TECHNOLOGY AND SOCIETY

TECHNOLOGY AND SOCIETY

Advisory Editor
DANIEL J. BOORSTIN, author of
The Americans and Director of
The National Museum of History
and Technology, Smithsonian Institution

AMERICA'S FABRICS

Origin and History, Manufacture, Characteristics and Uses

ZELMA BENDURE & GLADYS PFEIFFER

ARNO PRESS

A NEW YORK TIMES COMPANY

New York • 1972

Reprint Edition 1972 by Arno Press Inc.

Reprinted from a copy in The Wesleyan University
Library

Technology and Society
ISBN for complete set: 0-405-04680-4
See last pages of this volume for titles.

Manufactured in the United States of America

Publisher's Note: For this edition, the twelve
color plates have been reproduced in black and white.

———

Library of Congress Cataloging in Publication Data

Bendure, Zelma.
    America's fabrics.

    (Technology and society)
    Bibliography: p.
    1. Textile industry and fabrics. 2. Textile
fibers. I. Pfeiffer, Gladys (Bendure) joint author.
II. Title. III. Series.
TS1445.B395 1972   677   72-5260
ISBN 0-405-04685-5

AMERICA'S FABRICS

AMERICA'S FABRICS ARE COLORFUL

Many of America's fabrics receive their inspirational prints from local color as typified by patterns such as the southern palms, the western life, and America's flowers and foliage. This group of rayon fabrics, woven and knitted, represents different methods of dyeing and printing.

Starting at upper left, counterclockwise, the fabrics and coloring methods include the following.

The taffeta was first dyed yellow and then the pattern was roller printed on by the direct overprinting method.

The crepe with the green ground was first dyed and then, by roller printing, the color was discharged. Part of the discharged pattern was permitted to remain white while other parts were printed in color.

The foulard was first dyed black and then white and color discharged. The fabric was roller printed.

The crepe also was first dyed black. It was printed by the same method as the previous fabric except that the color was applied by screen printing rather than by roller printing.

The bold western print on the knitted acetate rayon is another screen print. The color was applied by the application method.

The next fabric also illustrates an application print made by screen on acetate tricot.

The green fabric, woven with thick-and-thin yarns, was roller printed to give a white discharge. It is commonly called a monotone.

This knitted fabric was first dyed gray, then the gray in the pattern was discharged and the colors were printed over the discharged pattern.

The last fabric was also printed by the discharge method in a roller printing machine. Here some of the discharged pattern remained white while other portions were color printed. The dark blue was overprinted.

Other colored illustrations of fabrics, printed by various methods, are facing pages 449 and 513.

AMERICA'S FABRICS

Origin and History, Manufacture, Characteristics and Uses

ZELMA BENDURE · GLADYS PFEIFFER

Photographic Layout by Crystal Stephen

Fabric Photographs by Nat Messik

THE MACMILLAN COMPANY · NEW YORK · 1946

To
Mother and Father
Cora Ann and Bert Elmer Bendure

FOREWORD

America's fabrics occupy a most important place in the scheme of our every day life. They add to our comfort and appearance, and to our very happiness, in countless ways, by their exceptional versatility.

Raw materials from which fabrics are made come from most parts of the world; plants, animals, and minerals all contribute. Man has improved natural fibers and synthesized new ones, and has designed ingenious machines that perform operations a thousandfold more rapidly and perfectly than were ever performed by hand. New dyes, new finishes, new methods, and even new fibers are being developed constantly to increase the utility, wearability, and beauty of our vast range of fabrics. This is the story of America's fabrics—an important story of men and machines.

This book is for those who work with fabrics, whether manufacturing or advertising, distributing or selling, and for students and consumers. It does not stress technical data essential for the fabric technician or for the student preparing to become a technician.

The purpose of the book is to give as complete a story as space permits of all fabrics in consumer use, from their beginnings to finished fabrics. First it tells the fascinating story behind fabrics, tracing through centuries of use of natural fibers down to recent discoveries of synthetic fibers. Next it describes the machines and methods used in making fabrics: from bobbins and needles to huge looms; from the artist who draws the design to the machine that prints the fabric. Third it explains fibers, yarns, and fabrics and their construction, finishing and coloring. Fourth it points out distinguishing characteristics of each fabric.

vii

Natural animal fibers, wool and silk, are discussed in two separate chapters as are the natural fibers, cotton, linen, and the minor vegetable fibers. Rayon, a cellulose synthetic fiber, claims one chapter followed by a chapter on all other synthetic fibers. Asbestos, the fiber from the mineral kingdom, ends the fiber section.

The first part of these chapters on the natural fibers discusses the fibers in detail up to the time they are ready to be spun into yarn. The rayon and other synthetic chapters describe completely the development of filament yarn ready for constructing fabric, and discuss staple fiber ready to be spun into yarn. From there on all fiber chapters summarize the procedure and processes from the fiber or filament to the finished fabric. In this way each of these chapters is complete within itself while the more detailed description of yarns, construction, coloring, and finishing is given in later chapters. The fiber chapters illustrate most fabrics made of each fiber. The fabrics are grouped and discussed by related types.

The next five chapters are concerned with spinning yarns and constructing fabrics by weaving, knitting, twisting, and felting.

The first chapter in the third section is an elementary discussion of the chemistry essential to a clear understanding of the chapters on pretreatments, dyeing, printing, and finishing.

An independent chapter is devoted to home fabrics, because of their wide range of individual types and textures. Following this is a chapter of definitions, descriptions, and illustrations of most fabrics in use today. These are arranged alphabetically for easy reference.

For consumer information and reference, the book ends with a resumé of the progress being made by manufacturers, stores, and national organizations, in advancing fabric knowledge.

Effort has been made to avoid oversimplification; an explanation of how and why fibers, yarns, and fabrics pass through certain processes is regarded as essential. Effort has also been made to interpret the language of the trade wherever its use is essential. The more intricate technical and chemical analyses have been omitted.

Each illustration has been selected for the definite purpose of supplementing graphically the text matter. Thus, with the picture of each raw material, each yarn, each method of construction, each fabric, and each machine, is told the story of America's fabrics—one of our nation's greatest industries.

WE APPRECIATE

We express special appreciation to those who gave time and attention to reviewing and correcting sections of the copy or complete chapters.

Eugene F. Ackerman, American Wool Council, Inc.

H. M. Chase, A. B. Emmert, Ira Hurd, and P. E. Smith, Riverside and Dan River Cotton Mills.

James Bunting; George Lawrie, Fairbairn Lawson Combe Barbour, Ltd.; Richard Pfefferkorn.

Clifford D. Cheney, Cheney Brothers.

Rene Bouvet and Joseph Leeming, American Viscose Company; W. O. Holmes, Hercules Powder Company, Inc.; Madelyn Kurth, Rayonier, Inc.; Ruth E. K. Peterson, United States Tariff Commission; Theodore Wood, American Bemberg Company.

T. H. Andrews, American Viscose Company; Allen F. Clark, Bakelite Corporation, Unit of Carbide and Carbon Corporation; E. I. du Pont de Nemours & Company; W. C. Goggin, Dow Chemical Company; R. H. McCarroll, Ford Motor Company; Bates Raney, Johns-Manville Corporation; Tyler S. Rogers, Owens-Corning Fiberglas Corporation; Stanley H. Rose, Aralac, Inc.

Richard F. Eshleman, American Viscose Company; Gustave Gerstle, Native Laces and Textiles, Inc.; Charles B. J. Molitor, North American Lace Company, Inc.

L. K. Fitzgerald, R. E. Henderson, and B. B. Howard, Riverside and Dan River Cotton Mills; J. B. Mellor, Cone Export and Commission Company.

Orval J. Boekemeier, Chief Specialist "T," USNR; Lloyd Eikenberry, Union High School, Willcox, Arizona.

C. A. Amick, W. H. Peacock, Dr. A. L. Peiker, and Dr. G. L. Royer, Calco Chemical Division, American Cyanamid Company.

Bernice Bronner, American Standards Association; Robert J. Painter, American Society for Testing Materials; Louis A. Olney, American Association of Textile Chemists and Colorists.

We appreciate the time and interest of the following companies who permitted us to visit their plants and explained to us the machines and processes in operation.

Riverside and Dan River Cotton Mills, Danville, Virginia; Botany Worsted Mills, Passaic, New Jersey; Cheney Brothers, Manchester, Connecticut; E. I. du Pont de Nemours & Company, Wilmington, Delaware; American Viscose Company, Marcus Hook, Pennsylvania; Liberty Lace and Netting Works, New York, New York; Scott and Williams, Inc., New York, New York.

The book is a mirror of the many photographs so generously supplied by manufacturers, advertising agencies, associations, government bureaus, and individuals.

In addition we gratefully acknowledge the assistance given us by many other executives of woven, knitted, and lace manufacturing companies; dyeing, printing, and finishing companies; textile machinery companies; and advertising agencies and textile associations.

Here also the authors wish to recognize the aid of manufacturers and associations who have supplied fabric information for manuals prepared in the past by the authors as well as those who encouraged the writing of this book.

THE SILENT PARTNERS

Many manufacturers bulletins and manuals, government bulletins, and the following books supplied valuable data:

Encyclopedia Britannica.

Encyclopedia Americana.

Matthews, Joseph M., *The Textile Fibers,* John Wiley & Sons, 1924.

Fink, Eugene D., *Applied Textiles,* New York State Vocational and Practical Arts Association, New York, 1941.

Lewis, Ethel, *The Romance of Textiles,* The Macmillan Company, New York, 1938.

Von Bergen, Werner, and Mauersberger, Herbert H., *American Wool Handbook,* American Wool Handbook Co., New York, 1938.

Murphy, William S., *The Textile Industries,* The Gresham Publishing Company, London, 1912.

Anderson, Arthur L., *Introductory Animal Husbandry,* The Macmillan Company, New York, 1943.

Bowman, Frederick Hungerford, *The Structure of the Wool Fiber,* The Macmillan Company, New York, 1908.

Merrill, G. R., Macormac, A. R., and Mauersberger, H. R., *American Cotton Handbook,* American Cotton Handbook Co., 1941.

The Cotton-Textile Institute, Inc., *Cotton from Raw Material to Finished Product,* New York, 1944.

Carter, H. R., *Flax, Hemp and Jute Yearbook,* H. R. Carter and Sons, Belfast, Ireland, 1932.

Mauersberger, Herbert R., and Schwarz, Dr. E. W. K., *Rayon and Staple Fiber Handbook,* Rayon Handbook Company, 1939.

Peterson, Ruth E. K., United States Tariff Commission, *The Rayon Industry,* Washington, D. C., 1944.

United States Tariff Commission, *Broad Silk Manufacture and the Tariff*, Washington, D. C., 1926.

Von Bergen, Werner, and Krauss, Walter, *Textile Fiber Atlas*, American Wool Handbook Co., New York, 1942.

Lemkin, Ph.D., William, *Visualized Chemistry*, Oxford Book Company, New York, 1943.

Des Jardins, Russell T., *Vitalized Chemistry*, College Entrance Book Company, New York, 1944.

Lewis, Ph.D., John R., *An Outline of First Year College Chemistry*, Barnes & Noble, Inc., New York, 1944.

Whittaker, Croyden Meredith, and Wilcock, C. C., *Dyeing with Coal-tar Dyestuffs*, D. Van Nostrand, New York, 1939.

American Association of Textile Chemists and Colorists, *Year Book*, New York, 1944.

American Society for Testing Materials, *A.S.T.M. Standards on Textile Materials*, Philadelphia, 1944.

CONTENTS

COLOR PLATES

AMERICA'S FABRICS—AN INDUSTRY OF MEN AND MACHINE

Hundreds of manufacturing plants, throughout the nation, are constructin fabrics by weaving, knitting, twisting, and felting. These and other mills ar finishing, dyeing, and printing fabrics. Huge chemical plants are making th essential chemicals.

Still other companies produce synthetic fibers to add to the natural fibers Elements from the air, water, and earth are all utilized. Nitrogen, hydrogen oxygen, carbon, and many others are combined to form often complex com pounds that are used to make synthetic fibers, and chemicals for dyestuffs an finishes.

Numerous plants and vast arrays of intricate machinery go to make up th physical part of this tremendous branch of commerce which includes th great wool, cotton, and beginning linen industries—and the silk industry to certain extent. The millions of Americans who produce and distribute Amer ica's fabrics attest to the full social significance of this great industry.

(Illustration, Pacific Mills.)

AMERICA'S FABRICS

One of our greatest industries

A book like this is the work not of one person or two, but of thousands of men and women. It turns back the pages of centuries to nomads tending their flocks, to monks stealing the eggs of the silk moths and hiding them in their canes before trudging homeward to their native land, to ancients who buried their dead swathed in linens, and to men who jealously guarded the secret of the mysterious trees that grew tiny white "sheep." It progressed from the blind acceptance of the contributions of animal and plant life to the inquisitive stage when man began to ask why and how nature performed its wondrous miracles. He studied the busy silkworm and then produced a viscous solution that, spun through a man-made spinneret, extruded a long filament resembling that spun by the silkworm through the tiny spinnerets in its head. He analyzed the wool, cotton, and flax fibers and determined why each contributes to man's comfort. He experimented with the earth's elements, he used minerals, vegetation, and even gases, and, by tearing apart and putting together, created fibers that by their very newness and distinction added more enjoyment to personal living.

A book like this reveals human development, from the simple inventions of ancient man and his handicraft to the machines that seem almost human as they perform with meticulous precision a multitude of intricate processes—spinning machines winding millions of yards of yarn on thousands of little bobbins until the whole room purrs like a giant cat; looms weaving back and forth with a throbbing clang as the harnesses rise and fall, rise and fall, while the shuttles shoot across, weaving each slender yarn into place, creating millions of yards of fabrics; cloth, gray, drab, limp, swishing through tubs of dyes, then over rollers, under rollers, being washed, dried, ironed, brushed with wires, often swirling from one machine to another like a banner, waving itself onward to become a beautiful, colorful, lustrous fabric. Truly, our modern fabric plants are a symbol of man's and America's ingenuity.

In 1941 came Pearl Harbor and war. It touched every part of our nation and one of America's greatest industries changed almost over night. Cotton plants wove canvas; lace machines turned out mosquito netting; wool mills made fabrics for military uniforms, silk mills wove miles of nylon parachute cloth, linen mills contributed airplane cloth, and rayon plants made yarns for tires—all a far cry from fashion's fleeting fancies.

Glass fibers were no longer used for noninflammable draperies but went to war wherever noninflammable or insulating fibers were needed. Other synthetics were detoured to help equip the nation for war.

But on this road to victory, again necessity became the mother of invention. To fulfill specific requirements some fabrics were given functional finishes and special features—contributions that could become later an everyday part of postwar fabrics. Many war fabrics were precision made and manufacturers acquired valuable knowledge that could be used in civilian production. Because of shortages in some fibers, new combinations were developed and consumer fabrics of unusual appearance and quality were manufactured. Thus, in a fast-moving world, the fabric industry kept pace.

A book like this has had to consider the immediate past and present, but has made few forecasts, because who knows? Will China, now that she has a great opportunity, pick up her lost silk culture? If so, will she be able to compete with rayon, nylon, and other popular fibers? Will linen become an American industry and will it be produced on a large scale? Will rayon make inroads into the markets of the cotton and wool industries? Who knows?

A quarter of a century ago, a book on fabrics discussed wool, cotton, silk, linen, and, briefly, rayon. Today silk is not available and linen plays but a small role among America's fabrics. The development of spun rayon and the improvement of filament rayon have raised rayon fabrics second to cotton only. Just as rayon first made its bid for attention in American markets, so today many new fibers and fabrics are being "discovered."

This does not mean, however, that the natural fibers are doomed. Who would trade a soft all-wool blanket or a beautiful wool tweed coat for a blanket or coat made from other fibers? Who would want to substitute a sheer linen handkerchief or a lustrous linen damask tablecloth for one made of other fibers? What house wife would want to do away with baby's cotton garments, with bath towels, sheets, and pillowcases that can be boiled and bleached without harm? Who would not desire sheer silk lingerie, so soft to wear and so easy to wash and iron?

There is no substitute for the natural fibers, nor is one natural fiber a substitute for another. True, they have been made into similar fabrics with identical weaves, but each has its own charac teristics that contribute to man's comfort and enjoyment.

Neither does our vast and varied field of synthetic fibers and filaments create substitutes for the natural fibers. Each synthetic by its special construction, has unique properties that are im parted to the yarns and fabrics made from it.

Rayons have added new beauty and versatility to fabrics. Nylon has created a definite place for itself. Soft casein and soybean fibers lustrous glass fibers, and the polymer resin fibers are making strides toward wider recognition. In fact, there are no substitutes no imitations among fibers. There are natural fibers and synthetic fibers. And each fiber in its own right is distinct and individual.

In former days consumers could readily identify fabrics. Today, fiber content is often obscured to the extent that only chemical analysis will reveal all of the fibers used in one fabric. Federal trade rulings have been passed to lessen the confusion, but fiber identification is still inadequate and will remain so until some rulings are withdrawn and one general ruling is made to cover all fabrics, stipulating that every fabric must be labeled with the percentages of all fibers contained in it, except perhaps decorative fibers not exceeding 5 per cent.

The trade-practice rule of 1937, requiring identification of rayons without specification of the kind or percentage of fiber content, does not take into account present-day conditions. Other parts of these rulings—requiring that satin, taffeta, and all such fabrics formerly made of silk be called "rayon satin," "rayon

4

affeta," and so forth—are completely antiquated. There should e a distinction between regenerated cellulose rayons (viscose and uprammonium) and cellulose acetate rayon. The only similarity s in the raw materials from which they are made. They are manufactured differently, react differently to dyes, and require different are. They should be identified just as are wool, reprocessed wool, nd reused wool.

The wool industry made the first great stride forward with the passage of the Wool Products Labeling Act of 1939 which became ffective July 14, 1941. This act is discussed on page 60.

In 1941 the linen industry followed suit with the passage of a imilar ruling which, however, falls short in some respects.

There are no rulings requiring other fiber designations. The nly rulings of any note regarding silk and cotton were two passed n 1938 on weighting of silk and on shrinkage of cotton. Today, ince most fabrics may be made from many different fibers or ombinations of fibers and are given many new and different nishes, the necessity for labeling has become paramount. If wool an be labeled, so can all other fabrics. But content is not the only nformation necessary. Functional finishes require special care; legrees of fastness of dyes to water and sunlight are not the same; nd some fabrics have controlled shrinkage while others have not. The knowledge of these details is essential to proper care and to atisfactory wear. This information cannot be passed on to the ltimate consumer unless there is standardized labeling of all abrics.

This book takes a definite point of view: it explains clearly nd in detail that practically any fabric can be made of more than ne fiber and that no fabric is the exclusive property of one fiber. f certain fabrics are made of only one fiber, there is always some pecial reason. For example, filament yarns cannot be napped. Therefore, yarns spun of short fibers are used for napping such as wool, cotton, rayon staple fibers, waste silk, casein, and soybean fibers. Linen fibers do not nap well and do not contribute to warmth. Many fabrics are made exclusively of cotton not because other fibers could not be used but because the price and adaptability of cotton make it most suitable for the purpose.

Naturally, fabrics were first made of fibers that were available at the time and seemed best suited for the purpose. But it does not necessarily follow that, today, a fabric woven from a different fiber but with the same weave, given the same type of finish, and having practically the same final appearance cannot be called by the original name of the fabric.

It is impossible for the average consumer to remember the details of long outdated rulings; as, for example, that the names "gabardine" and "flannel" mean wool gabardine and wool flannel and that, if the fabric is made of any other fiber, it must be stated, such as "cotton gabardine" and "cotton flannel." On the other hand, "batiste," if made of wool, must be called "wool batiste." Fabrics formerly made of any of the natural fibers must have the word "rayon" before the name of the fabric, if made of rayon. It seems logical to assume that gabardine is gabardine, flannel is flannel, and satin is satin, regardless of the fiber of which it is made, as long as the weave is the same, and the finish and final appearance are practically the same.

If a fabric is made of a fiber or fibers best suited to its purpose and price and if consumers are informed of the fiber content and what each contributes, we shall be on the way toward a better understanding between producer, distributor, and consumer.

Our world of fabrics is essential to our American way of life. In their texture and color they express our personalities in our dress and in our homes. By their fibers, yarns, construction, and finish they protect us from the elements. Dyed or printed they bring color into our lives; soft, absorbent, and sterilized they help care for our sick and wounded and give our babies a healthy start in life. Each, in its own way, adds to our comfort and enjoyment.

A book like this explains why and how our fabrics serve our ever-changing needs. It endeavors to emphasize and distinguish between each fiber, whether natural or synthetic, old or new; to explain each yarn and each method of construction; to analyze each dye and finish, and each method of application; and finally, to show what each fiber, construction, dye, and finish contributes to the individuality of the finished fabric. Truly, our today's fabrics have traveled a historically long and arduous, yet fascinating journey, from prehistoric simplicity to a vast business of men and machines—America's fabrics, one of our greatest industries.

Sheep grazing under tall stately pines on a cool summer range in the great Southwest (Arizona Highways).

WOOL—ANIMAL FIBER

The most unique

IN THE BEGINNING

AND in the beginning there were sheep—but lost in the obscurity of the past is the true tale of when and whence came the ancestors of today's wool-producing animals. Long before the dawn of recorded history, when primitive man lived in caves and lake dwellings, we find the bones of sheep already domesticated.

Later, when prehistoric man gave up his spear and club, he took up the shepherd's crook and exchanged his cave and lake dwellings for a nomadic life in which sheep invariably figured as the reason for roaming from place to place. Man's immediate wants were without a doubt supplied almost entirely by those mild and faithful animals. These primeval sheep provided him with flesh for his meat and with warm covering against the cold and damp. In addition, the care of the flock took little or no exertion; merely providing continued pasture for food and protection against the ravages of wild animals. As habits of the ancient gradually changed and he became more settled, his chief wealth consisted of flocks and land. It might be said that mankind roughly divided itself into two groups, shepherds and agriculturists.

The Scriptures and pagan literature are full of references to sheep, shepherds, and weaving. For example, in Genesis: "And Abel was the keeper of the sheep, but Cain was a tiller of the ground;" and in Deuteronomy (an edict settling an argument between the children of Israel wearing wool and the Egyptians clothed in linen): "Thou shalt not wear a garment of divers sorts, as of woolen or linen together."

8

A flock of sheep on a Navajo mesa at twilight wending their way homeward (W. M. Pennington, Durango, Colorado).

A Greek myth, originating in the seventh or eighth century before Christ, tells the story of a ram with a golden fleece, and the harrowing exploits of the venturesome Argonauts who braved the hazardous seas. Their leader, Jason, had previously fled from his country on the back of the ram, then, had sacrificed the ram and hidden the golden fleece. He and his Argonauts were later delegated to recover the golden fleece from the dragon-guarded cave in a foreign country. Conquering wars, perils, trials, and tribulations, Jason and his faithful Argonauts finally returned home in their invincible Argo with the precious golden fleece.

But early accounts of wool were not only in the old biblical and mythical world. Herodotus (484-425 B.C.) tells us that Babylonian dress consisted of a shirt of linen reaching to the feet and over it a woolen tunic. Ovid (43 B.C.-A.D. 18), in *Metamorphoses,* describes the process of woolen manufacture in his time. Pliny (A.D. 23-79) explains that garments can be made of wool without spinning or weaving. He evidently meant felting. Also according to Pliny, in *Historiae Naturalis,* the Tarentine sheep of Italy produced the finest wool in the world at that time.

During the reign of the Roman Emperor Claudius, Columella, a Roman, took a flock of sheep and settled in southern Spain.

This flock founded one of the finest strains of wool-bearing sheep known to the world today. Columella crossed his Tarentine sheep with the pure white sheep of the nomadic tribes of northern Africa. The progeny of the Tarentine ewes and the African rams were the forefathers of the famous and important breed of sheep, the Spanish Merino.

About A.D. 70, the progressive Romans considered the weaving of wool cloth such an important industry that sheep and wool fibers were given very special and exact attention. Over their backs the sheep wore covers in order that their fleece would become soft, wavy wool with a beautiful luster. To keep the wool fine, soft, and untangled, the sheep were periodically washed, and the hair was parted, combed, and moistened with expensive oils.

In A.D. 80, according to Caesar, the Romans introduced wool weaving to England. These conquerors of England, finding the weather cold and damp, established a wool-weaving factory at Winchester to make warmer clothing for the invaders. The Britons were quick to learn, and soon the Winchester looms gained reputation far and wide. It was said, "The wool of Britain is often spun so fine that it is in a manner comparable to the spider's web."

In A.D. 711, the prosperous but barbaric Saracens gained a foothold in Spain. Their extravagant living fostered many luxuries little known to other European countries, chief among them the weaving of beautiful woolens. By the thirteenth century, wool manufacturing had grown to such extent that in the town of Seville alone there were 16,000 looms and it was estimated that 16 million sheep moved each year from the north to the south and back again, requiring some 40,000 to 50,000 caretakers.

A few wealthy men held a monopoly of Merino sheep, and it was against the law to export sheep from Spain. The only person exempt was the king who made a gift of Merino sheep to the king of France who kept them at the royal farm at Rambouillet. Thus began the Rambouillet strain. Another gift was made to the king of England. Some of these sheep found their way to Australia to become the forefathers of the famous Australian Merino. When later, the Saracens were driven from Spain, with them went the industry they had so extravagantly supported.

During the period of the migration of nations, the principal occupation of all peoples was war, which left little time for the pursuit of peaceful arts. Textile activity in Great Britain seemed o be at a standstill until about the eleventh century when the barbaric Saxons were overthrown by the Normans led by William the Conqueror. In 1068 William allowed Flemish weavers, who fled continenal wars and persecutions, to settle in Carlisle under he protection of the queen. These skilled immigrants aised the standard of the English textile industry to hat of the continent, and weavers' guilds, the first of vhich was formed in 1080, became a strong power in British affairs. From the eleventh to the eighteenth entury, English rulers gave increasing attention to he wool industry. The continued immigration to the British Isles of competent machine weavers plus such nventions as the spinning jenny in 1764, by James Hargreaves, the steam engine in 1782, by James Watt, ind the water-power loom in 1785, by Edmund Cartwright, ushered in the industrial revolution.

Western-World Wool

The story of wool in the Western World is mostly n the form of speculation and comparison, but it is)elieved that simultaneously with their Eastern beginnings, wool and weaving were practiced on this side)f the world. In Mexico, Peru, and other South Amercan countries, peoples equally as advanced as those)f early Egypt and her neighbors were raising wool)earing animals and were weaving with such artistry hat, even today, their light, filmy, woolen fabrics 1ever been duplicated. Somewhere along the highway)f time, their unique methods of spinning and weavng were lost to the world.

These ancient Americans did not have the ordi1ary domesticated animals familiar to Asia and Africa: :heep, goats, and camels. America's lovely wool fabrics :ame from America's unique wool-bearing animals, :he vicuna, the llama, the guanaco, and the alpaca. These early peoples domesticated the mild tempered .lamas and alpacas and used them for food, for wool, ind for beasts of burden. They also tamed another :ypical American, the turkey, and used his meat for

11

their food and his feathers as blending fibers in their fabric

In North America, domesticated sheep landed almost as soo
as did the Pilgrim Fathers. The first small flock of sheep wa
brought from England to Jamestown, Virginia, in 1609. Th
growth of the flocks was so slow that in 1649 there were only 3,00
sheep in all of the colonies. Wool spinning and weaving were fror
necessity a household industry. Naturally, hand-spinning wheel
and locally made hand looms turned out uneven and often un
sightly cloth. But the products improved when the fulling mil
were established. In 1643, some twenty families from Yorkshire
England, skilled in carding and combing, settled in Roxbury
Massachusetts and here established the first fulling mill in Americ
—a very different institution from the present-day fulling industry

England, in order to protect her own wool industry, greatl
hampered the early colonial efforts by prohibiting the export o
sheep's wool and wool yarn from England. As a result of thes
stringent restrictions, the colonists were forced to build more sys
tematically their own industry.

In 1656, the Massachusetts General Court enacted the follow
ing order: "To assess each family for one or more spinners o
fractional part, that everyone thus assessed, do after this presen
year, 1656, spin thirty weeks every year a pound per week of linin
cotton wooling and so proportionately for halfte or quarter spin
ners under penalty of twelve pence for each pound short."

In 1766, all-day sessions of spinning were held in Providence
Rhode Island. In 1768, the senior class of Harvard graduatec
wholly clothed in American-made fabrics. As early as 1768, George
Washington manufactured 365 yards of wool cloth annually a
his Mt. Vernon home. In 1769, the first graduating class of Rhode
Island College wore at commencement clothes made entirely from
American fabrics.

In fact such was the patriotic enthusiasm and faith of the early
colonists in their America that the first several presidents of the
United States, as an expression of their own and their country's
loyalty to native industry, were inaugurated into office in suits
made from domestic fabrics.

The "Spirit of '76" which marked the beginning of political
United States also marked the beginning of industrial United
States, and now like our political strength, America's industrial
strength tops the world.

SHEEP

Sheep belong to the large family of Bovidae (hollow-horned ruminants), to the subfamily *Ovinae* (highland or mountain dwellers), to the genus *Ovis* (sheep), and to *Aries* (domesticated sheep).

Originally, sheep were covered with hair and only a soft, light down next to the skin. Through centuries of domestication the hair finally gave way to the wool we know today. Some sheep still have coarse hairs above the wool.

Today, there are some 698 million sheep in the world, of which the United States claims 55 million, contributing about 450 million pounds of wool annually to the fabric industry. In addition, the United States imports about 30 per cent more wool, even though she is second to Australia only in the size of sheep population. Other countries raising sizable amounts of wool are South Africa, Argentina, New Zealand, and Russia.

These millions of sheep roaming the earth are of a hundred or so different varieties, but of all the many breeds in existence today only about thirty have become fixed types adapted to conditions of their native countries and less than half of these types are well established in the United States.

Sheep are first classified according to the purpose for which they are produced, into wool types, and mutton types. Of course,

A little Navajo shepherdess makes pets of the lambs in the flock (Mrs. W. N. Searcy, Durango, Colorado).

Rambouillet ram, fine wool.

Merino ram, fine wool.

Hampshire ram, medium wool.

Shropshire yearling ram, medium wool.

Oxford ram, medium wool.

Southdown ram, medium wool.

Lincoln ewe, long wool.

Representative pure bred sheep (U. S. Department of Agriculture).

14

both types contribute to fabrics and both types are used for meat, but one is raised primarily for its wool and is improved to increase the growth and quality of the wool fiber. The other is grown for its flesh and is bred to increase the weight and the quality of the meat. About 90 per cent of the sheep in the United States are neither definitely the wool type nor definitely the mutton type but mixtures of various breeds. There are also definite crossbreeds, grown with the aim of producing the maximum poundage of meat and at the same time a medium length, fine quality wool. Because the wool of the crossbreed type sheep is similar to the fleece of the medium-wool mutton types, these sheep are considered mutton types.

Registered purebreds are less than 10 per cent of all sheep in this country; but because these registered sheep are used for breeding, there is an established proportion between the number of purebloods and the total number of sheep. For example, since 41 per cent of all registered purebreds were Rambouillet and Merino in 1930, roughly 41 per cent of the wool produced in that year would be Rambouillet and Merino types.

Wool is an important product of all sheep growers, contributing at least 14 per cent of the profit on even the largest mutton types.

The wool type sheep produce a quality of wool termed "fine wools." These sheep are usually of pure breeds. The mutton type wools are of two qualities: the "medium wools," produced by purebreds and by the most popular crossbreeds, and the "long wools," produced mostly by purebreds. In addition to these, there are a few "fur type" sheep.

Fine Wool Types

Fine wool is grown on sheep bred for wool production. The fleece of these sheep completely covers the body, grows very tight and dense (100,000 to 200,000 fibers per square inch), and is very heavy in oil or yolk—often 70 per cent of its entire weight.

The fiber of the fine wools is very fine (17 to 23 microns wide) with numerous distinct and uniform crimps (16 to 22 per inch). The length of the fiber is short, ranging from 1 to 4 inches. Less than one third of the fine wools are sufficiently long (2 inches or more) for combing. The fine wool long enough for combing is often termed "delaine."

Wool type sheep thrive on both farm and range, because they

Rambouillets on the Navajo reservation (Arizona Highways).

are adaptable and can withstand hardships. They shear from 10 to 25 pounds of wool annually.

The following are the fine wool types:

Merinos are of three distinct types, called A, B, and C types. The A type has heavy folds of deep wrinkles from head to tail. Its wool is the finest, densest, shortest, and crimpiest. The C type has almost no folds on the body and only a few on the neck. It is the largest of the three, with a strong constitution. Its wool is less dense and is longer and coarser. The B type is intermediate between the A type and the C type. Merino rams weigh from 140 to 225 pounds.

Rambouillets are the most popular sheep, judged by their numbers, in the United States today. They are larger than the Merino, more rugged, and have a coarser, longer fleece. Rambouillet may or may not have distinct folds and are graded into A and B types. Rams weigh from 225 to 250 pounds.

Medium Wool Types

Medium wool sheep, usually, are raised for mutton. They are larger and heavier; the rams weigh from 188 to 300 pounds. The fleece grows close and is dry enough for excellent protection. The wool is light in color, having less yolk than the fine wools. The fiber is longer (2 to 5 inches) and wider (23 to 27 microns) than fine wools. The annual shearing of these sheep produces from 8 to 12 pounds of wool.

The chief medium wool breeds in the United States today are *Southdown, Shropshire, Hampshire, Oxford, Dorset, Suffolk, Cheviot, Tunis,* and *Ryland.* With the exception of Hampshires and Suffolks, the medium wool breeds are more popular and better suited to farm than to ranch raising.

16

Long Wool Types

Long wool sheep are primarily mutton sheep. They are the largest breed, having big frames, square bodies, and broad backs. Rams weigh from 250 to over 300 pounds.

The fleece has a high luster, little yolk, and hangs in long, definite curls or locks. It grows open and loose on the body compared to fine and medium wool, which grows tight and close. The fiber is long (5 to 14 inches), coarse (35 to 40 microns), "lashy," and hairlike.

A sturdy little Southdown, a common breed on American farms competing for its dinner (U. S. Department of Agriculture).

The long wool types can be grown on land too damp or too low for other sheep, for the long, hard wool sheds the water more readily than does the shorter, closer fleece. However, these large sheep grow better on level land that requires little travel for pasture.

The most popular long wool breeds are *Lincoln, Cotswold, Leicester, Romney Marsh,* and *Blackface Highland.*

Crossbreed Wool Types

During the last fifty years, Merino and Rambouillet have been crossed with long wool sheep in order to achieve, as nearly as possible, types with mutton qualities of the long wools and wool qualities of the fine wools.

In the United States the following crossbreeds have been developed or are grown:

Corriedales have been developed in New Zealand since 1880 from Lincoln rams (and sometimes Leicester or Romney Marsh rams) and Merino ewes. The type is intermediate between the Lincoln and the Merino, smaller and less heavy than the Lincoln and larger and heavier than the Merino. The fleece possesses some of the softness and fineness of the Merino but it grows longer and has less weight and less yolk. Corriedales are popular with western range growers.

17

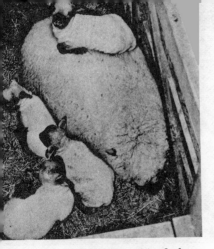

A young Shropshire finds a soft resting place on its mother's wide back (U. S. Department of Agriculture).

Columbia is a crossbreed type developed in the United States. The purpose was to produce a sheep suitable to western range conditions, which would breed true to type. After twenty years, the Columbia is the result of crossing Lincoln rams and Rambouillet ewes and then continuing to cross the crossbreeds without introducing new blood lines. This sheep is a large, vigorous, heavy-boned animal, the rams weighing about 275 pounds. The fleece is long staple, medium fine wool, shearing about 11 pounds annually. Columbias are increasing on the western ranges each year.

Panama is similar to Columbia but is the result of crossing Rambouillet rams and Lincoln ewes. This type also was developed in the United States to produce a more sturdy western sheep.

Romeldale, too, is an American production. It was developed by mating Romney rams with Rambouillet ewes. Like the other crossbreeds, the developed type is about half way between the two original types of sheep.

Targhee is a new American breed of approximately three-quarters Rambouillet and one-quarter Lincoln. This type has fine wool of longer staple than Rambouillet and quite acceptable mutton.

Fur Types

The Karakul sheep are grown not for their wool but for their skins. The lamb skins are valuable because they are suitable for fur. The pelts of the lambs are short haired, with tight, lustrous curls. Their natural colors are black, gray, or brown. The fur is termed "broadtail," "Persian lamb," or "Karakul," according to the age of the sheep.

SHEEP RAISING

Sheep are raised in one of two ways, by farm method or by range method. The divide between the two methods is approximately the 100th meridian. As a rule, east is the home of the farm flocks and west, the range sheep. Sheep on the farms are pastured, housed, and fed on a comparatively small acreage. The average farm flock has from twenty-five to fifty ewes. Flocks of less than twenty have proven uneconomical from the standpoint of investment in rams, equipment, and caretakers' time. Sheep eat 90 per cent of the commonly grown plants and can be raised on land unsuitable for other animals, but they do require continuous attention. About 30 per cent of the total number of sheep in the United States are raised by the farm flock method.

Range sheep are raised and cared for in the western states in three different ways. In the southwestern states, where the climate is warm, the sheep are ranged the year round on large ranches or lands of livestock producers. About 150 to 200 head can be raised on 640 acres.

In the central range areas, such as Colorado, Wyoming, and Utah, the sheep are moved in summer to government land in the high altitudes of the mountains and in the winter to lower, warmer ranges, either private or government owned, where plant life matures into cured feed during the winter.

A standard herd in the central section consists of 1,200 ewes with single lambs, but it may vary from 750 in rough country to 2,000 to 3,000 dry ewes in open country. These sheep must possess the flocking instinct if they are to be handled in large bands. They must be vigorous and sturdy to stand the hard trails. The fine wool breeds seem best suited for ranging.

In the northwestern area most of the sheep graze on government land in the spring and in summer, but in the winter they are cared for on leased or privately owned grazing lands. A few are ranged the year around. The number of sheep in the northwestern herds varies with winter range conditions. Of all the sheep in the United States, 70 per cent are range sheep.

Securing the Wool

Wool, in the form of clipped fleece, begins its trek toward the finished fabric shortly after the lambing season. In some countries the sheep are washed in fresh stream water and dried before shearing, but in the United States it is the usual custom to shear the sheep as they are.

The shearing season, like the lambing season, is dependent upon the climatic conditions of the land where the sheep are

raised. In Arizona and Texas shearing may start in February, whereas in Montana and North Dakota shearing may be in late June. As a rule, sheep are sheared once a year, although in some warmer places, if range conditions and demand for short wool warrants it, the flock may be clipped in the fall as well as in the spring.

Shearing is done by hand in many places, but mechanical shears, similar to the barber's electrical clippers, are widely used today, shearing from 100 to 200 sheep a day. Skillful shearers are able to slip the fleece from a sheep in one unbroken, continuous sheet that retains the form and relative positions of the mass almost as if the creature had stepped out of his skin. In unbroken condition each fleece is rolled up with the belly wool (which is clipped separately) and tied with its own wool or with paper twine. This facilitates sorting and grading later.

In some countries like Australia, where large numbers of sheep have uniform staple throughout, the fleece next may be "skirted" or trimmed and then

Above: Ready to shear and the first step in shearing. Below, left and right: Continuing and completing the shearing.

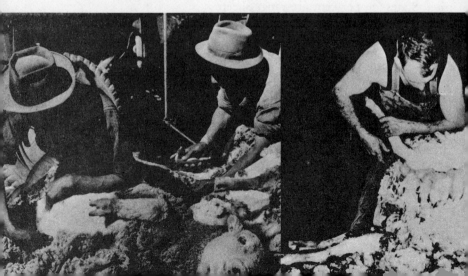

passed on to a "classer," who divides the wool into different classes, or counts, into which it will be spun later. The divided wool is then packed and transported to the manufacturer. But, since sorting and stapling of the fleece is done as a rule on the premises of the yarn manufacturer, the usual step after shearing is to pack the fleece "in the grease" into large bales, in which form they are sent to the manufacturer.

The wool clipped from lambs eight months old is termed "lambs wool"; from lambs twelve to fourteen months old "hogg" or "hoggett." All the rest of the shearings are called "wether" wool.

"Pulled wool" or "skin wool" are terms applied to wool taken from pelts of slaughtered animals. About 15 per cent of the annual domestic wool production is pulled wool.

After the skins are stripped from the bodies of the sheep, the pelts are transported to a pulling establishment and the wool is pulled by hand after the roots have been loosened. Pelts are treated in one of two ways to loosen the wool, (a) by sweating the pelts with moisture and heat, (b) by treating the pelts with sulfides and lime. If the pullery is not near the place of slaughtering, the pelts may need to be preserved for shipping. If so, they have to be softened again before treating. Pulled wool is considered inferior to clipped fleece because the roots of the fibers are damaged by chemicals and pulling, causing weakness and uneven dyeing.

Some wool has mixed with it some coarse, brittle hairs, which are dead white or opaque. These are termed "kemp hairs." They are especially conspicuous in lambs' fleece but are shed from the skin after several months of growth. Their presence in the matured wool cheapens the wool and is a sign of poor breeding.

Below: Folding the fleece ready for shipment. Right: Shorn sheep with fleece removed (American Wool Council, Inc.).

SPECIALTY FIBERS

Besides domesticated sheep, there are a number of other animals whose wool or hair can be made into yarns. While not exactly like wool from sheep, these animal fibers resemble wool both physically and chemically. Some are more nearly hair while others are definitely wool; but, in general, their treatment for yarn and fabric making is practically the same as that of wool from sheep.

The animals that contribute to specialty fibers as they are commercially known, are interesting, and the fiber from each imparts a unique characteristic to the finished fabric. They include the following:

The Angora Goat

The Angora goat hair, known as *mohair,* is the most available and the most commonly used of the specialty hair fibers. This goat was raised originally around Angora, a city and province in Asia Minor. Records show that it was domesticated in Turkey at least 2,000 years ago. The first Angora goats to reach the United States were seven does and two bucks brought from Turkey in 1849. Today that nucleus of nine has become over 4 million. It has been found that the climate and the vegetation of the southwestern United States are particularly suited for Angora goat raising.

The hair of the Angora goat grows long and heavy and, when shorn, the fleece weighs from 3½ to 5 pounds. It is the general practice to shear the Angora goat twice a year. The fall shearing is considered the best fleece. Mohair fiber grows in long locks of uniform length, varying from 4 to 6 inches for a half year's growth. The cleaned and scoured fleece is white.

There are three types of goats with mohair fleece, characterized by their different locks. The "tight lock" is ringleted throughout its entire length and is usually considered the finest textured mohair. The "flat lock" is generally wavy, although not in such tight ringlets as the first, and forms a more bulky fleece. The "fluffy fleece" as the name implies, has open locks, and is the weakest and least desirable.

One of the most trying problems in mohair production is presented by kemp. The Angora goat fleece contains these shorter,

22

coarser, inferior fibers, which dye differently and which are very difficult to comb out of the mohair. Breeders have been making an effort to rid the Angora fleece of as much kemp as possible. Some of the better mohair, especially the softer kid hair, now contains very little kemp.

The mohair fiber is soft and durable, with a distinct luster. It is smooth and somewhat slippery because the fiber has no deep serrations. It also lacks two other important characteristics of wool: the natural crimp and the ability to felt. Mohair fiber takes brilliant dyes readily and retains them well.

Mohair's reaction to moisture, dyes, and other treatments is different from that of wool. As a result, many interesting effects can be created when mohair is mixed with wool or other fibers. But it is equally usable when woven alone. Mohair is versatile. It is used successfully for materials with a heavy pile, as well as for both soft and hard textured fabrics.

Above: Cashmere goats (S. Stroock & Co., Inc.). Below: Angora goat (Goodall Fabrics, Inc.).

The Cashmere Goat

The Cashmere goat is today, as he was originally, a native of Tibet and of northern India. He receives his name from the province of Cashmere. It is here that the natives make the exquisite shawls from the fine down of the goat. To date any effort to transplant the Cashmere goat from its native land to any other part of the world has failed.

The Cashmere goat is smaller than the Angora goat. His outer or beard hair is straight, long, and coarse, while the undercoat or wool is exceedingly fine and downy. The fleece is white, gray, tan, or a mixture of tan and gray. The gray-tan mixture predominates.

The fleece is not shorn as is the Angora goat's but is obtained in the spring during the molting season. The hair and wool are either plucked off the animal, as they become loose, or are picked from brush and shrubs after the goat has rubbed them off. Each animal yields annually not more than 3 to 5 ounces of valuable wool.

The cashmere wool fiber is fine, soft, and has a crimp formed by wide waves. It has a distinct silky gloss and is smoother than sheep's wool fiber. Its length is from 1¼ to 3½ inches. Cashmere's wearing qualities are unexcelled, and it blends with other fibers well.

The term "cashmere" is often used incorrectly and indiscriminately, but only a fabric containing wool from the Cashmere goat can correctly be called cashmere.

The Camel

Strange as it may seem, the camel began his trek down through the centuries not in the old world, as might be expected from biblical and ancient literature, but in the new world. It was in Mexico that bones of the earliest camel have been found. Connection between these ancient Mexican camels and those of Africa and Asia has not yet been found, but it is assumed, because of similar characteristics, that the modern camel as well as the South American llama are descendants of this early camel of Central America.

Modern camels are of two distinct types. The dromedary, the one-humped variety, inhabits Arabia, Egypt, and Persia and is used chiefly as a burden carrier. But, since the dromedary contributes nothing to the fabric world (his hair being too short and too fine for weaving), he is mentioned only to distinguish him from his more versatile brother, the Bactrian.

The *Bactrian,* the two-humped camel, is found in all parts of Asia, from the cold north of Siberia to the sultry, humid Arabian Sea. That the Bactrian camel must live in such extremes of climate is probably why camel hair is a natural as a fabric fiber. The camel's hair must protect his body not only from extreme cold but from extreme heat. This remarkable fiber is soft, rich in luster and color, possessing warmth giving as well as cooling properties. Unlike sheep, which are shorn periodically, camels drop their hair continually in clumps and these are picked up by the caretakers of the animals.

Since there is probably no fleece with such an extreme range in texture as camel's hair, it is first divided into three grades. Grade 3 is the very tough, coarse outer hair, which has little commercial value and is utilized to make native tents, blankets, rugs, and rope. Grade 2, just beneath the outer hair, is a mixture of the tough hair and the finer fleece closer to the skin. This quality has recently come into use in the textile industry, giving a quality of camel hair fabric somewhat coarser than that made of grade 1. Grade 1 is the hair closest to the hide of

Two-humped camels
(S. Stroock & Co., Inc.).

the animal. This fiber is short, very thin, and extremely soft. It possesses a beautiful tan luster and, despite its extreme fineness, has great tensile strength. Since the fiber is short, grade 1 camel hair is spun only into soft woolen yarns because it takes long fibers to make hard worsted yarn. The word "noils" in camel hair means the choicest camel fibers. This is opposite to the meaning of the term "noils" as used in wool.

Camel hair in its natural state has a variety of color, ranging from pale to dark tan. There is some that is almost black and some pure white.

The Llama Family

The llama tribe is the New World branch of the camel family. True, they are smaller than the camel and have no hump, but they are camels nevertheless, having habits and characteristics similar to those of their Asiatic cousins.

The western portion of South America is the home of the llama family (except the guanaco, which inhabits Patagonia in Southern Argentina). Between two mountain ranges, extending from Southern Ecuador through Peru and Bolivia to Northern Argentina, runs a 300 mile wide mesa called the Puna. Here is the heart of llama land, especially in the higher parts ranging from 12,000 to 16,000 feet. In the Puna grows a grass called *ychu,* the food that

The South American llama
(S. Stroock & Co., Inc.).

seems to be a necessity to the llama family. Since this grass grows no place else on earth, it is probably the main reason why so far it has been impossible to transplant these animals to any other part of the world.

The llama family, today, consists of four distinct and two hybrid species. The distinct species are llama and alpaca, the domesticated members of the tribe, guanaco and vicuna, the wild members. The hybrids are huarizo, progeny of a llama father and an alpaca mother, and misti, offspring of a llama mother and an alpaca father. Brief characteristics of the individual members of the tribe are as follows:

The *llama* is the largest of the South American *camelidae*, weighing around 250 pounds. The llama's chief economic importance to the peoples of the Andes is his use as a beast of burden. There are some 3.1 million llamas in South America today, owned almost exclusively by Indians. The fleece of the llama is heavy. The outer coat is thick and heavy while the hair underneath is soft and silky. Most of the llama fleece is brown of one shade or another. The hair under the belly is generally white.

It is only the fleece of the female that is shorn and used for fabrics. That of the male, the burden carrier, is left on as a cushion for his pack and is only shorn after his death.

The *alpaca,* the other domesticated member of the llama family, is today, as he has been since at least pre-Inca times, a necessary part of the economic life of the great Andean Plateau. He is especially important to the fabric industry because of the large productivity of his fleece.

The alpaca is smaller than the llama. His hair, which at a distance looks black, owing to oil and dirt matting the long hairs, has a variety of colors. Gray and faun pelts are more prevalent but there are some darker brown, black, or pure white ones. The hair hangs down on the animal in long, shiny, tangled strands, measuring from 8 to 16 inches in length, or reaching even 30 inches if left

26

to grow for long periods. The fleece is shorn every other year and yields from 4 to 7 pounds. The fiber of the alpaca has great beauty. It is silky in texture, fine, and has a lovely luster as well as strength.

The alpaca is of two distinct types, the *huacaya* and the *suri*. The hair of the suri is finer, more glossy, thicker, and longer. It often completely conceals the animal's feet.

The majority of the 1⅛ million alpacas in South America are owned by Indians. There are some alpaca farms managed in a more or less scientific way, but, to date, any effort to raise alpacas on a large commercial scale, like sheep, has failed — probably because too little is known about the animal and his peculiar habits.

The hair of the two hybrids, *misti* and *huarizo,* is on the whole not as fine in quality and texture as alpaca hair but it has commercial value.

The *guanaco* is a very wild and timid creature. He is larger than the vicuna and the alpaca but smaller than the llama. The hair of the guanaco is straggly but of unusually fine texture. Its color is usually an attractive dull brown. It has good spinning qualities: softness and pliability. But, since the animal is scarce and exceptionally wild, guanaco fiber is limited and thus of small commercial importance. The young guanaco furnishes a fur called "guanaquito."

The vicuna, the second wild member of the tribe, is one of the most unique animals contributing to the fabric world for he yields what is thought

Members of the alpaca family. Above: Alpaca. Below: Guanaco (S. Stroock & Co., Inc.).

Vicunas, of the llama family (S. Stroock & Co., Inc.).

to be the rarest and finest fabric fiber known to man. He is smaller and more slender than his cousins and more gracefully proportioned. The color of his lovely coat ranges from golden brown to lustrous faun. His appearance is given distinction by an apron of long white hair, which falls down beneath his forelegs and continues along his flanks.

The rugged, barren, uninhabited, high vicuna region was once the home of the famous Inca nation and, before them, of the even more illustrious pre-Inca civilization. These peoples regarded the vicuna with great reverence and protected him, allowing only periodical and well-supervised hunts. Since the small animal was (and still is) very wild, he must be killed to obtain his precious coat. His fleece was considered so luxurious and valuable as to be fit only for the use of royalty.

With the coming of the Spaniard, the vicuna herds disappeared rapidly. So scarce had they become that Peru in 1921 was forced to enact laws to protect the vicuna and, with the help of other South American countries and the United States, is still making every effort to build up the herds to somewhat near their former glory.

The fleece of the vicuna contains two distinct types of hair, the outer, or beard hair, and the inner hair. It is only the soft inner hair that is used for fabrics. It is the finest of all natural fibers; it is less than 1/2000 inch in diameter, which is about one-half the diameter of the finest sheep's wool. Although exceptionally soft and silklike, vicuna is extremely strong and resilient. Since the fiber has a great resistance to dye, and since the luster and beauty of its natural color cannot be improved upon, this rich golden fleece is usually woven in its original color.

Vicuna is today insignificant in quantity in comparison to other fleeces, but as a result of present protective measures the day may be not too far distant when this royal fiber will find a recognized place in our growing fabric world.

Rabbits

Rabbit fur is not in the strict sense a fabric fiber but, since it has become one of the important furs used in combination with other fibers for yarns, it is rightly considered a part of the fabric family. The rabbit contributing the most fur for spinning is the Angora rabbit. France, Belgium, and England are producing sizeable quantities of Angora fur. In the United States, California leads in the number of Angora breeding farms.

White furred angora rabbits (S. Stroock & Co., Inc.).

The Angora rabbit is a small animal compared with the rabbit raised for meat. Its pelt has little value in the fur trade and the hair is used exclusively in fabric manufacture. The hair grows to the length of 2½ to 3½ inches in three months' time. The animal is usually clipped four times a year. A correctly cared for animal will yield 10 to 16 ounces of fur annually. Angora fur is white.

The hair from the more common rabbit, obtained mostly from Belgium, France, Russia, and Oriental countries, is also used in yarns. This hair is shorter than Angora fur, usually not more than ¾ inch in length, and has a higher percentage of guard hairs. The fur of the common rabbit is cut from the pelt while that of the Angora is shorn from the live animal.

Rabbit hair is of two kinds: the wool hair and the beard or guard hair. The wool hair is fine, soft, and very fluffy. The beard hair is shiny, stiff, coarse, and sharp. These tough hairs are often used to give a glint or other novelty effect to the surface of the fabric. Rabbit hair as a rule takes dye less readily and dyes lighter than the fiber with which it is mixed, giving a two-toned appearance to a fabric. Because of difficulties in handling, rabbit hair is almost always used mixed with other fibers. As a rule, it is blended with the fiber before spinning.

Furs from other animals are also making their appearance in the fabric world. Muskrat, raccoon, and mink are beginning to add to the variety of American fabrics.

The wool fiber, a miracle of nature, is one of God's greatest gifts to man. Its unique construction, designed to protect our ovine friends, enables it to give the same ideal protection to mankind.

The fiber's peculiar make-up forms a natural insulation, which tends to keep the body temperature normal. Thus, wool gives warmth in winter and protects the flesh from heat in summer. In fact the properties of this fiber are so different from all other fabric fibers, either natural or man-made, that to date no substitute has been found.

Like all hair, the wool fiber is the continuation of the epidermis, an appendage of the skin. Its root, or "hair follicle" as it is called, secretes an oil that bathes and feeds the fiber during growth and that acts as a lubricant, keeping it always elastic and pliable. In addition, a protective substance, known as "wool grease," is supplied to the fiber by numerous glands in the skin of the sheep. Wool grease serves to preserve the fiber from physical damage and to prevent the fibers from becoming matted and felted together. This grease in a refined form becomes lanolin, the basis of ointments to treat skin diseases and of the best of cosmetics and soaps.

The appearance of the wool fiber is likened to a scale-covered pine cone or to a scaly serpent's skin. A normal fiber from a healthy animal is oval in shape and nearly uniform in diameter throughout its length. "Evenness" in wool fibers denotes quality and serviceability. The wool fiber, in cross section, is composed of three parts: the epidermis, the cortex, and the medulla.

The Epidermis

If a single fiber of wool is taken from the tuft and drawn between the fingers, it is found that it draws very much more readily in one direction than in the other. This is due to the epidermal scales, which are flattened, pointed cells, forming the outside of the wool fiber as the bark covers the tree. This more or less hard covering gives body and resistance to the fiber.

Generally, these horny cells overlap each other like the scales of a fish and are free to two thirds of the depth of the cell. The ends of the scales are somewhat turned out, giving a serrated or notched appearance. These semifree scales have the ability to interlock, one with another, or to act like tiny springs, allowing the fiber to be stretched from one third to one half of its original length without harm.

In finer wools, one scale encompasses the whole circumference and the fiber gives the impression of tiny soup bowls, one inside another. In the coarser fibers with larger circumference, the scales go around the fiber only partially and overlap in both directions.

The marginal scales are most numerous in finer wools. On these fibers, there are as many as 2,900 scales or serrations to an inch, while the coarser wools have as few as 600. The dimensions, uniformity, soundness, and compactness of the epidermis determine the luster, firmness, and strength of the wool.

Above: Cross section of hair (× 500): E—Epidermis, C—Cortex, M—Medulla. Below: Epidermis structure of wool fiber of medium fineness (× 500) (Textile Fiber Atlas).

The Cortex

The major part of the fiber is that which lies just underneath the epidermis—the cortex. This interior part is composed of many elongated cells, the cortical tissue. The cells are spindle shaped, pointed at both ends, and larger in the middle. At the same time, they are more or less flattened. The peculiar shape of the cortical cells and their interesting arrangement, overlapping and interlacing, allow for bending and stretching without impairing strength.

These cells offer remarkable resistance to any rupture. Broken wool fibers, examined under a microscope, disclose the fact that the cortical cells themselves are never ruptured or deformed. They are merely pulled apart.

The Medulla

The pith or core of the fiber consists of several layers of oval cells, larger than those immediately surrounding them. The medulla or fibrocellular center is a more or less opaque portion. The cells are of granular matter resembling pigment. (It is thought that this pigment gives the fiber its color.) This is found particularly in wools that are very white when cleaned and lacking in luster.

This threadlike center portion may run the full length of the hair, may only occur at intervals, or may be absent altogether. The coarser the wool the more clearly defined the granular marrow. Some fine wools have no cellular center at all.

The medulla is living and, since it is nourished by the life of the sheep, it is affected by the same conditions as is the sheep—climate, food, water, and care. Its characteristics change with the changes in the life of the animal.

The medulla differs materially from the remainder of the fiber. Its function in the structure of the fiber has not been fully determined and it is believed by some authorities that its presence in a fiber impairs its strength for, as a rule, wool with conspicuous medulla is less strong than that which has little or none.

Wool Fiber Characteristics

The wool fiber has five physical characteristics: crimp, resiliency, felting ability, absorption, and insulation.

Crimp, says the United States Bureau of Standards, is the difference between two points on the fiber as it lies in an unstretched condition and the same two points when the fiber is straightened under specific tension, expressed as a percentage of the unstretched length. Since wool fiber normally relaxed is much shorter than when stretched, it is said to have a crimp. The wool fiber grows in a more or less wavy form and with a certain amount of twist. The waves in the fiber vary from flat waves to normal waves, to tightly bent waves.

The number of waves is usually in relation to the wool's fineness. As a general rule, the finer fibers have the greatest number of crimps, while the coarser fibers have fewer crimps per inch. On the other hand, one cannot say that the finer the wool the greater the number of crimps, for there are many exceptions. Crimps in wool fiber have a wide variation; they range from one to thirty crimps per inch. As a rule, uniformity and regularity of the crimps in the fiber indicates superior quality.

32

DYED WOOLS MAKE BEAUTIFUL FABRICS

Wool can be dyed from palest pastels to deep, dark or bright, rich colors. It is rarely printed. An exception may be a printed wool challis. The raw stock or fibers, the tops, the yarns, or the fabrics may be dyed.

In the left row, top to bottom, the following fabrics are included. The first is a yellow and black donegal woven in a herringbone twill weave. The fiber dyed yarns have thick yellow nubs or bunches at intervals. The soft doeskin was napped and fabric dyed. The colorful bouclé was woven in a twill weave with bouclé, fiber dyed yarns that were twisted to form loops. The mohair fleece was woven with long, lustrous mohair fibers. The fabric was deeply napped and dyed. The next is another colorful donegal woven with nubbed yarns that were fiber dyed before weaving. The last fabric in this row is a light weight, plain weave sheer made of wool and lustrous mohair. It was fabric dyed.

At the right, top to bottom, are the following fabrics. The first is a twill weave, softly napped fabric made of wool and rabbit's hair. It was fabric dyed. The light weight tweed, with the checked appearance was woven with colored yarns. The colorful, woven fleece was given a soft, deep napping and was fabric dyed. The clothy, loosely constructed, twill weave tweed was woven with dyed yarns. The duvetyn was napped and finished with a velvet-like surface texture and richly colored. The last fabric is a hard textured cassimere, woven in a twill weave with colorful dyed yarns. (All are Botany and Hockanum fabrics.)

Resiliency. The wool fiber has resiliency, which means the power or ability of the fiber to return to its original condition and length after elongation due to strain. Wool fiber mass also has springiness, which is the ability to return to the original volume after being compressed. This resiliency or elasticity in wool is due in part to the crimp of the fiber, but on the other hand the perfect coordination of the epidermal scales and the cortical cells also contributes largely to wool's resiliency. The entire fiber moves in unison and can be stretched 30 per cent or more beyond its own length without rupture and, when released, it returns to its former measurement.

Resiliency gives to wool fabrics the ability to hold their shape, to resist wrinkles, and to withstand wear. Like other physical characteristics of sheep, resiliency of the wool is strengthened or weakened by the physical condition of the sheep.

Felting is the ability of the wool fibers to interlock and contract when exposed to heat, moisture, and pressure. The rough exterior of the fiber is one contributing factor to felting. The pointed, open scales lock and interlock and seem to have an electric attraction for each other under certain conditions. Normally, wool fibers repel each other, but given moisture, softening heat, and pressure, the scales will cling, contract, and ultimately mat.

The hygroscopic inner tube of the fiber and the natural crimp also help, but no single factor causes felting. It is the result of many factors and the exact relationship of each to the others is not clearly known. All wool fibers do not have the same felting ability. In general wools that have higher felting ability are those with scales which are readily responsive to moist heat. Such scales inefficiently protect the fiber's sensitive hygroscopic inner tube. As a rule this wool has coarser, more open scales. Wools that are lower in felting power are those with scales which are hard and which have low hygroscopic power. Most of these wools are finer with more even closed scales. But there is no set rule for all rough scaly fibers do not felt readily nor do all fine fibers have low felting ability.

Wool grown on sheep raised in a more or less even climate seems to felt more readily. Fibers grown on sheep accustomed to extremes of weather seem to possess a harder, more protective surface. But, regardless of its cause, felting ability is one of wool's most important characteristics, peculiar to wool alone.

33

Absorption. Wool can absorb moisture from the surrounding air up to 50 per cent of its weight and can hold it more tenaciously than any other fiber. Wool can absorb moisture up to 20 per cent of its weight without feeling damp. When drying, wool gives off moisture very slowly, thus rapid cooling of the body of the sheep or of the wearer of wool fabric is prevented.

Insulation. Wool's construction also gives the fiber insulating qualities. That is, wool is a nonconductor of heat or cold. The serrations of the fiber and its numerous waves create natural air pockets. This fact, coupled with the fiber's reaction to moisture and air, allows wool to neutralize to some extent the body temperature with that of the surrounding air.

Hair

Hair, like wool and feathers, is in a way the continuation of the skin, having been formed by increased production of the epidermis. It is more or less flattened in shape. The outer end is usually pointed, while the lower extremity is bigger than the shaft and enlarges to form a conical bulb or bundle of cells, which secrete an oily substance that continually bathes the surface of the hair.

Like wool, hair also is composed of three parts: the center or medulla, the intermediate or cortex, and the outer or epidermal scales.

Although hair and wool are very much alike, so similar in fact that sometimes it is hard to tell whether a given fiber is hair or wool —there are marked differences between true hair and true wool.

True hair lacks the crimp of the wool fiber. It is more wiry. The scales of the outer part of the hair fiber have a more nearly rounded, regular form and are more closely fastened to one another, while the serrations on the outside of the wool fiber are irregular, pointed, and are free and turned out to often two thirds of their length.

Lacking the loose pointed scales, as well as probably other necessary physical and chemical properties of wool, hair does not have the felting characteristic of the wool fiber.

Hair, as a rule, is not as easily handled in fabric manufacturing as is wool. However, the hair that can be made into yarn contributes the same qualities to the finished fabric as it did to the animal pelt from which it was taken—light weight with natural insulation.

34

The wool reaches the fabric manufacturing plant in large sacks or bales weighing from 700 to 1,200 pounds. If the weather is cold, the bales are placed in a warm room from twelve to twenty-four hours before being sorted, so that the oil in the wool is soft, making handling easy.

These bales contain a great number of whole fleeces. The first step is to sort the fleeces as well as the wool on each fleece so that similar qualities are together. The fleeces differ in grades just as do the fibers on each individual fleece. The positions of these different grade fibers on the fleece are easy to remember when we keep in mind that nature provides the kind of covering the animal needs. The shoulders, sides, and back usually have the best grades, as these parts of the sheep have less contact with nature.

The fibers on the neck, head, throat, britch, belly, chest, and lower part of the legs are the coarsest, because these parts protect the sheep as he walks and grazes. The upper part of the legs, the lower part of the back, and the loin contain the middle grades. Often the middle grade in one fleece may be as good or better than the best grade of wool from an inferior fleece. Therefore the fleeces must be carefully sorted and the fibers from each section of one fleece be examined and judged.

As one watches a group of sorters quickly examining, pulling apart, and grading a fleece, it seems a very simple procedure; but, actually, the sorting of the many types of fleece and grades of wool fibers calls for years of experience and expert knowledge. The fibers are judged according

Each fleece has many qualities of fibers. The best qualities are found in sections 1, 2, and 3, the coarsest in sections 9, 10, 11 and the intermediate qualities are in sections from 4 to 8 inclusive (Goodall Fabrics, Inc.).

Sorting the grades of fiber in each fleece (Goodall Fabrics, Inc.).

to their length and fineness as well as by their color and quality. Careful sorting is essential because each type of fiber and each mixture of the various types is used to make yarns most adaptable to certain kinds of wool fabrics.

Cleaning the Wool

At this stage the wool is a dirty, oily mass; for the most part, just as it was sheared from the sheep. Even the paint used to brand the sheep may not have been cut out before shipment. The wool contains the sheep's natural oils (that keep the wool fibers lubricated) and perspiration (suint) as well as dirt, sticks, and burrs, the solution in which the sheep is dipped and often paint and tar. All of this must, of course, be removed. Most of it is taken out before the yarns are spun and the last is removed after the fabrics are woven.

Removing the fats and oils. The sorters shake out some of the foreign substances but the real cleaning and scouring job calls for more energetic methods. First, the mass of wool fibers goes through a machine that opens it up and removes much of the dirt and other foreign substances. The wool is dumped into the machine and fed onto a roll through combs that keep it even.

It then moves between two large cylinders with pointed spikes, where it is crushed and beaten, separating the fibrous bunches.

There are various methods of scouring and cleaning wool, but the three principal processes are the soap and alkali process, the solvent process, and the frosted wool process.

The soap and alkali process utilizes soaps and oils combined with alkali (usually soda ash) to form a soapy solution. This is called "saponification." Soap is used to emulsify or break up the fats and oils into tiny particles. Water carries away the fats and oils as well as the dirt which has been acted upon by the soap.

The wool passes through three to six steeping bowls. In a typical five-bowl machine, the first bowl contains water, the second, the alkali solution, the third, soap and water, and the fourth and fifth, rinsing waters. In this long scouring machine, big forks or rakes push the wool through the water in one bowl; lift and place it on rollers that carry it to wringers, which squeeze out the liquid as the wool passes through to the next bowl. Each bowl in turn has rakes that repeat the process. From the last bowl the wool is passed into a huge drying machine.

The solvent process differs radically from the soap and alkali process. A solvent, such as naphtha, dissolves the fats and oils. Then this solution containing the fats and oils is removed from the wool, after which the wool is washed.

The wool is placed in large tanks, which are then sealed. Thousands of gallons of the naphtha solution are passed through the wool. After three such baths the final liquid is drawn off and hot gas is circulated to remove the final traces of the liquid naphtha. The wool then goes through a scouring machine in which it floats through bowls of clear water. It finally passes through rollers and into the drier.

The frosted process works on an altogether different principle from the first two processes. The oil in the wool is frozen and then dusted out, after which the wool is ready for a light scouring.

The wool, on freezing conveyors, goes through very low temperature in a large machine. It is then fed to a dusting machine where the frozen fats are broken into a powdered mass and freed from the wool. This cleaning operation also removes vegetable matter and other foreign substances. The wool is then scoured and dried.

Removing the vegetable matter. Wool contains vegetable substances, burrs, short twigs, straw, and leaves, that have adhered to the fleece of the sheep. All of these have to be removed either when the wool is in a fibrous state or after the fabric is woven. Animal fibers take dyes which do not affect vegetable fibers and, therefore, any vegetable matter left in the wool will usually show up undyed or a different color than the dyed wool. For many fabrics this foreign substance must be cleaned out right after the wool has been scoured. This may be accomplished by a burr-picking process that mechanically removes the vegetable matter, or the wool may be carbonized, which consists of burning the vegetable substance, to be beaten out later.

In the burr-picking machine the wool goes through various cleaning stages. The first opens up the wool and combs out part of the impurities. In the next, strong air currents blow through the fibers, forcing out more foreign substances. The wool passes down between two fast-rotating cylinders (one of which has wire projections) shedding on its way the vegetable matter, which is blown out of the machine. The wool then passes to other cylinders rotating at such high speeds that they eject the remaining vegetable substances. The wool is brushed from the cylinders, freed from most of its impurities.

The carbonizing or burning process may be used for either the wool fibers or the finished fabrics. The principle of the operation is the same for either. The wool fibers are first soaked in an acid bath, which changes the vegetable matter to carbon. The acid solution is then removed and the wool goes through a very large machine containing gradually heated chambers, from 180°, through 210°, to 245°. When it leaves the machine, the vegetable matter is charred. Later, the wool passes through rollers that crush the charred carbon, which is beaten out of the fibers.

A wool washer with a long train of iron bowls (James Hunter Machine Co.).

The wool being pushed through the bowl by large rakes to a carrier, which delivers it up the inclined perforated bottom where it is flushed to the squeeze rolls (American Wool Council, Inc.).

The wool is given a rinse and then a soda bath, to remove any remaining acid in which the wool was soaked. A final rinsing removes all traces of the soda, and the clean, scoured wool is ready for carding.

The fibers, now clean, are ready to start their long journey from a tangled mass of millions of fibers to a tailored topcoat, a colorful dress, a luxurious drapery, a soft, downy blanket, a child's warm snowsuit, a serviceable furniture covering, or a rug—to a fabric made suitable for each requirement.

The decision as to the purpose of the fabric is made before the path the particular wool fibers will travel, is chosen. The warm blanket or snowsuit requires differently processed wool fibers from those utilized for hard-textured suit and dress fabrics.

Woolens and Worsteds

There are two types of wool fabrics, woolens and worsteds. The steps in yarn making are different for each. All wool fibers are carded but already in this initial process the fibers are beginning to be laid parallel for worsted yarns, while this is not essential for woolen yarns. Fibers for woolens are carded only, while fibers for worsteds go through further gilling and combing operations before they are spun into yarns.

39

For the most part, woolen yarns are softer, and more loosely spun. They are used to make the softer, bulkier, and heavier woolen fabrics, such as soft doeskins and kerseys, the bulky tweeds, and all the heavy coatings. The short protruding fibers of the woolen fabrics can be easily napped, creating millions of tiny air pockets that retain warm air and provide insulation.

For the most part, worsted yarns, which are made only of the long, combed, parallel fibers, are tightly twisted, smooth, and strong. They are utilized for the clear-surfaced, hard-textured fabrics, such as gabardine, serge, poplin, and other twilled and ribbed worsted fabrics.

Some fabrics may be made of either woolen or worsted yarns, for example, broadcloth, covert, and flannel. These have different degrees of softness and more or less obliterated weaves, according to the yarn of which they are woven. Knitted fabrics are made of either woolen or worsted yarns. The soft and thick yarns are woolen; the smoother, harder yarns are worsted.

Preparing the Fibers for Spinning

The first step in the manufacture of any kind of fabric is to get the wool fibers into position for making the yarn. If soft or heavy woolens are to be made, shorter fibers are required, and so short (or a mixture of short and longer) fibers are used. If hard-textured worsteds are to be made, only the long fibers are employed.

In preparing the fibers for spinning into yarn, it is necessary to open up the fibrous mass, to remove as much as possible of the

The wool stock passes through hydraulic squeeze rolls and is fed to the dryer at left where it is quickly dried (James Hunter Machine Co.).

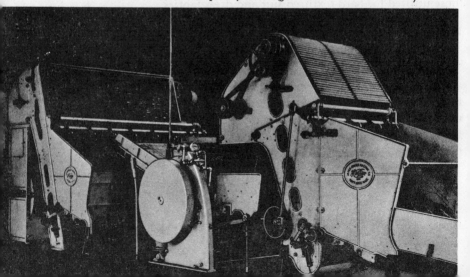

remaining vegetable matter, and to straighten out the position of the fibers. The fibers may or may not be laid parallel, depending on the final requirement.

The fibers are blended at this point. Different types or qualities of wool may be blended, or wool may be blended with other fibers, such as specialty fibers, cotton, rayon staple, waste silk, linen, or casein fibers.

Carding for woolen yarns consists of opening up the fibrous mass of wool, removing the burrs and other vegetable matter, blending and combing out the wool fibers, distributing them evenly in a wide, filmy web, but not laying them parallel. Leaving the carding machine, the wide sheet of long and short fibers is rolled into a loose ropelike form called a "roving" and wound onto spools, ready for spinning into yarn.

Carding for worsted yarns also opens up the wool, removes vegetable matter, blends and combs out the wool fibers, arranging them in a wide web. Leaving the carding machine, the wide web, usually combined with those from other machines is wound into a loose strand, called a "sliver." It may be given a slight twist. The worsted carding machine operates less strenuously than the woolen carding machine, as care must be taken not to break or damage the long fibers, which are later to be further separated from the short fibers and laid parallel for spinning into worsted yarns. This requires two further operations before spinning: gilling and combing.

Gilling further combs out and straightens the fibers and draws them out into thinner slivers. For example, six slivers may be combined into one sliver of the same size.

Combing for worsted yarns consists of combing out the short fibers and laying parallel the remaining long fibers. It also removes any remaining foreign substances. The short fibers, called "noils," are now again a tangled mass. They may be used to make certain types of woolen fabrics. The long fibers leave the combing machine in a soft, loose strand, called "worsted tops" or "combed slivers."

Drawing out and doubling of the combed slivers is the last operation before spinning. Drawing further straightens and parallels the fibers and thins the sliver. Doubling makes the sliver more uniform. These operations produce a strand called a "roving," which is ready to be spun into yarn.

Spinning. The woolen roving from the carding machine or the worsted roving from the drawing frame is next spun into yarn. While different types of spinning machines are used, they all perform three operations: final drawing out until the roving is the desired size, twisting, and winding the spun yarn onto a package such as a cone or a bobbin. These yarns may be used for making fabrics or may be further twisted singly or with other yarns to form ply yarns of many different types as explained in Chapter 10.

Weaving. Preparatory to weaving, the warp yarns are usually wound on large beams, which are placed on the loom to weave the lengthwise yarns of the fabric. (For weaving Wilton rugs the warp is wound on large spools.) The warp yarns are usually the stronger yarns as they must bear the brunt of the strain in the weaving operation. The filling yarns are wound on bobbins which later are placed in the shuttle on the loom. They weave back and forth, over and under the warp yarns to form the fabric's crosswise yarns.

Wools are woven on every type of machine and in practically every weave discussed in Chapter 11. The fabrics range all the way from the sheer, plain weave worsted chiffons and crepes to the heavy woolen and worsted rugs and carpets. The following pages show the important wool fabrics. Those used primarily in the home, as draperies and rugs, are discussed in Chapter 20.

Knitting. The yarns to be used for weft knitting are wound on spools or cones that fit the flat-bed or circular machines. For warp machines, the wool yarns are wound on warp beams similar to those used in weaving.

Knitting yarns must be strong and uniform because they are subject to more strain in elastic knitted fabrics, such as used for sweaters and underwear, than are yarns in woven fabrics.

Wool yarns of any type can be knitted on both warp and weft machines. Such items as underwear, sweaters, and hosiery are usually knitted on weft machines, while dress and lingerie fabrics are more usually warp fabrics. Representative knitted wool fabrics are illustrated on page 54 as well as in Chapter 12.

Felting. Since the wool fiber itself has felting properties, it is ideally suited to make felt fabrics. Wool fibers, alone or blended with other fibers, are felted into fabrics by the application of heat, moisture, and pressure. The fibers intermesh into a firm, durable fabric. Wool felt fabrics range from very thin to as thick as that used for rug cushions.

THE SPECIALTY FIBER FABRICS.

Above, left to right: Wool and rabbit hair, cashmere, vicuna, alpaca and wool, alpaca pile. Below, left to right: Llama, mohair, camel, Shetland.

Left to right: Gabardine, serge, sharkskin, cheviot.

HARD TEXTURED WOOL FABRICS

These are firmly woven, durable fabrics with a clear, hard surface, constructed to withstand wear. All but batiste are made of worsted yarns. Some may utilize the more tightly twisted woolen yarns as well as worsted yarns.

Plain weaves are used for batiste and, usually, for tropical worsted. *Batiste* is a softer woolen fabric, often printed. *Tropical worsted* has a firm, open weave that makes the fabric suitable for warm weather wear.

Rib weaves, which can be strong and durable, are used for the following fabrics, each quite different in appearance. *Bedford cord* (also the name of the weave) has pronounced ribs that run lengthwise in the direction of the warp. *Poplin* has crosswise ribs, very fine and closely spaced. *Ottoman* has broad, flat ribs running crosswise in the direction of the filling. *Bengaline* is light weight with fine ribs running crosswise.

Twill weaves are also employed for hard textured fabrics. *Serge* may have the finest twill or may have a heavier twill, easily discernible on both sides of the fabric. It is given a soft, smooth finish. *Gabardine,* being hard finished, clearly shows its fine diagonal lines on the face but not on the back of the fabric. *Hard finished worsted* is very similar to gabardine, but the twill is discernible on both the face and the back of the cloth. *Elastique*

44

Left to right: Elastique, cavalry twill, whipcord, unfinished worsted twill, covert.

is recognizable by its ribs in groups of twos on the face, not apparent on the back of the fabric. *Cavalry twill* has ribs heavier and wider spaced than those in gabardine. *Whipcord* has the most pronounced ribs of this group and is usually a heavier fabric.

Wool sharkskin is woven with a twill or a herringbone twill weave. Stripes and plaids are often woven into the fabric, making an attractive, closely woven, durable material for men's suitings. *Covert* has white and colored yarns alternating in both warp and filling. The twill is left to right, and the color lines are arranged from right to left, giving the fabric a speckled appearance.

Left to right: Bedford cord, poplin, bengaline.

Left: Kersey. Right: Duvetyn.

SOFT TEXTURED WOOL FABRICS

These are soft surfaced, lightly napped, or fulled and brushed fabrics of woolen or worsted yarns, mostly woven with a twill or satin weave. The weave is more pronounced in some fabrics than in others depending on the finish given the cloth. *Flannel,* usually woven in a twill weave, is given a light napping but the weave remains clear. *Unfinished worsted,* similar to flannel but heavier, is slightly napped and not closely sheared.

Another group of soft textured wools has the weave more obscured. *Broadcloth,* a smooth, dense fabric, is napped, sheared, and polished to a soft velvety texture. *Zibeline,* a highly lustrous fabric, with a somewhat hairlike surface, is napped and pressed flat, producing a silky velvet sheen. *Doeskin,* woven in a satin weave, has a fine, smooth surface with the nap laid down. *Duvetyn,* woven in a twill weave, has a silky, velvetlike, short nap. *Kersey,* a highly lustrous fabric, is very closely woven, napped, and sheared.

The fabrics with the more pronounced weaves may be plain colored but some, particularly flannel, are woven with dyed yarns, in stripes, checks, or plaids. The fabrics with the weaves more obscured are usually in plain colors.

Left: Chalk striped flannel. Right: Plain colored flannel.

Left: Plaid flannel. Right: Zibeline.

CREPE WOOL FABRICS

True wool crepes are woven with alternate right- and left-hand twisted yarns in the filling or in the warp or in both. The twisting of crepe yarns is discussed on page 288. The yarn and not the weave characterizes this fabric. Most crepe fabrics are woven in a plain weave as it permits the yarns to twist and form the crinkled, grainy surface texture.

Crepe fabrics range from the very sheer to the quite heavy but are usually light weight. They may be hard- or soft-textured, according to the type of yarns used and may be dyed in the yarn or in the piece, generally in plain colors.

Simulated Crepe Weaves

Some fabrics are woven to simulate true wool crepe without the use of crepe-twisted yarns. Usually, these fabrics have irregular weaves, sometimes a variation of the satin weave and sometimes a combination of plain and twill weaves. Or, thick and thin yarns may be loosely woven into a fabric to imitate those woven with crepe yarns.

Other crepy appearing fabrics may be woven in the leno or another loosely constructed weave, resulting in soft, spongy fabrics.

Left to right: True crepe, mock crepe, leno weave sheer.

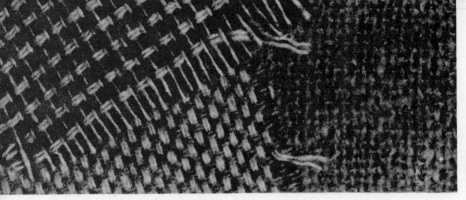

Left: Two hopsacking fabrics. Right: Homespun.

TWEED WOOL FABRICS

Fabrics in the tweed family range from the closely woven, rather smooth-surfaced tweed suitings to the spongy, rough-textured, loosely woven, heavier tweed coatings. They are woven in plain, basket, twill, herringbone twill, or novelty twill weaves. Most of these fabrics are of two or more colors. Some, called *monotone tweeds,* are of various shades of one color. Woolen yarns are generally used but some tweeds are woven with worsted yarns.

Homespun is a plain weave fabric, usually a loosely woven, spongy, woolen cloth, but it may be made of more tightly twisted yarns, in a more compact weave. At times, yarns of mixed colors are used to give more the appearance of the original homespun. *Hopsacking* is noted for its basket weave. Usually two filling yarns cross over and under two warp yarns, but it may be woven with three fillings crossing three warps. Sometimes different colored yarns are so woven in a basket weave that it simulates a twill

Left to right, tweeds: Twill, herringbone, and fancy twill.

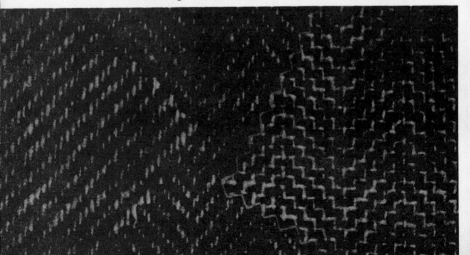

Left: Donegal tweed. Right: Cassimere.

weave. A *Donegal tweed* is usually a plain weave fabric, but it may be in a twill weave. It is distinguished by the colored nubs on the yarn that distribute bunches of different colored wool throughout the fabric. It may have a close weave, produced with firmly twisted woolen yarns, or it may be a heavier, more loosely woven fabric, made with large soft yarn.

A wide range of tweed suiting and coating fabrics are woven with thick irregular yarns that have nubs or slubs at intervals. These are somewhat like Donegal tweeds except that the nubs are of the color of the fabric and not of many different bright colors as those characterizing the Donegal tweeds. *The tweed fabrics* proper are woven in plain, twill, or herringbone twill weaves, or in novelty twill weaves that are combinations of plain, twill, and herringbone twill. Tweeds may be heavy, medium, or light weight. They range from rough textured fabrics to those with moderately smooth surfaces. Hard-finished herringbone twill weave tweeds are often called *cassimere*.

Left to right, tweeds: Fancy twill and plain weave.

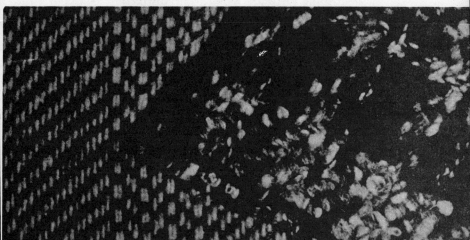

COATING WOOL FABRICS

These are heavy, warmth retaining fabrics. They are made of short as well as long wool fibers spun into rather loosely twisted woolen yarns. Coating fabrics may be divided into three groups; thick, pile or napped fabrics with raised surface textures; heavy, fulled and felted, smooth surfaced fabrics; and lighter weight, yet close, firm fabrics.

Group One. *Astrakhan,* a soft, thick, spongy fabric, usually with cotton back and wool or mohair face is made with lustrous, curled, loose, thick yarns. The pile is formed with extra warp (or filling) yarns curled before weaving. *Chinchilla* also has a raised or napped surface, but the pile has been rolled into close, thick tufts imitating chinchilla fur. It may be woven in a twill weave or knitted and then napped and finished; or it may be a double fabric, often with a plaid back. *Frieze* is a thick, sometimes double fabric, woven in a twill weave that runs from right to left. The face is napped and pressed down, sometimes in one direction and sometimes in another. *Fleece* is a thick pile fabric, with the pile erect. It may be woven or knitted and then napped, sometimes on both sides, or it may be woven with an extra pile yarn. *Velour* (the one illustrated here) is a soft fabric, with a smooth, napped face. It is woven in a satin weave and napped with a longer and more open nap than is duvetyn. Another type of velour is woven in a pile weave. It and frisé, another pile fabric, are discussed in Chapter 20.

Group Two. *Mackinaw cloth* is a thick, close, fulled and felted woolen fabric, woven in plaids with dyed yarns and slightly napped. *Melton* is similar in weight and closeness of weave to mackinaw cloth but is in plain colors. Its satin weave gives a smooth, soft surface texture. *Montagnac,* woven with floating filling yarns, is a soft, bulky, and rather lustrous fabric. The slack twisted filling yarns are napped into distinctive, small curly tufts.

Group Three. *Venetian cloth,* similar to melton in its smooth surface texture is produced by the satin weave, but is much lighter weight. These lighter weights are used for topcoats.

Opposite page, top to bottom, left: Frieze, velour, plaid-back chinchilla, melton. Right: Mackinaw cloth, Venetian cloth, knitted fleece (center), woven fleece. Below, center: Astrakhan.

NUBBY WOOL FABRICS

These fabrics are woven with plied yarns. That is two or more yarns are twisted together. The yarns are of different thicknesses and twisted together in such a way that, when woven, they produce nubs or bunches or small loops on the fabric. These yarns are illustrated and discussed in Chapter 10.

Bouclé fabric takes its name from the bouclé yarns with which it is woven. Two fine yarns are twisted together and this yarn is twisted with a thick yarn producing loops at intervals in the fabric. The fabric may be woven or knitted and has a kinky rough surface. Bouclé yarns are used in both warp and filling.

Ratiné fabric takes its name from the ratiné yarns with which it is woven. A fine and a thick yarn are twisted together and this yarn twisted with a fine yarn producing nubs on the surface of the fabric. This fabric also may be woven or knitted. In the woven fabric the ratiné yarns are used in the warp only and give the fabric a coarse, knotty appearance.

Éponge is similar to ratiné but the ratiné (or snarl yarns) are in the filling only. The fabric is spongy with an irregular surface caused by the nubs on the ratiné yarn.

Needle point is woven with very fine bouclé yarns making knots which closely cover the entire fabric. It is a close, firm cloth with a rough, finely nubbed appearance.

Other fabrics have similar textures produced by the weave rather than the yarns.

Top to bottom: Éponge, needlepoint, bouclé, ratiné.

SHEER WOOL FABRICS

These are openly woven fabrics, made with fine yarns. They are cool, light weight, and sheer. All are woven in a plain weave except some tropical weights, which may be in a twill weave.

Albatross is a soft, light fabric, made with soft, fine yarns. It is lightly napped and is dyed in plain colors. *Bunting,* used for flags, is a loose and openly woven, porous fabric, made with tightly twisted worsted yarns. *Challis* is noted for its small, usually printed floral or other design. It is light weight, firmly but not too closely woven. *Nun's veiling* is soft and thin but has a rather hard feel. It is woven with fine woolen yarns and is white or is dyed in plain colors. *Voile* is a sheer, openly woven fabric of highly twisted worsted yarns. *Tropical weights* may be made of woolen or worsted yarns in plain or twill weaves. Some are more closely woven than others. Frequently they are part mohair, which gives the fabric a crisp, firm hand. *Wool sheers* also include a group of light, open fabrics, woven with crepe twisted yarns or in novelty weaves that give a crepe surface texture. Many are woven in plain or basket weaves, and some in leno or twill weaves. Most of them are constructed of worsted yarns, but some are of the softer woolen yarns.

Left to right, top: Mock crepe sheer, albatross, printed challis. Bottom: Tropical worsted, herringbone twill sheer, plain weave sheer.

KNITTED WOOL FABRICS

Wool is especially desirable for knitted fabrics. Wool yarns for knitting can be as fine as those used for weaving. Worsted yarns may be knitted into firm textured warp knits that have sheer drapability but little elasticity. On the other hand, knitting machines are equipped to handle very heavy yarns. Thick woolen yarns or multiple ply yarns may be knitted into heavy cardigans that have the maximum amount of stretch and warmth giving qualities. Wool is especially desirable for fabrics that need napping to give added warmth or softness.

All varieties of wool yarns may be knitted but they should be spun strong, even, and smooth. Because of knitted fabric's elastic construction, especially in weft knits, yarns move, stretch, and return to normal length without the protection of yarns in close contact as are woven yarns.

Top to bottom: Plain, circular knit. Napped, plain knit with an extra yarn on the back. Plain, circular knit. Napped, tricot knit. Plain knit, napped sweater fabric with extra yarn on back. Lacy knit, made on Raschel machine.

FINISHING WOOL FABRICS

After the fabric has been woven or knitted, it passes through various finishing processes to attain its surface texture, its hand and feel, its color (if not woven of dyed yarns), in other words, its final personality.

As it leaves the loom, the wool fabric is in a limp and lifeless form. Woolens are loosely woven and have little character. While hard finished worsteds retain the appearance of their weaves, they too are without firm hand or clear surface.

The beauty of a hard textured worsted fabric is in its weave. Therefore, the finishes used bring the weave into clear relief and give a firm yet resilient body to the closely woven fabric.

The beauty of a soft textured worsted or woolen fabric is in its partly concealed weave, which remains evident as a part of the surface texture. Therefore the finishes do not obliterate the weave but slightly raise the

Top to bottom: Nailhead, Tattersall check, Glenurquhart plaid, chalk stripe, herringbone twill, shadow stripe, check, pencil stripe, hound's tooth.

55

fibers to give a light film, blending the colors and the weave.

The beauty of a tweed fabric is in its combination of colors in the weave. The finishes close up the weave, giving a compactness to the loosely spun yarns.

The beauty of a napped coating is in its soft, luxurious, deep pile. The finishes raise the nap and shear it to lustrous evenness.

The beauty of a wool fabric with a high sheen is in its velvety smooth surface, reflecting lights and shadows. It therefore requires finishes that raise the fibers and press them down in one direction.

All wool fabrics must have a finish, even though some worsted fabrics such as gabardine require little finishing in comparison to chinchilla (which requires a great deal).

Preparatory Treatments

Before undergoing the finishing processes that change its character it is necessary to prepare the fabric for these finishes. It must be carefully inspected and repaired if necessary, and it must be thoroughly cleaned. These are steps that contribute to the final appearance and make the finishing processes successful. These processes are more fully discussed in Chapter 16.

Inspecting, burling, mending. Before the fabric can be finished it must be inspected for such flaws as knots caused by tieing yarns together during spinning and weaving, for yarns missing in the weave, and for extra yarns woven into the fabric. The knots are cut out, the missing yarns are woven in by hand, and the extra yarns are carefully pulled out.

Scouring. At some stage in the wet finishing, oils and grease, sizing, and soil acquired during spinning and weaving, must be removed. The fabric is scoured with soap and alkali to remove these foreign substances.

Piece carbonizing. The fabric is carbonized if the wool fibers were not carbonized prior to spinning into yarns. This process burns out the vegetable matter. The fabric is soaked in an acid bath, which changes the vegetable matter to carbon. The fabric then goes through a machine that chars the vegetable matter, which is later shaken or beaten from the cloth.

The Finishing Processes

The following is a summary of the most important finishes

given to wool fabrics. The processes are fully explained in Chapter 19. They are not listed necessarily in continuity because the method of procedure for one fabric may be quite different from that for another. The processes are arranged here for convenience in discussions, and a fabric may not follow the sequence as listed here. Of course, all fabrics are not given all of the finishes described.

The first four finishing operations are applied when the wool fabric is wet or damp and are called "wet finishes." The fabric is dried before the "dry finishes" are applied.

Wet Finishes

Burning off loosened fibers and lint. This is called "singeing." Only worsted fabrics with a clear, hard texture are singed. The damp fabric, moving at a high rate of speed, passes over gas flames or red hot plates or rolls, and the loosened fibers and lint are burned off. As the fabric leaves the machine, it is plunged into water to remove any danger of burning.

Setting the warp and the filling. This is called "crabbing." As stated previously, the fabrics leaving the loom are in a rather limp state and the warp and filling may be irregular. The fabric passes through boiling water and then through cold water, which sets the warp and the filling in the weave.

Raising the fibers. This is called "gigging" or "napping." In this process the fibers are raised, either slightly to form a slight nap, or energetically to form a deep pile. Gigging utilizes small vegetable teasels from a thistle plant grown in Oregon and New York State. Napping is done by strong metal wires that give a deeper napping. The nap may be left upright or may be pressed down in one direction, as in broadcloth.

Setting the pile. This is called "wet decating," and it is usually the last operation in wet finishing, as it sets the pile so that it will remain the same through dyeing and dry finishing. At the same time it adds luster to the fabric. Heat and moisture is applied to the fabric while it is held at a certain tension on the roll.

Dyeing

Wool may be dyed at any step in manufacturing, from the raw fibers to the last wet finishing process applied to the fabric. The raw stock (fibers), the tops or slubbings (the combed worsted strand before being spun into yarn), the yarns, or the fabric may be dyed. The same dyestuffs may be used to dye wool in any of these four stages of processing.

Wool is an animal fiber and reacts favorably to acid dyes. When direct and chrome dyes are applied, a mordant must be used, which is some chemical that fixes the dye on the fiber. Wools are rarely printed.

Dyes are selected in consideration of the desired appearance and ultimate use of the fabric. As examples, some have a high degree of fastness to water and sunlight, others are susceptible to one but not to the other, while still others have a relatively low degree of fastness to both. Some dyes are bright and others are duller. The following are those dyestuffs used for coloring wool.

Acid dyes are customarily applied directly from an acid bath. Depending on the type and quality of the dyestuffs and the method of application, the final color varies from poor to excellent in its resistance to water, perspiration, light, and alkalies. Acid dyes have a wide variety of colors and can produce bright shades.

Mordant and chrome dyes are excellent for wool fibers. They give a wide range of colors that have a high degree of fastness to sunlight and a satisfactory degree of fastness to washing and alkalies if properly selected and applied. The wool must be mordanted with a metallic substance to fix the dye.

Basic dyes are used where high, bright shades are required. They have a poor degree of fastness to light and to washing and are employed when the fabric is not expected to be exposed to the sun. The dyestuff is applied directly to the fabric from a neutral or slightly alkaline bath.

Direct dyes have a wide color range and produce even, level colors but not as bright shades as acid and basic dyes. While they may vary from a poor to an excellent degree of fastness to light they have a poor fastness in washing, therefore direct dyes are used for fabrics that require infrequent or no washing. They are dyed from a neutral or an acid bath. Only a few of the direct dyes are used for animal fibers.

Vat dyes are used to a limited extent for wool. They have an excellent all-round degree of fastness. They include a wide range of shades and produce clear colors that are excelled in brightness by only acid and basic dyes. Vat dyes, being insoluble in water, must first be made soluble. After the wools are dyed, they are exposed to air or to chemicals to oxidize or fix the dyestuffs.

Dry Finishes

After the wool has been dyed and the wet finishes have been applied, the fabric is dried and given whatever is required of the following dry finishes.

Straightening and drying the fabric. This is called "tentering." During weaving, wet finishing, and dyeing, the selvage becomes irregular and the warp and filling yarns may slip out of line. At either side of the tentering machine are bars, adjusted to the width of the fabric. As the fabric passes through the machine, pins or clips on the bars hold the selvage of the fabric, even its width, and adjust the warp and filling yarns. The fabric then passes through hot to cool chambers and is dried.

Cutting off the raised and loose fibers. This is called "shearing." Most wool fabrics are sheared. It is not possible to singe worsted fabrics a second time, so the protruding fibers are cut off. Pile and napped fabrics are sheared to even the yarns. Pile fabrics may also be cut to produce patterns, such as checks and stripes. A large machine that operates much as a lawn mower (except that the blades are motionless and the fabric moves) shears off the protruding fibers or the raised yarns.

Brushing the fabric. Since the shearing machine leaves cut fibers in the fabric, the cloth is brushed to remove the short fibers.

Pressing the fabric. Moisture, heat, and pressure, the same as in pressing at home, are used for pressing the fabrics. This irons out the wrinkles and smooths the fabric, usually adding a luster.

Making the finish permanent. This is called "dry decating." This finish is necessary if the luster obtained from pressing is to be made permanent. Hot steam forced through the fabric sets the luster. The steam and pressure also set the yarns in the width and length of the fabric.

Steaming the fabric. Through most of these processes the fabric is exposed to moisture and heat, which tend to shrink it. A special steaming process removes the luster, with no effort made to shrink the wool fabric. A steaming process that includes pressure is for the purpose of shrinking wool fabrics.

Shrinking the fabric. The felting quality, peculiar to the wool fiber, causes shrinkage upon the application of moisture, heat, and pressure. If a shrinkage control finish is not given, the fabric will shrink. There are three methods of shrinking wool: hot steam, cold water, and chemical. With hot steam and cold water methods the fabric will shrink further when laundered. When given a chemically controlled shrinkage, the wool has little tendency to felt, thereby limiting future excess shrinkage.

Functional Finishes

While the wool fiber itself has inherent qualities that give it elasticity, resiliency, and absorbency, wool fabrics are frequently given special finishes to increase these properties as well as to give added features.

Wool fabrics do wrinkle and crease, but the elasticity of the wool fiber will, under normal circumstances—such as hanging the garment in circulating air—return the fabric to its original appearance. However, the fabric may pass through a special finishing process that makes it resistant to creases and wrinkles. A great many wool fabrics are being made water-repellent, wind, spot, and stain resistant. Some are rendered waterproof and others fireproof.

Wool, an animal fiber, is a favorite of moths. Today, wool fabrics can be given finishes that will make them resistant to moths. Some of these finishes are durable, others are less durable. While germs and mildew are not as active on wool fibers as on vegetable fibers, for certain uses, wool fabrics may be given finishes that make them resistant to germs or to mildew.

Fabrics or merchandise made of fabrics given these functional finishes usually carry a label stating the type of finish, what its effect is, and how to care for the fabric.

WOOL IS IDENTIFIED

In 1939 the Senate and the House of Representatives passed an act cited as the "Wool Products Labeling Act of 1939." It is for the protection of manufacturers, stores, and consumers.

Each and every product containing wool must be marked with means of identification bearing the percentage distribution of the total fiber weight of each fiber, exclusive of ornamentation (not exceeding 5 per cent of the total weight) .

The percentage of the kind of wool, whether new wool, reprocessed, or reused wool, must be designated.

The term "wool" means the fiber from the fleece of the sheep or lamb, or hair of the Angora or Cashmere goat (and may include the so-called specialty fibers from the hair of the camel, alpaca, llama, and vicuna), which has never been reclaimed from any woven or felted wool product.

The term "reprocessed wool" means the resulting fiber when wool has been woven or felted into a wool product, which, without ever having been utilized in any way by the ultimate consumer, subsequently has been changed into a fibrous state.

The term "reused wool" means the resulting fiber when wool or reprocessed wool has been spun, woven, knitted, or felted into a wool product which, after having been used in any way by the ultimate consumer, subsequently has been made into a fibrous state.

Merchandise marked "reprocessed wool" or "reused wool" is not necessarily inferior. There are many qualities of wool from many types of sheep, as well as different qualities of fibers in one fleece. A low grade of new wool may be inferior to a reprocessed or reused wool that was first made from a high grade of new wool. However, reprocessed and reused wool always make a product inferior to the one made of the same grade of new wool.

New wool fibers have more resiliency, because they were not exposed to damage, than do reprocessed and reused wools, and they can be woven into fabrics and made into garments that will retain their shape to a greater degree than those made of reworked fibers. Resiliency also aids in creating air pockets that are warmth retaining and new fibers, under ordinary circumstances will construct a fabric superior in warmth-giving qualities.

Many fabrics, constructed of reprocessed and reused wool by highly reliable manufacturers give satisfactory wear, appearance, and warmth (if required) at a price lower than is possible with the use of good-quality, new wool.

The terms "mohair" and "cashmere" or the name of any of the specialty fibers may be used for such fibers in lieu of "wool," provided the percentage of each such fiber is given.

The term "virgin" or "new" may be used if the fabric is composed of wool that has never been reprocessed or reused.

Pile fabrics may be marked to show the fiber content of the back and pile separately. Examples are, 100 per cent wool pile and 100 per cent cotton back (the pile constitutes 40 per cent of the fabric and the back 60 per cent). Linings, interlinings, padding, stiffening, trimmings, and facings must be identified separately if they contain wool, reprocessed, or reused wool.

SUMMARY

AN ANIMAL FIBER FROM SHEEP AND OTHER HAIR- AND
FUR-BEARING ANIMALS

Wool types: Fine, medium, long, and cross-breed wool.

Specialty fibers: Angora goat, Cashmere goat, camel, llama family
(llama, alpaca, guanaco, vicuna, huarizo, misti) and rabbit.

TO MAKE THE YARNS AND FABRICS

Shear, clean, sort, and blend.

Card, for woolens and worsteds: To remove foreign substances and
to partially lay the fibers parallel.

Comb and gill, for worsteds: To remove short fibers and to lay
the fibers parallel.

Draw and double: To further mix fibers and to lay them parallel
and to produce a strand (sliver), which is further drawn out
and slightly twisted into a roving.

Spin: To draw out and twist the roving into a yarn.

Construct the fabric: By weaving, knitting, twisting, or felting.

TO FINISH THE FABRICS

Preparatory treatments

    Burl: To inspect and mend.

    Scour: To clean the fabric.

    Carbonize: To remove vegetable matter by burning.

Wet finishes

    Singe: To burn off loose fibers.

    Crab: To set the warp and filling yarns.

    Gigg or nap: To raise the surface fibers.

    Wet decate: To set the pile and add luster.

    Shrink: To reduce the fabric in length and width.

    Bleach, dye, print: To make the fabric white or colored.

Dry finishes

    Shear and brush: To cut and brush off fibers.

    Tenter: To stretch, straighten, and dry the fabric.

    Press: To smooth the fabric.

    Steam: To shrink the fabric.

    Dry decate: To add luster and shrink the fabric.

Functional finishes: To render the fabric water-repellent, waterproof, resistant to creases, moths, germs, mildew, and fire.

Opposite page: A cotton field ready for harvesting (Bureau of Plant Industry,
Soils and Agricultural Engineering U.S.D.A.).

COTTON—VEGETABLE FIBER

The most plentiful

IN THE BEGINNING

Cotton as a source of human comfort first trickled into the dawn of history in far off India—say all the earliest chroniclers. Just how and why India, and only India, nobody knows, but all tales told by the first historians trace cotton's beginnings to this, then highly civilized, Asiatic nation.

Of course, since the exploration of the Americas, we have learned that cotton cultivation and use must have begun equally as early in wholly alien lands half a globe away, the lands now known as Mexico and Peru. Cotton has been found also in the pre-historic Pueblo ruins of Arizona. But new-world civilizations were unknown to us for many, many centuries, while India's cotton story has been passed along through countless entertaining channels since Sanskrit, the earliest known language of India, was recorded for posterity.

Just when did the use of cotton begin in countries so far apart and utterly foreign to each other? It is impossible to say. India's first records imply that cotton growing and weaving were at that time arts of long standing, while the origin of the use of cotton in Mexican and pre-Inca civilizations can be only partially determined by comparing the known works of art of the Western world at its assumed height with those of India of the same period.

As a fabric fiber, cotton may be as old as wool or older, but since our earliest historical records come from countries that did not know cotton at that time, it would seem that wool was used by man many centuries before cotton. This explains the fact that

COLORFUL COTTON FABRICS

Cotton fabrics, whether soft or crisp, thin or heavy, smooth or rough textured, lustrous or dull, can be dyed or printed with bright or subdued colors that have a high degree of fastness to water and sunlight. Below are a few of the many important cotton fabrics that add freshness and color to both wearing apparel and homes. Starting at upper left, counterclockwise, they include the following.

Gingham. A closely woven, plain weave fabric made with tightly twisted yarns that were dyed before weaving. The interweaving of the colored yarns forms a colorful plaid that is the same on both sides.

Chintz. This fabric has a surface glaze that remains after repeated washings. It was roller printed by a combination of methods.

Piqué. This is called "waffle piqué." The color was printed on by hand screening and was applied directly to the fabric in large, bold, colorful prints.

Lawn. This is a soft, thin fabric which was first dyed and the white pattern was obtained by discharging the color.

Plissé. Above the lawn is a checkered plissé. The color was printed on the fabric. The crinkled stripes were obtained by plissé printing.

Sateen. This smooth and softly lustrous sateen was first dyed and the pattern printed on after the ground color was discharged.

Poplin. This fabric was woven with crosswise ribs. The colored stripes were printed on by direct roller printing.

Organdy. This crisp, sheer fabric received its appearance from plissé printing, that caused the crinkled stripes, and from direct color printing.

Percale. While many percales have a colored ground, this fabric has colored flowers printed on a white ground by direct roller printing.

Batiste. This very small sample is a glazed fabric with colors printed on a white ground.

in most of the early accounts, cotton was described and explained in terms of the already familiar wool fiber.

Cotton's known history began about 3000 B.C. Specimens of cotton unearthed in ruins of the city of Mohenjo-Daro in the Indus valley of India show that cotton growing and manufacturing were a part of civilization even at that early date. One specimen found in the ruins is in the form of a small fragment oxidized in the handle of a silver vessel. It is interesting to note though, that the most ancient cotton fabrics now unearthed are the grave cloths of the pre-Inca civilization of Peru. It would seem from these early samples, that there was no connection between the Asiatic and American cottons—that the New- and the Old-World cotton were two distinct species, seemingly independent discoveries.

Early literature contains many references to cotton. A Hindu Rig-Veda hymn, some fifteen centuries before Christ, mentions cotton. The Bible refers to cotton numerous times. It is mentioned in King Solomon's time (1015-975 B.C.). The word "karpas," in the book of Esther, is said to mean cotton, and some authorities believe that Joseph's coat of many colors was hand printed cotton.

In 800 B.C., Manu, a writer of law books in India, states, "Let a weaver who has received ten palas of cotton thread give them back increased to eleven through rice water, which is used in the weaving. Whoever does otherwise shall pay a fine of twelve panas."

Herodotus (484-425 B.C.), the Greek chronicler, wrote, "The wild trees of that country (India) bear fleeces as their fruit, surpassing those of the sheep in beauty and excellence, and the Indians use cloth made from this tree wool."

When Alexander the Great invaded India in 327 B.C., he took back to Greece some of India's printed cotton, and records show that the "flowered robes" were greatly treasured by the Greek people. One of Alexander's generals wrote in his journal of the "linen from trees."

Greek Megasthenes in 300 B.C., an ambassador to the court of Chandragupta in India, wrote, "In contrast to the general simplicity of their lives, the Indians love finery and ornaments. Their robes are worked in gold and ornamented with precious stones and they wear also flowered garments made of the finest muslin."

Theophrastus (372-287 B.C.), a Greek disciple of Aristotle, said, "The trees from which the Indians make cloths have a leaf like that of the black mulberry but the whole plant resembles the dog rose. They set them in the fields, arranged in rows so as to resemble vines at a distance. They bear no fruit but the capsule containing the wool is, when closed, about the size of the quince. And when ripe, it expands so as to emit the wool which is woven into cloth either cheap or of great value."

The Roman recorders of 63 B.C. state that cotton awnings were used at the Apollonaris games held on July 6. But it was not until about the beginning of the Christian era that cotton left India in any quantities. About this time, Arabs were bringing Indian calicoes, muslins, and other cottons to ports on the Red Sea. Cotton plant cultivation was also beginning to spread to lands other than India.

In A.D. 77, Pliny wrote in his *Historiae Naturalis,* "In upper Egypt, toward Arabia, there grows a shrub which some call gossypium, and others xylon, from which are made the stuffs we call xylina. It is small, and bears a fruit resembling the filbert, within which is a downy wool, which is spun into threads. There is nothing to be preferred to those stuffs for softness and whiteness; beautiful garments are made from them for the priests of Egypt."

The first cotton grown on European soil was grown at Elis in Greece sometime between A.D. 100 and 200. These cotton yarns were manufactured into hair nets. Cotton cultivation and weaving were introduced into Spain by Abd-er-Rahman III (A.D. 891-961), and became established in Sicily while the Island was under the rule of the Arabs. Between A.D. 1200 and 1300, Barcelona flourished as a cotton manufacturing center, specializing in sailcloth.

In 1789, President Washington visited the cotton mill at Beverly, Massachusetts (New York Public Library).

The stormy career of cotton in England began about this time, and it is interesting to note that cotton's first use in England was not for clothing but for candlewicks. Very early also it was used for trimming doublets, for some sort of defensive pads, and as part of fortifications. In a Wardrobe Act in 1212, there is mentioned the price of 12 pence for a pound of cotton for stuffing the acton of King John. The English mixed cotton with flax and wool for cheaper fabrics. They also decorated their fabrics by embroidering with cotton threads.

A very entertaining story circulated through Europe during the thirteen, fourteenth, and fifteenth centuries as to the source of this wonderful fiber, cotton. Many who never traveled believed that in Tartary there grew a vegetable plant, the flowers of which were tiny lambs. The fleece of these lambs was used for fabrics and their flesh for food. The tiny lambs slept in the buds by day and leaned out at night to feed upon the grass beneath the plant.

In the fourteenth century an Englishman upon returning from Tartary wrote, "There groweth a manner of Fruyt, as though it were Gourdes; and when ther been rype, men Kutten hem ato, and men fynden with inne a lyttle Beast, in Flesche, in Bone and Blood, as though it were a lyttle Lomb with outer Woole. And men eten both the Fruyt and the Beast; and that is a great Marveylle."

Western-World Cotton

Cotton in middle America had been the basis of an elaborate fabric art for centuries. The clothing of America's ancients were made of true cotton fabrics instead of animal skins or bast fibers as in Egypt and other Old-World countries.

When Columbus landed in the New World he found wild cotton growing on Hispaniola and other West Indian islands. His journal dated October 12, 1492, describes the natives of Watling Island, where he first landed, bringing skeins of cotton thread out to his ship. In fact, the presence of cotton on the islands

67

was one of the main reasons Columbus thought he had reached India. On his return, Europe saw for the first time Sea Island cotton.

In 1519, Magellan found the natives of Brazil sleeping on beds of cotton down; and when Pizarro entered Peru in 1527, the natives were wearing cotton clothing. In fact, the early fabrics found in Peru and Mexico at this time were equal in beauty and perfection to the best that India produced.

Due to these discoveries of new and better cotton, many of the great explorations of the fifteenth and sixteenth centuries were undertaken in search of additional sources of cotton supply.

Cotton was first planted on the mainland of America, in the present state of Florida, by Spanish settlers, and the colonists planted cotton in Virginia in 1619. Cotton growing was an almost immediate success, and as early as 1621 England was complaining of the import of American cotton.

In 1700, England, to protect her wool industry, began passing laws prohibiting the use of cotton. An act of Parliament in 1721 imposed a fine of 5 pounds on the wearer and 20 pounds on the vender of cotton goods. England's restrictions on the manufacture of cotton materials forced the colonists to make their own cotton fabrics. For lack of machinery, these early efforts were of necessity household enterprises, and it is said that in 1700 cotton clothing was used by one fifth of the inhabitants of North Carolina.

In 1787, John Cabot and Joshua Fisher built the first American cotton factory at Beverly, Massachusetts. It was not a success. In 1788, a well-equipped factory was built in Philadelphia. In 1790, there were 3,138 bales (each 500 pounds) of cotton grown in America and 379 bales exported.

But it took the simple invention of the cotton gin in 1798 to really allow "King Cotton" to come into his own. In 1825, 533,000 bales of cotton were grown, and cotton plantations expanded to all the southern states. In spite of some major setbacks, such as the Civil War, cotton has continued to grow, until in 1945, some 21 million persons in the United States depend for their living on the growing, distribution, and manufacture of cotton.

THE COTTON PLANT

In botanical classification, the cotton plant is placed in the Mallow or *Malvaceae* family and under the genus or subdivision called *Gossypium*. The word cotton comes from the Arabic word *kutun* or *qutun* and from the old French word *coton*.

Our cotton plant of today is the result of thousands of years in the melting pot beginning with we know not what and continuing with selection, cultivation, and study through centuries until now we have a plant with so many assets, supplying human needs so vast, that it is almost unbelievable.

Henry W. Grady once wrote, "What a royal plant it is! The world waits in attention on its growth; the shower that falls whispering on its leaves is heard around the world; the sun that shines on it is tempered by the prayers of all the people; the frost that chills it and the dew that descends from the stars are noted. . . . It is gold from the instant it puts forth its tiny shoot. Its fiber is current in every bank, and when, loosing its fleeces to the sun, it floats a sunny banner that glorifies the fields of the humblest farmer that man is marshalled under a flag that will compel the allegiance of the world and bring a subsidy from every nation."

An entire cotton plant (Bureau of Plant Industry, Soils and Agricultural Engineering, U.S.D.A.).

One book on botany gives this description of the cotton plant: "The characteristics of the cotton plant are well marked; leaves, three- or five-lobed; flowers, corolla of five petals; calyx, fine toothed; stamens united in a column; fruit, a three-celled or five-celled capsule or pod, opening when ripe and displaying seeds embedded in the long fibers we name cotton."

All species are small tree or shrubby type and of tropical origin. Cotton is found in every region between the thirty-six parallels favoring a mild, warm climate. While the cotton plant botanically is listed as a perennial, in most portions of the United States the plant is killed each year by frost or cold weather; therefore replanting is necessary. The only exceptions are the Rio Grande Valley

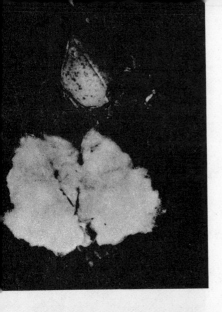

Above: Cotton pods before opening and after bursting open. Below: Fully opened cotton boll ready for picking (Bureau of Plant Industry, Soils and Agricultural Engineering, U.S.D.A.).

(where a small amount of cotton manages to survive the winter), Arizona, and California. The cotton that is allowed to become perennial, is referred to as "volunteer cotton." But, because only a very small amount of cotton will survive the winter and because it has inferior quality, practically all of the cotton in the United States is replanted each season. It grows from 3 to 15 feet in height, depending on variety and locality.

There have been, are, and will be many different kinds of cotton plants. The well-known tendency of the cotton plant to so readily respond to changes in soil, climate, and cultivation has made for countless changes and varieties since cotton first began to roam the globe along with civilization.

Cotton species come and go, change, improve, or deteriorate so fast that accurate descriptions of the varieties are often out of date almost before they are printed. For instance, a United States census report of 1880 listed fifty-eight well known varieties of cotton, but in less than fifteen years only six were still in common cultivation.

At present there are probably some fifty recognized species of the plant, but of these only five or six are of commercial value. Almost all cotton used today in the United States can be roughly listed under the headings of two of the distinct types of cotton, Sea Island and upland. Two other types, Asiatic and Peruvian, are imported in small quantities to the United States but are used to a greater extent in the countries where they are grown.

Sea Island cotton (Gossypium Barbadense) is supposed to have originated on the island of Barbados in the West Indies and spread to the surrounding islands, South America, and the lowlands on the coast of South Carolina and Georgia.

It is the finest of all cotton known. The plant has the appearance of a tree, growing from 6 to 15 feet tall, but is usually kept 6 or 7 feet by topping. The flowers are creamy yellow with a purple spot at the base of the petals; the seeds are small, smooth, and black, from which the lint separates readily; the pods or bolls are long, shapely, pointed, and full. The fiber is strong, silky, lustrous, regular, and grayish white in color. The staple is the longest of all cotton, ranging from

Sea Island cotton, shown in its natural size.

1½ to 2½ inches. It is slender and of a lustrous silky nature.

Modern Sea Island cotton is grown in the West Indies, on the islands along the coast of South Carolina, on a narrow strip of land close along the shoreline of South Carolina, and in sections of southern Georgia and northern Florida.

Egyptian cotton is original Sea Island transplanted to Egypt's climate and culture. The fiber grows yellow in color and is stronger and shorter than Sea Island, 1 ³/₃₂ to 1 ³/₁₆ inches. Its luster is unequaled, and it is especially good for all spinning yarns requiring high tensile strength.

There are some ten different grades or qualities, (with names such as *Ashmonni, Zagora,* etc.) each influenced by the soil and climate of the section of Egypt where it is grown. All of Egypt's cotton is dependent upon irrigation from the Nile.

Historically recent, Egyptian cotton found its way back to America. *Sakellarides,* a variety of Egyptian cotton, was transplanted to irrigated regions in states such as New Mexico, Arizona, and California. With different treatment, climate, and soil, a new cotton was born. It was given the name *Pima* because it was first grown in Pima, Arizona. The fiber of Pima is from medium to high grade, ranging from 1½ inches to 1⅝ inches in length, most of it being 1 ⁹/₁₆ inches. It varies from dark cream to light tan in

71

color, typical of its Egyptian ancestry. The cotton bolls of the Pima plant are small and therefore, as a result, the yield is low.

About 1930, another cotton strain was developed in Arizona, California, and New Mexico, by crossing Sakellarides and Pima. This new type is known as SXP. The staple is lighter in color and shorter in length than Pima but, even so, the yarn is slightly stronger. SXP gives a higher ginning output and a higher yield per acre than does Pima.

Upland (Gossypium Hirsutum) is the most extensively cultivated of all cottons. Its origin is thought to have been in Mexico, from where it spread to the southern United States. Today it is grown in all sections of the cotton belt except where Sea Island is grown. The cotton belt is the territory south of the thirty-seventh latitude, from Norfolk County, Virginia to the Pacific Ocean. In addition, a large quantity of cotton (over 2¼ million bales annually) from upland seed and other American strains is grown in Brazil.

The upland plant is shrubby and spreading, averaging about 3 feet in height. The flowers are white when opened, turning red with age. It produces large pods well packed with lint. The bolls are more numerous than Sea Island's, and the seeds are larger, greenish in color, and are surrounded by short fuzz beneath the longer, more valuable lint.

The upland cotton fiber varies in quality according to location. It is medium in diameter and fineness and is usually white, with a tendency toward the creamy. Its length is from ¾ to 1¾ inches. The fibers measuring 1⅛ inches or longer are termed "peeler" or "long staple upland."

Asiatic (Gossypium Herbaceum) cottons are grown commercially in India, China, Persia, Turkestan, Arabia, Russia, and Southern Europe. This cotton, which is thought to be indigenous to Asia, is botanically so different from the American that it is common belief that the two are of different origin. There are, however, some botanists who list this type with *Hirsutum.* The fiber is coarse and grayish white. *Lenant* cotton is the name given to one variety of *Herbaceum* grown in Arabia.

Peruvian (Gossypium Peruvianum) cotton is cultivated largely in Peru and Brazil. It has been introduced to Egypt and to the West Indies. The accounts of its origin vary with historians. Its beginning might have been the Sea Islands of the West Indies, from where it might have been taken to South America, or vice versa. The cottons of South America may have been the ancestors of the Sea Island cotton, for cloths made from such cotton have been found in the prehistoric tombs of Peru.

Peruvian cotton grows on trees 10 to 15 feet high. They are perennial and fruitful for ten years. The fiber is strong, rough, curly, or spiral shaped, and admirably suited for mixing with wool, giving strength, luster, and finish to the mixed yarn. Its length is 1¼ to 1½ inches. The smooth, black seeds adhere to an oval mass, a fact which gives this cotton the name "kidney cotton." Peruvian cotton will grow in the United States, but only as an annual.

THE COTTON FIBER

The cotton fiber is termed a "vegetable fiber" because it comes from a plant. It is also called a "seed hair" because it is the fluffy fibrous material which envelops the seeds of the plant. The cotton fibers are attached to the seeds and by nature are intended as parachutes to assist the seeds to be blown about by the wind.

Under the microscope, the longitudinal view of the cotton fiber is shown to be a long, continuous, single cell, like a twisted, flattened, or collapsed tube, or a ribbon with delicately thickened edges and a slight twist. The twists, convolutions, or "hooks" may run as high as 150 to 400 per inch.

The fiber length is composed of three parts: the root, which is irregular and cone shaped; the stem or center, which is more or less even in width; and the tip or free end, which tapers to a fine rounded point.

The cross section shows that the cotton fiber has three parts: the cuticle and primary wall, the secondary cellulose, and the lumen.

The cuticle and primary wall. The cuticle, which is the outmost covering, contains the fiber's natural waxes and oils. The wax not only protects but gives softness or harshness to the fiber. The primary wall is closely joined to the cuticle and with it covers the entire fiber. The primary wall is the main body of the fiber during its growth. It consists of cellulose chains, lying perpendicular or in a transverse direction to the long axis of the fiber. The primary wall is completed in about the first twenty days of the fiber's growth.

Secondary cellulose. This section makes up the major portion of the mature fiber. After the primary wall has finished growing, the secondary wall is formed by depositing a layer of cellulose

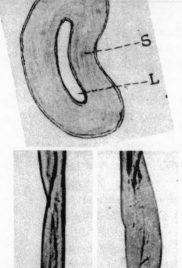

Above: Schematic drawing of cross section of wool fiber: P—Primary wall and cuticle; S—Secondary cellulose; L—Lumen. Below: Mature American upland fiber (× 300) (Textile Fiber Atlas).

within the primary wall each day for about twenty-five days. The secondary layers of cellulose are composed of fibrils, spiralling about the axis, at angles up to about 25°. These fibrils are more nearly parallel to the axis than are the fibrils in the primary wall. Sometimes the spiral goes in an S fashion and sometimes in reverse, in form of a Z twist.

If for some reason this inner section is not formed or is incomplete, the fiber appears as a thin-walled flat ribbon, with little or no twist. The quality of cotton is dependent upon the degree of maturity of the secondary cellulose.

Lumen. This is the central interior opening of the fiber. This narrow, rounded opening runs the full length of the one-celled fiber, giving it a hollow tube effect.

Chemically, the cotton fiber is (except for impurities), pure cellulose, which is a carbohydrate composed of carbon, hydrogen, and oxygen, represented by the formula $C_6H_{10}O_5$. It is considered a fiber which contributes coolness since cellulose is a very poor conductor of heat which prevents the transmitting of warmth to the body.

The cotton fiber is highly hygroscopic. It absorbs and releases large quantities of moisture. It takes the moisture into the pores of the fiber walls. Thus, the fiber is said to dry by absorption. The frictional hold is increased by water, making the cotton fiber stronger when wet. Due to this moisture absorbing quality, the fiber is also very susceptible to dyeing and finishing, resulting, if good quality dyes are used, in colors that are distinct and fast.

The peculiar construction and twist of the mature cotton fiber make it a natural for spinning and weaving. The fibers hook and cling one to another, allowing cotton to be made into yarns that have durability and strength.

There are many types and varieties of cotton fibers, ranging from ¾ to 2 inches in length, from 1/2000 to 1/1200 inch in diameter, and from pure white in color to dull yellow or gray. Some fibers are clear, lustrous, and soft; some thin, wiry, and smooth; others broad, large, and harsh; still others dull, irregular, and rough. These numerous kinds of fibers are selected and blended into yarns to best suit particular kinds of fabric.

Which cotton fiber or which blend is the best? True, there are many different grades and many different price values, but, regardless of size, color, or texture, each grade or staple is suited for a particular purpose. It is suitability that gives value; a sheer fine textured organdy frock would be out of place in a munitions plant; it would be hard to bandage a bleeding wound with sturdy, tightly woven duck; and soft canton flannel while wonderful for the baby's skin would be useless as a tent.

FROM COTTON SEED TO FIBER

The main steps of cultivating and securing the cotton fiber are:

Ploughing. After the last of the cotton is picked in the fall, the ground is ploughed as soon as possible for the spring planting. This accomplishes two things, first, fall or early winter ploughing loosens the soil so that it absorbs more winter rain. Since moisture is most necessary for seed germination and early growth, stored-up winter moisture is very helpful. Second, fall ploughing turns under the old stalks and leaves, which puts organic matter back into the soil and also helps to destroy weevils hibernating in the stalks.

Seed selection. Cotton seed is selected by the grower in a variety of ways. He may bring back from the gin seed from his own cotton plants or he may save seeds from cotton planted in special sections solely for seed. He may buy seed from the gin, or select and purchase special types from seed companies. It is estimated that about 500,000 tons of seed are required for the 22 million acres of cotton planted in the United States each year.

For many years, federal, state, and county governments, some cotton mills, and some cooperative groups have furthered a movement known as "single variety community." After detailed research and study, a particular variety of cotton is selected for a given community and each grower is urged to plant the same type of cotton. This not only gives each grower better selected seeds,

75

especially suited to his soil and climate, but prevents the cross-breeding of types which would lower the value of his cotton yield.

The United States Department of Agriculture states, "Ideal planting cotton seed may be described as seed selected from cotton that is true to type and pure of variety; well matured; free from disease and insect pests or insect injury; delinted, recleaned, and graded; and with a minimum germination of 88 per cent."

Bedding up. The beds for the cotton seed are made by turning furrows parallel to each other, so close that the deep furrows may lap. This may be repeated to build up the beds to the right height. The spacing of the furrows varies with the type of soil and the variety of plant. The average distance is 4 feet.

Busting out the middles. An alley is ploughed in the middle of the furrow with a "middle buster," in preparation for the seed. Fertilizer may be added before ploughing or it may be dropped at this time. Today, a modern touch is given to some of the well-known American cotton species. By adding differently colored dyes to the soil in which the cotton is planted, the planters are growing cotton of various colors, such as blue, green, red, and black.

Planting the seed. Most of the seed is planted by machinery, although some irregular or small patches may be sown by hand. The machine usually deposits the seeds at regular intervals and covers them at the same time. The seed is planted in different localities at different times, ranging from January to the middle of May.

Chopping out. Normally the seed germinates within a week. As soon as the plants have reached a few inches in height and have formed three or four leaves, the field is thinned or chopped out, to leave space between plants and to get rid of undesirable plants. Here, too, the distance between plants depends upon plant types and the character of soil. The cotton field will, after chopping, receive about three hoeings and three or four ploughings before picking time. This cultivating is necessary to rid the cotton of grass and weeds and to keep the furrows deep and the soil damp around the plants.

Picking. The first blossoms appear about two and a half months after planting. About six weeks later, the first bolls open, and cotton appears. From then on until frost, the bolls continue to mature and burst open.

In order to keep the fields clean of open cotton and to avoid injury by rain or storm, the cotton field is picked three or four times, and sometimes oftener, during the season. Cotton picking is done either by hand or by machine. Hand picking gives a higher grade of cotton. Cotton picking by hand is really a memorable sight: men, women, and children picking cotton, filling large canvas sacks, dragging them to be weighed, dumping them into wagons or trucks—and then starting over again. Picking time is often festival time.

There are today mechanical cotton pickers that harvest a high percentage of the cotton and at the same time, leave the plant and immature cotton bolls unharmed. They can gather the lint without damage to the fiber and without much trash.

Dragging a long "pick-sack," the picker goes up one row and down another (Cotton Textile Institute, Inc.).

There have been many types of cotton pickers advanced and tried out through the years, but experimentation has narrowed the types, as best for all-round commercial purposes, to machines employing spindles and doffers. A typical spindle and doffer type has two drums that revolve as the cotton plant passes between them. Each drum is provided with cam-actuated picker bars on which are fastened rotating spindles with tiny needles or barbs that catch the lint as the spindles roll through the plant. Before the spindles reach the plant they are dampened by passing under moistened rubber pads. This makes doffing easier. The picker bars carry the cotton laden spindles from the plant to the rubber doffers, which rotate close to the spindles and remove the cotton from them.

The cotton is then conveyed by vacuum to a separating chamber, which removes considerable trash. More foreign substances are ejected as the cotton passes along a grating and is blown into the storage basket. The cotton is dumped from the picker machine by a hydraulic lift. Tests have shown that by installing additional cleaning machinery at the gin, the grade can be brought approximately to that of hand picked cotton.

Ginning. The growers deliver their cotton to the gins. Some gins are no more than small wooden sheds housing the ginning machinery, while others are large, modern ginning plants that dry and clean the cotton as well as separate the fibers or lint from the cotton seeds. Ginning has become a specialized business. The gin may be operated for the public, for one plantation exclusively, or for both the public and a plantation. A four-gin establishment with forty saws, each in constant operation, will turn out forty to sixty bales of cotton a day.

The yield per acre is determined by the ginned yield; in other words the amount of long lint resulting from ginning. Today an average yield is a little over 260 pounds per acre—due to better soil control and especially to better seed—as compared to 1920 when it was around 130 pounds per acre. Lint ought to weigh one third of the weight of the seed cotton.

Frequently the cotton is damp because of recent rains, or dew, or picking the cotton before it is thoroughly dried out. If it is ginned in this damp condition, the long lint will not cut off

Cotton picking by machine. A picker at right is picking a field of tall and rank leaf growth cotton. Left, below: The rear of the machine, showing the basket well filled. Above: A partially picked field (International Harvester Company).

clean and more short lint will result. After picking, the cotton should be thoroughly aired and dried and then stored so that the lint may absorb some of the oil in the seed. However, this latter is not always done.

Drying the cotton is performed in several ways. If no plant equipment is available, the cotton may be stacked in small piles in sheds, or it may be placed on trays and moved into the sun each day until dried. These are slow processes, and today a substantial percentage of ginning plants clean, dry, and gin the cotton mechanically.

The wagons or trucks filled with cotton, as it is picked in the cotton fields, are driven to the gin. Pneumatic elevators draw the cotton from the trucks or wagons and suction tubes convey it to a drier. After drying, the cotton is forced against a screen and some of the dirt and foreign substances are removed. Spiked rollers then tear apart the cotton, again throwing it against a screen, removing additional dirt and trash.

A conveyor belt carries the now dried and cleaned cotton to the feeders, which deliver it to the gin stands, containing long saws with hookline teeth. The cotton drops down on these fast revolving saws and is torn apart. The seeds fall out, and are usually collected in a trough at the bottom of the gin stand. The lint remains on the saws and is conveyed to a press box by a blast of air or by brushes. As the linters are delivered to the press

Right: Interior view of cotton gin, showing stands and overhead cleaners, and press in background. Below: A cotton gin building, showing a load of cotton under suction pipes, and bales of cotton on the platform (Bureau of Plant Industry, Soils and Agricultural Engineering, U.S.D.A.).

box an overhead press compresses the cotton into uniform layers. When the press box is filled, bagging is drawn over the layers of cotton and they are securely pressed together, usually by a plunger that moves upward. The bale is then tied with bands, ready for shipment.

Another type of gin is the roller gin, which is used less frequently and primarily for long staple fibers. In place of the combs it has a leather covered roller with a knife set parallel to the roller. The cotton falls on the roller and is carried to the knife, which pulls the cotton apart, permitting the seeds to fall below.

After the lint cotton is packed into bales averaging about 500 pounds, it is sent to the yarn manufacturer.

FROM FIBER TO FABRIC

Opening and cleaning the cotton. The large bales of cotton arrive at the manufacturing plant. Before starting the process of yarn making, the fibers must be separated and the remaining foreign substances, such as leaves, sticks, and dirt, removed. To accomplish this, the cotton passes through separate machines or through one continuous machine that performs the following operations: bale breaking, cotton opening, and picking.

The cotton is blended. Thin layers from several bales are put into one batch so that the cotton fibers will be mixed. The blended batches may be stored in large bins, but to save time and handling, usually the cotton is transferred from various bales immediately to the first machine.

The bale breaker, or the feeder, breaks up the large bunches by means of spikes on rolls. The cotton passes over aprons (which look much like a series of slats) and is carried to the top of the

Left: A sample of cotton is cut from the bale, now at the warehouse. It will be used for checking and grading (Cotton Textile Institute, Inc.). Opposite page: The bales are opened and cotton from different bales are blended (Pacific Mills).

machine by elevated aprons. At the top of the aprons, rolls with spikes pull the cotton off the aprons. The rolls deliver the cotton to revolving doffers, which carry it out of the machine ready for the next operation. Some dirt drops out as the fibers are torn apart.

The opener further breaks up the cotton bunches and removes heavy dirt. This machine is an airtight tank or drum with a shaft down the center, which has attached to it a series of metal plates with small bent metal bars. The cotton enters the bottom of the machine and the revolving shaft or cylinder beats and whirls the cotton with its prongs. When the fibers are sufficiently blown apart, they rise to the top of the machine and are delivered to a screen, ready for the next operation. Much of the dirt is beaten and blown out during this process and drops to the bottom of the tank.

The picker continues the opening and cleaning, and delivers the cotton in a wide sheet or "lap" ready for carding and yarn making. Previously three pickers were used, then two, and now picking may be a continuous operation utilizing one picker with three beaters. The new machines, in doing a better job of opening and breaking, eliminate two picking operations. The cotton passes through the machines ("hoppers") on aprons, is delivered to two screens and then to the next beater section, and, again, to two screens and to the finisher section. The cotton leaves the last screens passing through rollers, where it is formed into a large, round, soft roll called a picker "lap." In this form it is ready for carding.

Types of yarns. There are two types of cotton yarns, carded and combed. The carded yarns are usually thicker and have protruding short fibers. They are made of only the short fibers, or of

short and long fibers. The combed fibers are the long fibers, remaining after the short ones have been "combed out." Combed yarns are smoother and usually more tightly twisted. Diameter for diameter, the combed yarns may be stronger, but a great many strong, heavy, and durable yarns and fabrics are made of carded cotton. Fine organdy, batiste, broadcloth, percale sheeting, and all such fabrics are woven with the tightly twisted smooth yarns made from combed cotton.

All cottons are carded. In this operation the bunches of cotton are further opened up, and dirt and foreign substances are removed. While this is primarily a straightening and cleaning process, many of the shortest fibers are removed. The big round roll or lap of cotton fibers is fed into the carding machine and, after being cleaned and carded, leaves the machine in a wide, gossamerlike sheet that is gathered into a soft, continuous strand called a "card sliver."

If yarns are to be spun of these short and long fibers, the card sliver is passed on to the drawing and roving machines. If only the long fibers are to be used, the card sliver must pass through additional processes.

Some cottons are combed. To prepare the card sliver for combing it must again be made into a wide sheet. Several card slivers are combined into what is called a "sliver lap." This is further drawn out into a wide ribbon lap, in which the fibers have been laid more parallel. Combing the ribbon lap removes the short fibers and noils, leaving only the long fibers called "combed slivers." It also further lays parallel the long fibers.

Drawing and roving. These are necessary final operations prior to spinning the yarn. The slivers are drawn out, to straighten the fibers and to lay them as parallel as possible and at the same time to reduce the strand in diameter. Slivers may be doubled and drawn out together. Roving continues the drawing-out process and gives a twist to the sliver. Leaving the roving machine, the strand is called a "roving" and is ready for spinning into yarn.

Spinning the yarn. The rovings from the roving frames are spun into yarns, usually on a ring spinning machine. This operation further draws out the yarns, twists them, and winds them on bobbins or other packages. In this form they may be used for weaving, knitting, or lace making, or they may be further twisted on twisting machines that look very much like the spinning machines. One, two, or more yarns are frequently twisted together to form what are called "ply yarns." Warp yarns are usually more strongly twisted than filling yarns, as they must withstand greater strain in weaving and finishing.

Weaving. The yarns to be used for the warp (lengthwise yarns) are wound on large beams or rolls that are later placed on the loom. The yarns to be used for the filling (crosswise yarns) are wound on bobbins which are later placed in the shuttles. The warp yarns and the filling yarns interlace at right angles to construct a fabric.

Cotton fabrics are woven on all types of looms and in all weaves (except for some carpet looms and weaves). Weaves are discussed in Chapter 11. The following pages show most of the important cotton fabrics. Others used for the home, are described in Chapter 20.

Knitting. Yarns to be used for knitting on weft machines are generally made into large hollow-center balls. The cotton yarn is cross-wound on machines constructed especially for balling yarn for knitting. The yarns for warp knitting machines are wound on beams the same as the warp yarns for weaving. If cotton yarn is to be used for Milanese machines, it is rewound from the warp beams on special sectional beams used on Milanese machines. Most cotton yarns for knitting do not have as much twist as the usual filling yarns used for weaving. The soft, "low twist" is multiple ply and unmercerized and has a fuzzy appearance. Both carded and combed yarns are used. Since knitted fabrics are produced by forming loops with needles and are made to give elasticity, it is very important that knitting yarns be made as smooth and uniform as possible. Therefore, cotton yarns for knitting are usually made with considerable care. Cotton can be knitted on all types of knitting machines and in all stitches discussed in Chapter 12, but the majority of knitted cotton fabrics are made on weft machines. Representative cotton knitted fabrics are shown on page 98.

Twisting. Yarns to be used for lace fabrics are wound on large beams, on smaller beams, and on bobbins. Any and all types of lace are made of cotton, as discussed in Chapter 13.

Felting. Cotton fibers, alone or blended with other fibers, may be felted into a fabric. Heat, moisture, and pressure intermesh the fibers and press them firmly into a fabric.

SHEER COTTON FABRICS

These are the sheer, crisp or soft, plain colored, printed, or decorated fabrics, used when a cool, dainty, and fresh appearance is desired. All but the last two of the group discussed here are woven in a plain weave. Those two are woven in a leno weave. *Organdy,* woven with fine, tightly twisted yarns, is the most crisp and transparent. *Dotted Swiss* is next in crispness, and is noted for its dot that may be woven-in, printed, or electrocoated on the fabric. It usually has a somewhat open weave. A true dotted Swiss has a woven-in dot. *Dimity* is a crisp fabric, identified by its fine cords, produced by thicker, tightly twisted yarns. The cords may run lengthwise or both lengthwise and crosswise forming checks. *Voile* is a soft draping, sheer fabric. It is light weight and transparent and woven with highly twisted yarns. *Challis* is a light weight, very soft fabric, usually with a printed design. *Cheesecloth* is an open, very loosely woven, soft fabric.

Top to bottom: Dotted Swiss with woven-in dot, dotted Swiss with electrocoated dot, dimity, embroidered organdy, voile, challis, cheesecloth.

Gauze appears similar to cheesecloth, but is more closely woven with tightly twisted yarns. *Marquisette* is a fine, open weave fabric, firmly woven in a leno weave. It may be soft or crisp and is lacy in effect. *Leno weave shirtings* usually have novelty stripes or checks.

CRINKLED COTTON FABRICS

The two following fabrics, while similar in appearance are produced by two different methods. *Seersucker* has a permanent lengthwise crinkle formed by holding certain warp yarns at different tension during weaving. Those that are not held taut draw up forming a crinkle. *Plissé* receives its puckered effect after the fabric is woven. A caustic soda solution is printed on the fabric which causes certain portions of the fabric to shrink, making a crinkled effect on the portions not printed. The crinkle may be in stripes or checks, or in a patterned effect. The plissé printing method now produces fabrics that retain their crimp after laundering.

Top to bottom: Marquisette, leno weave sheer gauze. The next two are seersucker, the last three are plissé.

MEDIUM WEIGHT COTTON FABRICS

These are fabrics used extensively for women's and children's garments, and men's and boys' shirts and shorts. They do, of course, have many other uses. All are plain woven.

One group has two major uses. In white, they make infants' wear, handkerchief, and lingerie fabrics. In plain colors or printed, they are soft fabrics suitable for children's wear, women's dresses, and blouses. *Batiste* is a sheer and light weight, highly mercerized fabric, noted for streaked effects running lengthwise. *Lawn* is basically a light fabric, softer and more closely woven than batiste. Other sheer fabrics, more crisp and lustrous, are often referred to as lawn. *Nainsook* is slightly lustrous and soft, similar to batiste but heavier and somewhat coarser. *Longcloth,* also made of soft yarns, is similar to nainsook but heavier. *Cambric* is a fine, soft fabric with a slight

Above: Batiste (upper) and lawn. Below, counterclockwise: Calico - print muslin, chintz, airplane cloth, cambric, percale, balloon cloth.

luster on the surface due to calendering. A lining cambric has a stiff, glazed surface. *Airplane cloth* is a very closely woven fabric, made of fine strong, usually two-ply, yarns. *Balloon cloth* is a soft, firm fabric, similar to airplane cloth but with a higher thread count.

The second group consists of three printed fabrics, heavier in weight than the first group. *Muslin* (illustrated with a calico print) is a soft, closely woven fabric, made with firm, strong yarns. *Calico* is rarely manufactured today. *Percale* is a strong, closely woven fabric that has a dull luster. *Chintz* may have a close and fine weave or a coarser weave. It is noted for its gay print and highly glazed surface. Muslin and percale sheetings are discussed in Chapter 20. Sheeting is often dyed or printed and used for apparel fabrics. It also may be glazed or given a soft finish.

The third group is not printed but is woven with colored yarns. *Gingham* is a finely woven fabric, made of combed

Above, top to bottom: Gingham, corded chambray, chambray. Below, left to right, top row: Oxford cloth, longcloth, end-and-end cloth. Bottom row: Longcloth, madras.

cotton. Its colors may be in stripes, checks, or plaids. *Chambray,* woven in various weights, is identified by changeable colors achieved by weaving white filling with colored warp yarns. By alternating the yarns, it may have white or colored stripes or checks. *A corded chambray* has raised lengthwise stripes. *End-and-end* cloth also has a variegated surface color. This is produced by white filling yarns and alternating white and colored warp yarns. The fabric seems to have tiny checks, an effect of the white yarns. *Oxford cloth* is a heavier fabric. Like chambray, it has colored warp and white filling yarns, but the filling yarns are heavier and thicker and the weave is over two and under two warp yarns, giving a two-by-two basket weave appearance. *Madras,* a closely woven fabric, white or yarn-dyed, is noted for its corded stripes or checks. Small dobby patterns are frequently woven in. It may be fine, as shown, or heavier, as the so-called curtain madras discussed in Chapter 20.

NUBBED OR SLUBBED COTTON FABRICS

These fabrics have a rough, irregular surface texture caused by uneven yarns.

Éponge is a loosely woven, spongy fabric, made from thick yarns unevenly twisted with a very fine yarn, forming nubs. *Ratiné* is another spongy fabric, made with yarns twisted together to produce a knotty, irregular yarn. The nubs or knots are scattered throughout the fabric. *Bouclé,* with small loops on the surface of the fabric, is made with yarns twisted together to form loops. (See Chapter 10.)

A second group consists of fabrics woven with yarns spun with short or long, thick slubs. These yarns are introduced into various weaves to produce such fabrics as so-called *slubbed broadcloths* or *slubbed poplins.* Since the slubs are produced during spinning or twisting and are not part of a natural yarn, the fabrics are different from pongee or shantung.

Left to right: Fabrics woven with slubbed yarns, thick-and-thin yarns, nubbed yarns in warp and filling, nubbed yarns in filling woven with leno weave.

HEAVY WEIGHT COTTON FABRICS

These are durable cotton fabrics, made with strong yarns and usually woven to resist strain and wear.

Osnaburg is a coarse, plain weave fabric, similar to crash. It is in medium to heavy weights and is woven with rough, uneven filling yarns. *Duck* is a strong, heavy, plain weave fabric, often woven with two-ply yarns. It is made in many weights and may be bleached, dyed, or printed. *Canvas* is a firm, heavy fabric. It may be bleached, dyed, printed, or woven in stripes with colored yarns. *Monk's cloth* is loosely woven in a basket weave, with thick, heavy yarns. It may be a two-toned natural color (as the sample shown), or dyed in plain colors, or woven in stripes. *Crash* has a rather open weave, like Osnaburg. It may be white, dyed, printed, or woven in stripes or checks.

Interlining cotton fabrics include *crinoline,* an open, mesh fabric, made of fine, low-grade yarns. It is highly sized and stiff. *Buckram,* a meshy fabric woven with thicker ply yarns has a stiff, thick sizing. Heavier buckram is made by gluing two sized fabrics together.

Top to bottom: Osnaburg, duck, canvas, monk's cloth, crinoline, buckram. Below, left to right: Yarn dyed glass toweling, plain crash, printed crash.

Left to right: Printed poplin, yarn dyed poplin, plain poplin, rep.

TWILLED COTTON FABRICS

These are strong, long wearing fabrics, identified by their horizontal or twill lines. *Gabardine* has a fine, distinct, steep twill that is scarcely discernible on the back of the fabric. *Tricotine* (the same as wool cavalry twill) has a steep twill with raised, double diagonal lines. *Whipcord,* often napped on the back, has a heavy, more reclining twill, usually from right to left. In wool whipcords, the twill runs left to right. *Denim* is a very strong, twill weave fabric, woven with colored yarns in plain colors, stripes, or checks. The twill is right to left or vice versa. *Jean* is a strong twill or herringbone twill weave fabric. *Coutil,* woven in a herringbone twill weave, may be fine or heavy. It frequently has a dobby pattern. There are many strong, durable, novelty twill weave cotton fabrics.

While the twill in wool fabrics runs from left to right, it is often reversed in cotton fabrics, and runs from right to left.

Left to right: Gabardine, whipcord, tricotine, plaid twill.

Left to right, Bedford cords: Fine ribbed, wide ribbed, and heavy ribbed.

RIBBED COTTON FABRICS

All the ribbed weave fabrics have the ribs running length-wise or crosswise except rep, which may have them in either direction. These ribbed fabrics are formed by weaving groups of yarns as one yarn. (A fabric woven with some thicker yarns to form ribs is not a ribbed weave fabric.) These weaves make firm durable fabrics. *Poplin* has a fine rib and may be white, plain colored, or printed. *Broadcloth,* a lustrous fabric, has the finest of all ribs. It may be white, dyed, printed, or woven with colored yarns as the illustrated striped broadcloth. *Rep* has a heavier, more distinct rib. *Bedford cord* has wider, more pronounced ribs, with warp yarns floating on the back. In the largest ribs stuffing yarns may be added between the back warp yarns and the ground weave. It may be white, dyed, or woven in stripes with colored yarns. *Piqué,* a ribbed fabric, is discussed on page 92, as there are various types of piqué today, that have a raised, textured surface.

Left to right, top: First two, striped denim, coutil. Bottom: Blue denim, jean.

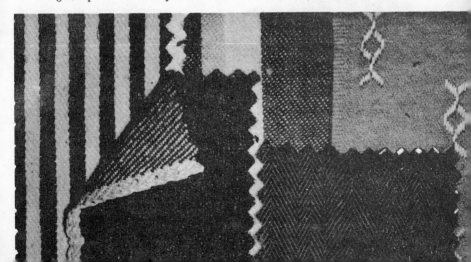

LUSTROUS COTTON FABRICS

These fabrics have a smooth, lustrous surface achieved by the weave and the finish. Either the yarns or the fabrics are mercerized to produce a sheen. *Sateen* is woven in a satin weave, with more filling yarns on the surface. It may be plain colored, or printed in checks, stripes, floral, or other designs. *Venetian cloth* is similar to sateen but is a more compact fabric with more warp yarns on the surface. Its luster and finish give it the appearance of satin. *Foulard* is a fine, reclining twill weave fabric with a printed design. *Damask* has a ground usually in a twill or satin weave and a Jacquard woven pattern in a lustrous satin weave, frequently with more warp yarns on the surface.

TEXTURED COTTON FABRICS

These are a group of fabrics that are individual in their appearance and construction. *Piqué* forms the widest range of fabrics in this group. Today, piqué has wales or ribs running lengthwise, formed

Top to bottom: Sateen, satin, foulard, damask. Below, left to right: Piqué, waffle piqué, bird's-eye piqué.

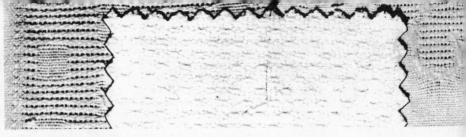

Left to right: Mock leno, huckaback, poplin with leno design.

by floating yarns on the back. The wales may be very fine or quite heavy, when stuffer yarns are added between the back and face warp. *Bird's-eye* is a fabric that has a small geometric pattern with a dot or "eye" in the center. The filling is heavier and loosely twisted, giving the fabric good absorbency. At times, a more firmly woven fabric may be woven and printed and called *birds-eye piqué*. A fabric similar to bird's-eye is *honeycomb* or *waffle*. It is a rough fabric, with long floating yarns forming raised squares. The loose yarns make a soft, absorbent fabric for towels. When made of fine yarns and printed, it is called *waffle piqué*. *Matelassé* is a double fabric, with a raised surface joined to the back fabric in such a way that it has a quilted appearance. The back of the fabric shows long floating yarns. *Huckaback* is another fabric with loosely twisted floating yarns that increase its absorbency. Both the warp and filling yarns float to form a small geometric pattern. The *Jacquard patterned* fabrics have usually a plain, twill, or satin ground weave (plain in the sample shown) and patterns of unlimited sizes and designs woven in on a Jacquard loom. The *dobby patterned* fabrics have geometric designs of limited size woven in with a dobby attachment.

Left to right: Bird's-eye, waffle piqué, matelassé.

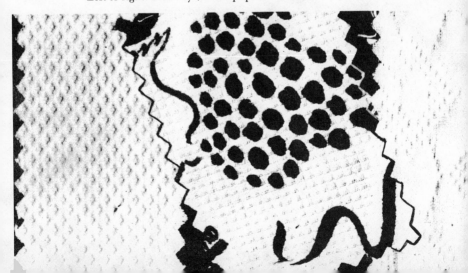

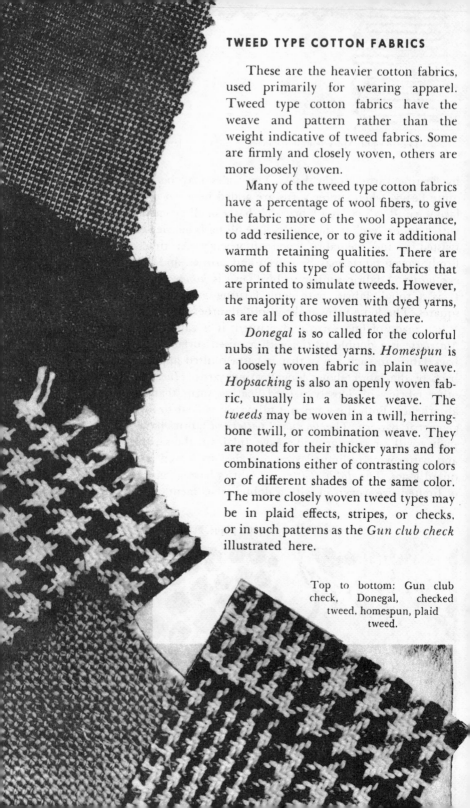

TWEED TYPE COTTON FABRICS

These are the heavier cotton fabrics, used primarily for wearing apparel. Tweed type cotton fabrics have the weave and pattern rather than the weight indicative of tweed fabrics. Some are firmly and closely woven, others are more loosely woven.

Many of the tweed type cotton fabrics have a percentage of wool fibers, to give the fabric more of the wool appearance, to add resilience, or to give it additional warmth retaining qualities. There are some of this type of cotton fabrics that are printed to simulate tweeds. However, the majority are woven with dyed yarns, as are all of those illustrated here.

Donegal is so called for the colorful nubs in the twisted yarns. *Homespun* is a loosely woven fabric in plain weave. *Hopsacking* is also an openly woven fabric, usually in a basket weave. The *tweeds* may be woven in a twill, herringbone twill, or combination weave. They are noted for their thicker yarns and for combinations either of contrasting colors or of different shades of the same color. The more closely woven tweed types may be in plaid effects, stripes, or checks, or in such patterns as the *Gun club check* illustrated here.

Top to bottom: Gun club check, Donegal, checked tweed, homespun, plaid tweed.

NAPPED COTTON FABRICS

These fabrics are napped on one or both sides to give added warmth. They may be white, dyed, printed, or woven of dyed yarns. Some are very lightly napped while others have a deep nap.

Most of the napped cotton fabrics belong to the flannel family. *Flannelette* is a plain-weave fabric, printed and napped on the face only. *Outing flannel,* a plain or twill weave (the sample shown is plain weave), is usually woven in stripes with colored yarns. It is a soft fabric, napped on both sides. *Flannel,* usually a twill weave, is closely woven of dyed yarns and is napped on one side. *Suede flannel,* also called *duvetyn,* is a thick, smooth fabric. It is deeply napped on both sides and sheared, and the fibers are pressed into the fabric, giving it the appearance and feel of felt. *Cottonade* is a firm twill weave fabric, napped on the back. It has peppered light stripes, formed with ply yarn made of white and colored yarns. *Interlining flannel,* napped on one side, is woven in a plain weave with stiffer yarns. *Blankets* are discussed in Chapter 20.

Top to bottom: Flannelette, outing flannel, flannel, suede flannel, duvetyn, cottonade.

PILE SURFACE COTTON FABRICS

These fabrics are woven and finished in various ways to produce a pile surface. Some are cut pile and others have uncut pile loops.

One group has the pile woven-in when the fabric is woven and the loops may be cut or uncut. Examples are velvet (cut) and frisé (uncut). Another group is woven flat and yarns are cut, after weaving to produce a cut pile. Examples are corduroy and velveteen. A third fabric, terry or turkish toweling, has loops woven in by holding certain warp yarns slack. The more detailed explanation of these weaving methods are discussed in Chapter 11. A pile can also be formed by weaving with chenille yarns.

Corduroy and velveteen are both woven flat, with floating filling yarns which, after weaving, are cut to form a pile. *Corduroy* is woven in a plain weave so that, when cut and finished, the pile runs lengthwise in cords. According to the type of yarn and the length of the floating yarns, the cords are finer or heavier. *Velveteen*

Above: Corduroy, from fine to wide wale. Bottom sample, printed corduroy. Below: Velveteen. Sample at left is hollow-cut velveteen.

has the yarns floated so that, when cut, they form an all-over pile. *Hollow-cut velveteen* has the pile cut in ridges so that it gives the appearance of some types of corduroy.

Velvet may be woven face to face with connecting pile yarns that are cut to form the pile on each fabric. Or the pile yarns may be woven over wires with a knife fastened to one end which cuts the pile. *Hollow-cut velvet* omits the pile warp for short spaces, forming wide ribs of pile.

Terry cloth has loops on the face, or the back, or on both. It is woven with warp yarns forming the pile. The uncut loops may be on one side only, on both sides, or with the pile in stripes or checks.

Frisé is woven with warp pile yarns weaving over a wire to form loops, the same as velvet except that the loops are not cut. The sample shows some pile loops omitted in the weaving to form a pattern.

Chenille is constructed quite differently from all the other fabrics mentioned. First a fabric is woven, which is then cut into strips to produce a chenille pile yarn, which, in turn, is used as the filling of the final fabric. The pile on the chenille yarn forms the pile of the fabric.

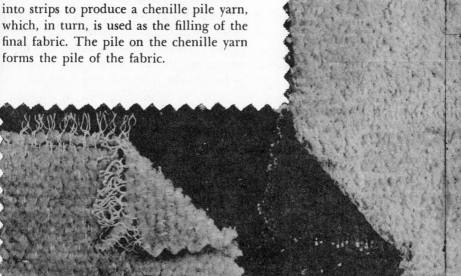

Above: Velvets. Top is printed velvet, middle is a heavy velvet, lower right is cut velvet. Below, left to right: Chenille, frisé, terry.

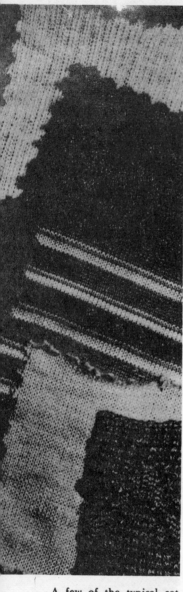

A few of the typical cotton knitted fabrics. Top to bottom: Two-by-two rib, one-by-one rib, plain stitch, balbriggan, links-and-links.

Cotton's qualities make it excellent for knitting. Its twist ensures elasticity, one of the main requirements for plain knitted fabrics. It takes long staple combed cotton to make even, smooth, strong yarns suitable for knitting. Because of the open and loose construction of most knitted fabrics, especially the plain and ribbed knits, there is more strain on the yarn than there is in woven fabrics. Cotton is especially suited for knitting by machines with needle arranging devices that knit shaped garments such as hose, sweaters, underwear, and gloves. Cotton is admirable for knitted fabrics that require heavy napping such as underwear and outer winter fabrics. It is equally desirable for lighter, open knit fabrics that must stretch to fit the body and absorb moisture rapidly.

Most cotton knitted fabrics are made with the plain or ribbed stitch on weft flat-bed or circular machines but some are made on warp machines, especially the Simplex machine which knits a double fabric used mostly for gloves, and the Raschel machine which knits openwork lacy fabrics. Cotton also is knitted on links-and-links machines making both the purl and the links-and-links stitches.

The two-by-two rib shown is loosely knitted with light, soft yarns while the one-by-one rib is more closely knitted with heavier, harder twisted yarns. Dyed yarns can be handled by special needle manipulation in order to form a pattern in the plain knitted fabric. Balbriggan is always weft knitted while a variety of fancy ribs are possible on a links-and-links machine.

FINISHING COTTON FABRICS

The gray, limp cotton fabric is delivered from the looms to large, modern finishing plants. Often, the cloth is whirled full width high under the ceiling of the washing room through a slit in the wall to the finishing room. Here it begins its journey through fire, water, and chemicals to a victorious finish. According to the type of fabric required it passes through all or many of the following finishes.

Preparatory Treatments

Burning off the surface fibers, by singeing the fabric, is one of the first treatments. The cotton cloth is passed over flames or red hot plates or rollers. Frequently, both sides of the fabric are singed at one time. As the fabric leaves the singeing machine, it is plunged into water to extinguish any sparks that may ignite the fabric.

Removing the sizing that was applied to the warp yarns (to make them stronger and more manageable) is another treatment prior to coloring and finishing. A chemical, such as sulfuric acid, may be added to the bath after singeing, to soften the sizing material. The fabric is left in bins for a time, permitting the desizing solution to act. Then, the fabric is washed and boiled, and it is now clean, ready for bleaching.

Bleaching the fabric makes it white. Even if cotton is to be dyed, it is generally bleached first to remove the fiber coloring matter. Cotton is usually bleached after the fabric is woven. The fabric is kier boiled and bleached with chemicals, such as sodium hypochlorite, sodium chlorite, or peroxide.

Mercerizing is a preparatory finish given to some fabrics prior to dyeing. Cotton yarns can also be mercerized. Mercerization gives the fabric increased luster, a soft feel, added strength, and better dyeing qualities. The fabric is immersed, under tension,

A textured hand-woven cotton fabric with rows of uncut pile made of loosely twisted yarns (Dorothy Liebes).

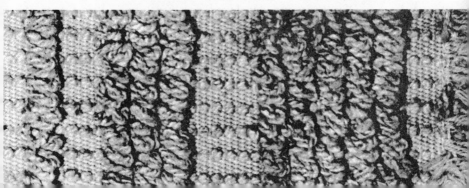

in a caustic soda solution. The fibers swell, shrink in length, and become almost cylindrical in shape, acquiring a sheen.

Dyeing

Cotton fibers may be dyed while still in the raw state, or the yarns or the fabric may be dyed. For some dyestuffs, mordants are applied that aid in fixing the dyestuffs to the fiber. Cotton fibers are little affected by volatile organic acids, but mineral acids will destroy them. They are not affected by alkalies.

Cotton can be dyed readily but has less affinity for dyes than has wool or silk. One of the first requisites for most dyed cotton fabrics is a satisfactory degree of fastness to water and to sunlight. Resistance to perspiration, to chlorine, and other chemicals used in laundering is essential also.

There are various classes of dyes used to color cottons. They are selected according to the requirements of the particular fabric.

Vat dyes, used on cottons give the best all-round fastness of any cotton dyestuffs to light, washing, perspiration, acids, and alkalies. Many can be washed with chlorine and some can be boiled. They produce a wide range of clear shades, second in brightness only to basic dyes. Vat dyes, being insoluble in water, are made soluble with chemicals, which fix the dyestuff in the fiber. After the fabric is dyed, it is exposed to air to oxidize the dyestuffs, or it is oxidized by further use of chemicals.

Azoic dyes also have an all-round degree of fastness, second only to vat dyes, and will withstand chlorine and boiling. Since the dyestuff is precipitated on the fiber rather than absorbed by it, the color has a tendency to crock or rub off. The cloth is prepared with a developer. The color is diazotized and the fabric dyed. The diazotized dyestuffs and the developer couple and the color is precipitated on the fiber.

Direct dyes give even, level colors. They have a relatively low degree of fastness to washing. According to the type of direct dyestuff used and its application, the color may have from a poor to an excellent degree of fastness to sunlight. The dyestuff is applied from a neutral or an alkaline bath.

Diazotized and developed dyes are usually direct dyes. After these are applied, the fabric is treated with chemicals that give it an excellent degree of fastness to washing. By treating the dyed fabric with acids (called "diazotizing") and then developing the solution, a new dye is really produced within the fiber. These dyes are commercially known as "tub fast."

Basic dyes give high, bright shades but, since they have a poor degree of fastness to light and washing, they are least used to dye cottons. The fabrics should not be exposed to the sun or should not be laundered frequently. While basic dyes are usually applied directly to animal fibers, an acid mordant is required to fix them to the cotton fiber.

Sulfur dyes produce a wide range of rather dull colors, but as they have a very good degree of fastness to frequent and hard washing, they are excellent for fabrics such as those used in work clothes. However, these dyes are not fast to chlorine, which is usually used in commercial laundries. The degree of fastness to perspiration and light is fair. Sulfur dyes are applied in an alkaline solution. The dyestuff is made soluble and the fabric dyed, after which it is given an oxidizing treatment to develop the dye and to produce a faster and brighter color.

Pigment dyes. According to the property of the pigment and its method of application, pigment dyes may have from a fair to an excellent degree of fastness to washing, light, acids, and alkalies. One of the main problems of pigment dyeing is crocking and resistance to rubbing. Most pigments are bonded to fibers by means of synthetic resins or other film-forming materials. The dyeing mixture, containing among other materials, the pigment and resin, is applied to the fabric, usually by padding. After application, the resin is dried and the pigment is bonded to the cloth with a permanent binder.

Printing

A great many cottons, from the sheer, crisp organdies to the heavy drapery fabrics, are printed. All types of printing are used —screen, block, and roller printing. Cottons are printed by direct, resist, and discharge methods as well as by plissé and warp printing.

Finishing

After the fabric has been bleached white, dyed, or printed, it is dried and is given added finishes in the dry state.

Drying is usually done on a tentering machine and is called "tentering." The fabric, while damp, is passed into a long frame that has clips on each side, which grasp the selvage of the cloth and

A hand-loomed fabric with a loosely woven cotton ground decorated with loops of lustrous metallic yarns (Dorothy Liebes).

carry it along, stretching it to its normal width. While still in the frame, it passes into enclosed chambers where hot circulating air dries the fabric under exact temperature control.

Starching is required for fabrics such as cambric and percale and for many other cotton fabrics. Softening compounds such as waxes and oils are added to the starch to keep it from peeling off. Other substances, like glucose, may be added to prevent the starch from becoming too dry and harsh. Materials may be added to make it resistant to mildew. The starch and added ingredients are poured as a thick paste into a mixing machine, and the fabric is passed around rolls in the paste, which is forced into the fabric. This starch washes out, however, and must be replaced every time the fabric is washed. Today, some fabrics such as organdy are given a crisp finish that remains after laundering.

Weighting may be given to some fabrics not starched. Such materials as clay, sugar, and salts are used with softening oils. While this makes a more presentable fabric, it also can be washed out. Today, there is a special finish that adds weight to cotton fabrics and remains in the fabric after laundering.

Ironing usually follows starching. This is called "calendering." The fabric passes under pressure around heavy rollers in a calendering machine. This gives it a smooth and usually a lustrous appearance. If required, as in chintz, a highly glazed surface may be produced.

Embossing or watery moiré effects can also be given to cotton fabrics on a calendering machine. Fine engraved lines on the rollers and special ways of running the fabric around the rollers produce the embossed or moiré effects.

Raising the fibers to produce a light, soft, or very deep nap is necessary for some fabrics. Flannelette or cotton flannel has a slight nap, while cotton blankets have a deep nap. The fabric passes through machines, which have either vegetable teasels or wires that brush out the fibers and raise them to an upright position on the surface, thus forming a pile. This is a different method than that used to produce a pile on fabrics such as velveteen and corduroy (which are woven with extra yarns that are cut to form a pile) , or on fabrics such as plush (where the pile is woven in) .

Shearing and brushing are necessary for many fabrics, either to produce a clear surface, free from short fibers, or to even the fibers on a pile surface. After shearing, the fabric must be brushed to take out all the cut-off fibers.

Shrinking is the last finish given to cotton fabrics. Processes have been developed that add so little to the cost of the goods and so much to their usefulness that all cotton fabrics for general use should have controlled shrinkage. A shrinking process using water and steam and holding the fabric under tension in the machine, controls further shrinkage in length and width to ¾ per cent. Before leaving the machine, the fabric is given a smooth, ironed surface and with this, its long journey from the cotton fields to the finished fabric is completed.

Functional finishes. In addition to the previously described finishes, cottons may be given other finishes that add to their beauty, their suitability, and their durability. They may be made waterproof or water-repellent, resisting both wind and rain. They may be made crease- and crush-resistant. Some are given finishes that make them crisp and smooth after laundering, without starching. Some are given a high glaze that will remain on the fabric. Others obtain increased power of absorbency. They may be rendered resistant to perspiration, mildew, and germs, and some are made resistant to fire. These finishes may be applied to the yarns but are usually applied to the fabric. The fabrics may be given these special finishes before or after dyeing, or during the wet or dry finishes.

Today there are a relatively great number of such finishes, many of them reliable. Some are durable, others semidurable. Some will remain after washing or dry cleaning, others can be washed and not dry cleaned or vice versa, while still others need to be reprocessed after washing or cleaning. Most manufacturers of these finishes supply consumer good's labels which explain the result produced by the finish and the necessary care of the finished fabric.

SUMMARY

A VEGETABLE FIBER FROM THE COTTON PLANT

Commercial cotton species: Sea Island, upland, Asiatic, Peruvian

TO MAKE THE YARNS AND FABRICS

Plant, cultivate, and pick.

Gin: To remove the cotton fibers from the seeds.

Open and clean: To open bunches and remove foreign substances

Card: To remove dirt, to partially lay the fibers parallel, and to remove the shortest fibers. All cottons are carded.

Comb: To remove short fibers and to lay the long fibers parallel

Draw and double: To further mix fibers, to lay them parallel, and to produce strand (sliver), which may be further drawn out and slightly twisted to make the roving.

Spin: To draw out and twist roving into yarn.

Construct the fabric: By weaving, knitting, twisting, braiding knotting, or felting.

TO FINISH THE FABRICS

General finishes

Inspect and mend.

Singe: To burn off protruding surface fibers.

Desize: To remove warp sizing.

Bleach, dye, print: To make fabric white or colored.

Mercerize: To increase luster and strength.

Starch: To give body to the fabric.

Calender: To press the fabric.

Tenter: To stretch, straighten, and dry the fabric.

Nap: To raise the fibers.

Shear and brush: To cut off and brush surface fibers or to even nap.

Moiré: To give watery surface texture.

Emboss: To give a crepe effect.

Shrink: To reduce the fabric in length and width.

Functional finishes: To render the fabric crease-resistant, water-repellent, spot- and stain-resistant, waterproof, absorbent, crisp, and resistant to fire, flame, perspiration, mildew, and germs.

Opposite page: Shocks of flax ready for processing and converting into linen yarn (The Irish Linen Guild).

LINEN—VEGETABLE FIBER

The Oldest

IN THE BEGINNING

To linen—honor and respect. Far in the lead of the colorful pageantry of fabrics, down through the ages, comes aristocratic linen, the ancient of ancients.

Long before anyone thought to clip the skins of sheep or long before the silkworm was recognized as useful, flax was contributing her varied qualities to the welfare of man. History tells us that at least as far back as 5000 B.C., and perhaps earlier, barbarians and predynastic Egyptians were using flax, or at least a flaxlike fiber. Wild flax was made into cordage, fish lines, and fish nets.

To the valley of the Nile is given the honor for the discovery of the use of flax and for the origin of linen weaving. The fertile soil surrounding this historical river grew flax so abundantly and of such quality that it was only natural that the prehistoric Egyptians living along the Nile should find the possibilities of this stately but fragile plant. It is also understandable that they should elevate their linen fabrics to the pedestal they reserved for the most perfect. Linen, made from the cleanest plant of the field (not "untidy wool from a profane animal"), must cover the most sacred, clothe the most spiritual, and be a part of the most cultural. The dead, in order to be presentable for entrance into the "Court of the Sun," had to be clothed in the most perfect of costume— they had to be wrapped in linen. The tomb of Beni Hassan gives proof of flax as used by early man. On the wall of this tomb were painted, some 2,500 years before Christ, a series of graphic pictures depicting the entire process of flax culture.

The Swiss lake dwellings, brought to light in recent years, gave up many specimens of flax and linen fabrics. There were looms, combs, and equipment that showed exactly how these early people, who built their homes out over the lakes, wove their linen into fabrics. At least four or five distinct weaves had been perfected and were part of the everyday life of these prehistoric peoples of the Stone Age. These fabrics give evidence of culture long practiced and enjoyed.

Homer, the blind poet of about 1000 B.C., alludes to linen manufacturing in the islands around Greece.

Linen was an everyday word to the writers of the Bible. They tell us that the high and mighty priestly order of the Pharaohs were allowed to wear garments made only of linen. During the time of Moses, a plague of hail destroyed the flax crop. The ten curtains of the Tabernacle of Israel were to be "fine twined linen." When the Israelites departed from Egypt they carried "fine twisted linen."

Herodotus (484-425 B.C.) the "father of history" writes, "The Egyptians wear a linen tunic fringed about their legs, called 'calasure,' over which they wear a white woolen garment, nothing of woolen taken into the temple or buried with them as their religion forbids it." Xerxes twisted linen into rope to build a bridge across the Hellespont and the Phoenicians used it for their first sails to harness the winds. Julius Caesar of imperial Rome wore beautiful linen robes.

Linen manufacture sprang up in other countries as civilization spread. Flax growing appears to be coexistent with recorded agriculture in all countries climatically suitable. Irish records show that linen was manufactured and exported from Ireland as far back as the thirteenth century.

As a fiber crop, flax continued to expand throughout the sixteenth century and became commercially important in Europe. Many linen guilds and monopolies came into existence and grew strong. Laws were passed protecting linen and allowing the linen mer-

chant and others connected with the industry to become powerful.

The last half of the eighteenth century saw the linen industry government controlled with a sharp class system established. Linen markets, with their seal masters, who certified that the linen fulfilled strictly official requirements, rose to importance and the standard of living of the weavers fell to that of paupers.

Arkwright's invention of roller drawing in 1775 speeded up cotton spinning, thus making cotton yarn more available, but the low wage of linen weavers delayed the general introduction of power looms for linen until long after they were in use for cotton.

In 1810, Napoleon offered a prize for the best invention of flax spinning machinery. Philippe de Gerard obtained in that year a patent for such machinery. He never received his promised prize, but his invention is still the basis of modern flax machines. He was forced to take his invention to Russia, where it was successfully put to use.

The first Irish flax spinning mill was started in 1827. The life of the weavers became intolerable until, between 1845 and 1852, there started a mass emigration of Irish linen weavers to the United States. The problems of the linen weavers were not confined to Ireland, for Germany and other sections of Europe showed the same unhappy conditions.

Western-World Linen

North American history of flax also dates back to prehistoric times. Although no actual pieces of linen fabric have withstood the ravages of time, weaving tools have been found in the earliest excavations. There is also in existence a great deal of pottery—such as jars and bowls—which bear fabric marks. Some ancient American tribes had a method of decorating pottery by wrapping the damp clay in a piece of woven cloth, which left a patterned design. These prehistoric bits of fabric were made not only of grasses, buffalo hair, and kindred fibers, but of cotton, flax, and hemp as well.

The early North Americans probably did not cultivate the plant as did the ancient Egyptians but used the wild flax that grew on their river banks. In Mexico, Central America, and Peru the earliest excavations contain no flax. Linen seemed to be unknown to these peoples. This was probably due to the fact that flax grows very poorly in tropical dry countries.

Cultivated flax seed and European knowledge of flax growing and weaving came to North America along with the first white settlers. Flax was grown in all the colonies north of the Carolinas, and linen was hand spun and woven by the colonists for their own use from the beginning of the first settlement. But, as in Europe, cotton soon overshadowed linen in America. With the invention of the cotton gin in 1793, which allowed easy, inexpensive handling of cotton, the American home linen industry died a natural death. There was, though, a comparatively short boom in the linen industry during the Civil War because of the scarcity of American cotton at that time.

At different times, from 1890 down to 1932, the United States Department of Agriculture had carried on experiments in an attempt to produce varieties of flax and retting processes suitable to the United States. The early experiments were carried on in Michigan, Minnesota, and Oregon.

Oregon had a sort of revival of the linen industry in 1915. At that time it was thought that the flax industry would be assisted if it were relieved of the task of retting. For this reason, the inmates of penitentiaries were given this job. The prisoners still ret some flax, but most retting and other processing are carried on in especially equipped plants.

In 1932, government experimenting in breeding, culture, and retting of flax was moved to Oregon. By 1935, cooperative plants began to make their appearances, but it took a war and the curtailment of foreign imports to bring to the fore the real possibilities of the linen industry in Oregon. In 1938, the government began research for developing fiber-flax machinery to reduce the cost of pulling and processing. It has also helped design processing plants. Machinery is being developed, and constantly improved, for pulling flax, cleaning tow, deseeding, scutching, and hackling. Cotton machinery is being better adapted to linen drawing, spinning, twisting, and weaving. Worsted combing machines have been utilized for combing linen fibers.

Over 2½ million acres are producing seed flax for linseed, which is used in making oil and other commodities. This straw is available for a lower grade of linen, usually combined with cotton. Statistics show that in 1936 Oregon grew 210 tons of linen fiber. By 1944 this state's production had increased to 3,000 tons of fiber flax, with fourteen retting and scutching plants in operation. Prac-

tically all domestic line flax is grown in Oregon. Seed and some fiber flax is grown in Minnesota, North and South Dakota, Wisconsin, and other central northwest states. A small amount of seed flax is being grown in Oregon and Georgia. It is considered that flax can be produced where oats and wheat can be grown for the harvesting times would be about the same.

The flax plant in bloom (The Irish Linen Guild).

The State Engineering Experiment Station at the Georgia School of Technology has been carrying out a project to develop an economical method of extracting the fiber from flax strain and to determine the best methods of utilizing the flax fiber in cotton machinery, whether alone or blended with other fibers.

America's linen industry at present is dependent on foreign markets, and a tariff, and the ability of American growers and manufacturers to produce fabrics comparable to those made of imported linen. During the war the Army and Navy used America's linen for parachute harnesses, marine fire hose, yarns for shoes, and for other such necessary items. With better growing and retting facilities and with the building of plants and machinery, the American linen industry seems to have a future. What will it be? Only time will tell. But regardless of man's shifting loyalties and changing values, with each new turn of a century, still on the throne sits—"Linen—the Cloth of Kings."

THE PLANT

Flax is the plant from which linen is made. The botanical name for common flax is *Linum usitatissimum*. There are many species of flax, but only this botanical type has attained commercial value.

There are two types of flax grown, fiber flax and seed flax. One is grown mainly for fiber purposes, with the seed crop secondary, the other is grown for its seed and the fiber qualities are secondary. Fiber flax is made into linen yarns and flax seeds are used for replanting and for linseed oil and other products.

Flax is an annual plant, growing relatively short, from 12 to 40 inches. The average height is 18 inches. It grows from seed only and is self-pollinated. The plant grows tall and erect, with a rather shallow taproot system. The leaves grow alternately on the stem and are narrow and pointed.

The flowers are blue or white. Their construction is comparatively symmetrical: five sepals, five petals, ten stamins, five double-celled carpels, and ten seeds. The seeds, small, flat, and brownish, are grown in a small round capsule attached to the end of the branches. These tiny seeds weigh about 56 pounds to the bushel.

The fibers for the linen yarns grow in the bast or woody part of the stem of the flax plant; thus they are termed "bast fibers."

A cross section of the stalk of the flax under the microscope shows that it has six sections: the outside layer of rind or cuticular cells; a thin layer of living cells; a layer of bast fibers (used for linen) bound with resinous and pectinous material and associated with lignins and hemicelluloses; a cambium layer which forms new tissue; a woody layer; and a pithy center core.

The plants of the two types, seed and fiber flax, are distinctly different and are grown and cared for differently.

Fiber flax grows relatively tall, slender, and sparsely branched. It produces a high yield of good quality, long fiber, but the yield of seed is very low. The plant grows from 3 feet 6 inches to 4 feet tall. In order to make sure that the flax has little chance to branch out, and thus ruin the long fibers, the seed of the fiber type is sown very thickly so as to have as "close stand" as possible.

Seed flax is short stemmed and numerously branched. It yields abundant seeds, but the fiber is short, coarse, and strong. In order to make room for branching out and for blossoms and seeds

Left to right: A package of flax seed and above it the top portion of flax stalks showing the seed bolls; flax fiber; unbleached linen yarn; bleached linen yarn (The Irish Linen Guild).

111

on the ends of each branch, the seed is sown thinly.

There are many agricultural varieties of each type improved to suit different soils and conditions.

THE LINEN FIBER

The linen or bast fiber is composed of small elementary bast cells, combined to form a single fiber. Each of these cells in the fiber is joined to the others by a thin membrane and tapers to a sharp or somewhat rounded or forked point on either of its ends.

Chemically, the linen fiber is mostly cellulose but never in as pure a condition as cotton, because on a large scale it is impossible to remove all the encrusting matter.

Two stained linen fibers showing lumen and nodes (× 500) (Textile Fiber Atlas).

The fiber, ranging from 12 to 36 inches in length, has the appearance of a cylindrical tube with a minute channel down the center, which is open at both ends. The tube, however, is not continuous (as in cotton, which is a single cell) but is separated by distinct joints, swellings or "nodes," which appear throughout the length of the fiber at irregular intervals varying from two to six times the diameter of the tube. The fiber remains tubular, for it is prevented from collapsing (as does the cotton fiber) by the frequently recurring joints.

The cross section of the fiber shows a rather solid mass with a very tiny opening in the center. The diameter of the fiber varies from 1/1100 to 1/1800 inch. There can be a great difference in linen fiber quality. It may be as fine as the finest silk or as coarse as the coarsest cotton.

The fiber is cylindrical, straight, firm, smooth (except for the nodes), and semitransparent. It is next to silk in strength and is stronger than wool or cotton. The natural color is gray with a brownish tinge. It may be bleached to any degree of whiteness but loses strength with bleaching. Of all the natural fibers, linen is the least receptive to dyes.

The flax fibers, or elongated filaments, grow in the stalk in the form of bundles. A bundle is composed of from three to twenty fibers, held together by pectin. Before spinning, these fiber bundles need to be subdivided.

The length of the fiber and its uniformity make spinning easy, but the chief feature that allows the flax fibers to cohere and cling together to form yarn is the swellings or "nodes" spaced at intervals throughout the length of the fiber. The nodes of one fiber hook onto and cling to the nodes of others, thus forming yarns. The linen fiber has unique characteristics giving value to the finished fabric.

Rapid absorption of moisture. It has been stated that the fiber is composed of tiny, sharp-pointed cells or "fibrils." Each of these minute fibrils has between them what is known in science as "capillary attraction." A simple experiment shows how this works. If a long, narrow glass tube is placed open end down in a tumbler of water, the water will quickly rise in the tube much higher than the surface of the water in the tumbler. In the same way, the linen fiber quickly draws up into its cells the moisture that touches the sharp ends of the cells. The moisture retention in linen is 8½ per cent, the same as cotton, but linen draws the moisture much more rapidly.

Linen dries by *absorption* whereas cotton dries by *surface attraction.* For this reason, cotton towels are given more surface by terry loops while in linen a smooth surface does just as well. Linen gives off moisture more rapidly by evaporation than other natural fibers. In other words, wet linen becomes dry more quickly. Linen's reaction to moisture is such that its natural luster and smooth, attractive appearance improves with a reasonable amount of washing and ironing.

No shedding of fluff. The linen fiber does not shed fluff from the ends of the fiber. The smooth, solid fiber has no fluffiness nor any tendency to felt. It remains firm, thus never giving a linty surface to a fabric. Such hardness contributes gloss and smoothness to the finished linen.

Strong with little elasticity. Linen is the strongest fiber grown, with less elasticity than other natural fibers. Some textbooks give this table on relative strength and elasticity, taking flax as 100.

	Flax	Silk	Cotton
Tensile strength	100	52	48
Elasticity	100	605	152

Thus, linen is a natural for fabrics that need to be strong and taut, that do not tear easily, and that do not appreciably expand or contract with changes in atmospheric conditions. Such are the requirements for coverings for airplane parts, types of net, fire hose, and other fabrics.

Good conductor of heat. A fiber that is a good conductor of heat contributes to coolness, while a poor conductor of heat contributes to warmth. According to this, linen, the best conductor of heat, gives the most coolness while silk, the poorest conductor of heat, is the most warmth giving. The fibers contributing from warmth to coolness, are ranged by most authorities as follows: silk, wool, rayon, cotton, flax. Linen also has good ventilating qualities.

Pliability. Pliability is the property of remaining in any position in which an object is placed. Resiliency is the opposite of pliability. On bending, a resilient fiber springs back again to its original position.

Flax is the most pliable of the natural fibers. The solid fiber has no twist as does cotton, or wave as does wool, therefore it lacks the ability to spring back into its original position. The pliability of the flax fiber causes linen to crease readily and to remain creased. On the other hand it gives draping qualities.

FLAX GROWING

Flax is adapted primarily to the temperate zone but it grows reasonably well over a great range of temperature, moisture, and soil conditions. Although a certain amount of humidity is always essential, hot tropical countries are unsuited for flax growing. The plant seems to grow best in alluvial deposits of rivers, as in the valley of the Nile, the banks of Irish rivers, of rivers in Holland and Belgium, and in the Willamette Valley in Oregon.

The greatest yield per acre and the best quality of line flax is grown in Belgium, Holland, France, and Russia. Here, flax grows abundantly and the fiber is long and fine. Northern Ireland grows fine flax, but due to uncertain weather conditions during harvesting time in August, the quality is irregular. The flax grown in these countries is fiber flax. Little or no seed flax is raised and in some places like Ireland the seed from the fiber plant is usually discarded.

Russia produces by far the largest amount of flax, growing both fiber flax and seed flax. This country produces more than all other countries together, about 90 per cent of all flax grown. Russia's acre yield is low and the fiber is medium quality.

The United States, Canada, and the Argentine raise large quantities of seed flax. Some of this flax, after the seed is harvested, is made into yarns. The United States grows over 4 million acres of seed flax annually; in addition there are over 15,000 acres of fiber flax raised, mostly in the state of Oregon.

FIBER FLAX CULTIVATION

For the best success of flax growing a rigid rotation of crops should be followed, that is, flax should be planted on the same plot of ground only about once in seven or eight years.

Flax seed is so small that the soil has to be carefully prepared with as fine a tilth as possible. If available, a corrugated roller is used to smooth the thoroughly cultivated soil. Silt or clay loams in moderate state of fertility seem best for flax growing.

The crop is seeded early in the growing season. The seed may be sown by a drilling machine, which distributes the seeds evenly and correctly, thus providing a better stand. Hand broadcasting of the seed is customary in most sections of Europe. Often the seed is sown by a small seeding apparatus called a "fiddle broadcaster." The rate of seeding varies. In Europe 84 to 140 pounds are used per acre, while in the United States 80 to 90 pounds are sown per acre.

After the seed is sown, it is very advantageous to firm the seedbed with the corrugated roller so that the seed may become embedded in the soil. As soon as the flax is a few inches high, it is thoroughly weeded, as weeds or any foreign growth injure the quality of the fiber. The plant is ready to harvest about four months after planting or just before the plant reaches maturity.

Harvesting the Flax

Fiber flax, used to make linen yarns, threads, and fabrics, is always pulled and never cut as is seed flax, which is used primarily for linseed oil and seed production. For the best fiber the flax is pulled just before it reaches maturity. When the lower third of the stalk turns yellowish and the leaves drop, a desirable state of ripeness is reached. A test is made by cutting and squeezing a seed pod. If the sap oozes out, the flax is not ready to pull, but if there is no sap, all of it has been absorbed by the seeds and the crop is ready for harvesting.

While a considerable amount of fiber would be lost if the

Pulling flax by machine and laying it in bundles in the field to dry (The Irish Linen Guild).

root were not utilized, the primary reason for not cutting fiber is that it would leave a blunt end that would seriously affect the retting and spinning quality of the flax.

Flax is now generally pulled by machine. In some countries, particularly Ireland, flax is grown on many small farms and is hand pulled because machine pulling is not feasible. However, hand pulling is a long and arduous task. Whether pulled by hand or by machine, the stalks are grasped as near the ground as possible. Care must be taken that the entire fiber from its top to root be intact and injured in no way. For this reason, machine pulling does not always result in as perfect flax as does hand pulling. If the flax straw is broken in machine pulling this can never be entirely corrected in future processing.

Machine pulling. There is more than one type of flax puller, but in operation they perform the work in much the same way. Gripping belts, traveling in opposite directions, pull the flax out of the ground and convey it upwards to a binding unit that ties it into bundles. A machine pulls a swath about 3 feet wide.

One machine has three gripping belts and each pulls a 1-foot swath. Another has two rubber covered wheels with rubber belts, which grip and pull out the flax in two 18-inch strips. As machines of this type move along the flax, gripping belts running over pulleys grasp the flax much as it is grasped in hand pulling and pull it out of the ground. The pulleys pass it on to the binder, which binds and deposits it on the ground in sheaves. These machines can pull from 8 to 10 acres a day.

Drying. After the flax is pulled, the bundles are built into shocks, called "wigwams," to permit the stalks and seeds to dry and be seasoned. In some countries a number of shocks, some thirty-eight to forty, are arranged in rows with other shocks on top. After drying the flax is stored, ready for deseeding, retting, scutching, and making into linen yarns.

Deseeding. Usually the first step in the preparation of flax for linen yarns is to remove the seed bolls. The seeds are most valuable in the countries where they are used for linseed oil.

Today, practically all flax is deseeded by machines which perform quickly and effectively the centuries old hand operations. Various types of deseeders are used. Some whip out and others comb out the seed ends only; still others comb out both the seed ends and the root ends. This last method lays the fibers more parallel for drying and for future scutching. In one hand operation, the flax straw is drawn through combs or spikes called "ripples." Frequently deseeding is called "rippling." As the straw is drawn through, the seed bolls fall off, because they are too large to pass through the ripples or teeth of the forklike comb. Later the seeds are extracted from the bolls. In another hand operation the bundle of flax is gripped and extended against two rollers which crush the seed bolls.

Preparing the Fiber

Retting. As stated previously, the fibers used to make linen are beneath the outside cells. The inner portion of the stalk consists of a woody core. Therefore, one of the first steps in producing linen is to separate the bast fibers from the remainder of the stalk. In order to loosen and detach the fibers, the pectin (gum) by which the fibers are attached to the rest of the stalk must be removed. This is accomplished by retting.

Warm water-retting is most frequently used in the United States. It is also widely used in Belgium, in Northern Ireland, and in some other countries. Warm water-retting is considered the best method from the point of view of application, length of time required, and results, which are second only to retting in the River Lys. In the United States, large open tanks are filled with warm water and the flax straw is placed in the water, which is maintained at a temperature from 85° to 95° F. The bundles are set in an upright position and are stacked two bundles high in the tank. A timber grid or grating is locked over the top of the straw, to prevent it from floating. The straw swells and the water becomes a golden yellow color and large bubbles rise to the surface and burst as fermentation starts. The bacteria organisms then begin swarming in the water, which changes to an opaque yellow color. Gaseous bubbles rise to the surface and burst, forming white froth that turns to a brownish scum on top of the water.

The bacteria begin their work on the skin and cortex, finally reaching the pectin that holds the flax fibers to the skin and cortex. Retting is stopped at just the right time, before the fiber is reached. This method requires four to five days. In Belgium, France, Northern Ireland, and England closed tanks are used.

Cold water-retting requires from seven to twenty days and was used for centuries in Holland, Ireland, and Belgium. It may be done in ponds, lakes, or streams. In Holland and Ireland the flax straw is submerged in water-filled dams, into which more water is permitted to trickle during the retting process. In Holland, clay is placed over the dam to hold down the flax and to exclude air and regulate the temperature.

The most famous cold water-retting was done in the River Lys in Belgium, which produced the famous Courtrai flax. The straw is laid in burlap-lined wooden crates which are placed in the water and weighted down by stones. From April 15 to October 15 is the time allowed for retting in the river. Today, many warm water-retting plants are established along the River Lys but some flax is still retted in the river.

Dew-retting is a primitive method and is most extensively used in Russia. It may take three weeks and more, depending upon weather conditions. A layer of deseeded flax straw is spread out thinly on the ground, usually on grass. Dew and rain provide the moisture. The straw is turned from time to time, to expose all sides to the moisture and to the drying sun. Its most favorable aspect is that it produces a softer fiber, but the results are uncertain because of changes in the weather. In some countries, where labor costs are lower it is considered cheaper than other types of retting. The chemical action in dew-retting is different from that in water-retting. A mold or fungus develops on the straw, which also works through the skin and cortex, loosening the pectin so the fiber can be separated later from the woody core and skin. Dew-retting produces a soft, strong, lustrous fiber.

Chemical-retting has been attempted to shorten the retting period by chemically retting flax, but it has not proven entirely satisfactory to date. There seems to be little advantage. The chemicals add to the cost, and the extra handling required in washing the flax to remove the chemicals consumes any time that may have been saved in the actual retting process.

Drying. After retting the flax straw is dried. The bundles of straw are spread on the ground for a short time and then separated into two or three parts and again set up in wigwam fashion. They must be turned to permit the straw to dry evenly. If not, the straw unexposed to the sun will be harsher and less even in color than the exposed straw. This will in turn cause uneven scutching. After drying for about twenty-four hours, the straw is lifted and tied into bundles with the root ends even and is stored for several weeks to allow moisture from the air to be distributed evenly throughout the straw.

Breaking. The seeds have been removed, the flax is retted, dried, and seasoned, and the straw is now ready for the removal of the woody portions. To do this, it is necessary to break up and beat out the woody waste straw.

Three operations are necessary to remove the woody portions: evening up the root ends of the stricks (a handful of straw),

Top: Cold water-retting in a linthole. Bottom: Warm water-retting in tanks in Oregon. Heavy timbers on grates hold the flax under water. (Bureau of Plant Industry, Soils and Agriculture Engineering, U.S.D.A.).

crushing and breaking the straw, and beating the flax to obtain long fibers.

Scutching. While there are various types of machines in which details of the operation differ slightly, the following is a typical method.

The bundles of straw are opened out, and the straw, with roots leveled, is laid on a table in a thin layer, and is fed to the breaking machine. Here, it passes through rollers that break and crush the straw. From the breaking machine the flax passes to the scutching machine, which is combined with the breaker.

Straw is grasped near the middle by a gripping device and passed through the scutching machine. Blades beat and scrape the root ends of the flax, removing the woody portions. Half way through the machine the gripping device changes and a second set of belts grip the flax, and the branch ends are scutched. The woody portions, called "shives," are conveyed by suction to a fuel shed as they are usually used for fuel. The short fibers, called "scutching tow," drop down through the bottom of the machine and pass through a shaker. The tow is treated further and is fed through a tow-cleaning machine before being baled. The long fibers are deposited at the other end of the scutching machine. They are hand dressed, gathered into stricks, and baled ready for roughing and hackling.

FROM FIBERS TO FABRICS

The scutched line flax is now clean, the woody portions are removed, and the flax is ready to be made into linen yarn and woven into fabrics. The following is a brief resumé of the different processes through which it passes. The more complete explanation of these operations is given in later chapters.

Roughing. This is a preparatory hand operation. The scutcher dashes the ends against a board (a rougher) covered with steel spikes. This lays the fibers parallel and removes any foreign matter, cleans off small loose fibers at the end of the piece or strick, and breaks up the long fibers.

Hackling. This used to be a hand operation similar to roughing except that the boards had finer spikes. In machine hackling the flax is clamped in holders and moves through an ingenious machine which has steel bars with pins that comb the flax. On the first trip through the machine the flax is held with the root ends exposed, and the roots and middle are combed. At the end of the machine, the clamps are unscrewed automatically, the flax is reversed, and the tops are combed as the flax passes through the

Right: Deseeding the flax in a "whipper" deseeding machine. Below: Breaking and crushing the straw to remove woody portion (Linens & Domestics).

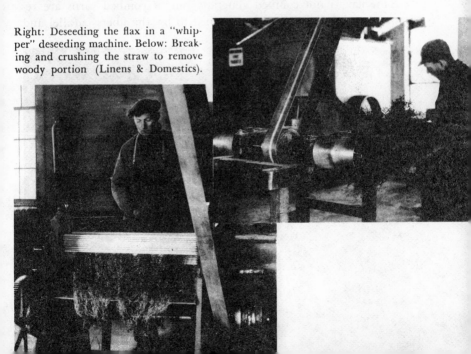

second machine. The short fibers, called "hackling tow," are removed from the hackles by means of a slow revolving brush. The brush is cleaned by an oscillating comb and the tow is deposited into wooden boxes. The long fibers, called "line," are delivered at the end of the machine by the attendant or are passed to an automatic spreader, which delivers the flax in the form of a sliver or ribbon.

The line fibers are now finely separated, smooth and glossy, and ready for drawing, doubling, and spinning into yarns. The tow used for coarser linen fabrics is a tangled mass of millions of short fibers, which must be put into a more or less parallel position before drawing, doubling, and spinning.

Carding and drawing tow. The mass of tow fibers are passed through a carding machine that breaks up and separates the neps or bunches and, to a limited degree, lays the fibers parallel. Leaving this machine, the strands are called "ribbons" or "slivers" of carded tow. These slivers are then passed through a drawing frame. Tow flax is not combed generally but, if combed yarns are required, the slivers are now combed to lay the fibers parallel and

to eliminate the shortest fibers. The carded or combed slivers are now ready for drawing out and spinning.

Drawing. The long line fibers are sorted according to length and then laid on a spreading frame by hand or by an automatic spreader, attached to the hacklers with ends overlapping, and are fed to the drawing frames in the form of a sliver or ribbon. The slivers are drafted and doubled on each machine, and the operation is repeated on several machines until the sliver is the size and quality desired.

Roving. On the roving frame the sliver is drafted and in addition it is given a slight twist, leaving the machine as a rove, ready for spinning.

Scutching. Top of opposite page: The bundles of straw are opened up and laid on the table in a thin layer (left) and the flax is fed into the scutching machine. Bottom of opposite page: The straw is crushed, beaten, and scraped in the scutching machine. Below: The long line fibers leave the machine. Above: The fibers are taken off the machine by workers (left) and gathered into stricks ready for roughing and hackling. (Illustrations top of pages, The Irish Linen Guild. Bottom of pages: Fairbairn Lawson Combe Barbour, Ltd., Leeds.)

Spinning. The rove is drawn out to its final required size, twisted into a linen yarn, and is now ready for being made into a fabric.

Weaving. The linen yarns to be used as warp, or lengthwise yarns, are wound on large beams. Those to be used for filling, or crosswise yarns, are wound on cops, or pirns, or paper tubes. Due to the characteristics of linen yarn and the uses of linen, it is woven in fewer weaves than are other fibers. The beauty of linen is in its own texture rather than in the weave. The uneven yarns give an interesting texture to the plain weave, and many yarns of other fibers are made to imitate this original linen feature. The fine, lustrous linen yarns woven on a Jacquard loom with, for example, more warp face yarns in the pattern and more filling face

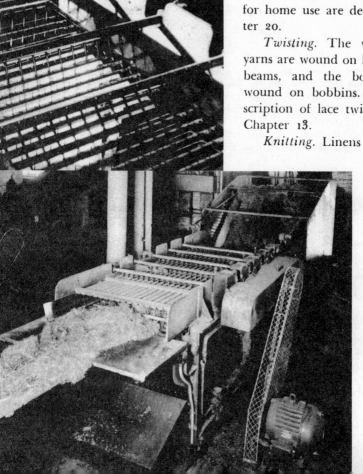

yarns in the ground give a soft two-tone white effect to a smooth, lustrous linen tablecloth. Some of today's linens are illustrated and briefly discussed on the following pages. Others for home use are described in Chapter 20.

Twisting. The warp and beam yarns are wound on large and smaller beams, and the bobbin yarns are wound on bobbins. A complete description of lace twisting is given in Chapter 13.

Knitting. Linens are rarely made

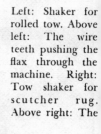

Left: Shaker for rolled tow. Above left: The wire teeth pushing the flax through the machine. Right: Tow shaker for scutcher rug. Above right: The

into knitted fabrics, because of the fiber's lack of elasticity. The other yarns are more adaptable and easier to handle for such fabrics.

FINISHING LINENS

Linen requires fewer finishes than do other fabrics except those made of filament yarns such as silk, rayon, and nylon. The linen fibers are longer and smoother and so there are fewer short, protruding fibers. For this reason linen is not napped. It is a naturally lustrous fiber and woven into fabrics has an attractiveness of its own without requiring many finishing touches.

Preparatory Treatments

Boiling removes from the fabric the dressing or sizing used on the warp yarns and the oil accumulated during yarn manufacture and weaving.

Bleaching is required whether the fabric is to remain white or is to be dyed or printed later. Some authorities believe that, for the finest linens, the best result is still obtained with the ancient method of grass bleaching. Sometimes lawn bleaching is combined with chemical bleaching. However, most linen is chemically bleached with hypochlorite or such other solutions, much the same as is cotton. The coloring matter in the

flax leaving the machine. (Lower illustrations: Fairbairn Lawson Barbour, Ltd., Leeds. Upper illustrations: The Irish Linen Guild.)

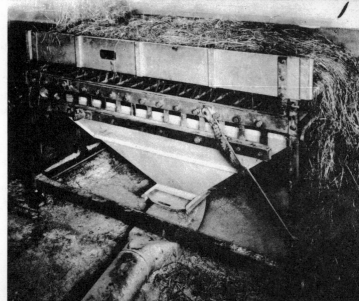

wax and the intracellular substances contained in flax usually require a longer process for removal. After bleaching, the linen fabric is frequently mangled on a very heavy hydraulic mangle, the purpose being to flatten the linen and make the weave closer.

Dyeing

Linen usually is visualized as a snowy white fabric, probably because most linen goods are preferred white due to their uses, and because of linen's great affinity to bleaching. But it is also true that of all the natural fibers, linen is the most difficult to dye.

Linen fibers are little affected by volatile organic acids but can be destroyed by mineral acids. The fibers are not affected by alkalies and are affected very little by heat. White linens can be boiled. The dyestuffs used for colors are those that have affinity for the linen fiber and are not injurious to it.

Linen, being a vegetable fiber, can be dyed with the same dyestuffs and in the same manner as is cotton. However, only a few dyes are used for linen. Considering the uses of linen, high, bright shades are seldom required. Furthermore, since linens are always laundered, the dyes must have a high degree of fastness to washing.

Top to bottom: Linen damask, linen huck toweling, linen glass cloth, linen sheeting, handkerchief linen, dress linen, linen suiting (The Irish Linen Guild).

Dress and suit linens. Left to right, top: Sheer, medium, heavy slubbed, medium weight with thick-and-thin yarns. Bottom: Variegated colors made with slubbed yarns, yarn dyed checked fabric, combination twill weave.

For most linens, a satisfactory degree of fastness to light and perspiration is also essential. While many of the dyes discussed in Chapter 17 may be used to dye linens, vat dyes are most extensively used.

Vat dyes. If the dyestuff is well selected and properly applied, the color will have excellent resistance to washing, sunlight, acids, alkalies, chlorines, and caustic soda. Heat does not affect these dyes. They produce a wide range of clear colors that may be made very bright shades. The vat dyestuffs are reduced and made soluble to dye the goods, after which these are exposed to air or chemicals are added to oxidize the dyestuffs.

Top to bottom, left: Coarse plain weave; herringbone twill weave, American fiber flax and cotton parachute webbing. Right: Linen crash with thick slubs; coarse, plain weave, linen crash.

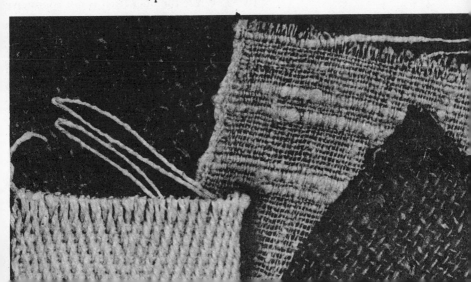

Printing

Many linens are printed. Formerly block printing was used, as the large, clear designs suit the purposes of linen fabrics. Now, linens are mostly screen and roller printed by the direct printing method.

Finishing

Drying and evening by tentering is necessary after bleaching, dyeing, or printing. The damp linen fabric is run into a long, flat tentering frame. There are clamps on either side of the frame, which grasp the selvage of the cloth, even it to its normal width, and carry it into drying chambers where it is dried by controlled hot air.

Ironing or calendering gives a smooth, lustrous surface to the linen fabric. The cloth is run under pressure between rollers.

Pounding or beetling is a process used for some fine linens to impart to them compactness and luster. Large wooden blocks pound the fabrics in the machine, flattening the fibers. This closes up the weave and adds luster due to the greater surface of the fibers. Today, beetling is used infrequently and the fabrics are mangled to flatten the fibers and close up the weave.

Shrinking the fabric is the last finishing process. While the fiber itself is not conducive to shrinking as are the wool and cotton fibers, the weave, particularly if not a close weave, will cause the fabric to shrink when laundered.

The same shrinking process as applied to cotton is now given linens. A fabric controlled this way will not shrink over ¾ per cent.

Top to bottom: Sheer handkerchief linen; fine huck toweling; lustrous Jacquard woven linen damask; damask toweling; yarn dyed, plain weave crash.

While the fabric is held at a certain tension water and steam shrink the cloth to predetermined dimensions. Upon leaving the machine, the fabric is given a smooth, ironed surface. It is now a finished fabric, sleek or rough textured, lustrous, crisp or soft—depending on the finishing processes through which it has passed.

SEED FLAX

The fiber from seed flax is coarse and quite different from that obtained from fiber flax, as these plants are grown for seed first and only incidentally for fiber. Experiments with seed flax have been going on in the United States for some time. Seed flax fiber has been combined with cotton for yarns and woven into fabrics. However, only a very small proportion of seed flax fiber can be mixed with cotton.

The seed flax crop is harvested with combines that mow the flax, cut off the seed heads, and lay the straw in the field where it is dew-retted. After retting, the straw is delivered to the decorticating plants. Here "tow-breakers" break up the straw into short lengths called "tow," which is sent to the cotton plant. The shive and short fibers are removed in centrifugal-air cleaners. The remaining fibers, alone or blended with cotton, are processed on cotton waste machinery. They are carded into a sliver; several slivers are drawn into a roving; and the rovings are spun into yarns by additional drawing and twisting. The damp yarns held under tension are further twisted in a twister machine and are woven into fabrics.

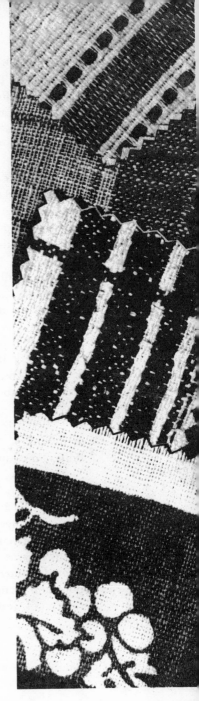

Top to bottom: Coarse huck toweling, yarn dyed; plain weave, yarn dyed luncheon cloth; plain weave, yarn dyed toweling; plain weave, printed luncheon cloth. These are all woven of American flax and cotton.

SUMMARY

A VEGETABLE FIBER FROM THE FLAX PLANT

The Flax plant. There are two types of commercial flax plant, fiber flax, grown for the fiber, and seed flax, grown for the seed. Fiber flax is pulled, seed flax is mowed.

FIBER FLAX

Pull and dry.

Ret: To separate the fibers from the bark.

Dry: To dry and condition the retted straw.

Deseed or ripple: To remove the seeds.

Scutch: To remove the woody portions.

Rough: To lay the flax straw parallel and to remove neps and foreign matter.

Hackle: To comb and separate the long fibers (line) from the short fibers (tow), and to lay parallel the line fibers. Tow is carded and combed, to separate and lay parallel the short fibers.

Draw and double: To further straighten, mix, and lay parallel the fibers and to produce a strand or sliver. Roving further draws out and very slightly twists the sliver into a rove.

Spin: To draw out and twist the rove into a yarn.

Construct the fabric: By weaving, twisting, braiding, or knotting.

TO FINISH THE FABRICS

General finishes

Boil: To clean the fabric and to remove the dressing used in the warps.

Bleach, dye, print: To make the fabric white or colored.

Mangle: To flatten and close in the yarns in the cloth.

Tenter and dry: To make the fabric even.

Iron: To give a smooth, lustrous finished surface.

Shrink: To reduce the fabric in width and length.

Functional finishes: To render the fabric crush- or crease-resistant, spot- and stain-resistant, crisp, absorbent, water-repellent, waterproof, resistant to moths and germs, and fire-resistant.

Opposite page: Hemp stalks showing the leaf pattern around the female flower (U. S. Department of Agriculture).

MINOR VEGETABLE FIBERS

The strongest

There are some twenty or more different vegetable plants capable of producing fibers that can be made into yarns, but most of them are negligible as far as commercial production is concerned. Many are made into fabrics for local use; still others are woven into rope or twine only. The vegetable fibers of present commercial value can be divided into three groups: seed hairs, such as cotton, kapok, and pulu; bast fibers from the stems of plants such as flax, hemp, jute, and ramie; and structural fibers from the leaves of plants, such as sisal and coir.

Besides the two outstanding vegetable fibers cotton and linen, there are a few others playing a more or less important part in the fabric world today. They are hemp, jute, ramie, sisal, and kapok.

HEMP

There is but one species of true hemp—*Cannabis sativa*. Many other fibers such as sisal hemp and New Zealand hemp are incorrectly named Manila hemp. These are not hemp and have no botanical relationship to it.

Hemp is adapted to a great many countries throughout the temperate zones. As a cultivated crop, it is grown chiefly in European countries, Japan, Turkey, Chile, and the United States. Russia produces as much hemp fiber as do all other countries combined.

It is thought that hemp originated just north of the Himalaya mountains. Ancient Chinese writings indicate that hemp was used for fiber in China in 2800 B.C. Like linen, it was used by man long before he could write. Hemp supposedly spread from Asia to Europe about A.D. 500. It came to Chile in 1545 and to the United States in 1645.

The plant. Hemp is an annual plant grown from seed. It is cross pollinated and has a deep taproot system. The seed-producing flowers are grown on separate plants from those that produce pollen. Thus, there are two types of plants: the pistillate or "female" seed plants, and the staminate or "male" pollen-producing plants. Under normal conditions the crop will produce about equal amounts of each. The male plant dies soon after the pollen is shed, while the female plant continues to live until the seeds are mature.

The plant's physical appearance varies with the type of cultivation. If sown thickly for fiber yield, the plant is tall, slender, whip-like, and free from branches except at the top. The usual height of fiber hemp is 5 to 10 feet. When the seed is spaced for seed crop, the plant produces long, coarse branches from nearly every node, and the main stalk becomes very thick. The seed type may grow as tall as 16 feet. There are a great many varieties and types of hemp, ranging from short dwarf plants to tall, rank growing tree types. Some are used for fiber, some for oil, and others for drugs.

The stem of the hemp plant is hollow except at the base. The thickness of the tissue around the center varies greatly. The fiber, like the flax fiber, is found between the cambium and the epidermis.

Growing. Hemp is seeded early in the growing season in very fertile soil. It is thickly broadcast by hand or by machine. Usually no cultivating or weeding is practiced after the seed is sown. The crop is ready for harvesting as soon as the male plants are in full bloom and are freely shedding pollen.

Harvesting hemp with a hemp harvester that cuts and spreads the stalks on the ground to ret (U. S. Department of Agriculture).

It is generally believed that for the softest, finest (though weakest) fiber, it is necessary to harvest the plant as soon as the shedding of pollen begins and for the strongest fiber and the greatest yield it is advisable to wait until the plant is completely matured.

In some countries the plant is pulled, but in most places the crop is cut a few inches above the ground with a hand sickle or a hemp harvesting machine. The acre yield ranges from 500 pounds in some countries to 1,800 pounds in others. The United States yield is about 1,600 pounds.

The fiber. Hemp is a bast fiber. The fiber shows cross markings but usually has a wider lumen than does flax. Its length is almost as long as is that of the flax fiber, but its thickness is somewhat greater. The cross section of the fiber is irregular, varying from triangular to polygonal shapes. The corners are much more rounded than those of flax.

From the Fiber to the Fabric

The preparatory processes are the same as for flax, namely, retting, drying, breaking, and scutching. The last two may be hand or machine operations.

Making the yarn. These processes are the same as for linen. The operations include hackling, which combs out the short fibers, the "tow," leaving the long fibers, the "line;" drawing out and doubling a number of slivers into one sliver; further drawing and slightly twisting the sliver into a roving; and spinning the roving into hemp yarn.

Uses. Hemp has been used, at one time or another, for a wide range of products, from fine fabrics to coarse ropes. Its characteristics, such as strength, pliability, and tenacity, make it ideal for certain types of thread, twine, and rope.

JUTE

Jute is obtained from two plants of the *Corchorus* family. They have physical differences, but from the standpoint of fiber they are essentially alike.

Jute has been grown and processed in Bengal since very remote times, although it probably did not originate in Northern India. Today, the cultivation of jute is still most extensive in Bengal, (83 per cent of the total world production) but it has spread to other sections of India. The manufacture of jute is carried on not only in the East but to some extent in European countries as well. Brazil is making a successful effort to grow jute.

Jute growing has been tried out in the United States but no industry has developed. This is probably due to the lack of adequate machinery and to the impossibility of competing with the cheap labor markets of India.

The plant. Jute is an annual plant, growing from 5 to 10 feet high. It has a cylindrical stalk as thick as a man's finger. There are no branches except near the top. The light green leaves are from 4 to 5 inches long and 1½ inches wide; broad at the base and growing to a sharp point. The two lower teeth of the leaf are drawn out into bristlelike points. The flowers are small, yellowish white, and growing in clusters of two or three opposite the leaves. Plants are grown not only for the fiber but also for the leaves, which are used as a pot herb.

Growing. Successful growing of jute demands a hot, moist climate with a moderate amount of rain. Too much rain is detrimental and a very dry season is disastrous.

Like flax, the seed of jute is broadcast in the spring, from March to the middle of June, depending upon locality and weather. Cutting depends upon the time of planting; it may be as early as June or as late as October. The crop is ready for cutting when the flowers begin to fade. If gathered earlier, the fiber is weak; if left until the seed is ripe, the fiber, although stronger, is coarser and lacks the characteristic luster.

Left: The hemp, cut and left in bunches, is being spread for dew retting. Right: The hemp grower holds in his right hand stalks of fiber hemp that have been properly weathered and are now ready for breaking. In his left hand he holds hemp fiber after the stalks have been broken (U. S. Department of Agriculture).

The fiber. The individual bast cells of the jute are very fine and much shorter than flax fibers. The lumen varies through the fiber, narrows in some places to a thin line or disappears altogether. Nodes or cross markings are usually absent. The cross section shows the cell to be polygonal with a pronounced oval lumen. The fiber is always found in bundles, even after manufacturing. The best quality of jute fiber is clear yellowish in color, with fine, silky luster. It is soft and smooth to the touch.

Jute fiber has decidedly less strength and tenacity than has flax or hemp. This is due to the construction of the walls of the separate cells that make up the fiber. The walls vary in thickness at different points, making the fiber unequal in strength through its length. Fiber quality also changes with age and exposure. Recently prepared fiber is stronger, softer, whiter, and more lustrous than that which has been stored awhile. The older fiber is brown, harsh, and brittle in quality. Air changes the fiber's color and quality.

Jute fiber is more woody in character than is flax or hemp. It is highly hygroscopic. In dry atmosphere it may have no more than 6 per cent of moisture, but in damp conditions it may have as high as 23 per cent.

From the Fiber to the Fabric

Retting of jute is very much like that of flax and for the same purpose—to remove the bark from the jute fiber. After retting is completed, a native operator, standing in the water, grasps a number of stalks of jute and rips off some of the bark next to the roots. The stalks then are wrapped in bales, ready for softening.

Softening. Since the jute fiber is woody and brittle, it must undergo this preliminary process in order to facilitate the yarn making processes. The bales are opened and the fibers are sorted. The bale opener or jute crusher partly crushes and softens the fibers which are then fed between a series of fluted rollers in the softening machine and water and oil are poured over them. Upon being delivered from the machine, the roots are cut off and the fibers are allowed to rest for twelve to twenty-four hours, permitting the oil and water to spread through and soften the fibers.

Preparing the yarn. The jute fibers are delivered to a carding machine and leave the machine in a long, round sliver. The slivers are then drawn out by combining a number of slivers into one, the size of an individual sliver. During the drawing out process the fibers are combed parallel. This doubling and drawing out is repeated until the desired sliver is obtained.

The sliver is delivered to the roving frame where it is drawn out to about eight times its length and is given a slight twist, and then, as a roving, is wound on bobbins. It is now ready for weaving, unless a finer yarn is desired. If so, the roving is further drawn out and twisted on the spinning frame, to produce the required jute yarn ready for weaving.

Jute bleaches easily up to a certain point, sufficient to enable it to take brilliant or delicate colors, but it is difficult to bleach it pure white. Most colored jute has a poor resistance to light.

Jute is sometimes used alone to weave fabrics; but more frequently it is combined with other fibers. It may be combined with wool for carpets, with cotton or linen for draperies, household cloths, and padding. It is often used with wool for horse blankets. The coarser yarns are used for sacking.

RAMIE

Ramie is obtained from the plant genus *Boehmeria nivea*, which is closely related to the stinging nettle genus, although it lacks the stinging hairs. The term "ramie" is used by English-speaking people to designate the plant with snow white leaves on the under surface, and the fiber obtained from it. "Rhea" is the term used in India for not only the white leaf type but for one with leaves all green. "China grass" means the hand cleaned, but not degummed, fiber as it comes to the market, never the plant.

Ramie grows in warm temperate countries, chiefly in China, but it is also grown commercially in Formosa, Japan, and the Philippine Islands. It is grown in Central America, and South America, but here no industries have developed. There have been

Jute is combined with thick spun, softly twisted wool to weave this original fabric created and hand-woven by Dorothy Liebes. See for contrast the heavy durable jute (right) used as part of the back of many carpets.

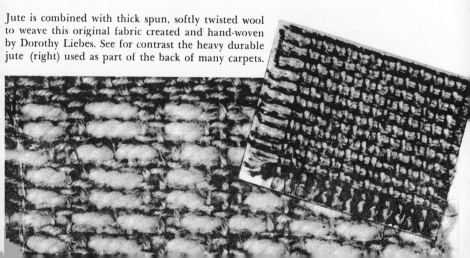

some experiments made to grow ramie in the United States, but the results have been inconspicuous.

The plant. Ramie, a shrubby plant, grows from 3 to 8 feet tall, with straight shoots that are sent up each season from perennial underground root stock. It also grows from seeds, cuttings, or layers, or by division of the roots.

The leaves are nettle shaped and have a silver-colored downy substance on the back. Minute greenish brown flowers grow closely along the slender axis.

Two to four crops may be cut in a single season if conditions are favorable, for ramie grows easily—4 tons or more of moist stems are obtained per acre. When the plant is ripe, it is cut and the leaves and small branches are removed. The outer cover and the layers of fiber are then stripped off like ribbons. These strips contain the bark, the fiber, and a quantity of very adhesive gum. The bark and as much of the gum as possible are removed before drying. After drying, this fiber is termed "China grass." The China grass yield is approximately 2½ per cent of the green stem ramie.

The fiber. Ramie, like linen, is a bast fiber. The longitudinal view of the long, coarse ramie fiber appears microscopically an irregular, knotty, and often ribbonlike fiber, with pronounced diagonal cracks on the surface. The cross section is similar to cotton except much larger, with fissures or cracks running from the outer edge towards the lumen. The cracks reduce the strength of the fiber and lower its spinning qualities.

The fiber varies in length from as short as 1½ inches to as long as 12 inches. Its thickness is about three times that of linen. The variability of the ramie fiber gives it a low drawing and spinning quality. Since it is impossible to make perfect yarn with varied fiber lengths, ramie must be limited to a low grade of yarn, or must be separated into reasonable groups, or cut into even lengths.

From Fiber to Fabric

Removing the bark. The bark and as much of the gum as possible are removed by hand before the plant dries–a slow and tedious process.

Degumming. As in processing silk, it is necessary to remove the gum. This is usually accomplished by immersing the fibers in hot caustic solution or other chemical solution. After the gum is removed, the alkali or chemicals used are neutralized, the fibers are washed, the moisture removed, and the fibers dried. The fibers may be softened the same way as described for jute.

Making the yarns. The fibers pass through the same processes as does linen, namely dressing, carding, drawing out the slivers, further drawing and twisting to produce a roving, and spinning into yarn. Since the fibers vary in length, they handicap preparing and spinning unless they are sorted or cut into even length groups.

Ramie is easily dyed and is little affected by moisture. The fabrics have a hairy texture but the superfluous hairs may be removed by singeing or burning.

Properties. Ramie is stronger than any known fiber, it has the luster of some silk or mercerized cotton, it resists atmospheric changes, dyes easily, and is affected little by moisture. It lacks extensibility and flexibility. Ramie makes strong, lustrous fabrics. It can be made into as fine a fabric as that used for lingerie, table-cloths, or nets, or into as heavy a fabric as canvas.

KAPOK

Kapok are long, downy fibers from a tree, native to the West Indies and to tropical America. It is used for stuffing and padding and is not made into fabric.

SISAL AND COIR

Sisal, a hard fiber, is made from the leaves of a plant closely related to hemp. The East African colonies, the Netherland East Indies, and Haiti are the chief sources of sisal, but it is also grown in Mexico. The plant has long, straight stalk-like leaves growing out of a short trunk. The fibers are strong and flexible, from white to yellowish brown in color. Sisal is used alone and combined with kraft fiber to make attractive and often colorful rugs.

Coir is from the cocoanut tree grown in the South Pacific. It is a strong, fibrous material, brown to reddish brown, obtained from the outer covering of the husk of the cocoanut, the fruit of the tree.

Two spun rayon and ramie blended fabrics.

SUMMARY

VEGETABLE FIBERS FROM PLANTS

SEED HAIRS

From seeds of plants
  Cotton
  Kapok
  Pulu

BAST FIBERS

From stems of plants
  Flax
  Hemp
  Jute
  Ramie

STRUCTURAL FIBERS

From leaves of plants
  Sisal
  Coir

Opposite page. A full grown silkworm (shown here larger than actual size) is rearing its head looking for a place to attach itself on the mulberry leaf, so it can begin spinning its cocoon (International Silk Guild).

SILK—ANIMAL FIBER

The most luxurious

IN THE BEGINNING

Silk, the queenly multi-wound, brief coverlet of the tiny silk-worm, was spied by human eyes first in ancient China, some 4,000 years before Christ. It is not known exactly how the Chinese discovered that the countless silkworms after feeding on their mulberry trees protected themselves by yards and yards of beautiful silk yarn suitable for weaving, but there are entertaining myths on the subject.

One story is that a beautiful Chinese princess, while in her garden one day, dropped a cocoon into a cup of hot tea. Taking it out she discovered that she could unwind the strong continuous fiber from the softened exterior. Whether or not this is true, it is recorded that Empress Si-Ling-Chi, wife of Emperor Huang-ti, (2640 B.C.) experimented with the wild silkworms that lived on the mulberry trees. She learned how to feed and raise them and, more important, how to reel or unwind the silk filaments from their cocoons. It is to her that the Chinese give the credit for the invention of the loom. During the reign of Huang-ti, silk culture was encouraged and the peoples began an industry that became an important force in China, and only in China for centuries.

Cocoons, their growth and care, silk, its reeling and weaving, became so important that this precious art was kept a guarded secret from the rest of the world for hundreds of years. Tending the silkworms was one of the highest honors that could be granted to ancient Chinese ladies. Confucius. (500 B.C.) wrote that the emperor and his vassals maintained near a stream, a government-

142

owned nursery for silkworms and mulberry trees and that it was the custom for the ladies of the palace to draw lots to see which ones would be lucky enough to be sent to the nursery to care for the worms.

Chinese silk was well known throughout the ancient world long before people found out from where it came and how it was constructed. Caravans traveling from China to Persia, and thence west, carried silk cloth. Silk became the noble fabric of the Roman Empire. Pompey (106-48 B.C.), during his conquests, returned from China wearing a beautiful robe woven of silk.

The knowledge of silk culture began to trickle out of China along about A.D. 300. Through Korea the information reached Japan. One of the ancient Japanese books states that about this time Japan sent some Koreans to China to engage competent people to teach the art of making silk goods. Four Chinese maidens were brought back and these girls, by instructing the Japanese court in the art of plain and fancy weaving, started the Japanese silk industry. A little later the cultivation of silkworms began in India. The legend states that the eggs of the silkworm and the seeds of the mulberry tree were carried to India, hidden in the lining of the headdress of a Chinese princess. On the other hand, references in Sanskrit indicate that a silk industry existed in India as early as 1000 B.C. or perhaps even earlier.

Besides references to silk in the Books of Amos and Ezekiel in the Bible, the first mention of the silkworm in Western literature was by Aristotle (384-322 B.C.). He writes of "a great worm which has horns and so differs from others. At its first metamorphosis, it produces a caterpillar, then a bombylius and lastly a chrysalis. . . . From this animal, women separate and reel off the cocoons and afterwards spin them. It is said that this was first spun in the Island of Cos by Pamphile, daughter of Plates." Classical literature is full of references to silk but very vague as to its source. Even Pliny knew little more than Aristotle. In A.D. 273 Aurelian, who would not allow his wife to buy a silk shawl, tells that silk was worth its weight in gold.

About A.D. 550 two Nestorian monks who had long resided in China, learning the art of silkworm culture, were fortunate in being able to smuggle a few silkworms out of China and into Constantinople, by carrying them concealed inside their hollow canes. These few worms were supposedly the beginning of all the varieties that supplied the Western World for more than 1,200 years. Byzantine silks became famous; the Saracens mastered the industry; Venice, Florence, and Milan became known as silk centers.

By the seventeenth century the Mediterranean area had a well-established silkworm industry. Other countries made many efforts to grow silkworms but with little success. Many companies were formed in England for the introduction of sericulture (the growing of silkworms) but they were failures, and the rearing of the silkworm has never become a branch of the British industry.

Western-World Silk

To date, America's story is not much different from that of England. In 1522 Cortez appointed officials to introduce silkworms into Mexico. Acosta mentions the Mexican venture which died before 1600. In 1609, James I tried to compel the Virginians to take up sericulture instead of tobacco raising. They were encouraged by bounties and stimulated by rhymes.

> "Where Worms and Food doe naturally abound
> A gallant Silken Trade must there be found.
> Virginia excels the World in both—
> Envie nor malice can gaine say this troth."

In 1623 it was decreed that any Virginia planter should be fined 10 pounds if he did not cultivate at least ten mulberry trees for each 100 acres. All the threats and bounties failed, for silk culture in Virginia was not profitable in comparison to tobacco. Similar tries were made in Georgia, beginning in 1732. In 1739, 10,000 pounds of cocoons were received at a filature (silk-reeling establishment), which had been built in Savannah, for reeling the silk from the cocoons. But silk in Georgia could not compete with cotton after the cotton gin was invented.

In 1755, a Mrs. Pinckney of South Carolina carried with her to England enough silk of her own raising to make three dresses. One dress she presented to the Princess Dowager of Wales. One of the remaining was still in existence in 1809.

Different states made special effort to establish a silk industry. Mulberry trees were introduced into Pennsylvania in 1763 and 1790 by a Dr. Aspinwall, who had sent trees to Connecticut in 1762. In the first part of the nineteenth century, silk culture seemed to promise possibilities of development. In 1810, three counties in Connecticut produced $28,500 worth of raw silk. There was enough progress made in 1830 to engender tremendous speculation. A new mulberry tree gave promise, and there was a wild rush for purchase of the young plants.

In 1833, the Cheney brothers began experimenting with silk culture. Their first nursery was at South Manchester, Connecticut. In 1836 they received 15,000 mulberry trees from China. Three hundred of these Ward Cheney planted horizontally, and 3,700 shoots sprang up, the leaves from which fed some 6,000 silkworms. This multiplication of trees started a boom. In 1836 the Cheneys leased 117 acres in New Jersey for a nursery. In October, 1837, they sold $14,000 worth of trees. Silkworms were $5.50 an ounce. In 1839 the trees were $1 and $2 apiece and in some places $500 a hundred. But in 1840 a crash came. The new mulberry trees could not stand a severe northern winter. They died, and the silk growers had wasted their money. In 1844 the growers had another shock when a fatal blight affected almost all the rest of the mulberry trees in the United States.

There have been many attempts to revive interest in the production of raw silk, but none were successful. There never was much raw silk produced in the United States. The fortunes made and lost were in speculating with mulberry trees. Although the United States became the world's greatest user of silk, this was imported from Japan, Italy, and some from China and Russia.

Recently Brazil has begun to concentrate on the growing and spinning of silk. There are now in Brazil, over eighty spinning mills with more than 2,500 reeling basins, producing close to 800,000 pounds of raw silk annually. Of course, this is not a great deal if compared to average monthly consumption of raw silk in the United States, which was 4,000,000 pounds in 1939. Still, because it has favorable climatic conditions, Brazil has possibilities of becoming a large silk growing country. It is possible to harvest many more cocoon crops in Brazil than in the Orient, and the quality of silk compares very favorably.

The Brazilian Government Trade Bureau has this to say in

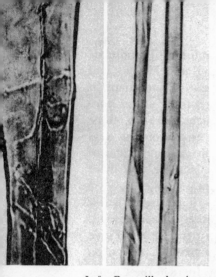

Left: Raw silk showing gum (sericin). Right: Degummed fibers (× 500) (Textile Fiber Atlas).

regard to their mulberry tree and silkworm cultivation:

"The fact that the world's silk market remained under Japanese control until recent years is due solely to the circumstance that no other country dared start such industry on a large scale, fearing inability to compete against the cheap wages paid by Japan to her silk laborers and the consequent unbelievably low cost of yarn.

Cheap wages, however, can easily be offset by higher production and increased output, and experiments carried out in Northern Brazil during recent years to ascertain the possibilities for cultivating the mulberry tree for culture of the silkworm yielded astonishing results."

Of course, the silk manufacturers in America, by application of inventions and power, have been able to produce much of the best silk fabric in the world. Silkworm culture and reeling are still essentially hand work. American human behavior is such that we lack the infinite oriental patience required to unwind correctly the tangled mass of silk filament from the tiny cocoons.

A reeling machine has been invented in the United States and the reeling of raw silk has begun. Texas is the seat of these beginnings of machine reeling. This state also grows mulberry trees and raises the cocoons for silk yarns. California has also become interested in the possibilities of future silk culture.

Will American inventive genius find mechanical means that can commercially compete with the accuracy of deft handwork? It may be that silkworm culture and silk reeling (like handmade lace) should be relegated to the sphere of art and left to those countries whose inherited interests and abilities are best suited to perform the intricate operations so foreign to the capabilities of modern Americans. Anyway, this is, at least, food for thought.

But regardless of complications and difficulties, there always will be a place for silk in the family of fabrics, for no other to date has ever filled the position held by silk—the aristocrat.

THE SILK FIBER

Silk fiber used for fabrics is the fiber from the cocoon spun by the larvae of the mulberry silkworm *(Bombyx mori)*. The silk fiber is in every respect one of the most perfect natural substances known for yarn making.

Silk, a smooth, structureless fiber, is semitransparent. Its luster depends upon the uniformity of the outer layer that reflects the light without dispersion. Under the microscope, the raw silk fiber (the fiber as it is reeled from the cocoon) appears a long rod of consolidated gum. In reality, it consists of two filaments (two secretions from glands on two sides of the body of the silkworm), which are cemented together and enveloped by silk glue, called "sericin." The surface of the raw fiber is very irregular, with fissures, creases, and folds, caused mainly by deforming, breaking, or rubbing off of the sericin during reeling.

The cross section of the silk fiber is triangular in shape, with rounded corners. In raw silk the two filaments are normally joined, facing each other with the flat side of the triangle. There is a difference in the cross sections of the three layers of silk fiber forming the cocoon. The inner layer, or the layer closest to the chrysalis, is flat shaped, the middle, more uniform and rounded, and the outer, rougher and more irregular.

The cross section of the fiber consists of three parts: the central silk cylinder composed of a substance called "fibroin," which forms

Silkworms feeding on mulberry leaves (National Federation of Textiles).

the principal part of the fiber; a layer of silk albumen or sericin; and a very thin coat of wax or gelatine. Both outer layers are eminently fitted to receive various vivid coloring matters. About 20 to 30 per cent of the fiber is sericin. After degumming or removing most of the sericin, the fiber is soft, smooth, and rodlike, with a pearly luster.

The silk fiber is the longest of all natural fibers, ranging from 800 to 1,200 yards. It is a very fine fiber, averaging from 10 to 13 microns in diameter (Chapter 22). Each fiber varies in diameter through its length, being thicker and stronger toward the middle than at its extremities. It is said that to produce 1 pound of raw silk about 3,000 cocoons are required.

The perfectly homogeneous structure of silk renders it stronger in proportion than any other natural fiber. It is said to have the tensile strength of iron wire. It has elasticity and can be stretched one fifth of its original length after which it will return to its natural state.

Silk has great power of absorption. Its weight can be increased as much as 30 per cent with moisture, without any change in appearance. Its ability to take dyes of vivid color is, to a large extent, due to this absorbing ability. It is a low conductor of heat, therefore, though it is fine and light, it is a warmth giving fiber.

Silk's characteristics of strength, lightness, elasticity, and absorption, as well as its beauty of color and its high luster are such that to date no one fiber, either natural or synthetic, has been able to imitate it entirely.

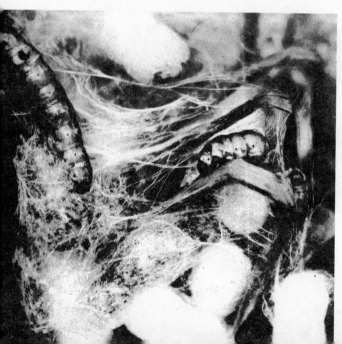

Silkworms spinning their cocoons. Some are preparing to spin, others have only a thin web around them, while still others have finished their cocoons and are completely covered with a long, continuous, silk filament (National Federation of Textiles, Inc.).

A silkworm room where the worms have been transferred to branches to spin their cocoons (Cheney Brothers).

THE LIFE OF THE SILKWORM

A silkworm is a larva of a moth. After centuries of experimenting it was found that the *Bombyx mori* species produced the best silk. It gets its name from Bombyx and from the mulberry tree, the *Morus alba*. In the beginning, the Chinese searched and gathered the spun cocoons from the mulberry leaves, but today they are highly pampered and carefully watched through every step from the eggs to the completed cocoon.

One moth lays from 350 to 500 eggs and soon dies. After about thirty days, the eggs hatch into tiny worms, called "ants," which begin to feed on mulberry leaves. The feeding lasts for forty days after hatching. During this time the worms have four twenty-four hour periods of sleep, on the sixth, twelfth, eighteenth, and twenty-sixth days. In the entire time they consume about fifty times their own weight. They now look for a place to spin their cocoons. They are no longer little insects but are grayish white caterpillars about 3½ inches long. As soon as the caterpillars begin to rear and move their heads, looking for a place to attach themselves preliminary to weaving their cocoons, they are removed from the trays upon which they were fed and are placed on branches ready for spinning.

The worm attaches itself to straws or branches placed upon shelves, then doubles itself on its back like a horseshoe, with its

The cocoons are removed from their mountings (National Federation of Textiles, Inc.).

legs on the outside and, by moving its head in a figure eight, extrudes the filament and spins itself inside a cocoon. It is estimated that the head describes about one ellipse a second, and some 300,000 in making the cocoon.

As the silkworm digests the mulberry leaves, two fluids are formed in its body, fibroin and sericin. From two tubelike glands in the body, the fibroin (a viscous fluid) is ejected through one tiny spinneret, which is an opening in the head of the worm. At the same time, from two other glands, the sericin (a gumlike secretion) passes through the spinneret. As the secretions contact the air, the two fibroin filaments coagulate and are cemented together.

Around the eighth day after spinning, the cocoons are gathered and made ready for reeling. Some cocoons are left to mature in order to produce moths to lay eggs. The moth appears from ten to twelve days after the cocoon is formed and mates immediately. The female lays her eggs from four to six days later on a sheet of paper arranged for her, and the silkworm growing starts again.

The cocoons are carefully sorted and placed in large bins (Cheney Brothers).

SECURING THE RAW SILK FILAMENT

Preparing the Cocoons for Reeling

The cocoons from which the silk is to be reeled are treated to kill the chrysalis inside the cocoon before it escapes and destroys the filament. Since the cocoon has been wound in a figure eight, the breaking through of the escaping chrysalis would cut the silk filaments into thousands of short fibers.

The most effective method of destruction is by use of hot air. A large quantity of cocoons are placed in a dryer, and hot air is circulated through the chamber, suffocating and killing the chrysolides and, at the same time, drying the moisture in the cocoons. This method keeps the cocoons clean and eliminates exposing them to wind and air.

Hot sun also will suffocate the chrysalides, but it hardens the gum, making it difficult to unwind the cocoons. It also withdraws some of the color. A third method is to force hot steam from eight to ten minutes through the cocoons that have been placed on trays in a steam cupboard. After steaming, the cocoons are moved to canvas beds and are allowed to dry in the air, away from the sun. The drying takes from eight to ten weeks. This is a tedious method. First the steaming must be timed accurately. If too long, the fiber may be damaged, if not long enough, the chrysalides may recover and later escape. Whichever method is used, the dried cocoons are then put into sacks, ready for sorting and reeling.

Sorting. The cocoons are carefully sorted, as only the perfect cocoons can be used for reeling. Often adjoining worms do not have sufficient space to spin and therefore interspin with other cocoons. These "doupion" or double cocoons and the pierced cocoons are used for spun silk. In gathering the "fresh" cocoons,

In the boiling room the cocoons are immersed into very hot water to soften the silk for easier reeling (National Federation of Textiles, Inc.).

Top: Ready to begin to reel silk from cocoons in pans to reels over heads of workers. Above: Silk reeled on six-armed reels. Below: The skeins of silk are steamed (National Federation of Textiles, Inc.).

some are crushed, smashing the chrysalis, which stains its cocoon as well as others near by. These cocoons can be used, however, if not otherwise damaged but must be reeled at once to prevent deterioration.

Reeling

Reeling is the process of unwinding the silk filament from the cocoon. To do this, the sericin that holds the filaments together must be softened. The cocoons are boiled, steamed, or soaked in hot water. They are also kept in heated water during reeling.

Brushing. A certain number of cocoons are placed in a deep basin of water. In the bottom of the basin are heated coils that keep the water at a certain temperature. A broom with a circular base is placed over the basin. It rotates first one way and then another way until it catches the outside fluff of the cocoons. The broom is raised and the cocoons, with the end of the filament exposed, are transferred on a strainer to a basin of water beside each reeler's reeling machine. The reeler's basin of water is maintained at a temperature of from 180° to 200° F. After the reeler has received her supply of cocoons, the reeling operation begins.

Hand reeling. In the past, reeling has been a hand operation, as the filatures (reeling machines) are not really machines but rather the equipment for the reelers. A filature provides for four, six, and eight skeins to be reeled at one time. In this operation, filaments from two to twenty cocoons are reeled simul-

152

taneously to make one silk yarn. The usual number, however, is four to six. The reeler takes the filaments from the cocoons and twists the ends together to form one yarn. This is then laced through an apparatus just above the water level. From there it is wound around a tiny glass wheel about 18 inches above the basin; then down around another glass wheel; upward again, twisted several times; through a glass hook above the reeler's head; and through a slit in front of the reel, to be attached finally to the reel.

When all the filaments that are to be wound on one reel are collected, the reel is set in motion. The filaments unwind, causing the cocoons to bob around in the water. Since each cocoon supplies but 800 to 1,200 yards of filament and many more yards are wound into one skein, the reeler is constantly attaching filaments from new cocoons to these filaments. If a filament breaks, an expert reeler can attach a new filament before half a yard has passed.

The girls who do the reeling hold perhaps the most important position in the entire production of silk. The work requires constant judgment, a developed sense of sight, touch, and timing, as well as dexterity. It is the one reason why in the past silk has never become an American industry—American wages could not compete with the low pay of the Orient and Europe, where girls receive some 25 to 50 cents a day and keep.

In reeling, the yarn must be kept as nearly even in diameter as possible. As an example, in a yarn with five filaments, two may be coarse and three fine. As these must be continued, a filament with similar fineness or coarseness must be joined to them. A yarn made up of five filaments is scarely visible, and judgment of a very fine individual filament requires experience. Expert and quick joining of like filaments adds greatly to the value of the raw silk and the finished product; and it is essential if the yarns are to have a comparatively uniform diameter.

In some instances the filaments are rereeled to another frame into skeins of fairly uniform weight. In either case, the skeins are tied, inspected, and twisted into standard skeins. After sorting according to denier (size) and quality, the skeins are packed into bundles of 30 or more each, called "books." The books are made into bales of about 133 pounds each, and are ready for shipment to yarn manufacturing plants, where they are degummed, and made into yarns and fabrics.

A reeling machine designed to eliminate hand reeling (Anscowain Manufacturing Company).

Machine reeling: As stated previously, the only process (from the silkworm to the finished fabric) that was not possible in America, because of competitive cheap hand labor, was reeling the filaments from the cocoons. The first big forward step has now been made to remedy this situation. A semi-automatic cocoon finding and reeling device has been invented.

This reeling machine must first be threaded with the proper number of filaments, one from each cocoon, before the machine begins its operations. This is much the same as lacing the warp yarns through the loom before weaving starts. Once the machine is started all subsequent cocoons are automatically joined as preceding cocoons run out. There is no stoppage of the continuous reeling.

The number of filaments required (from as many cocoons) for different denier yarns, are entered in the eye of the reeling device and then are reeled onto a first-time spinner bobbin. While the machine is in operation and the filaments pass from the cocoons to the bobbin, the yarn is given two and a half turns to the inch.

The machine also performs another essential operation in silk manufacture. During the reeling, the silk filament is automatically degummed, and is made ready to be sent to the throwster for twisting into single or ply yarns. The degumming takes place as the silk passes from cocoon to bobbin. Raw silk is now being reeled in the United States from cocoons raised on scientifically planted mulberry groves. This operation is now on a small scale and still in the experimental stage.

154

TYPES OF SILK

While silk is often believed to be only the long, continuous filament spun by the cultivated silkworm, there are four types of commercial silk.

Raw silk. Long continuous silk filaments are reeled from the cocoon of the domesticated silkworm *(Bombyx mori)*. Most of this silk is white or yellow.

Wild silk or tussah. Long, continuous silk filaments are reeled from the cocoon of the wild silkworm. The filaments are coarser, more irregular, and have less gum than do the domesticated silk filaments. They include silk from the *Pernyi* and *Mylitta* varieties, brown in color; the *Theophila* and *Rondotia,* white or gray with more gummy substance; and *Antherea Yamamai,* green in color.

Doupion silk. Long filaments are reeled from double, triple, or more cocoons that have been spun over and become entangled with other cocoons. These are rough, irregular yarns.

Waste silk. Short fibers obtained from many sources as damaged unreelable cocoons, and reeling and rereeling waste. When spun into yarns it is called "spun silk," or "schappe silk."

FROM FILAMENTS AND FIBERS TO FABRICS

Raw Silk

When the bales of raw silk are received by the yarn and fabric manufacturers, they are opened and inspected. The skeins are then wound on bobbins. From the bobbins they may be wound on beams, ready for making fabric. However, the yarns usually are twisted first.

Throwing is the term applied to the twisting of raw silk that has been reeled into multiple filaments. (It is also applied to twisting filament rayon.) It includes not only twisting but also doubling, which consists in combining two or more silk filaments or yarns and twisting them together into ply yarns. Any number of raw silk yarns from one to fourteen, and any amount of twist up to one hundred turns per inch may be used, but certain standard twists are used most commonly. The bobbins are placed on the first spinning frames, where the yarns receive their first twist. The doubling and further twisting are on twister machines.

155

Bales of raw silk being weighed. The bales are made up of thirty skeins each, called "books" (Cheney Brothers).

Spun Silk

This silk is commonly called "waste' silk, because until the introduction of silk machinery only the long reeled filaments were used. It consists of silk that cannot be unwound from the cocoon and reeled into skeins. These short fibers are spun into yarn much the same as are cotton and wool fibers. Spun silk is a far cry from being wasted, as it is used for making many woven and knitted fabrics as well as laces. It is particularly adaptable for pile fabrics, such as velvet, and napped fabrics, such as duvetyn.

It is obtained from the following sources: (*a*) *The inside layers of the cocoon.* When the silkworm begins to spin, it emits a lusterless, uneven filament, as it attaches itself to the twigs or straws. This cannot be unreeled and is often filled with twigs, leaves, and straws. (*b*) *The outside layers of the cocoon.* As the spinning is being completed, the cocoon spins a finer and finer filament that is too weak to be spun into yarn. Just before the filament begins to taper off the filament is coarsest. It is too coarse to use in reeling. Therefore these two layers are not reeled. (*c*) *Pierced or damaged cocoons.* Some moths are permitted to emerge. Other cocoons are damaged. (*d*) *Reeling waste.* In joining filaments there may be a waste in finding and joining the ends. (*e*) *Gum waste.* If the skeins are rereeled, there is always a certain amount of waste. (*f*) *Manufacturer's waste.* During throwing, some broken yarns are accumulated. The name "gum waste" is often applied to these fibers also.

Raw silk is shipped to the manufacturers to be reeled into skeins, ready for yarn making. Spun or waste silk is a tangled mass of fibers, broken cocoons, twigs, sticks, worms, and chrysalides. First, the fibers must be separated from all these foreign substances as well as from the sericin. There are two methods by which this is accomplished and each produces a different kind of spun silk.

One is called "discharged spun silk" and the other "schappe silk."

Discharging. The fibrous mass is put in open meshed cotton bags, which are placed in large tubs, covered with soapy water, and boiled for about two hours. They are then put through a hydro-extractor. A second soapy boiling of about one and one-half hours is given. The fibers are again hydroextracted and dried. This method is used in America. It produces fibers with clear brilliance.

Schapping. The fibrous mass is placed in a large vessel. It is covered with hot water, and the tank is sealed. After a few hours, the gum ferments and loosens. The fibers then are soaked for a few hours in vats of hot water, after which they are washed, hydro-extracted, and dried. The fermentation causes a most offensive odor. This method is used in Europe. It produces a fiber with a clear level color, less lustrous than those cleaned by the discharg-ing method. They are suitable for making velvet.

After degumming, the silk waste is taken to a "cocoon beater" if it contains many worms or chrysalides. The silk, on a revolving disk, is beaten by a leather whip, which loosens and knocks out the foreign substances. Still a tangled mass, freed from all foreign substance and gum, the fibers are ready to be spun into yarns.

Spun silk, unlike the raw silk filament which requires twisting only, must pass through several processes before the mass of fibers can be spun into yarn. The following is a brief resumé of these processes, more fully explained in Chapter 10.

Opening the silk. The fibers pass through a machine which opens up the bunches and begins to lay the fibers parallel. They leave the machine in a wide sheet, called a "lap."

Filling. The lap passes through a second machine similar to the opening machine, and the fibers are further opened up and straightened. An additional operation in the machine fastens the

The gum silk being wound from skeins onto spools preparatory to throwing and fabric making (Cheney Brothers).

fibers in a strip between two boards (book boards). This strip of fibers is cut an even length.

Dressing. The strips of silk are combed and the short fibers (noils) are eliminated, leaving only the long fibers (drafts).

Drawing and doubling. The drafts, still in a fibrous state, are fed to a large revolving drum, where they wrap evenly around the drum. When the amount is of the desired thickness, it is drawn into a wide filmy sheet, again called a "lap." A second drawing machine forms the sheet into a soft round strand called a "sliver." Eight or more of these slivers are run through a third machine, are drawn out and combined into one sliver. This is repeated on three more drawing machines. Doubling and drawing lay the fibers parallel and even the sliver.

Roving. Twenty to forty slivers pass through a roving machine and are again doubled and drawn out into a "slubbing roving." Passing to other roving machines, the yarn is further doubled and drawn out into a fine roving.

Spinning. The fine roving is given its final drawing out, is twisted and wound on spindles or cones.

Raw and Spun Silk

Weaving. The warp yarns, filament or spun silk, are wound on large warpers or reels and then rewound on warp beams. The filling yarns are wound on bobbins. Most known weaves have been used for silk fabrics. In fact, many of the most complicated weaves were developed for creating the exquisite, long lasting silk fabrics, softly colorful, with intricate and unusual patterns and designs.

On a following page a few representative silk fabrics are illustrated. Today, most fabrics formerly made of silk are woven with rayon and are illustrated in the chapter on rayon.

Knitting. For warp knitting the yarns are wound on beams as for weaving. For weft knitting they are wound on cones or spools. For Milanese machines the

A raw silk and wool drapery fabric hand-loomed by Dorothy Liebes.

158

yarns are wound on small, sectional warp beams. Silk fabrics can be knitted on both warp and weft machines and in all stitches. See Chapter 12 for an explanation of knitted fabrics.

Twisting. Yarns for twisting are wound onto two types of beams and on bobbins. Beautiful, fragile appearing, yet durable laces can be made with silk. Laces are discussed in Chapter 13.

Preparatory Treatments

Before silk can be colored and the final finishes applied, the silk is degummed and often bleached. As previously stated, silk is an animal fiber consisting of proteins—about 75 per cent fibroin and 25 per cent sericin.

Degumming of waste silk takes place during discharging or schapping by fermentation. Raw silk may be degummed before weaving, in skeins, or after the fabric is woven. The sericin, usually yellow, is "boiled off" in a hot soap solution, leaving the raw silk with a natural soft luster. At times only a part of the gum is removed, producing what is known as "souple."

The removal of the gum also decreases the weight of the silk and the "boiled off" liquor is usually added to the dyebath later.

Bleaching is required for silk that is to remain white and often for silk that is subsequently colored. Raw silk, later dyed a light color, is usually first bleached. If dyed a dark color, bleaching is not necessary. Most wild silk is a dark color and requires bleaching to be white or to be dyed any color. Silk may be bleached with such chemicals as sulfur dioxide or hydrogen peroxide. Bleaching is done practically always before degumming.

Dyeing

Silk has an excellent affinity for certain dyes. Being an animal fiber, the dyes used for silk are the same as those for wool. Each dye is selected with consideration of the ultimate appearance and use of the fabric. The following are dyes used for silk; a more complete explanation of these and their application is given in Chapter 17.

Acid dyes produce brilliant shades on silk. Their degree of fastness to water, light, perspiration, and alkalies depends on the type of acid dyes used and on their application. As a group, they have a good fastness to washing and an excellent fastness to light. The fabric is dyed directly in an acid bath or from a neutral bath.

Basic dyes produce colors of great brilliance but have a poor degree of fastness to washing and light and are used less frequently than other dyes. They are applied from a neutral or slightly alkaline bath.

Mordant and chrome dyes are another class of dyes with satisfactory brightness, although less than acid and basic dyes. They produce a wide range of colors and have a high degree of fastness to sunlight and are often used for curtain and drapery fabrics. Their degree of fastness to washing, perspiration, and alkalies is much better than acid or basic dyes. The goods must be mordanted with metallic salts to fix the dye that is applied later.

Diazotized and developed dyes are less frequently used for silk. These are usually direct dyes to which chemicals are applied to add to their fastness to light and water. Acids are added to the dyestuffs, which change them to a new class of dyes. These are then developed and become soluble in water, dyeing the fabric.

Vat dyes have an all-round excellent degree of fastness to washing, light, perspiration, acids, and alkalies. Their use in silk, however, has been limited. They produce clear colors, brighter than any except those achieved by acid and basic dyes. Vat dyes are insoluble in water and a reducing agent is added to make them soluble. After the goods are dyed, they are oxidized by chemicals or by exposing to air.

Printing

Silk fabrics may be printed by any method—roller, screen, or block. They also may be printed by direct, resist, discharge, or any other type of printing. However, silks are usually dyed and then printed, which means that the resist and discharge methods are used more frequently than is direct printing.

Finishing

Since the silk fiber itself contributes beauty to the silk fabric, with its natural luster or sheen and with its soft drapability or crispness, most silk fabrics require fewer finishes than do fabrics made of cotton or wool. The following is a summary of the finishes more fully discussed in Chapter 19.

Singeing. The fabric traveling at a high rate of speed is passed over red-hot plates or rolls, or through flames, to singe off any surface fibers or lint.

LUXURIOUS SILK

Silk has always produced luxurious fabrics with a feel and appearance exclusive to silk alone. The fabrics shown here are all pure silk. Starting upper left, counterclockwise, they include the following.

The crosswise ribbed poplin has been fabric dyed. The lustrous, soft, black satin has a pattern produced by the resist printing method. The heavy twilled serge was fabric dyed. The next is a flat crepe that was paisley printed which is the name given to a design that imitates paisley shawls that were woven with different colored yarns. The fine ribbed faille was fabric dyed.

Above the faille is a crepe that has a Jacquard woven pattern. The color was printed by the direct method on a white ground. The fine taffeta has a changeable color because of the interweaving of the previously dyed blue and gray yarns. The softly lustrous, closely constructed satin was fabric dyed. The luxurious blue brocade has a Jacquard pattern woven with gold metallic threads.

The four fabrics from right to left at the top are typical of the very crisp or very soft silk fabrics. All were piece dyed. They include a crisp organdy; a sheer, semicrisp leno-weave marquisette; a soft, thin chiffon; and a very fine voile.

Shearing. Pile fabrics such as velvet and plush are sheared to cut the fibers even. After shearing the fabric is brushed to remove any loose clinging fibers.

Steaming and pressing. Such fabrics as velvet may be steamed and pressed, laying the pile fibers down and giving the fabric a sleek, lustrous surface.

Sizing. A stiffening solution is applied to some fabrics, such as to the back of some heavier satins.

Calendering. Many fabrics, particularly the more firmly woven and crisp silks such as taffeta, are pressed and given a luster by passing the fabric around and between rollers in a calendering machine.

Tentering. Fabrics pass through the tentering machine where the selvages are evened and the warp and filling yarns are restored to their proper positions.

SUMMARY

AN ANIMAL FIBER FROM THE SILKWORM

Kinds of commercial silk: Raw silk from cultivated silkworms, wild silk or tussah from wild silkworms.

Two types of yarn: Long, continuous silk filaments, reeled from cocoons; spun silk yarns, spun from short fibers called "waste silk."

TO MAKE THE YARNS AND FABRICS

Silk filaments

Raise the silkworms that spin the cocoons.
Kill the chrysalide: To protect the filament.
Reel: To unwind the filament from the cocoon.
Throw: To twist multiple filaments into yarn.

Spun silk

Discharge: To open up fibrous bunches, to remove foreign substances and sericin, and to clean by boiling.

Schappe: Same as discharge but by soaking and fermentation.

Open and fill: To further open up the bunches and to partially lay the fibers parallel.

Dress: To comb out the short fibers (tow), leaving the long fibers (drafts).

Draw and double: To further mix and lay fibers parallel and to produce strand (sliver), which is further doubled, drawn out, and slightly twisted into a roving.

Spin: To draw out and to twist roving into yarn.

Construct the filament silk or spun silk fabric: By weaving, knitting, twisting, or braiding.

TO FINISH THE FABRICS

General finishes

Singe: To burn off protruding surface fibers.

Degum: To remove the gum (sericin).

Bleach, dye, print: To make the fabric white or colored.

Nap: To raise surface fibers.

Shear and brush: To cut and brush off surface fibers or even nap.

Steam and press: To lay down napped fibers and to produce lustrous surface.

Size: To give body to the fabric.

Weight: To add weight and give body to the fabric.

Calender: To press and give luster to the fabric.

Tenter: To stretch, straighten, and dye the fabric.

Functional finishes: To render the fabric crease-resistant, water-repellent, resistant to spots, stains, and snags, and fire- and germ-resistant.

Opposite page: Cellulose is changed into a viscous liquid solution which pours through large tanks, and ultimately extrudes through spinnerets to harden into rayon filaments (Tennessee Eastman Corporation).

RAYON—SYNTHETIC FIBER

The Most Versatile

IN THE BEGINNING

If fifty years ago the "world of tomorrow" could have been foreseen, the prophets would have predicted an era filled with so many strange discoveries as to seem almost supernatural. They would have described our radio, airplanes, and submarines; foretold the birth of radar and electronics; predicted the atomic bomb; envisioned a new field of synthetic plastics and synthetic fabrics as well as many other wonders of the age in which we live—all so new, different, and startling as to be unbelievable to ordinary man.

Even today, many people do not realize that cellulose which produces beautiful, soft, rayon fabrics also makes explosives and films, is a part of enamel and lacquer, and composes many solid plastics with a multitude of uses. These plastics are hard and smooth. They may be transparent, translucent, or opaque, and can be pigmented with any color desired. The plastics, the same as rayon fabrics, have a color and texture that are distinctive. They are used for packaging foods, medicine, cosmetics, and other merchandise, they are parts of automobiles and airplanes, are made into clear, colorful belts and buttons, cooking utensils, dental plates, hat straws, door knobs, and thousands of other items used daily. Their versatility is as manifold as scientists have had time to make them.

The fact is there were many successful uses for cellulose long before rayon yarns were made for fabrics. Rayon began shortly before the middle of the nineteenth century; but there had been some scientists who foresaw its potentialities 100 to 200 years previously. In 1664 Dr. Robert Hook, an English scientist, discussed the possibilities of spinning fibers similar to those spun by the silkworm. In 1734, René de Réaumur predicted the making of filaments from resins similar to those being used at that time for varnishes. It was not until 1844 in England that a chemist treated cellulose with caustic soda and produced an alkali solution, called "alkali cellulose," which half a century later was to become the basis for the largest production of rayon in the United States, the viscose process.

In 1845, in Germany, cotton cellulose was mixed with sulfuric acid and nitric acid, and guncotton, an explosive used as a substitute for gunpowder, was discovered. This cellulose nitrate was the basis of the solution that was to be used, some forty-five years later, for the production of nitrocellulose—the compound that produced the first rayon yarn.

Experimenting further, it was found that cellulose nitrate was soluble in alcohol and ether and from this, pyroxylin was born with its many well-known uses for such products as films and lacquers.

In 1857, in Germany, it was discovered that by dissolving cellulose in a copper ammonium solution a hydrated solution, called "cuprammonium" was obtained, which in the twentieth century was to be used to make cuprammonium rayon yarn.

In 1865, again in Germany, it was discovered that acetyl (an acetate radical) combined with cellulose and formed cellulose acetate, which thirty years later was used as the basis for making cellulose acetate rayon yarn.

Another important step was the discovery of the effect of camphor solvents on pyroxylin under heat and pressure, resulting in a solid product called "celluloid." This was manufactured into toilet articles of all kinds, umbrella handles, picture frames, piano keys, and a multitude of other articles of everyday use.

The discovery that amyl acetate produced a solvent action on pyroxylin was very important. This action changed pyroxylin so that it could be used in enamels and lacquers, could be mixed with oils and pigments for dyeing, and was suitable for waterproofing and coating fabrics and other materials.

Patents were granted for making rayon filament yarn by four different processes. The first process, nitrocellulose, patented in 1884, is not being used in America today. Patents for the other

processes were granted as follows: for the viscose process in 1892; for the cuprammonium process in 1890 and 1897, and for the acetate process in 1894 and 1902.

The Nitrocellulose Process

During the latter part of the nineteenth century, a Swiss chemist developed the idea that, since the silkworm fed on mulberry leaves, the cellulose for filaments could be made from them. He produced filaments by dipping a needle into the solution and drawing them out. The experiment, however, was not successful commercially. One invention often leads to another. About this time Thomas Edison invented the electric light and with it came a demand for satisfactory filaments. Two Englishmen, working on carbon filaments for electric lamps, produced the first filament by the method of forcing the liquid through tiny holes (orifices) in a spinneret and coagulating (hardening) the filaments in alcohol.

It was in 1884 that a French chemist, Count Hilaire de Chardonnet, began the first commercial production of rayon yarn. His patent, based on the nitrocellulose process, was described as "the squirting of an ether-alcohol solution of nitrocellulose through apertures into the air." Chardonnet is known as "the father of the rayon industry," because he worked out the mechanical details of transforming the liquid into a solid filament and because he operated these methods on a commercial scale, opening his plant in 1889.

A Belgian company began making rayon by this method in the United States in 1920, but it was discontinued in 1934.

The Viscose Process

As stated previously, this process was based on a discovery made in 1844. However, it was not until 1892 that three chemists in England patented the process of treating cellulose with carbon disulfide in the presence of caustic soda. This, when dissolved

166

in water becomes a golden viscous or gelatinous substance, from which the viscose process derives its name. From 1892 until 1900 the viscose solution was given a wide variety of uses. It was used for finishing fabrics, in rubber goods, for solids as door handles, for cellulose films, and many other uses. As with the products of other processes, carbon filaments for electric lamps were the first adaptations of the extruded viscose rayon filament.

In 1902 in England the centrifugal spinning box was invented. This great contribution made it possible to spin the viscose solution into filament yarns for fabrics. (This is described under methods of spinning viscose yarns.)

By 1905, the manufacturing of viscose yarns was begun in England on a commercial scale. The English company established subsidiary branches in Europe, Canada, and, in 1903, in the United States. Today, a number of plants are producing viscose rayons in the United States. The viscose process is responsible for about 88 per cent of the world's production of rayon.

The Cellulose Acetate Process

This process was based on the discovery, in 1865, that cellulose combined with acetyl resulted in cellulose acetate. But it was not until 1894 that the same English chemists who developed the viscose process, patented the process of making carbon filaments for electric lamps by the cellulose acetate method. As with the products of other processes, the cellulose acetate compound was used for fabric finishes, paints and lacquers, films, and many solid plastics long before it was made into a filament.

The adaptation of this process to the making of rayon yarns for fabrics was developed in the United States. The English chemists passed on their experiments to Boston chemists who, in 1902, patented the process of spinning yarn from cellulose acetate.

A company was formed at the beginning of World War I, but little was accomplished in the

United States until 1924. In the meantime, two chemists with laboratories in Switzerland started a plant in England, and in 1924 they bought the American company, established in 1914, and made it into a subsidiary of their English plant. Since then, other large rayon manufacturers have been making cellulose acetate rayon. This process is responsible for about 9 per cent of the world's production of rayon, but approximately 23 per cent of the United States production.

The Cuprammonium Process

This process was based on the discovery, in 1857, that cellulose dissolved in a copper ammonium solution produced a hydrated cellulose, cuprammonium.

In 1890, a French chemist patented the process for making rayon filament by this method, but the results were not successful. It was again the use of the rayon substance for making carbon filaments for electric lamps that brought about this rayon fabric yarn. In 1897, in Germany, patents were granted and successful production of rayon was begun by the cuprammonium process. Factories were built in Germany, France, and England. In 1927, a plant was established in the United States. Today, only one plant in the United States is manufacturing cuprammonium rayon yarn. This process accounts for about 3 per cent of the world's production of rayon.

Rayon Staple Fiber

Rayon staple fibers are short lengths (usually no more than 6 inches) cut from long, continuous rayon filaments. Today, viscose and cellulose acetate filaments are used to make rayon staple fiber.

The production of short fibers was first attempted in France and a patent was granted in 1910. Germany then took up the idea and made great strides during World War I by using the short fibers as a substitute for wool and cotton. Then, Italy, England, Japan, and the United States began producing rayon staple fiber. The greatest advance has been made since 1930.

In that year 6 million pounds were produced. In 1936 this amount had risen to 300 million pounds, and in 1940 to 1,237 million pounds.

Rayon Is a Synthetic Fiber

Rayons are called "synthetic fibers" because chemists constructed them by synthesis (discussed in Chapter 8). The original compounds may be changed in physical form only and not in physical structure. If so, they are called "regenerated fibers" which means to produce anew. In making other synthetics there is a chemical as well as a physical change in the original substances. Viscose and cuprammonium rayons are regenerated fibers as there is a physical change only in the cellulose and it remains chemically cellulose. When making cellulose acetate rayon the cellulose is changed both chemically and physically and becomes a cellulose ester with a chemical formula different from cellulose.

CELLULOSE

Cellulose used for making rayon is a component part of plant tissue. As found in plants it is not pure but is mixed with many other components of the tissue, such as some tiny waxy or fatty bodies that repel the water and moisture and protect the plant tissue, and other bodies that are sticky, mucilagelike substances. Some plants, particularly trees, have the cellulose mixed with an encrusted substance. All of this must be removed as only pure cellulose can be used for making rayon. The noncellulose parts must be destroyed chemically without harming the cellulose. Therefore, the best plants for manufacturing purposes are those with a type of cellulose that resists these chemicals to the greatest extent.

While experiments are going on constantly to find other plants that will yield suitable cellulose, wood from western hemlock and southern pine, and cotton linters are the only raw materials that have been used successfully to date for rayon production.

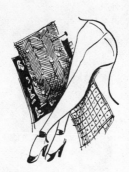

WOOD PULP

Bleached sulfite wood pulp, often referred to as "dissolving pulp," is used to make rayon. In the early days of the industry, spruce was considered the most suitable pulpwood for the manufacture of dissolving pulps. Research later developed two additional raw material sources, western hemlock and southern pine. Today, these two are used exclusively by the major manufacturers of dissolving pulps in the United States.

Logging trees from 3 to 6 feet in diameter in frequently inaccessible mountain regions calls for engineering skill and a large outlay of equipment. Logging camps, railroads, and truck roads must be built. Often substantial steel bridges are required to cross rivers and chasms. As logging progresses, spur or branch tracks and roads are built to the advanced timber stands. Steam, electric or Diesel engines, and caterpillar tractors convey the huge logs to the loading spur.

Two "fallers" fell the huge trees. They must judge how and where a tree is to fall to prevent it from being damaged and from injuring other trees. Then the tree is "limbed," and "buckers" saw it into standard log lengths for convenient transportation to the sawmill.

Breaking Down the Wood into Chips

The first step in the conversion of pulpwood to dissolving pulp is breaking down the logs. A tree is cut into 6 foot lengths. The real breakdown of the logs take place when they reach the sawmill or "breakdown plant." The large western hemlock logs are conveyed directly to a hydraulic barker. The smaller logs are placed inside revolving drums where dirt and bark are removed. Water sprays at high pressure wash the logs after they have been barked. Cut-up saws then reduce the logs to smaller sizes for con-

Cellulose is obtained from the wood pulp of giant trees (American Viscose Corporation).

venient removal of any remaining bark, knots, and defective spots.

When the wood has been thoroughly cleaned and inspected for imperfections, it is conveyed to a "chipper" that reduces it to uniform sized chips about 1 inch long and ⅛ inch thick. Chips are run over an oscillating screen to sift out all oversized chips and slivers. Now the pulpwood, in the form of uniform sized chips, is ready for conversion into wood pulp.

Dissolving the Wood Pulp

The raw materials required for the manufacture of bleached sulfite wood pulp are wood, limerock, sulfur, water, and bleaching chemicals.

Sulfur is burned in a rotating burner to form sulfur dioxide, and the gas is blown into the bottom of an acid tower filled with limestone, while water is sprayed into the top to form calcium bisulfite. This chemical solution is pumped

Wood chips being conveyed from the chipper to storage bins (Rayonier, Incorporated).

from storage tanks into the bottom of large digesters filled with approximately 100 tons of wood chips. Steam pressure is then applied to "cook" the chips. The cooking liquor reacts with the lignins, etc., to make them water soluble and to separate them from the individual wood fibers. Cooking or digesting time varies, depending upon the grade of pulp desired.

When the cooking cycle is completed, the digesters are discharged under pressure into "blowpits." During the release from pressure in the digester to atmospheric pressure outside the digester, the wood chips tend to break apart and as they are blown against the blowpit target, they are completely broken up into separate cellulose fibers.

The excess of used cooking liquor is drained off in the blowpits after which the cellulose fibers are dispersed into the water suspension and flow over vibrating screens to remove any portions not completely disintegrated. From these "knotter" screens the pulp then passes over washers, where any remaining traces of cook-

Riffling or washing and screening the liberated cellulose fibers to prepare them for subsequent bleaching (Rayonier, Incorporated).

ing liquor are removed. Millions of gallons of pure water are used in a single day in these washing operations.

Bleaching the Wood Pulp

Bleaching is the final chemical purification process, which removes the remaining impurities and incrustations and whitens the pulp. This is accomplished in several stages, the degree of bleaching depending upon the grade or quality of pulp desired.

Bleached sulfite wood pulp for conversion into rayon and other chemical uses has usually a higher chemical purity than wood pulp used in the manufacture of paper. This is because in the manufacture of rayon the chemical properties of the wood fibers are more important than their physical properties. The alpha cellulose content is a measure of the degree of chemical purity and in dissolving pulps it ranges between 91.5 per cent and 96.5 per cent.

Cellulose fibers are being thickened into a heavy pulp mat prior to the final stage of bleaching (Rayonier, Incorporated).

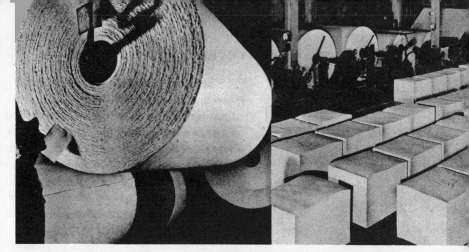

Left: The mat of cellulose fibers is wound on a large reel to form a roll weighing 4 to 5 tons. Right: The rolls in the background are cut into smaller rolls or sheets ready for conversion into rayon at a rayon plant
(Rayonier, Incorporated).

Control of the composition of the cooking liquor, time and temperature of cooking, control of the composition of the bleach liquor, and time and temperature of bleaching, all are factors in achieving the required degree of alpha cellulose content. When extra high alpha cellulose content is desired (around 96 per cent), a special caustic treatment is given to dissolve out a major portion of the beta and gamma celluloses.

Forming the Sheets

After the final bleaching operation, the stock is run over a wet machine, where in one continuous operation the pulp is formed into a wide sheet of matted fibers, which first passes between numerous sets of rolls to remove excess water and then passes over steam-heated driers. This continuous sheet of pulp is dried to the proper moisture content and, when it emerges from the dry end of the machine, is wound on a reel to form a jumbo roll, weighing about 5 tons.

During the various manufacturing steps, constant check samples are taken and sent to the laboratory for analysis. Pulp which does not meet the rigid standards for chemical characteristics, sheet formation, color, and cleanliness is diverted to other uses.

Wood pulp is shipped to rayon manufacturers in specified sheet or roll sizes.

COTTON LINT AND LINTERS

Cellulose is obtained from cotton linters.

The part of cotton that is used for rayon are the short fibers adhering to the cotton seeds after they have been ginned to remove the spinnable fibers. Thus, fibers from cotton seed are of two types: lint fibers, the long fibers removed by ginning and used for yarn making; and fuzz or linters, a short, thick undergrowth, found on the seeds of some types of cotton.

Cotton seeds, a byproduct of cotton lint fibers, were used to build a thriving industry of cottonseed oil. Since the short fibers or fuzz absorbed much oil, they had to be removed; and so there developed another byproduct, cotton linters. These were first used for gunpowder and later for stuffing of mattresses and upholstery. Since there still remained a surplus the chemists began experimenting with it for use in rayon. With the invention of the delinting machine, cotton linters became available for rayon and other uses, supplementing the wood cellulose.

Chemical cotton is used as a basic ingredient in the production of cellulose derivatives used in plastics, photographic films, paints and lacquers, as well as fibers and fabric finishes.

Delinting the Cotton Seed

The delinting is performed in the cottonseed-oil mill. The cottonseeds are first screened to remove foreign substances as dirt, leaves, stalks, and bolls. The seeds are then fed into a linter machine, which is much like a cotton gin. First they contact a magnetic plate that removes any iron left by the screening operation. Then they contact saws which detach the linters from the seeds, permitting them to fall through a grate in the machine. A revolving brush removes the linters from the grate and blows them up through a flue. An open condenser receives the linters and forms them into a soft roll.

The machine may be adjusted to cut off only the longer linter fibers (first cut) and the seeds are put through the reset machine a second time to remove the remainder (second cut). If the seeds are delinted but once, both the first-cut and second-cut fibers are removed at one time (mill runs).

Cotton linters for rayon are graded in much the same general manner as cotton lint for cotton fabrics—for the amount of long fiber and fuzz, smoothness, color, elasticity, uniformity, and amount of foreign substance present. In general, first cuts are grades 1, 2, and 3; mill runs are 3, 4, and 5; and second cuts are 5, 6, and 7.

The rayon industry uses the second cuts and a certain amount of so-called hull fiber. This latter is obtained by crushing the hulls and sifting out the fiber residue. Regardless of the length of cotton fibers, the cellulose content remains the same.

The linters leaving the delinting machine are baled and delivered to the plants that convert them into chemical cellulose, freed from foreign substances and containing 99 per cent alpha cellulose.

Producing the Chemical Cotton

The first step is the selection of the linters. This is usually done in the cottonseed-oil mill but may be in the cotton field. Linter standards for chemical cotton used for rayon are carefully

Cotton pulp, chemically treated, leaving the drying tunnel
(Hercules Powder Company).

set and samples are constantly tested and analyzed for quality.

The bales of cotton linters are opened in a bale-opening machine. This has a cylinder (with spikes) that rotates, opening up the bales and delivering the fluffy linters, buff to brown in color. The linters may be mechanically cleaned, removing foreign material.

Digesting the Cotton Linters

To obtain purified cellulose, it is necessary to remove the fatty and waxy substances and any remaining hulls. The brownish cotton linters are placed in machines called "digesters." A mild solution of caustic soda (an alkaline solution) and a detergent (wetting-out agents) such as soap are used. The wetting-out agent may consist of sodium salts of sulfuric acid esters of a saturated aliphatic alcohol.

After the cotton linters and the liquor are mixed, the air is taken out of the digester and the temperature is raised. The steam pressure is held for from two to six hours. This cooking, without air, saponifies the fats and waxes and destroys the pectates, hulls, and other substances.

After the linters are cooked, a valve in the bottom of the machine is opened, and the steam blows the linters through a separator which removes the steam from the linters. From the separator the now yellowish linters drop into a tank, where they are washed to remove the caustic soda. They are now pure chemical cellulose.

Bleaching the Linters

Bleaching of the linters is very much like the bleaching of cotton. Chlorine is usually used, but hypochlorite, peroxides, or other bleaching chemicals may be applied.

The linters are placed in a large bleaching vat or tub. The chlorine may be in the form of a liquid or of a gas; or a sodium or calcium hypochlorite solution may be used. Sulfuric acid is used as a souring agent and oxalic acid is added to reduce the iron content.

The linters are washed after bleaching. This process of bleaching and washing may be repeated. While wet, the linters from various batches are blended.

Drying the Linters

The linters are now purified chemical cotton. To be delivered in fluffy fibrous form, they are passed through squeeze rollers and are carried on a traveling apron through a long drier maintained at controlled temperature. In this fluffy, fibrous form the linters are baled, ready to be delivered to the acetate rayon yarn plants.

If they are to be used for viscose or cuprammonium rayon, the purified linters are sheeted on a paper machine. The linters are taken from the bleaching machines and, while in a wet state, are blended. They are passed to a paper-making machine to be formed into long sheets of a predetermined thickness, and are rolled into large rolls. Or the sheets may be cut into desired lengths and baled, ready for shipment.

Opposite page: Sheeting chemical cotton is being rolled from the paper machine to the lower roll, while the upper roll is being unrolled and fed to the cutting machine. Inset: Weighing bale of chemical cotton (Hercules Powder Company). Right: Weighing cellulose sheets made from wood pulp (American Enka Corporation).

CHANGING THE SOLID INTO A LIQUID

The Viscose Process

Pure cellulose fibers from either bleached sulfite wood pulp or purified cotton linters, or a mixture of the two, are used for making viscose rayon. It is received in the yarn plant in large packages containing sheets looking much like thick blotting paper. The sheets are cut into about a 12- by 18-inch size and sorted so that sheets from many different packages are in one batch. This assures a more even quality in the finished product.

The cellulose passes through three major steps as it is changed from a solid (the cellulose) to the liquid viscose spinning solution:

Forming the alkali cellulose—white sawdustlike
    crumbs—   *a solid*
    with caustic soda
Forming the cellulose xanthate—orange colored plastic
    crumbs—   *a solid*
    with carbon disulfide
Forming the viscose spinning solution—a golden honey-
    like solution—   *a liquid*
    with caustic soda

Forming the alkali cellulose. Groups of sheets are placed on edge in long vats or open tanks that have metal divisions separating

Forming the alkali cellulose. Left: Placing the cellulose sheets in a vat (American Enka Corporation). Below: Long rows of vats steeping or mercerizing the cellulose sheets (American Viscose Corporation). (Robert Yarnell Richie was photographer for all American Enka Corporation pictures).

the groups. A solution containing about 18 per cent caustic soda is added, and the sheets are steeped. This is called "mercerizing." The mercerization of cotton, which gives cotton luster and strength, also uses caustic soda, but that is the only similarity between the two processes, as here the caustic soda is used to take out impurities and to form the alkali cellulose. The sodium molecule combines with the cellulose and forms sodium cellulose.

After about two hours the steel divisions are pressed together by a hydraulic ram. This squeezes out the caustic soda solution, leaving the moist sheets weighing about three times more than when dry.

Shredding the alkali cellulose. Before the alkali cellulose sheets are ready for the next major step they are shredded into crumbs. The sheets are lifted from the vats and dumped into shredding machines with heavy revolving blades that beat and tear the sheets, breaking them into crumbs resembling white, damp sawdust.

Ageing the crumbs. The crumbs are now sent to the ageing cellars. There they remain under a controlled temperature of 18° for about fifty hours. A large circular hollow pipe, down the center of the tank, aids in controlling the temperature of the crumbs. During this ageing the caustic soda further acts on the cellulose to complete the change to alkali cellulose.

Shredding the alkali cellulose. Left: Placing the steeped cellulose sheets in the crumbing or shredding machine (American Viscose Corporation). Right: Unloading the crumbs or alkali cellulose from the machine (The Du Pont Company).

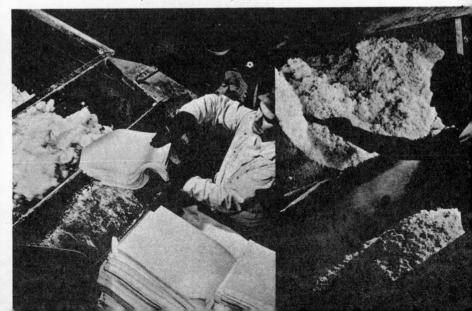

Forming the cellulose xanthate. After it has aged, the alkali cellulose is dumped into large churns and liquid carbon disulfide is added. The churn slowly rotates, for about two hours, thoroughly mixing the crumbs with the chemical solution constantly sprayed over the mixture. The result is a spongy mass, now orange in color (because of the sulfur), called "cellulose xanthate." If the rayon is to be delustered, the chemical needed for this purpose is added to this mixture.

Forming the viscose. From the churns, the cellulose xanthate is delivered to large mixers containing a dilute solution of caustic soda and a small amount of sodium sulfite. The mixers have powerful revolving blades or stirrers, which beat and stir the mixture for five to six hours. The xanthate is thoroughly dissolved, forming a liquid the consistency and color of honey. This is the viscose, the cellulose solution used for spinning the rayon filaments. Here, as throughout the entire process, the temperature is carefully controlled.

Filtering and ageing. Before the viscose solution is ready for spinning it must be filtered and aged again. Drawn from the mixers into tanks it is run through the meshes of filter presses to remove any foreign substance or undissolved cellulose xanthate and is delivered through pipes into a second tank. This process, from tank to filter to tank, is repeated three more times before the viscose is delivered to the last tank, awaiting its turn to be spun into fila-

Forming the cellulose xanthate. Below: The crumb and the liquid carbon disulfide are mixed in the slowly revolving churns (American Viscose Company). Left: Removing the plastic, orange colored cellulose xanthate (American Enka Corporation).

Filtering. Two types of filter presses which remove foreign substances and undissolved cellulose xanthate, as the viscous solution is run through meshes of filter material (American Viscose Company, and American Enka Corporation).

ments. It remains in the last tank for about one hundred hours, until it obtains the right consistency for spinning.

A strong vacuum on this last tank extracts all the air bubbles, and the viscose solution receives a final filtering before it passes through pipes to the spinning machine. The air pressure and vacuum are applied alternately. These two mechanical treatments of filtering and deaerating are necessary so that the viscose solution will not break or clog the spinneret openings. The solution is now a dark, buckwheat-honey color.

Ageing. The viscous solution is aged for many hours in huge tanks to obtain the right consistency for spinning (American Viscose Company).

The Cuprammonium Process

Pure cellulose from cotton linters or wood pulp is used for making cuprammonium rayon. It is received in the yarn plant in sheet form, either flat, like large desk blotters, or in rolls, like thick wrapping paper.

The cellulose passes through but one major step in being changed from a solid to the liquid cuprammonium solution:

Forming the cuprammonium spinning solution—dark blue
    and honeylike—   *a liquid*
    with copper sulfate and ammonium hydroxide

The washed and bleached cotton linters or wood pulp leaving the washing machine are fed to large solution mixing tanks. Copper sulfate and ammonium hydroxide (forming an ammoniacal copper oxide) are added. This solution dissolves the cotton linters or wood pulp and forms a dark blue spinning solution with a honeylike consistency.

The spinning solution is filtered several times as it passes from the mixing tanks to the storage tanks in the same manner as is the viscose solution. The impurities are filtered out, leaving a pure solution. It is then deaerated to take out all the air bubbles and thinned to obtain the proper cellulose content and the right degree of viscosity for spinning.

Cellulose acetate rayon. Left: Cellulose is obtained from wood pulp or cotton linters. Right: Cellulose fiber from cotton linters are being loosened to prepare them for steeping (American Viscose Company).

Forming the cellulose acetate liquid. Left: The pulp is steeped with acetic acid. (American Viscose Company). Below, left: The linters are being placed in mixers. Right: The chemicals have been added and the material is now being kneaded into a mass known as cellulose acetate (Tennessee Eastman Corporation).

The Acetate Process

Purified cotton linters for making acetate rayon are received in the rayon manufacturing plant in large bales. The cotton is dumped from the bales into a machine, which opens it up and loosens the fibers preparatory to making the rayon yarn.

The cellulose passes through three major steps in being changed from a solid (the cellulose) to the liquid acetate spinning solution. While in the viscose and cuprammonium processes the change is from a solid gradually to a liquid and back to a solid, the acetate process has two additional steps. It goes from a solid to a liquid, back to a solid, then to a liquid, and in the final spinning, back to a solid.

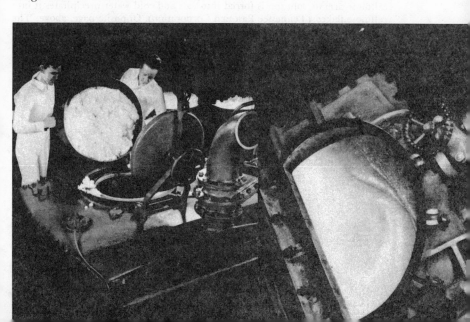

Forming the cellulose acetate liquid—water-white
  and syrupy—   *a liquid*
  with acetic acid, acetic anhydride, glacial acetic
  acid, and sulfuric acid
Forming the cellulose acetate flakes—white and ricelike—
                                                       a solid
  with water
Forming the acetate spinning solution—clear and
  water-white—   *a liquid*
  with acetone

Forming the cellulose acetate liquid. The fluffy cotton fibers are placed in large pans and are steeped in a solution of acetic acid. The mixture is then dumped into a kneading machine and acetic anhydride and glacial acetic acid are added, with a small quantity of concentrated sulfuric acid included as a catalyst. A powerful stirrer kneads the mixture into a glutinous mass, known as "cellulose acetate." After the chemical reactions are completed, the liquid is poured into huge storage tanks and allowed to age until the chemical properties of the cellulose are changed.

Forming the cellulose acetate flakes. The acid that is in the cellulose acetate solution must be removed. From the storage tanks

Ageing. Left: The acid dope is now placed in tanks or jars to age until the chemical properties of the cellulose have changed (American Viscose Company).

Forming the cellulose acetate flakes. Right: From the storage tanks or jars the cellulose acetate solution is forced into vats and cold water precipitates it into cellulose flakes (Tennessee Eastman Corporation). Opposite page, above: The controls that operate the precipitating tanks (American Viscose Company).

or jars the solution is poured into large precipitating tanks of cold water. Rotating paddles stir up the solution, and the cellulose acetate is washed free of the acetic acid and forms solid white particles. The acid is dissolved in the water and drawn off. This washing process is repeated several times until all traces of acid are removed. The material is then passed to a drier that whirls off the water and dries it, in flaky form, in warm filtered air. Flakes from a number of different lots are blended.

Forming the cellulose acetate spinning solution. The cellulose acetate flakes are dissolved in acetone by constant stirring and mixing in large mixing machines. This takes from four to twelve hours. The final acetate spinning solution is almost colorless, thick, and syrupy. It is run from the mixer into a blending machine, which holds a solution from a number of batches. In this way, each batch is thoroughly blended with many other batches.

From the blending machine the solution is forced through a series of filters where impurities are removed; after which it is deaerated by a vacuum to remove air bubbles and is stored in tanks ready for spinning.

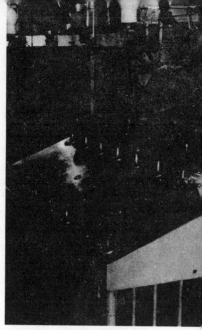

Forming the cellulose acetate spinning solution. Below left: After drying and blending, the material is passed into dissolvers and dissolved (in acetone) into the acetate spinning solution (American Viscose Company). Below, right: The solution is aged in huge tanks or jars ready for spinning (Tennessee Eastman Corporation).

The caps on the spinnerets through which the spinning solution is extruded. The larger is used for Bemberg and the smaller for viscose and cellulose acetate. Approximately actual size.

PRELIMINARY TO SPINNING

To understand what is meant by rayon spinning it is necessary to remember the spinning of silk, because the meaning of this "spinning" is altogether different from that applied to spinning fibers into yarn. The latter means the drawing out and twisting of a roving into a yarn. Spinning to produce the rayon filament derives its meaning from the silkworm spinning its cocoon. The worm feeds on the mulberry leaf. As it digests the food, a glutinous liquid, made up of fibroin and sericin, is formed in its internal glands. The liquid is extruded through two spinnerets below the mouth. As the liquids leave the two tiny spinnerets and come in contact with air, they combine and harden into one filament.

Man, in making rayon, has attempted to reproduce the marvel of spinning the silk cocoon. He uses plant cellulose, the same as does the silkworm, mixes it with a chemical, and delivers the solution from storage tanks through filters and pipes, to extrude it through man-made spinnerets into air, or through a chemical solution, where it is hardened into a filament. The spinneret, therefore, was named after the silkworm's spinning device.

Types of Rayon

There are two general types of rayon—filament and spun rayon. The long, continuous filaments produce only the sleek, smooth fabrics. Therefore continuous filaments may be cut into short lengths and spun into yarns (as are the natural fibers) to produce fabrics that resemble those made of the natural fibers. With these short fibers any desired texture may be obtained.

Types of Spinnerets

For the viscose and cellulose acetate processes the spinneret is a small metal cap, looking very much like a tiny, high silk hat. The illustration shows its actual size. It is made of hard, light weight, noncorrosive precious metals, as platinum and rhodium.

or platinum and gold. The minute circular holes cannot be seen unless the spinneret is held to the light. There may be as few as thirteen or as many as several hundred openings, and each tiny hole is diamond cut and polished.

For the cuprammonium process the spinneret, made of nickel, is flat and larger, with rather large holes as illustrated in natural size. The coarse rayon strands, the size of the openings, are thereafter elongated in a stretch-spinning device, as explained later.

For rayon staple fibers (short lengths cut from groups of long filaments and used for spun-rayon yarns), some spinnerets have a larger surface area and may have as many as 3,000 tiny holes.

For thick, round filaments, used for making such materials as artificial horsehair and bristles, the spinneret has round, comparatively large openings.

For ribbonlike filaments, used for such items as hat braids, there may be an L-shaped or rectangular opening from

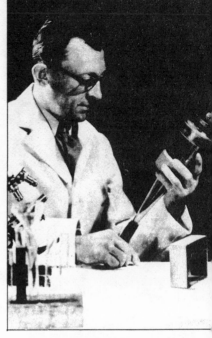

Above: The Bemberg (cuprammonium) spinneret (American Bemberg Corporation). Below left: The viscose spinneret (American Viscose Company). Below right: The acetate spinneret (American Viscose Company).

1/10 inch to an inch in width. Wide spinnerets with long, narrow slits are used for making wide plastics for such uses as packaging.

Types of Filaments Produced

Spinning in all three processes—viscose, cellulose acetate, and cuprammonium—and for rayon staple fiber produces a multifilament yarn. That is, all the filaments leaving the openings in one spinneret are combined to make one yarn. This is similar to silk spinning, where two filaments combine into one.

Spinning for round, heavy, artificial horsehair or for the flat bands produces a monofilament. That is, the filament leaving each opening becomes one filament and is not combined with others.

Summary of the Chemicals

It has been explained that chemicals are required to change wood or cotton cellulose into liquids. In like manner, chemicals are necessary to change the liquids back into solid filaments. The following outline shows not only the chemicals used in spinning but repeats, for clarity, the chemicals employed in changing the solids into liquids. In general, it is the use of different chemicals and the variations in number and kinds of holes in the spinnerets that produce different kinds of rayon. There are some other, individual differences that are explained in the discussion of each process.

The viscous, syrupy rayon solutions which are spun into filaments are chemical compounds (American Viscose Company).

Chemicals used to change the solid to a liquid	Viscose	Processes Cuprammonium	Acetate
Sodium hydroxide (caustic soda)	x	x	
Carbon disulfide	x		
Copper sulfate		x	
Ammonium hydroxide (ammonia)		x	
Acetic anhydride			x
Acetic acid			x
Sulfuric acid			x

Chemicals used to change the liquid back to a solid			
Sulfuric acid	x	x	
Salts	x	x	
Carbohydrates	x	x	
Acetone			x

Remembering that man endeavors to duplicate nature's silk in making rayon, and understanding the all important spinneret, it should not be difficult to grasp how cellulose is changed into a liquid and from a liquid back to a solid. The production is similar in all three processes; the viscose, the cuprammonium, and the cellulose acetate.

All three:

I. Start with a solid:

The wood or cotton cellulose

II. Change the solid into a liquid to produce:

a. The viscose spinning solution

b. The cuprammonium spinning solution

c. A cellulose acetate *liquid*

Cellulose acetate has two additional steps:

(i) Change the liquid into a solid to produce:

The cellulose acetate flakes

(ii) Change the solid into a liquid to produce:

The cellulose acetate spinning solution

III. Change the liquid into a solid to produce:

a. Viscose filament yarn

b. Cuprammonium filament yarn

c. Cellulose acetate filament yarn

SPINNING OR CHANGING THE LIQUID BACK TO A SOLID

The previous section has covered the manufacture of rayon from the solid, a plant cellulose, to a liquid, the spinning solution. As stated this was accomplished with chemicals.

The next step is to change the liquid after it is extruded through the spinneret back to a solid. Just as the method for making the acetate spinning solution differed from the other two processes so the spinning is different. Chemicals are used for viscose and cuprammonium rayon, while warm air is used for the acetate process.

For all three processes, the spinning solution is fed from storage tanks through main feed pipes to branch pipes, and from there through a filter to tubes, on one end of which are the spinnerets. As explained previously, each spinneret has a number of tiny holes and all the filaments extruding from one spinneret are hardened together to form one yarn. The method of hardening, however, is different for each process.

Above: Viscose bobbin spinning (American Enka Corporation). Below: Viscose continuous spinning (Industrial Rayon Corporation).

Viscose Spinning

The tube that holds the spinneret is bent so that the spinneret is immersed in an acid bath. As the solution leaves the spinneret, it enters a circulating, coagulating bath of warm water, containing sulfuric acid, sodium sulfate, zinc sulfate, and glucose. The tiny individual strands coming from one spinneret harden as they strike the acid bath and form a single filament. The many filaments from one spinneret form multifilament rayon yarn.

This continuous filament or yarn is carried through the bath, up and out, to be collected in one of three ways: the bobbin method, the centrifugal pot spinning or cake method, and the continuous method.

In the bobbin method, the filaments are wound parallel on perforated bobbins, with no twisting, and are given the finishing treatments in this form.

In the centrifugal pot spinning or cake method, the filaments are drawn out of the acid bath almost vertically, wound twice around a wheel, passed up to be wound three times over a second wheel, from where they are dropped through a funnel into a spinning pot that rotates 3,000 to 10,000 revolutions per minute. This force causes the filaments to wind from the inside of the pot to the center, forming a "cake." A representative cake contains some 40,000 yards of rayon yarn. The filaments are parallel until they leave the bottom of the funnel, but since the rim speed of the spinning pot is greater than the speed at which the filaments enter, they are twisted together, forming yarn, before they are laid on the side wall of the spinning pot.

In the continuous method, the filaments pass downward from the coagulating bath over a series of revolving spools, rotating several times around

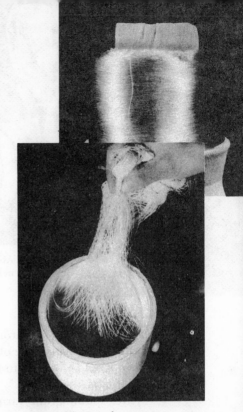

Below: Viscose cake or centrifugal pot spinning (American Enka Corporation). Above: Close-ups of cake, after drying (top) and before drying (American Viscose Company).

191

Above: Cuprammonium spinning, the glass funnel method (American Bemberg Corporation). Left: Cellulose acetate spinning (American Viscose Company).

each spool. As they pass around the spools they receive the various finishing treatments. One of the last spools is heated and dries the filaments, which pass on to be wound on a twisting spindle, ready for fabric making.

After spinning, the yarns are placed in closed cabinets where they remain about six hours. During this time, the viscose and sulfuric acid further combine, resulting in complete coagulation of the filament.

Cuprammonium Spinning

Because only one important manufacturer in the United States makes cuprammonium rayon, the following describes the method used exclusively by this company. For this process the spinneret is attached to a glass funnel immersed in an acid bath. The cuprammonium solution is pumped through the spinneret into the funnel through which flows soft deaerated water. On its way through the funnel, the now plastic filament is slightly stretched and it slowly coagulates as the water absorbs most of the ammonia and some of the copper from the spinning solution. Leaving the funnel, it passes through a mild sulfuric acid bath which further hardens the filaments and absorbs most of the remaining ammonia and copper.

Leaving the acid bath, the filament passes over a roller which

LUSTROUS TO DULL,
DYED AND PRINTED RAYONS

The luster of rayon can be controlled and the yarns, from those with a high sheen to a very dull, produce a wide range of color values when the fabric is dyed or printed. These are all filament rayon fabrics. Starting upper left, counterclockwise, they include the following.

The firmly woven taffeta has beautiful, soft colors printed by the direct method on a white ground. The next fabric was woven with thick-and-thin yarns. It was fabric dyed and the white was produced by discharging the ground color. The dark shade was printed over the ground color. The blue flat crepe was fabric dyed, then the color discharged, and the pattern color was printed on. The sheer crepe received its color by direct printing on a white ground. The lustrous blue satin was fabric dyed.

The next two fabrics have bold, colorful printed designs. The first one was dyed yellow, then the color discharged, and the pattern color printed. Portions were permitted to remain white. The top fabric had its color printed directly on a white ground. Some of the green was discharged and the red color printed on these portions.

The next is a very sheer fabric, soft and draping. It was fabric dyed in a pale pastel color. The last is a typical rayon sharkskin, dull in luster, firmly woven and suitable to tailoring. It was fabric dyed.

pulls and stretches it to a very fine yarn. It is this process and the stretching in the glass funnel, called "stretch spinning," that produce the fineness of cuprammonium yarns.

In this method the acid neutralizes the alkali changing the liquid back to a solid. The filaments are drawn off, and they are reeled into skeins, ready for the finishing processes.

Acetate Spinning

In this process, the spinneret is at the top of a tall shaft or cabinet. As the spinning solution leaves the spinneret, it travels down inside a cylinder, through which a constant stream of warm air rotates. The acetone in the acetate evaporates in the air, which coagulates and hardens the filament. At the bottom of the shaft the filaments pass over a guide and are wound on a bobbin. They are then ready to be twisted nd wound onto cones, spools. warp beams, or other forms.

PREPARING THE FILAMENT YARNS

Viscose. The yarn on the cakes may be wound into skeins. In this form or in cake form it is washed, dried, worked in a sulfur solvent to take out the sulfur, washed, bleached, washed again, treated with oil or soap to soften it, and then dried. If bobbins are employed, the yarn is thoroughly rinsed and dried while still on the bobbins. It is wound from the spinning bobbins to small spools, and then rewound to other spools on rapidly revolving spindles that put a twist in the yarn. After being twisted, the yarn is reeled into skeins and given the finishes described above. The yarns from the continuous process need no further treatment. They may be wound on cones, spools, tubes, quills, or warp beams.

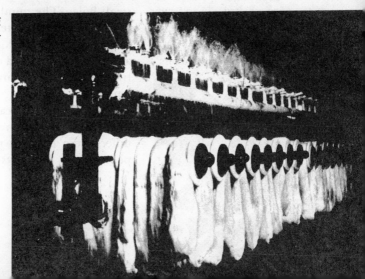

Washing skeins of cuprammonium rayon (American Bemberg Corporation).

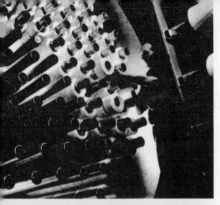

Top: Rayon packages being placed in vacuum dryer. Above: Closing the dryer. Below: Drying rayon in a hydro extractor (American Enka Corporation).

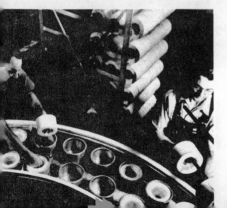

Cuprammonium. The skeins are sprinkled with water to remove any remaining acid and copper sulfate. They are then washed in a soap bath, dried, given a second soap bath and final drying and humidifying. The yarn may be twisted or untwisted and is wound on cones, bobbins, or any other desired form.

Cellulose acetate. Cellulose acetate rayon does not have to be made into skeins, as the yarn requires no finishing after spinning. It may be twisted during rewinding to suitable packages.

SPUN RAYON
FROM RAYON STAPLE FIBER

Rayon staple fiber may be made by any of the three rayon processes—viscose, acetate, or cuprammonium—but at present only the first two are utilized.

Up to spinning, the manufacture of rayon staple fiber is exactly the same as that of filament rayon. The same raw materials are used, the same spinning solution is manufactured in the same way, and the filaments are formed by the same methods.

However, for rayon staple fiber the spinnerets are larger than for filament rayon. They may have from 500 to 3,500 small holes. The number of holes determines the final diameter of the yarn. This can be controlled, which is not as easily achieved for wool and cotton yarns made from fibers of many diameters and lengths.

It is in the physical handling of the filaments that spun rayon manufacture differs from filament rayon. As was explained, all filaments from one spinneret combine to form the filament yarn, which is wound on a package or is reeled into a skein. For rayon staple fiber, all the yarns from *many* spinnerets are collected into one, heavy ropelike strand of tow.

For acetate rayon staple fiber, the tow is then cut into the desired short lengths. The acetate yarns require no further finishing and the short fibers are now ready to be spun into yarns.

For viscose rayon staple fiber further finishing is required and the tow may be cut at various stages of the finishing process. It may be cut right after it leaves the spinning machine while the tow is still wet. It is then washed, disulfured, washed, bleached, washed again, dried, and lubricated. Or the tow may pass over rollers before cutting and receive the finishes mentioned above. The tow is then cut after it is dried. Or the tow may be given all the finishing treatments except the final drying, and then be cut while still wet, and then dried.

Whichever method is used, the short fibers are later placed in a machine that blows them apart, after which they are pressed into bales. In this form the rayon staple fiber is ready for spinning into spun rayon yarns.

They are called "spun rayon yarns" because the yarns are made by spinning rayon staple fiber into yarn. This is done on any of the common spinning systems, such as cotton, wool, or spun silk.

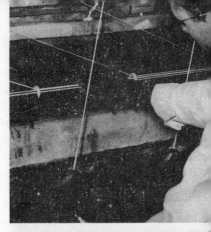

Spun Rayon. Top to bottom: Spinning the viscose rayon staple fiber; tow of filaments from spinning machines being led to the cutter; a mass of fluffy rayon staple fibers after drying (American Viscose Company).

195

Length of fibers. The tow is cut into the lengths required for each type of yarn to be spun. They vary from ½ inch to 8 inches depending on the system by which they are to be processed. Lengths up to 1½ inches and some 2½ to 3 inches are spun on cotton machinery. Wool machinery uses from 4 to 6 inches, and some use 3 inches. The spun silk manufacturers use lengths from 3 to 5 inches. Rayon staple fibers, cut into the lengths of the natural fibers, are spun into yarns and made into fabrics that resemble in appearance those made of the natural fibers.

The diameter is also important as it determines the weight or thickness of the yarn. In the silk system 1½ to 5 deniers rayon staple fiber is used, in the cotton system 3 to 5½ deniers, and in the wool system 5 and 5½ deniers. Acetate yarns are often made as coarse as 16 and 20 deniers when they are to be used as trimmings.

DELUSTERING

This process is applied to rayons only. Filament rayon has a brilliant luster. To change its appearance and increase its uses it was thought necessary to subdue its sheen. This may now be accomplished by two methods, called "internal" and "external delustering." In the first method, the delustrant is added to the spinning solution before it is extruded through the spinnerets, and delustering takes place before the spinning solution is coagulated into a filament. In the second method, the delustrant is applied to the surface of the coagulated filament or yarn.

There are several chemical compounds with which rayon may be delustered, but the most frequently used is titanium dioxide (an insoluble pigment), which is a brilliant white compound obtained from titanium ore. Yarns delustered with a pigment in the spinning solution are called "pigmented yarns."

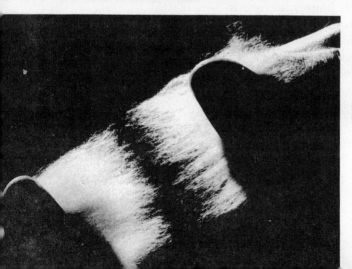

Spun rayon is made of fibers uniform in length and diameter. These illustrated cellulose acetate rayon staple fibers have a crimp and when stretched as shown relax to their average length (Tennessee Eastman Corporation).

Viscose and cuprammonium rayons may be bright, semidull, or extradull (matt), while acetate rayons are classified as bright and dull. However, bright acetate has less luster than does bright viscose. Rayons with a bright luster account for about 50 per cent of the production and the other half is about evenly divided between semidull and extradull.

Attractive patterned fabrics are produced by weaving bright and dull rayons, as for example using the dull for the ground and the bright for the pattern. The delustering compound is also used in discharge printing. The white pattern is printed with the pigment or delustering compound on a colored ground.

FROM FILAMENTS OR FIBERS TO FABRICS

The long, continuous filaments leaving the spinning machines are ready for fabric making. The rayon staple fibers, according to the lengths in which they are cut, may pass through all or most of the yarn-making processes used for woolen, worsted, cotton, and spun silk yarns. This may include carding, combing, drawing, roving, and spinning.

Weaving. The yarns, filament or spun rayon, are wound on large beams to be used as the warp or lengthwise yarns in the fabric. Those to be used for filling or crosswise yarns are wound on bobbins, which are placed in the shuttles in the looms. Rayons are woven on all looms and in all the weaves discussed in Chapter 11. On the following pages are illustrated and discussed the most important rayons. Others, for home use, are described in Chapter 20.

Knitting. The yarns for weft knitted fabrics are wound on cones that are placed on holders in the knitting machine. For warp knitting the yarns are wound on warp beams similar to those used for weaving. Yarns for Milanese machines are wound on smaller sectional warp beams. Knitting yarns must be flexible and pliable because most types of knitted fabrics are even and smooth. Both filament and spun rayon yarns meet these requirements and are used for knitting.

Fabrics are knitted on both weft and warp machines and in all stitches as explained in Chapter 12. Representative rayon knitted fabrics are illustrated and discussed on pages 212 and 213.

Twisting. The yarns are wound on large beams, on smaller beams, and on bobbins. Filament rayon makes fine, lustrous lace. A more complete description of laces is in Chapter 13.

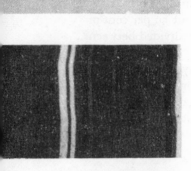

CREPED RAYON FABRICS

These fabrics are woven with alternate right- and left-hand twisted yarns. Employing these tightly twisted yarns in both warp and filling produces a balanced fabric. A true crepe is made in only this way. Some fabrics are woven to simulate these fabrics.

Rayon crepe fabrics cover a wide range of weights and textures. They may be sheer fabrics woven with fine yarns; or firm and closely woven; or more loosely woven with thicker yarns into fabrics with a clothy feel. Some have a somewhat smooth surface texture while others have a pebbly or mossy texture. Crepe fabrics may be woven with filament or spun rayon crepe yarns.

They are most frequently woven in a plain weave but other weaves may be utilized. Satin back crepe is a fabric woven in a satin weave with crepe yarns.

Crepe fabrics may be woven with dyed yarns, or the fabrics may be dyed or printed. Only a few crepe fabrics are illustrated here.

Top to bottom: Crepes. Fine and close texture, ribbed and printed, light weight and printed, loose and clothy texture, wavy ribbed and printed, satin back. Below, top to bottom: Sheer and yarn dyed, sheer and printed.

RIBBED RAYON FABRICS

These fabrics have crosswise ribs except piqué and Bedford cord, which have lengthwise ribs. The ribs of piqué and Bedford cord are formed by yarns floating on the back of the fabric. All of the others are formed by heavier filling or by groups of warp yarns that weave as one. Ribbed rayon fabrics may be woven with filament or spun rayon yarns.

Poplin has the finest rib next to broadcloth. The ribs are closely set. *Faille* has fine ribs, flatter than poplin. *Bengaline* is similar to poplin but heavier, with more rounded ribs. *Grosgrain* is still heavier, with clearly defined round ribs. *Ottoman* has wider ribs, often at intervals, producing a striped effect. *Gros de Londres* has one heavy, then one or two fine ribs. *Marocain* has closely set creped ribs, being woven of creped yarns. *Bedford cord* has wide, flat ribs, with floating warp yarns in the back of the fabric. *Piqué* in the market today is the same as Bedford cord only finer, with narrower ribs.

Top to bottom: Poplin, faille, bengaline, gros de Londres, grosgrain, ottoman. Below, left: Both are Bedford cords.

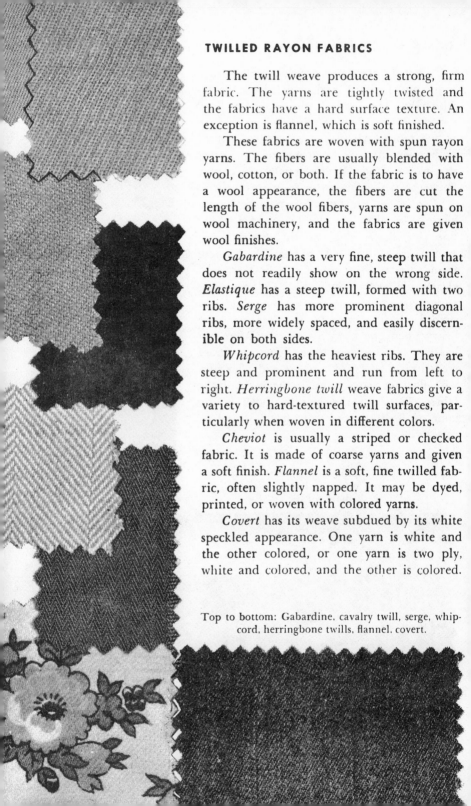

TWILLED RAYON FABRICS

The twill weave produces a strong, firm fabric. The yarns are tightly twisted and the fabrics have a hard surface texture. An exception is flannel, which is soft finished.

These fabrics are woven with spun rayon yarns. The fibers are usually blended with wool, cotton, or both. If the fabric is to have a wool appearance, the fibers are cut the length of the wool fibers, yarns are spun on wool machinery, and the fabrics are given wool finishes.

Gabardine has a very fine, steep twill that does not readily show on the wrong side. *Elastique* has a steep twill, formed with two ribs. *Serge* has more prominent diagonal ribs, more widely spaced, and easily discernible on both sides.

Whipcord has the heaviest ribs. They are steep and prominent and run from left to right. *Herringbone twill* weave fabrics give a variety to hard-textured twill surfaces, particularly when woven in different colors.

Cheviot is usually a striped or checked fabric. It is made of coarse yarns and given a soft finish. *Flannel* is a soft, fine twilled fabric, often slightly napped. It may be dyed, printed, or woven with colored yarns.

Covert has its weave subdued by its white speckled appearance. One yarn is white and the other colored, or one yarn is two ply, white and colored, and the other is colored.

Top to bottom: Gabardine, cavalry twill, serge, whipcord, herringbone twills, flannel, covert.

UTILITY RAYON FABRICS

While these fabrics are woven and finished the same as identical fabrics made of cotton, they have a soft luster and feel peculiar to rayon. They are spun rayon fabrics, made from rayon staple fibers that have been cut the length of cotton fibers and spun into yarns on cotton machinery.

There are many other fabrics belonging to this group that have been made of such spun rayon but are manufactured infrequently.

Gingham is a yarn dyed fabric, woven in stripes, checks, and plaids, in a close, plain weave. *Madras* is a fine, plain weave fabric, identified by woven corded stripes or checks, or by small dobby patterns. It may be white or yarn-dyed, or white with colored stripes or checks. *Broadcloth* has a fine cross-ribbed tight weave. It may be white, or piece or yarn dyed.

Batiste is the lightest weight among these fabrics. Woven in a plain weave with a high count, of very fine yarns, it gives soft draping qualities. It may be white, dyed, or printed. *Challis* is a very soft and pliable fabric in a plain weave. A typical challis has small printed floral designs but may be dyed a plain color, printed with all-over designs or woven with colored yarns. There are many *leno weave sheers* used for shirtings and other garments. They make thin, cool, and strong fabrics. *Crash* is a heavier utility fabric used for apparel that must be durable. It has a strong, usually coarse, plain weave, woven with colored yarns that are frequently uneven.

Top to bottom: Gingham, dobby pattern broadcloth, plain broadcloth, batiste, challis, leno weave sheer, striped crash, crash.

SLUBBED AND NUBBED RAYON FABRICS

These are fabrics woven with uneven yarns. The yarns may be plied (twisted together) in such a way that they form nubs, knots, or loops on the fabric. Very short slubs are called "seed slubs." Other yarns are so twisted together that elongated slubs or thicknesses are formed in the yarn. These yarns may be used one way of the fabric or in both warp and filling. Some have thick and some have finer elongated slubs.

When rayon is being spun the viscous solution may be made to extrude intermittently through the spinneret and cause thicknesses at intervals in the filament.

Some fabrics give the appearance of being woven with elongated slubbed yarns but they are thick yarns, in either warp or filling, woven in at intervals. These are called "thick-and-thin yarns."

Ratiné fabric takes its name from the ratiné yarns with which it is woven. This yarn is made by twisting together a thick and thin yarn and then twisting the resulting plied yarn with a thin yarn, forming small nubs. The ratiné fabric has nubs at intervals throughout the surface. *Bouclé* fabric takes its name from the bouclé yarns with which it is woven. This yarn is made by twisting together two thin yarns and then twisting this with a thick yarn, forming small loops at intervals. The bouclé fabric can be distinguished by spaced loops covering its entire surface.

Left to right, top: Slubbed poplin, slubbed yarns in warp and filling. Bottom: Thick-and-thin filling yarns, heavy slubbed poplin.

Shantung is woven with yarns that have elongated slubs. It was originally woven with silk filaments that had natural differences in thickness considered imperfections. The rayon slubbed filaments produced during spinning imitate this natural silk feature.

Fabrics with a slubbed or nubbed surface texture include sheer fabrics woven with tighty twisted yarns, open weave crash, tightly woven ribbed fabrics, and many other fabrics with various yarn weights.

The distinguishing characteristic of each fabric is not the type of rayon of which it is made but of the surface texture produced by the knots, slubs, nubs, or loops in the yarn. Fabrics made of either filament or spun rayon yarns are included in this group; and of course each type does contribute to the individual appearance of the fabric. If spun rayon, the yarns are usually made of rayon staple fibers cut the length of cotton fibers, although they may be cut other lengths.

Top to bottom, right: Bouclé, slub crash, ratiné, sheer seed slub, thick-and-thin yarns in warp and filling, sheer, thick-and-thin. Below, top: Slubbed sheer. Bottom: Long fine slubs on ribbed fabric.

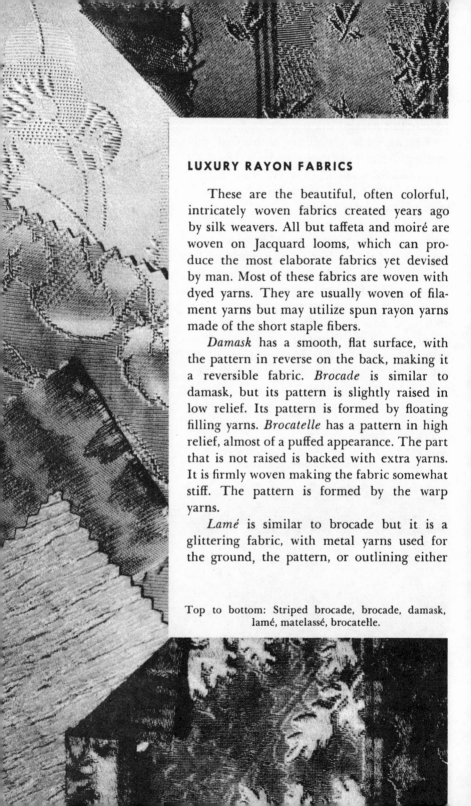

LUXURY RAYON FABRICS

These are the beautiful, often colorful, intricately woven fabrics created years ago by silk weavers. All but taffeta and moiré are woven on Jacquard looms, which can produce the most elaborate fabrics yet devised by man. Most of these fabrics are woven with dyed yarns. They are usually woven of filament yarns but may utilize spun rayon yarns made of the short staple fibers.

Damask has a smooth, flat surface, with the pattern in reverse on the back, making it a reversible fabric. *Brocade* is similar to damask, but its pattern is slightly raised in low relief. Its pattern is formed by floating filling yarns. *Brocatelle* has a pattern in high relief, almost of a puffed appearance. The part that is not raised is backed with extra yarns. It is firmly woven making the fabric somewhat stiff. The pattern is formed by the warp yarns.

Lamé is similar to brocade but it is a glittering fabric, with metal yarns used for the ground, the pattern, or outlining either

Top to bottom: Striped brocade, brocade, damask, lamé, matelassé, brocatelle.

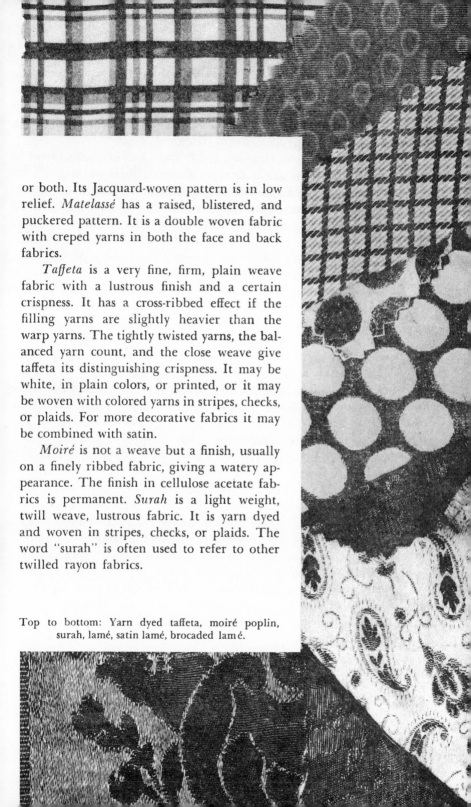

or both. Its Jacquard-woven pattern is in low relief. *Matelassé* has a raised, blistered, and puckered pattern. It is a double woven fabric with creped yarns in both the face and back fabrics.

Taffeta is a very fine, firm, plain weave fabric with a lustrous finish and a certain crispness. It has a cross-ribbed effect if the filling yarns are slightly heavier than the warp yarns. The tightly twisted yarns, the balanced yarn count, and the close weave give taffeta its distinguishing crispness. It may be white, in plain colors, or printed, or it may be woven with colored yarns in stripes, checks, or plaids. For more decorative fabrics it may be combined with satin.

Moiré is not a weave but a finish, usually on a finely ribbed fabric, giving a watery appearance. The finish in cellulose acetate fabrics is permanent. *Surah* is a light weight, twill weave, lustrous fabric. It is yarn dyed and woven in stripes, checks, or plaids. The word "surah" is often used to refer to other twilled rayon fabrics.

Top to bottom: Yarn dyed taffeta, moiré poplin, surah, lamé, satin lamé, brocaded lamé.

These are smooth, bright, or softly lustrous fabrics, ranging from the light, easily draping, to the heavier, rather stiff, fabrics.

Lustrous rayon fabrics are usually woven with smooth filament yarns. Sharkskin may be woven with filament or spun rayon yarns. If spun rayon, the fibers are the length of cotton fibers.

Satin, in all its various weights and kinds is predominant in this group. There are many types of satin. All are woven in a satin weave and have a lustrous surface and dull or semilustrous back. Some have more warp yarns on the surface and are called "warp faced satins." Others have more filling yarns on the surface and are called "filling faced satins." *Crepe satin* is woven with alternating right- and left-hand twisted crepe yarns and is distinguished by its crepy surface and its thick and thin yarns.

Satin Canton is a heavier satin, with a crepe back that is finely ribbed. The ribs give a pebbly appearance to the lustrous face. *Bridal satin* is a medium to heavy weight very closely woven, and rather stiff fabric. *Panné or slipper satin* is stiff, highly lustrous, and one of the heaviest satins. *Lingerie satin* is a light weight, soft satin, often with designs printed in pastel colors.

Dull lustered satin does not have a sleek surface and the weave is more easily noted. *Printed satin* is usually light to medium weight. Frequently, dyed yarns are used to make colorful *striped satins*. Interesting surface textures are acquired in *Jacquard woven satins* by combining bright and dull yarns. Other fabrics, such as some velvets, may have a satin back.

Radium is a smooth, supple fabric, closely woven in a plain weave, with slightly twisted, soft yarns. It may be dyed in plain colors and is frequently printed. *Foulard* is another light weight, soft fabric. It is woven in a twill weave and printed.

Sharkskin is a less lustrous fabric and has a "springy" feel. Today typical rayon sharkskin is woven in a plain weave. It receives its dull tone and texture from delustered yarns which are twisted to produce the characteristic sharkskin appearance. This and another type of sharkskin are discussed in Chapter 21.

Top to bottom, left: Printed foulard, printed lingerie satin, bridal satin, Jacquard satin, dull yarn satin, sharkskin. Right: Striped satin, crepe back satin, printed satin, slipper or panné satin, radium.

SHEER RAYON FABRICS

These are the fabrics of gossamer sheerness, soft and clinging or crisp and fresh. Some are soft, light weight, and well-draping. They may be white or dyed and some are printed.

These fabrics may be woven with long, continuous filaments or with yarns spun from rayon staple fibers. If spun yarns are used the rayon filaments are cut the length of the cotton fibers. The yarns are spun on cotton machinery and the fabrics given finishes applied to the same fabrics woven with cotton yarns. Ninon and chiffon are woven with filament yarns, organdy with spun rayon yarns, and marquisette may be woven with filament or spun rayon yarns. The fabrics made of filament yarns are given many of the finishes applied to silk fabrics.

The *sheer crepe* illustrated is woven in a plain weave with exceedingly fine twisted yarns. The crepe is obtained by weaving alternate yarns of left- and right-hand twisted yarns. *Chiffon* is a sheer, softly draping fabric woven in a plain weave with

Left to right: Romaine crepe, sheer slubbed gauze, parachute cloth, metallic printed marquisette, organdy.

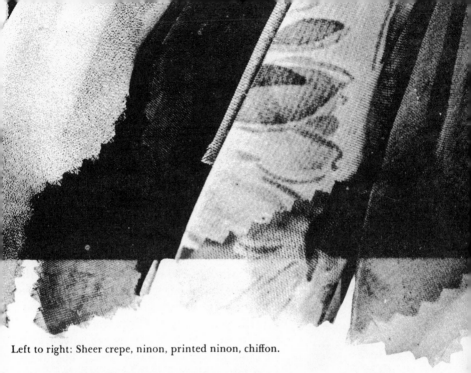

Left to right: Sheer crepe, ninon, printed ninon, chiffon.

fine tightly twisted crepe yarns. It may be piece dyed or printed. *Ninon* is somewhat more crisp than chiffon; it is often called *voile*. It also is woven in a high count with fine yarns. These three very soft fabrics may be white, dyed plain colors, or printed usually in soft, pastel colors.

Organdy is a crisp transparent, almost wiry fabric, woven with fine, tightly twisted yarns. It is frequently given a crisp finish that is retained after laundering. *Marquisette* is a strong yet very sheer fabric, soft or crisp, openly woven in a leno weave. It may have woven dots, as sample shown with metallic dot, or it may be decorated with a dobby or Jacquard pattern.

Romaine crepe is a semisheer fabric with a very light, clothy feel. It is woven with crepe yarns in a plain weave and dyed plain colors. The *novelty gauze* illustrated is an open, plain weave fabric with thick and thin yarns in both warp and filling. It is woven with tightly twisted yarns in an open mesh weave that results in a sheer clothy fabric. *Parachute flare cloth* is a soft, fine, "feather-weight," highly lustrous fabric. It is usually woven, as shown, in a plain weave.

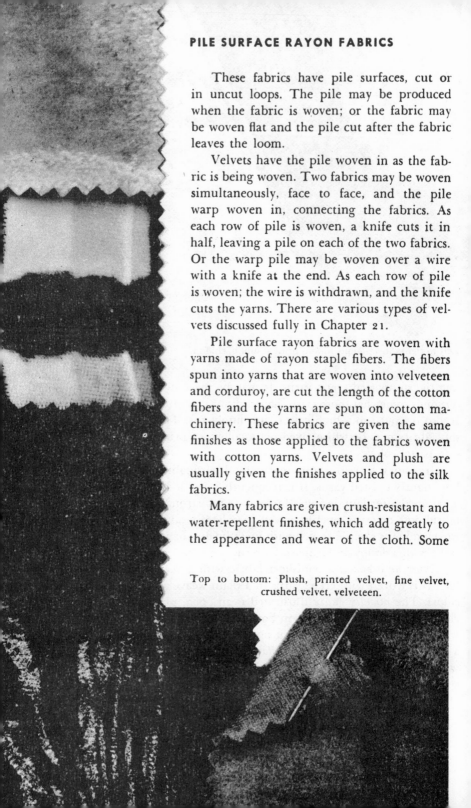

PILE SURFACE RAYON FABRICS

These fabrics have pile surfaces, cut or in uncut loops. The pile may be produced when the fabric is woven; or the fabric may be woven flat and the pile cut after the fabric leaves the loom.

Velvets have the pile woven in as the fabric is being woven. Two fabrics may be woven simultaneously, face to face, and the pile warp woven in, connecting the fabrics. As each row of pile is woven, a knife cuts it in half, leaving a pile on each of the two fabrics. Or the warp pile may be woven over a wire with a knife at the end. As each row of pile is woven; the wire is withdrawn, and the knife cuts the yarns. There are various types of velvets discussed fully in Chapter 21.

Pile surface rayon fabrics are woven with yarns made of rayon staple fibers. The fibers spun into yarns that are woven into velveteen and corduroy, are cut the length of the cotton fibers and the yarns are spun on cotton machinery. These fabrics are given the same finishes as those applied to the fabrics woven with cotton yarns. Velvets and plush are usually given the finishes applied to the silk fabrics.

Many fabrics are given crush-resistant and water-repellent finishes, which add greatly to the appearance and wear of the cloth. Some

Top to bottom: Plush, printed velvet, fine velvet, crushed velvet, velveteen.

velvets are given a finish that makes them washable. A crisp finish, applied to the back of the fabric, imparts a stiffness to some pile fabrics.

The fabrics shown here are the very sheer, *transparent velvet;* a *Lyons velvet,* heavier and crisp; *delustered (matt) velvet;* a *crushed velvet;* and a *printed velvet.*

Plush, which usually has a higher and less dense pile, is woven the same as velvet.

Velveteen is woven flat, with floating filling yarns intermittently but uniformly woven. After the fabric leaves the loom, the floating yarns are cut to leave a pile over the entire surface.

Corduroy is woven like velveteen except that the filling yarns float across the fabric in a straight line. When cut, they leave cords, wide or narrow, depending on the length of the floating yarns.

Pile fabrics imitating fur fabrics are woven in a pile weave and then dyed and finished to imitate fur fabrics. The length of the pile, the fabric coloring, and the finish contribute to the fabric's appearance and texture. The pile may be long and straight, curled, or pressed down, according to the fur it is imitating. These fabrics may be made to imitate many furs such as leopard, pony, Persian lamb, broadtail, Hudson seal, moleskin, kid, and beaver.

Top to bottom: Three imitation fur fabrics and below a wide wale corduroy.

KNITTED RAYON FABRICS

Rayon yarns for knitting may be either viscose, acetate, or cuprammonium rayon and they may be filament or spun rayon. Filament rayons are best suited for fabrics that need smoothness, luster, sheerness, and elasticity. Filament yarns knitted in a plain stitch give sheerness and resiliency to such garments as hosiery and foundation garments.

Filament yarns knitted in the warp stitch produce fabrics that are soft, light weight, and firm, making them suitable for lingerie and dress fabrics.

Spun rayon knits into fabric that can be soft and light weight or heavy and firm. Spun rayon is particularly suited to napping, whether it be a very light nap giving only a soft appearance or a heavy nap such as is used for some types of sweaters or bed jackets.

Top to bottom: Two-bar tricot knitted with fine filament yarn; two-bar tricot of heavy lustrous filament yarns; plain knitted fabric designed with a tuck stitch; yarn dyed, two-bar tricot; printed Milanese; yarn dyed, one-bar tricot; yarn dyed, two-bar tricot; yarn dyed, one-bar tricot.
Opposite page: Printed, two-bar tricot; spun rayon, one-bar tricot, design formed by floating yarns; spun rayon, one-bar tricot; plain knit with extra napped yarns on the surface; printed jersey; plain knit with extra napped surface yarns forming a pattern; printed and napped plain knit.

FINISHING RAYON FABRICS

Many more finishing processes may be applied to rayon than to fabrics made of any other fiber. This is because filament rayon yarns and spun rayon yarns are made into such a wide range of fabrics. However, with the exception of delustering, there is no finishing process exclusively applied to rayon, and it utilizes many of the finishes applied to cotton, woolen, worsted, silk, and linen.

The lustrous or dull (delustered), long, continuous rayon filaments are made into bright or subdued fabrics. They may be silky smooth, crepy, finely ribbed, gossamer sheer, soft, or crisp. They may have a velvety pile, a luxurious Jacquard pattern, or a permanent, sleek, watery surface. These fabrics are given many of the finishes applied to cotton or silk fabrics.

The spun rayon yarns are made from the short rayon staple fibers, cut from groups of rayon filaments. These yarns are not brilliant as are the rayon filaments before delustering. The short fibers, cut the lengths of natural fibers, duplicate to an extent the luster of the natural fiber of the same approximate length. Luster, bright or

213

dull, is caused by the reflection of light on the fibers. One long, continuous filament, as silk or rayon, reflects more light and is therefore shiny, while a multitude of short fibers spun into yarn reflect less light.

If the filaments are cut the lengths of the woolen or worsted fibers and are spun into yarns on woolen or worsted spinning machinery, the fabrics woven from these yarns are given many of the finishes applied to woolen or worsted fabrics. These fabrics may be firmly woven twills or ribs, hard textured and clear surfaced; or light weight, closely woven, with a softly napped film; or sheer or crepy, with a spongy feel. They may be heavier fabrics, with a soft or harsh tweed texture; or with a rough surfaced, nubby appearance; or napped to a soft, deep pile, which on some fabrics is pressed down to form a smooth, softly lustrous, velvety texture.

If the filaments are cut the lengths of cotton or linen fibers and are made into yarns on cotton spinning machinery, the fabrics woven from these yarns are given many of the finishes applied to the cotton and linen fabrics. Most of these are utility fabrics. The majority are woven with a close, plain or twill weave; many, with colored yarns. They may be of fine or coarse yarns, producing a fine, smooth appearance, or an uneven texture, or a rough surface.

Since the finishes applied to rayon fabrics are many of the same as for fabrics made of natural fibers, they are but briefly summarized here. Following chapters give a more complete explanation of each finishing process, and the chapters dealing with the four natural fibers give a brief description of the finishes applied to each.

General Finishes

The finishing processes used for fabrics woven from filament rayon yarns, and for fabrics woven from spun rayon yarns (processed on cotton machinery) are the same, with the one exception of napping, as it is not possible to raise short fibers from the long continuous filaments.

The processes include most of those applied to cotton, linen, and silk as the following chart shows. Filament silk and spun silk have not been separated, as all finishes apply to both with the exceptions of gigging and napping which apply to spun silk only.

The Finishing Process	Applied to				
	Cotton	Linen	Silk	Filament Rayon	Spun Rayon*
Singeing	x	x			x
Calendering	x	x	x	x	x
Moiréing	x		x	x	x
Embossing	x		x	x	x
Gigging	x		x		x
Napping	x		x		x
Shearing	x	x	x	x	x
Brushing	x	x	x	x	x
Weighting			x		
Tentering	x	x	x	x	x

The finishing processes used for fabrics woven from spun rayon yarns, processed on woolen or worsted machinery, are the same as some finishes used for woolens and worsteds, as the following chart shows.

The Finishing Process	Applied to		†Spun rayon the length of:	
	Woolens	Worsteds	Woolen fibers	Worsted fibers
Singeing	x	x	x	x
Gigging	x		x	
Napping	x		x	
Dry decating	x		x	
Tentering	x	x	x	x
Shearing	x	x	x	x
Brushing	x	x	x	x

Burning off surface fibers or lint. This is called *singeing.* The fabric is passed rapidly over a gas flame or hot plate.

Giving a smooth surface. This is called *calendering.* The damp fabric is run around heavy, heated rollers that press or iron it.

Moiréing is also a calender operation. It produces a water-mark effect and, if applied to ribbed rayon acetate fabrics, the finish is permanent. Two fabrics are face to face with the ribs

---

* Made from rayon staple fiber cut the lengths of cotton or linen fibers.
† Made from rayon staple fibers cut the lengths of woolen and worsted fibers.

not parallel. As the dampened cloth is run between the rollers in the calendering machine, the heavy pressure flattens the weave, producing the watery or moiré pattern.

Embossing is another calender operation that produces a crepy effect. One roller on the calendering machine has lines or a pattern engraved on it. There may be one or more cotton rolls on which the design is impressed from the engraved roller. As the moist fabric passes between the steel and the cotton rolls, under pressure, the creped effect is produced on the fabric. It is permanent only on acetate or part acetate rayon fabrics.

Raising the surface fibers. This is called *napping* if steel wires are used on the machine; a method applied usually when a deep pile is desired. Or it may be called *gigging* if vegetable teasels are used on the machine. This is not as strenuous an operation as napping and is applied usually when a slight nap is required.

Cutting the uneven or loose fibers. This is called *shearing*. Many fabrics are sheared to give them a clear surface, others are sheared to even the nap or pile.

Brushing is necessary after shearing to remove all the loose fibers.

Straightening the fabric and setting the width. This is called *tentering.* The damp fabric is fed into the machine, on both sides of which are pins or clips that hold the fabric by its selvage, straightening and setting the width as it travels through the frame of the machine.

Lustering and protecting the nap. This is called *dry decating.* This is used for only spun rayon fabrics made of yarns processed on woolen machinery. By means of heat, moisture, and pressure, the nap is set. At the same time, luster is imparted to the fabric. This is partly accomplished by the straightening out of the surface.

For fabrics that have a mixture of rayon and natural fibers, the finish is determined by the proportion of each fiber in the fabric. For example, if a fabric is a mixture of wool, cotton, and spun rayon, with wool predominating, the fabric will be given wool finishes.

Functional Finishes

In addition to the wide range of finishes that may be given to filament and spun rayon, the fabrics may be given additional functional finishes that add to their appearance, serviceability, and uses. These include all of the functional finishes applied to wool and cotton, with the possible exception of the glazed finish, which is more suited to cotton fabrics.

These finishes are applied to spun rayon, rather than to filament rayon fabrics. Some are made crease resistant, others are given a crisp hand. Some are made water-repellent or waterproof, and wind, spot, and stain resistant, while others are rendered flame and fire resistant. Rayon fabrics may be given finishes that make them resistant to mildew and germs, particularly fabrics that have some cotton content. Rayon and wool mixtures are frequently finished to be resistant to moth attack.

Many of these finishes contribute other features. They may make the fabric lintless, nonwilting, and aid it to retain its new appearance—crisp and firm, or soft and draping, without becoming sleazy. Some of them may render the fabric resistant to perspiration and to odors, and others contribute some shrinkage control or prevent slippage and stretching. Many add to the durability as well as to the beauty of the finished fabric.

Fabrics or merchandise made of fabrics that have been given these finishes usually carry labels that explain the type of finish, its effect, and how to care for the fabric.

Dyeing

Rayons can be readily dyed brilliant colors and with a relatively high degree of fastness, when the proper dyes are used. Filament rayons may be dyed in the yarn on in the fabric. Spun rayons are dyed in the rayon staple fiber stage, or the yarns, or the fabrics are dyed.

Cuprammonium and viscose rayon (regenerated cellulose rayon) take much the same dyeing as does cotton. Cellulose acetate is an acetate ester (carbon, hydrogen, and oxygen), and its dyeing properties are not the same as those of regenerated cellulose. Therefore the dyeing of most cellulose acetates is with different dyestuffs and different dyeing procedures. Vat, pigment, and especially prepared acetate dyes are used to dye acetate fabrics.

For the most part, the same dyestuffs have a higher degree of fastness on regenerated cellulose rayon than on cotton.

Acid dyes have very poor affinity for rayon. It is this dye that gives the bright shades to silk. They can be used on regenerated cellulose rayon with the application of a mordant (chemicals that fix the dye, which is applied later). The dyes, while bright, have a poor degree of fastness to water and light and are seldom used.

Basic dyes also give bright shades. They have a moderate degree of fastness to water but a poor degree to light and rubbing. As a result they are used for fabrics that will not be exposed to sunlight or washing. They are used at times to redye fabrics when a brighter shade is required. They are applied to viscose and cuprammonium rayons with an acid mordant.

Direct dyes are used most often for regenerated cellulose. They produce a wide range of colors, not as bright as acid and basic dyes. As direct dyes have a great affinity for these fibers, they have a higher degree of fastness than when applied to cottons. They have a very good degree of fastness to light, some practically as effective as vat dyes. However, there is not too high a degree of fastness to washing. These dyes are applied from a neutral or alkaline bath.

Diazotized and developed dyes are usually direct dyes with added acid compounds that make the dyes faster to light and washing. They are used for all three types of rayon. During the diazotizing and developing of the dyes, a new dyestuff is formed that has higher color fastness.

Sulfur dyes have a rather limited use for rayons and are not used for cellulose acetate rayons. The colors are somewhat dull. They have a good degree of fastness to light and washing but none to chlorine used in laundries.

The dyestuffs are applied in an alkaline solution. The dyestuff is made soluble and the fabric dyed, after which it is given an oxidizing treatment to develop the dye and to produce faster and brighter colors.

Vat dyes are used for all three types of rayons. They have the highest all-round degree of fastness to light, washing, and perspiration. When applied to rayons, it is difficult to get an even level of color, due to the dyestuffs high affinity for rayon. Vat dyes are insoluble in water and a reducing agent is required for the dyeing. After dyeing, the dyestuffs are oxidized with chemicals or by exposing the goods to the air.

Azoic dyes produce brilliant shades on the regenerated cellulose rayons. They have an affinity for rayon and have good fastness to light, washing, and chlorine. Since the dyestuff is precipitated on the fiber the color has a tendency to crock. The cloth is prepared with a developer. The dyestuff is diazotized and the fabric dyed.

The diazotized dyestuff and the developer couple and the color is precipitated on the fibers.

Dispersed dyes are a group of dyestuffs used for cellulose acetate rayon. The dispersed dyestuffs, in insoluble form, are absorbed by the fiber when dyed. These dyes have a high degree of fastness to light and washing. Since they dye no other fiber, they can be used to produce interesting color effects when acetate rayon is combined with other fibers. The dispersed dye colors the acetate rayon and a dye, with no affinity for acetate rayon, colors the other fiber.

Cellulose acetate rayons can be dyed by adding the dyestuffs to the spinning solution. However, this is not highly satisfactory as the dyestuff will stain the spinning equipment.

Pigment dyes. Rayons, dyed with pigment dyes, may have from a fair to an excellent degree of fastness to light, washing, acids, and alkalies, according to the property of the pigment and its method of application. Since the color is bonded on the fabric the color may crock or rub off. Most pigments are bonded to fibers with synthetic resins or other film-forming materials. The dyeing composition, containing the pigment and resin plus other ingredients, is usually applied by padding. After application, the resin is dried and the pigment is bonded to the fabric with a permanent binder.

Printing

A wide range of rayon fabrics, both woven and knitted, are printed. The great majority are roller printed by machine. All methods of roller printing are used—direct, resist, and discharge. Many times the results produced are so similar that it is difficult to distinguish by which method the fabric was printed.

Rayon fabrics are also printed by the screen, block, and batik methods. Large, bold, and clear prints can be produced by the screen and block methods. Plissé printing, which produces a puckered effect, gives rayon fabrics an attractive surface texture and, if properly printed on either filament or spun rayon, the crinkle will not wash out.

Direct, resist, and discharge printing, as well as the roller printing and the hand printing methods are discussed in Chapter 18. The color photographs which are included in the chapter, illustrate most types of printing used today on rayons, as well as on other fabrics.

SUMMARY

SYNTHETIC FIBERS FROM WOOD PULP AND COTTON LINTERS

Kinds of rayon: Viscose, cuprammonium, and cellulose acetate.

Types of rayon: Long, continuous filament rayon; spun rayon, spun from viscose or cuprammonium short staple fibers.

TO MAKE THE YARNS AND FABRICS

Filament rayon

Prepare the wood pulp or cotton linters: To obtain the cellulose.

Convert the cellulose by use of chemicals: To obtain a viscous solution ready for spinning. For viscose and cuprammonium rayon, the solution is cellulose in a physically changed form. For cellulose acetate, the cellulose has undergone a chemical change.

Spin the solution: To obtain long, continuous filaments.

Spun rayon

Cut groups of long continuous filaments into rayon staple fibers: To obtain short fibers the length of cotton, woolen, worsted, or linen fibers.

Make the yarns: According to the length of the fibers, they are carded, combed, doubled, drawn, and spun on cotton, woolen, or worsted machinery and by their respective methods.

Construct the filament rayon or spun rayon fabrics: By weaving, knitting, twisting, braiding, or felting.

TO FINISH THE FABRICS

General finishes

Filament rayon: May be given the same finishes as those for silk or cotton.

Spun rayon: Rayon staple fibers cut the length of woolen, worsted, cotton, or linen fibers, may be given the same finishes as those for woolens, worsteds, cottons, or linens.

Bleach, dye, print: To make the fabric white or colored.

Functional finishes: To render the fabric crease-resistant, absorbent, water-repellent, spot- and stain-resistant, waterproof, crisp, and resistant to perspiration, mildew, germs, flame, and fire.

Scientists are constantly researching and experimenting to produce new and different synthetic fibers as additions to our fabric world
(Celanese Corporation of America).

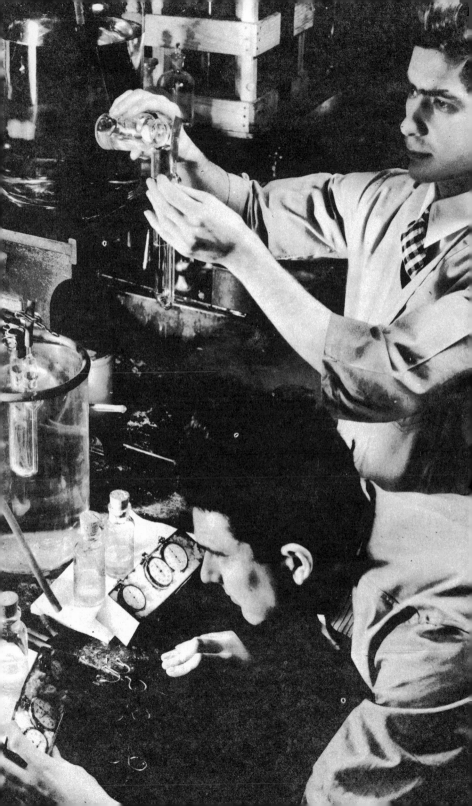

The Newest

IN THE BEGINNING

Fabrics made from linen, silk, wool, and cotton fibers became part of man's life thousands of years ago—in a past so remote that only legends handed down through the ages tell about their beginnings.

About the middle of the nineteenth century scientists began experimenting with cellulose and were able to change it so it could be used as part of paints and varnishes or be made into solid plastic items. Some experiments resulted in a physical change only in the cellulose, while one caused a chemical as well as a physical change in the cellulose.

It was not until around the dawn of the twentieth century that scientists successfully produced a cellulose filament which was later given the name of rayon—our first synthetic fiber.

Synthetics include all filaments and fibers that are produced by synthesis, which means the process of forming or building up of a complex substance or compound by the union of elements or the combination of simpler compounds or radicals.

Chemicals are used to make all synthetic fibers. Some of them, the polymer fibers, are made entirely of chemicals. Others are made from a cellulose, protein, or mineral base with the use of chemicals, which may change the original substances in physical form only, with no change in their chemical structure. These synthetics are called "regenerated" which means to "produce anew" or to "bring into existence again." On the other hand the chemicals may change the original substances chemically as well as physically and thus produce a new chemical compound, different from that of the original compound. Therefore these fibers must be distinguished from the regenerated fibers.

As stated in the rayon chapter, viscose and cuprammonium are regenerated cellulose rayons, while cellulose acetate rayon is a cellulose ester, chemically different from cellulose.

Casein and soybean fibers are both synthetic protein fibers. Casein is an animal protein fiber, made from milk. Soybean is a vegetable protein fiber, made from soybeans. While the protein is physically changed with chemicals, the fiber remains chemically protein.

The twentieth century has brought many improvements in fabrics made of both the natural and synthetic fibers but the great contribution of this century's scientists has been the introduction of the polymer fibers, which are made from complex chemical compounds. These chemical fibers, the same as the rayon fibers, are a part of our advancing age of plastics for the same compounds used for filaments may be used for solid plastics and for fabric finishes. The polymer fibers that are produced today for fabrics, include nylon, Vinyon*, Saran*, and Velon*.

The polymer fibers are so new and so revolutionary that their future importance cannot be estimated at this time. Even the potentialities of nylon, which made a terrific inroad into the silk and rayon fields just before the war, cannot be foretold. Others now contributing to one field may expand into another and some may not be successful in the ultimate competition. There is a stimulating, exciting era of fabrics just over the horizon and each discovery adds to the imaginative range of expansion—the future of which is unpredictable.

NYLON

Nylon is considered a product of "pure" science, as the chemists who began experimenting in 1928 in a laboratory, set up by E. I. du Pont de Nemours & Company, were searching for knowledge, attempting to fill in some of the missing links in certain chemical processes. They were not necessarily looking for a new fiber.

The study was being made on polymerization; they were trying to find out how small molecules unite to form large molecules

* Trade-mark names.

with new and different physical properties. As the muscles in man grow, the small molecules unite with others to form larger molecules. The growing tissue of plants is formed the same way, by forming larger molecules, called "polymers."

Early experimenting was with certain types of acids and alcohols. When heated, it was found, the acids and alcohols combined to form polyesters, called "superpolymers" in the laboratory. Some of these substances could be drawn into a strand, and this, when cooled, could be further drawn out, several times its length, into a transparent, lustrous filament that was strong and elastic.

Then, in 1930, began the experiment to find the most adaptable and practical chemicals for making this filament. Adipic acid and hexamethylene diamine proved to be eminently satisfactory. The polymer from these was called "polymer 66," because each of the two components—diamine and adipic acid—has six carbon atoms.

Steps in Manufacturing Nylon

Nylon is made from four elements: carbon, hydrogen, nitrogen, and oxygen. Basic raw materials are a hydrocarbon obtained from coal; nitrogen and oxygen from the air, and hydrogen from water. As in making rayons, the material changes in the process in a certain sequence, from solid to liquid to solid again, and so forth.

Left: Molten polymer is extruded on a huge casting wheel and sprayed with water, which helps solidify the nylon into a strip. Right: Nylon flakes, shown pouring from a hopper, are made by chopping the strips.

Manufacturing nylon, from the chemical to the finished filament, includes the following steps:

Combining the basic chemicals
    to form the intermediate chemicals—
    adipic acid and hexamethylene diamine   *solids*

Combining the intermediates
    to form a nylon salt; reaction is carried out
    in solution, but the resulting hexamethylene-
    diammonium-adipate is precipitated as   *a solid*

Polymerization or linking together the molecules
    to form a polyamide—viscous liquid when
    hot, but on cooling   *a solid*

Hardening and cutting the polymer
    to form the flakes   *a solid*

Melting the solid flakes
    to form a syrupy polymer   *A viscous liquid*
                                                (when hot)

Spinning the molten fluid nylon
    to form the nylon filament   *a solid*
                                                (when cool)

Left: The flakes are melted and the molten material is extruded through a spinneret and the yarns wound on bobbins ready for fabric making. Right: Close-up of the nylon spinneret (The Du Pont Company).

The chemical compound. The coal-tar hydrocarbon and other intermediates are formed into chemicals known as "adipic acid" and "hexamethylene diamine." This is accomplished by means of high-pressure synthesis.

Combining the chemical compounds. Measured amounts of the acid and the diamine solutions are run together, and the acid and base combine to form a salt, which has a long scientific name—hexamethylene-diammonium-adipate — and a highly complex formula.

Polymerization. This is the all-important step in the manufacturing of the nylon filament—the linking together of small molecules into large molecules. It is through this step that production of this unique fiber was made possible.

The nylon salt is made into a water solution and is piped to evaporators where it is concentrated to a certain salt percentage. This solution is then run into a metallic gas-tight vessel, called an "autoclave," and heated. The nylon salt in the solution is "ionized," that is, the salt is broken up into electrically charged diamine and dibasic acid parts. Each part now may be considered as a short chain with a hook on either end. Heat is applied, and a diamine molecule hooks up with a dibasic acid molecule to form a larger new molecule with a diamine hook at one end and a dibasic hook at the other. These in turn hook up with others in like manner, forming a longer and longer molecule—so-called molecular chains or polymers. Since this polymer contains many amide groups it is known as a "polyamide."

The linking of the chains is controlled by controlling the temperature and by adding at the right time a chemical that closes the hooks, eliminating further linking together.

The substance is now a thick, viscous liquid, ready to leave the autoclave.

Hardening and cutting the polymer. Under the autoclave is a large, slowly rotating wheel. The molten, viscous polymer leaves the autoclave through a slot in the bottom of the machine. As it pours down over the wheel, a shower of water causes it to harden into a wide, translucent ivory-white solid—nylon. As the ribbon of nylon leaves the wheel, air blows off the water.

Nylon flakes are produced by cutting up the strips of nylon into fine chips on a large rotary cutter. This is done to facilitate mixing and blending of different lots of material.

Nylon fabrics, left to right: Satin, chiffon, patterned sheer, taffeta.

Melting the flakes. Prior to melting, flakes from many batches are blended and poured into a hopper. The oxygen is removed by passing prepurified nitrogen through them. The nitrogen is then eliminated by a vacuum process.

The flakes fall through an opening in the bottom of the hoppers onto a melting grid made of heated coils. From the grid, the then molten polymer flows into a funnel-shaped melt chamber, ready to be spun into a nylon filament.

Spinning the molten nylon. This operation is very much like that of acetate rayon. The spinneret is much larger and has from ten to thirty-four holes.

From the melt chamber the viscous solution is pumped to the spinneret. At the same time, the pump squeezes out any bubbles that may prevent an even flow of the solution through the spinneret; and the solution is filtered through a layer of sand.

The thick, syrupy polymer is extruded through the holes in the spinneret. All filaments from one spinneret form one strand of nylon yarn. As they leave the spinneret, they are air cooled and pass into a conditioner, where they are moistened so that they adhere together as one single strand. The strand of yarn then passes over a lubricating roll, which further aids the sticking together of the individual filaments. It is next wound on a cone, ready for stretching and twisting.

Drawing out the yarn. This process is the great discovery that led to the final construction of the unique nylon yarn. Up to here, there has been no change in the molecules since they were linked together in polymerization. Now the yarns are stretched by drawing them out over rolls revolving at different speeds. They are spun at a denier approximately four times the denier size of the final yarn in order to allow for this necessary drawing out.

As the yarn is drawn out, the molecules line up parallel and close to each other, producing a yarn with high tensile strength and true elasticity. This manipulation leaves the yarn with a definite amount of inherent shrinkage and a tendency to elongate and contract with changes in humidity.

Nylon yarns may be twisted together in the same manner as are other yarns. A highly twisted yarn may be set in a steam oven to prevent it from kinking.

Sizing of nylon yarns, to protect them during subsequent operations, is by the application of a water soluble material, such as polyvinyl alcohol.

Setting the fabric. In the course of twisting the yarns, preparing the warp, and weaving or knitting the fabric, the nylon yarns may be deformed slightly under the tensions employed. This means that minimum tensions must be used in the construction operations, as when handling yarns containing rubber. Nylon yarn can also be set to permit no further stretching or shrinking. This method of setting is applied to nylon only.

Usually, the fabric is subjected to boiling water or steam under pressure. Experiments have been made in dry heat finishing. This gives more softness, a better hand, improved resilience and draping qualities, and a greater widthwise shrinkage. Just what causes nylon to set under heat and pressure is not known unless it is that the molecules shift into "more comfortable" positions and remain there.

Dyeing. Acetate dyestuffs have been used to dye nylon. However, it is thought that acid, chrome, and other dyestuffs can be adapted to nylon yarns and fabrics.

The nylon family. The term "nylon" includes a whole family of chemicals that may be compounded and spun into yarns. Only one nylon is discussed here. Different chemicals, finer and coarser yarns, and fabrics of different inherent qualities and textures are planned for future nylons.

The nylon fabrics. To date the great bulk of nylon yarns for civilian goods has been used to produce hosiery and lingerie. Cer ain other fabrics, as marquisette, have been manufactured in limited quantities. Experiments have been made in adapting nylon to such widely varied items as men's neckwear, crush-resistant velvet, upholstery fabrics that would be moth resistant, and pile fabrics resembling fur. Nylon can be given a water-repellent finish, and nylon fabrics coated with synthetic resins have been used as war materials.

While the short nylon staple fibers for spun nylon have not been produced commercially, they may be introduced in the future. Also, nylon may be combined with other fibers to construct fabrics with different qualities.

Nylon features. Nylon has a high tensile strength and can be made into fine denier yarns, sheer and lightweight fabrics, with an individual clarity of weave. It has great elasticity and the ability to take a set preventing any appreciable further shrinkage. Nylon is easy to wash and dries quickly. It produces smooth fabrics with high abrasion resistance. The yarn itself is resistant to mildew, moth, and soil-rot.

A variety of nylon fabrics. Left to right, top row: Sheer marquisette, curtain netting that needs no ironing or stretching, nonsag lace. Bottom row: Striped marquisette, crush-resistant velvet, light weight foundation garment fabric. Center: Fine, twilled fabric made of nylon staple fiber.

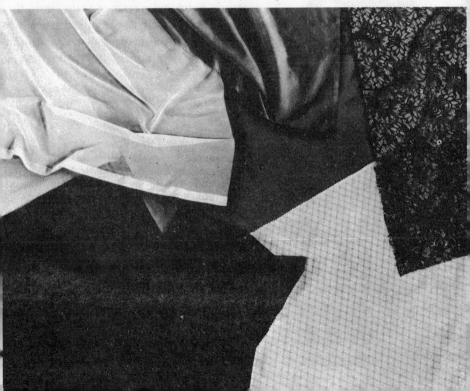

COPOLYMERS OF VINYL CHLORIDE AND VINYL ACETATE

VINYON*

The Vinylite resins used to make Vinyon yarns for fabrics are very similar to those that can be molded into plastics or films or are used as finishes for coating fabrics.

These resins, like those of nylon, make true synthetic fibers. They are obtained by the copolymerization of vinyl chloride and vinyl acetate. These belong to the group of resins known as thermoplastic resins, which means they may be formed and re-formed many times by the application of heat and pressure. (See resins in Chapter 15.)

The vinyl resins have been known for over a hundred years, but it was not until 1927 that they were produced on a commercial scale. After extensive research by the Carbide and Carbon Chemicals Corporation and the American Viscose Corporation, the present fiber was produced in 1938. These resins are known today by the trade-mark name of "Vinyon" and for simplicity, the fiber will be so identified hereafter.

Polyvinyl chloride alone is a hard, tough, water soluble resin with a high softening temperature. Polyvinyl acetate is a clear resin that softens at a relatively low temperature. When these two in a mixture are polymerized simultaneously they produce remarkable characteristics.

* Trade-mark name, registered in the United States by the American Viscose Corporation.

Top to bottom: A very sheer Vinyon fabric. A heavy, lustrous, chain twill weave Vinyon fabric. An unusual drapery fabric created and hand-woven by Dorothy Liebes. It combines yarns with small Vinylite solid plastic rods.

Manufacturing Process

The steps in the manufacturing process, from the chemical compounds to the final filament, consist of forming and re-forming the resin from a solid to a liquid, to a solid, and so forth; just as in the making of rayon, nylon, and other synthetic fibers. Polymerization, or the linking together of the molecules is the same general chemical reaction as used for making nylon. The following are the major steps in Vinyon production:

Processing the basic chemicals
    to form the compound chemicals,
    vinyl chloride and vinyl acetate   *a gas and a liquid*
Polymerization or linking together the molecules
    to form a powder   *a solid*
Dissolving the Vinylite resin powder
    mixed with acetone to form a syrupy solution   *a liquid*
Spinning the solution (extruding and drying)
    to form the Vinyon filament   *a solid*
Drawing out the yarn

Processing the basic chemicals. The first step is the forming of the chemical compound, vinyl chloride and vinyl acetate.

Polymerization. The vinyl polymers of the two compounds, vinyl chloride and vinyl acetate, form long chain molecules. Through the introduction of light, heat, and oxidizing agents, the molecules "hook up" in an end-to-end manner into longer molecules, called "molecular chains" or "polymers." Polymerization is controlled so that further linking together of the molecules may be stopped at any time. The substance becomes a white powder, ready to be dissolved and spun into filaments.

Dissolving the Vinylite resin powder. The white, fluffy powder, colorless and odorless, is mixed with acetone in a large mixing machine, in which paddles thoroughly combine the powder and acetone into a viscous, syrupy solution.

Preparatory to spinning. After ageing, the dope is filtered under pressure through a filter cloth. It is then put into a large chamber where it is deaerated. A vacuum is created in the chamber above the liquid and the air bubbles are drawn to the surface. This step is necessary if the solution is to spin smoothly and evenly. It it now ready to be spun into filaments.

Spinning the filament. From the deaerating chamber the

liquid travels through pipes to the spinning machine that is the same as the one used for acetate rayons. The viscous solution is forced through spinnerets at the top of the machine and is dropped down through a pipe about 10 feet long. As the strands drop through the pipe, warm air circulates and evaporates the acetone causing the strands from one spinneret to form into a multifilament which is wound on a bobbin. The yarn is weak and nonelastic and must be drawn out.

Drawing out the yarn. The molecules linked together through polymerization are not in uniform arrangement. Drawing out, stretching, and lining up the molecules in parallel formation are necessary to increase elasticity and strength of the filaments. They are stretched under steam. Previous to stretching the Vinyon filaments are irregular and much larger than after stretching.

At an approximate stretch of 300 per cent, Vinyon becomes "Vinyon S.T."—meaning Vinyon with medium stretch or medium strength. This is used for certain materials not requiring great strength—such as filter cloth. Under steam at 800 per cent or 900 per cent stretch, the yarn becomes "Vinyon H.S.T.," which means high-stretch or high-strength yarn. An outstanding characteristic of this fiber is that it is thermoplastic, and at high temperatures the stretched filaments will shrink up to 75 per cent of their stretched length. For this reason the yarn is "set" after it is stretched by heating it under tension at 90° to 100° C.

Aged again. The stretched yarns are now aged for about a week. No special conditions are necessary, merely the passage of the required time. This ageing creates a static tension within the filament and permits a better alignment of the molecules. After ageing, the yarns are twisted three and one-half turns, then wound on spools or cones.

Finishing the yarns. The yarn may have either an oil finish or a soap finish. In either case it is applied by drawing the yarn over a saturated wick just before spooling. The finish makes it easier to handle the yarn in the loom. Later, the fabric is scoured to wash out the soap or oil. It can be satisfactorily dyed with water-soluble cellulose acetate dyestuffs with the addition of organic chemicals that act as assistants. The yarn has a naturally bright finish but can be dulled by adding pigments to the spinning solution.

The filaments may be combined to make heavier yarns. The thread count for Vinyon fiber is in multiples of forty, such as 40 denier, 80 denier, 120 denier, etc. Each twenty-eight filaments increase the denier by forty. In its stretched form, it is usually a multifilament denier, considerably finer than silk. Vinyon staple fiber (short fibers) are produced by cutting long, continuous filaments into predetermined short lengths.

Features

Vinyon fiber has many excellent features. It is unaffected by water and has resistance to shrinking unless exposed to temperatures over 65° C. Unlike rayon, its wet strength equals its dry strength. Vinyon is actually water-repellent but can be surface wetted. It is a flexible, strong, durable, and nontoxic yarn, resists mineral acids and alkalies, is noninflammable and is not affected by oxidizing agents or by most inorganic acids. It will not support mildew bacteria, mold, or any fungi. It is insoluble in gasoline, mineral oils, alcohols, and glycols.

Uses

Vinyon's uses range from chemical-resistant hosiery to heavy tents and awnings. It is particularly adaptable to fire-resistant curtains, draperies, and upholstery and certain work clothing. Being water-resistant, its uses include fabrics for umbrellas, shower curtains, bathing suits, and waterproof clothing. It has been made into sturdy fabrics for tents, awnings, sails, fishnets, and other nautical equipment. When blended with cotton, wool, or rayon, Vinyon produces a fabric that holds its pleat and shape. Experiments are being conducted to fuse Vinyon fibers on cloth, thus producing a pile fabric not created in the weave. Vinyon staple fibers, used for felts, shrink from heat, form around the wool fibers, and bond the felting quality of the wool fiber.

Vinyon has a low heat sensitivity. However, future Vinyon will withstand the boil and will dye better. It is unsuitable for undergarments as it accumulates static electricity.

VINYLIDENE CHLORIDE POLYMERS—SARAN*

The development of vinylidene chloride by The Dow Chemical Company began actively over a decade ago. Research men investigated the material while working on chlorinated aliphatic compounds. This original study was followed up by a thorough

* Trade-mark name registered in the United States by The Dow Chemical Company, Midland, Michigan.

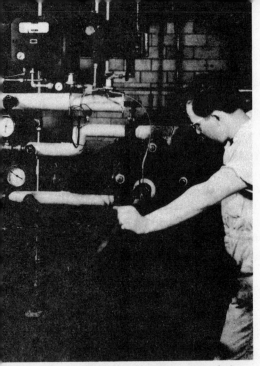

Saran monofilaments are being extruded into a cooling water bath (The Dow Chemical Company).

research program on vinylidene chloride. As a direct result, it was possible early in 1940 to introduce commercially the first vinylidene chloride polymers.

Chemical and physical structure. Vinylidene chloride has petroleum and brine for its basic raw materials. Ethylene (made by cracking petroleum) and chlorine (from the electrolysis of brine) combine to form trichloroethane, which is converted to vinylidene chloride. It is a clear, colorless liquid, having a boiling point of 31.7° C.

Vinylidene chloride can be readily polymerized to form a long, linear, straight chain polymer. By selection of copolymers and by control of the polymerization conditions, polymers may be formed that have softening points ranging from about 70° C. to at least 180° C. Soft, flexible materials to hard, rigid materials can be obtained. These polyvinylidene chloride plastics are known today by their trade name, "Saran," and for simplicity, will be so identified hereinafter.

Saran has three physical conditions: amorphous, that is, with no specific shape or form; crystalline, or crystal-like, transparent solid, with the polymers in an unorganized array; and oriented, after the filament is stretched and hardened and the polymers have lined up in parallel rows.

Most organic thermoplastics exist in an amorphous state and do not exhibit crystallinity. To a greater or lesser degree, Saran exhibits regions of crystal structure. Under special conditions, Saran can be made amorphous but when allowed to return to room temperatures, Saran gradually changes to its normal crystalline state. Amorphous Saran can be converted by mechanical working to an oriented crystalline state.

234

General properties of Saran. One of the outstanding characteristics of Saran is its resistance to chemicals and solvents. At room temperature, it is extremely resistant to all acids and to all common alkalies except concentrated ammonium hydroxide. Slight discoloration with little change in mechanical properties will occur when Saran is exposed to concentrated sulfuric acid or caustic over long periods. It is substantially unaffected by both aliphatic and aromatic hydrocarbons, alcohols, esters, ketones, and nitroparaffins. It is swelled or softened only by oxygen-bearing organic solvents. The resistance to chemicals or solvents decreases with a rise in temperature. The resistance of Saran to any chemical is in part a function of the crystallinity of the polymer. It is chemically more resistant in the crystallized form than in the amorphous state.

A second important characteristic of Saran is its extremely low water absorption. Saran is thermoplastic and has a definite softening point, which limits the temperature at which it can be used. Since softening points vary with composition, the upper limit of the operating temperature can be varied from 150° F or lower, to 250° F. From the standpoint of fire hazard, exposures to much higher temperatures are not dangerous, since Saran is not inflammable.

Manufacturing Process

Saran is supplied to extruders as a fine powder. This powder is fed to a standard, modified, screw type plastic extrusion machine. A screw in a heated chamber forces the powder through a second heated zone, where it becomes quite fluid and is forced through several small orifices in a die. The plastic emerges as a number of thin streams that have the consistency of a thick syrup. These go directly into a water bath, where they are supercooled and emerge as soft amorphous filaments. The plastic strands are then stretched about 400 per cent with suitable equipment and are wound on spools. The Saran filaments, after stretching, have very high strength, good flexibility, and other desirable properties.

Right: Saran screening.

235

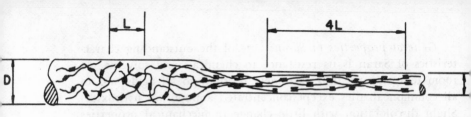

Orientation of the molecules. At the left is the arrangement before stretching. Note how they line up in parallel arrangement after stretching. The diameter (D) is greatly reduced (D2). The length (L) is stretched to the length (4L). Unoriented or unstretched the filament has low flexibility. When oriented it has high flexibility (The Dow Chemical Company).

Extruded and oriented sections are now being produced for fabric uses, ranging in diameter from 0.004-inch to 0.100-inch circular monofilaments and in other shapes, having maximum dimensions up to 0.200. These materials have shown adaptability to standard fabric-manufacturing operations and have been constructed by braiding, weaving, knitting, and twisting. While there are many uses now for monofilaments in the commercially available sizes, the potential fields for single and multiple, fine fibers, with the properties of oriented Saran are even greater.

Many uses of Saran monofilaments are in fields formerly supplied by imported natural products, such as hemp, reed, rattan, horsehair, and linen. The following are typical examples:

Articles of apparel. Its general attractiveness and its range of possible colors fit Saran for such accessories as belts, suspenders, handbags, and shoes.

Upholstery fabrics. Saran makes useful and attractive upholstery fabrics because of its durability, ease of cleaning, abrasion-resistance, and flexibility, as well as color possibilities. Saran fabric used on subway seats shows no material wear and appears like new after a hard service of several years. Saran upholstery fabrics are now used on train and bus seats, as well as for household furniture and automobile seat covers.

Softly lustrous Velon brocades with luxurious draping qualities. These fabrics are woven with multifilament yarns on a Jacquard loom.

VINYLIDENE CHLORIDE POLYMERS — VELON*

Velon is made of the same basic raw materials as is Saran. Ethylene (from petroleum) and chlorine (from brine) combine to form trichloroethane which is converted into vinylidene chloride. The desired polymers are formed by the selection of copolymers and by control of polymerization. The polymer, which is in an amorphous state at time of polymerization, is converted into an oriented crystalline state by mechanical means. This process is more completely explained in the preceding Saran section. In the Velon method, other chemicals are added to the basic raw materials before the viscous solution is extruded through the spinneret to form the filament.

If the filaments are to be colored, the dyestuffs are also added to the original formula and become an integral part of the filament itself. The colors may be deep and rich, brilliant, or a pastel shade. They may be transparent, translucent, opalescent, opaque, or a mixture.

Features: Velon has all of the features discussed under Saran. The colors are not affected by sun or water. Velon fabrics are resistant to dust and dirt and to certain stains. Water and perspiration have no effect on the fabric.

Uses: Velon is produced as a monofilament or a multifilament. The monofilaments are used for upholstery fabrics and for such durable fabrics as those required for awnings, luggage, handbags, golfbags, shoes, and belts. The finer multifilaments are made into such fabrics as brocades suitable for apparel, millinery, shoes, handbags, and curtains. Extremely fine multifilament yarns will give an even wider range of possibilities for future Velon fabrics.

* Trade-mark name registered in the United States by The Firestone Tire and Rubber Company.

Velon fabrics. Left to right: A multifilament sheer fabric is superimposed on a heavy monofilament upholstery fabric. The next two are rough textured fabrics for draperies and table mats, made from multifilament yarns.

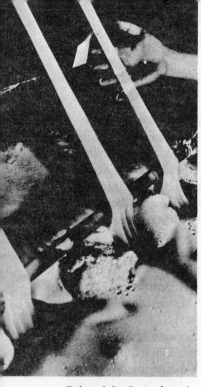

CASEIN FIBER

The use of casein dates back hundreds of years, when it was used as a binder for paint. European cathedrals built in the fourteenth and fifteenth century still show the bright, unfaded casein paint applied half a thousand years ago. This is similar to the history of rayon. For years prior to the development of rayon filaments, cellulose was used for paints, varnishes, and plastics.

Casein fibers are blended with other fibers to make attractive fabrics—the casein fiber contributing to the fabric its own particular features. It is a synthetic protein fiber made from casein obtained from milk. The chemical compound consists of carbon, hydrogen, oxygen, nitrogen, sulfur, and phosphorus. (Wool and silk are natural protein fibers.) Casein occurs in milk as a calcium salt combined with calcium oxide (lime). To free the casein, it is necessary to break it up with an acid which leaves the casein (now a base) in form of curds.

It was not until the twentieth century

Below, left: Ground casein is poured into trap door, mixed in machine, in foreground, treated with chemicals, and heated in huge vats shown in left background. Left: Thousands of strands of casein fiber coming out of three spinnerets. Right: Close-up of the spinneret (Aralac, Inc.).

238

Left: Chemicals are recovered from the tow. Right: From the revolving drum at the top, cut and finished fibers drop down, ready to be baled and shipped to fabric mills (Aralac, Inc.).

that casein was developed for use in fabrics. Some 120 billion pounds of milk are produced annually. About half is used for whole fluid milk consumption. The other 50 per cent is used to supply cream, butter, and other fats, leaving some 57 billion pounds of skim milk of which about 13 per cent is marketed commercially. The other 87 per cent was used for feed purposes, fertilizers, etc. It is from this skim milk that Aralac* derives its casein, of which there some 3 pounds in every 100 pounds of milk.

Atlantic Research Associates, a scientific research division of National Dairy Products Corp., in 1937 began investigating the possibilities of using casein for fibers and by 1940 had developed a usable fiber. It was first blended with rabbit fur for making hat felts and it is still being used successfully for this fabric.

As it has great ability to blend, casein fiber is blended with wool, mohair, cotton, and rayon. It adds a soft draping quality, resiliency, and beauty to cotton and rayon fabrics. The casein fiber has a smooth surface, similar to rayon. It is durable and has a certain elasticity. The soft and lively fibers are particularly suited for interlining fabrics.

Manufacturing Process

The production of the fiber from the skimmed milk to the finished fiber is briefly as follows:

* A trade-mark name registered in the United States by Aralac, Inc.

The curd is precipitated	*a solid*
Skimmed milk is heated and acid added; the curd is washed, broken into particles, and dried.	
The curd is dissolved	*a liquid*
with an alkaline solution	
A viscous spinning solution is formed	*a liquid*
by heating	
Casein strands are spun and cut into short fibers	*solids*

Precipitating the curd. Skimmed milk is heated from 95° to 118°. An acid, such as dilute sulfuric acid or hydrochloric acid, is added to the milk. The precipitated curd separates from the whey. The curd is washed in cold water, to remove the acid and salt, and is pressed or passed through a centrifugal separator to eject the moisture. The casein is then broken into small, ricelike granules and is dried in air or by other means. It is then ready for delivery to the fiber-making plant.

Dissolving the curd. Upon reaching the plant, the ground curd is poured into a huge mixing machine. After mixing, it passes to a large vat where it is further mixed with an alkaline solution.

Spinning solution is formed. The dissolved curds in the alkaline solution are now heated until they form a viscous, honey-like solution, ready for spinning into filaments.

Spinning the casein filaments. From the vat, the viscous solution is fed to the spinning machine, which contains spinnerets. The solution is forced up through the spinnerets. All of the thousands of strands, extruding through the tiny holes in one spinneret, coagulate and harden, by the action of a formaldehyde solution, forming a ropelike tape called "wet tow."

Soft casein yarns, combined with yarns spun of other fibers and with brilliants. are hand-woven into a colorful and unusual drapery fabric (Dorothy Liebes).

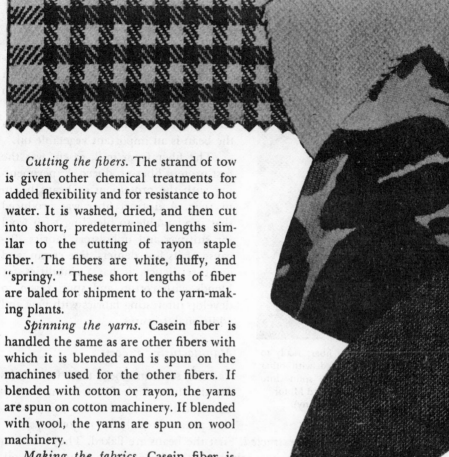

Cutting the fibers. The strand of tow is given other chemical treatments for added flexibility and for resistance to hot water. It is washed, dried, and then cut into short, predetermined lengths similar to the cutting of rayon staple fiber. The fibers are white, fluffy, and "springy." These short lengths of fiber are baled for shipment to the yarn-making plants.

Spinning the yarns. Casein fiber is handled the same as are other fibers with which it is blended and is spun on the machines used for the other fibers. If blended with cotton or rayon, the yarns are spun on cotton machinery. If blended with wool, the yarns are spun on wool machinery.

Making the fabrics. Casein fiber is usually blended with other fibers, for both woven and knitted fabrics.

Finishing the fabrics. A fabric made from blended casein fibers may have the same finishes as those given to the fiber with which it is blended. However, it should not be given any of the finishes that require an alkaline bath and should not be dried with high temperatures.

The affinity of casein for direct dyes is similar to that of wool and it has a great affinity, as does wool, for acid dyes. It can be bleached with hydrogen peroxide.

Top to bottom: Spun rayon and casein yarn dyed check; spun rayon and casein herringbone twill; spun rayon and casein printed twill; spun rayon, casein, and wool mixture, plain weave.

Soybean fibers ready to be blended with other fibers and spun into yarn (Ford Motor Company).

SOYBEAN FIBER

The soybean fiber is a synthetic protein fiber. Soybeans are easily raised in the United States. They are harvested in abundance, as soybean meal is used extensively for feeding cattle. The oil from the bean is an important vegetable oil.

The fiber from milk casein was the first protein fiber produced. Experimenting with the protein in soybeans began in 1937. A distinct fiber was produced that is adaptable to blending with wool for suitings and upholstery fabrics. Because of its fine deniers, fine crimp, and great resiliency, it has possibilities for blends with cotton and spun rayon to develop interesting fabrics with new and additional features.

Manufacturing Process

The soybean fibers used for fabrics are the short staple fibers. From the soybeans to the finished fibers the production passes through five major steps.

The oil is extracted. First the beans are flaked. They are then conveyed through a solvent bath (hexane) to remove the oil. After the oil is extracted, the meal passes through pipes, where steam removes the solvent.

The protein is extracted. Uniform batches of the meal are blended, after which the protein is extracted. There are various methods, some of which are very complex. One method is to leave the meal for a half hour in a weak alkaline solvent solution. The solution from this mixture is clarified by filtering. Then acid is added, the protein in the solution is precipitated, and a protein curd results. The curd is washed and dried, and is then ready to be made into the viscous spinning solution.

Changing the protein curd into the viscous spinning solution. The dried curds are dissolved in suitable chemicals to form a syrupy, viscous spinning solution. The solution is then deaerated to remove all air bubbles so that the spinning is uniform.

Spinning the solution into filaments. The spinning solution is forced through spinnerets. All the filaments from one spinneret combine to form one strand. The spinneret has 500 tiny holes, which means that 500 filaments unite mechanically to form one fibrous strand. Leaving the spinneret, they go through a precipitating bath which congeals and hardens them. The bath usually consists of dilute sulfuric acid, a formaldehyde solution. A salt such as aluminum sulfate or sodium chloride is added to facilitate dehydrating the fibrous strands.

As the strands are pulled through the acid bath, they travel over two pulleys, the second rotating faster than the first. This stretches the individual fibers, increasing their elasticity and strength.

Lustrous Fiberglas fabrics: Coarse and sheer.

Finishing the filaments. The protein in these fibrous strands is set by a bath in a formaldehyde solution. To obtain the short staple fibers, groups of the long, wet, continuous strands are cut into short lengths, ranging from 1½ to 6 inches. When dried, they are loose fluffy fibers, white to light tan in color, with a medium luster. They are soft and resilient and have a natural crimp. They can be dyed during spinning by adding the dyestuffs to the spinning solution or can be dyed when blended with other fibers and woven into fabrics.

GLASS FIBER

Glass fiber is an inorganic, mineral material made of raw materials more or less common to the production of all forms of glass. Glass has been made for centuries. The Venetians used drawn glass strands to decorate their glassware. It was not until 1893, however, that Edward D. Libbey succeeded in drawing from the heated ends of glass rods, coarse glass fibers that could be combined with silk fibers to form a fabric. This was not a true glass fabric, but it was used for lamp shades. Patents were issued in Germany and England early in the twentieth century for glass-fiber processes, but the fibers were too coarse for weaving.

In 1931, the Owens-Illinois Glass Company and the Corning Glass Works began individual research in this field. Rapid progress was made in producing finer and more pliable fibers. In 1938, the Owens-Illinois Fiberglas Corporation was organized, and Owens-Illinois Glass Company and Corning Glass Works turned over to it all the assets they had devoted to the development of Fiberglas*. Very soon, however, almost the entire production of the new company was concentrated upon turning out war materials. Glass fabrics were urgently required for war necessities, where their incombustibility, high strength, and resistance to decay made them indispensable.

Glass fibers. Glass fabric fibers are fine as gossamer, so thin that they are almost invisible. Each strand of continuous filament fiber is made up of more than one hundred individual filaments. Several strands are combined to make a yarn. One small ball of glass, ⅝ inch in diameter, can be drawn into 97 miles of filament. The fibers are like tiny glass rods, highly lustrous and exceedingly smooth. They will not shrink. At tension approaching breaking strength they show an elongation of about 3 per cent.

It is because the fibers are so incredibly thin that they have great flexibility. Though so fine, they also have great strength. Fibers averaging 23/100,000 inch in diameter have a tensile strength of more than 250,000 pounds per square inch. Since they have no cellular structure, they cannot absorb moisture. They

* Trade-mark name registered in the United States by Owens-Corning Fiberglas Corporation.

Left: Machine for forming the glass marbles used to make glass filament.
Right: A close-up of the marbles (Owens-Corning Fiberglas Corporation).

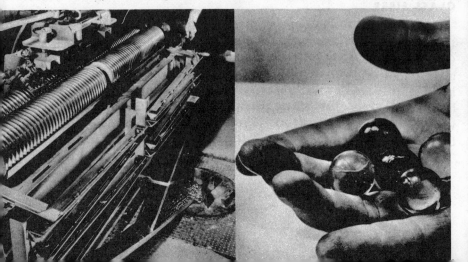

will not rust or decay. Moths cannot eat them. They are unaffected by acids, except hydrofluoric acid. Like asbestos fibers, they will not burn. Glass fabrics have been used for such items as draperies, curtains, bedspreads, table covers, shower curtains, and lamp shades. Used as a drapery material, the fabric falls into draping folds. It is not affected by salt air or by humidity, which often cause sagging and stretching in other fabrics.

Due to their incombustibility the fabrics are ideally suited to decorative uses where noninflammable properties are essential, such as in theaters, schools, and night clubs. This quality may also make them adaptable to such items as awnings. Since repeated flexing will damage the fabric, it cannot be used for furniture upholstery or for wearing apparel.

Manufacturing Process

There are two processes for producing fiber glass: the continuous filament process and the staple fiber process. The rayon staple fiber is made by cutting groups of long, continuous filament rayon fibers into short lengths. Glass staple fibers, on the other hand, are made as the molten streams of glass leave the perforated bushing. Like other man-made fibers, glass fibers are produced by changing the basic materials from a solid to a liquid, from a liquid to a solid, and so forth. With the exception of the handling of the streams of molten glass as they leave the bushing, the two processes for making glass fibers are the same.

The primary mineral ingredients used for the production of

Left: Long, continuous filaments are drawn from a chamber in which the glass marbles are melted. Right: A strand of filament is wound on a high-speed winder (Owens-Corning Fiberglas Corporation).

glass fibers are silica sand and limestone. All raw materials are native to the United States.

The steps in manufacturing are:

Changing the minerals (solids) into molten glass	*a liquid*
Forming the molten glass into glass marbles	*a solid*
Remelting the marbles into molten glass—the forming solution	*a liquid*
Forming, or changing the liquid into continuous filament or staple fibers	*a solid*

Preparing the molten glass. Accurate batches of the raw materials are collected in a traveling batch car and thoroughly mixed in a rotary batch mixer. The batch is then carried to a furnace, where it is melted and refined.

Forming the glass marbles. The molten glass flows to the marble-forming machine, where small glass marbles, about ⅝ inch in diameter, are made. The purpose of forming the marbles is to permit visual inspection for any impurities such as "stones" or "seeds" that may interfere with producing the fibers.

Remelting for the forming solution. The marbles are delivered to small electric furnaces where they are melted. The bushings are at the base of the electric furnace. Each bushing has more than one hundred orifices. The molten glass flows downward from the electric furnace through the bushings.

Forming the continuous filament. All the filaments from one bushing are gathered into one continuous strand. The strand of

A warp beam of fiber glass at the back of a loom showing the filaments laced through the heddles in the harnesses, preparatory to weaving (Owens-Corning Fiberglas Corporation).

Dyed and printed Fiberglas fabrics.

multiple filaments is carried down to the floor below, to a high-speed winding machine. As the winder draws the strand, it draws out each filament emerging from the furnace to a diameter much smaller than that of the hole through which it emerges. These continuous filaments are twisted (or thrown) and plied into yarns of various twists and sizes.

Forming the staple fibers. All the thin streams of molten glass leaving the bushing are yanked into fibers, varying in length from 8 to 15 inches, by air or steam jets. The fibers drop upon a revolving drum. Here they form a gossamer web. The web of fibers is collected from the drum into a strand, which is slightly drawn out so that the fibers lie parallel and is wound on a tube as an untwisted, soft strand. The strands are then twisted and plied into yarns of various twists and sizes. Continuous filament yarns produce fabrics that are smooth and lustrous; staple yarns produce fabrics that have a slightly fuzzy appearance and are less lustrous.

Dyeing. At first the dyestuffs were added to the molten batch, but when the glass was formed into fine fibers the colors became paler, so that only pastel colors were possible. Now glass fabrics can be dyed and printed with many pigment colors and in many shades, fast to sunlight and dry cleaning.

SUMMARY

FIBERS FROM POLYMERS

Nylon: Obtained from adipic acid and hexamethylene diamine.

Vinyon*: Obtained from copolymers of vinyl chloride and vinyl acetate.

Saran* and Velon*: Obtained from vinylidene chloride polymers.

PROTEIN FIBERS

Casein (Aralac*): Obtained from milk.

Soybean: Obtained from soybean.

MINERAL FIBER

Fiber glass (Fiberglas*): Obtained from silica sand and limestone.

* Trade-mark names.

Opposite page: The vein of ore can be noted at top and bottom with a mass of asbestos fibers between the ore (Johns-Manville).

ASBESTOS—MINERAL FIBER

The heat-resistant

Asbestos is a nonmetallic mineral. At first it was thought to be simply rock; but it was found that it was rock that had been changed by the superheated volcanic steam of prehistoric volcanoes. The grayish veins of asbestos twist and turn in a network of lines through the volcanic rock formations.

It is not a fossilized formation of wood or vegetable origin but an inorganic mineral, mostly composed of silicon and magnesium with a small percentage of iron and about 15 per cent of water. The asbestos mined in Arizona has some aluminum, a trace of sodium, and only ½ per cent of iron.

There are two major classes of asbestos, called "amphibole" and "serpentine." The fiber used for yarn is a member of the serpentine class and is called "chrysotile asbestos." The greater part of the asbestos used in this country is mined in Canada. Considerable quantities are also found in Arizona.

Asbestos in its natural state is slippery and soapy to the touch. The fibrous surface is so hard that it will dull a sharp knife and yet it can be scraped with the fingernails into a soft and fluffy, shiny, fine, fibrous mass, ranging from a pure lustrous white to shades of gray. The fine fibers have a delicate texture and a perfectly smooth surface. They have great flexibility and high tensile strength.

Perhaps its most important quality is its ability to resist heat—it will not burn. In fact, "asbestos" is a Greek compound word meaning indestructible, incombustible, unburnable. Fabrics made today from asbestos yarns are those that must be resistant to heat and chemicals. The fabric is soft, flexible, and very strong. It is widely used commercially where fire protection is essential, such as for curtains in theatres.

Asbestos makes excellent protective clothing for workmen who require fire protection in certain industries. Such items include aprons, helmets, arm protectors, gloves, leggings, gaiters and shoes, coats, and overalls which may be lined with duck.

Making Asbestos Fabrics

Sometimes workmen carefully pick the long fibers, ⅝ inch and over in length, from the shattered rock which they break with small hammers. However, most asbestos is mined with big steam shovels and leaves the mine combined with rock and dirt. It is dumped into huge crushers that break up the rock and the asbestos. The broken mass passes over screens. Huge "vacuum cleaners" suck up the asbestos fiber and the smaller rock particles fall through the screen. The asbestos fibers are then graded and those of suitable length are used to make fabrics.

The asbestos fibers pass through the same steps as does cotton. Usually a small percentage of cotton is blended with the asbestos fibers to facilitate spinning. The fibers are carded, leaving the carding machine in a soft, cylindrical roving, with no twist. Drawing and spinning draw out and twist the asbestos roving into a yarn, which is woven into a fabric.

Usually a plain weave is used, but some twill weaves are employed when a closely woven, tight, heavy cloth is desired.

Left to right, opposite page: By air suction the fibers are separated from the particles of rock which are released by crushing. The fibers are passed into the carding machine, where they are carded, and leave the machine in a soft roving. Right: Asbestos fireproof fabrics (Johns-Manville).

SUMMARY OF FABRIC FIBERS

NATURAL	SYNTHETIC
PROTEIN	**PROTEIN BASE**
Wool	Animal: Casein
Silk	Vegetable: Soybean
CELLULOSE	**CELLULOSE BASE**
Cotton	Regenerated: Viscose rayon
Linen	Cuprammonium rayon
Hemp	Esters: Acetate rayon
Jute	
Ramie	
Kapok	
Sisal	
Coir	
MINERAL	**MINERAL BASE**
Asbestos	Fiber glass
	POLYMERS
	Polyamids: Nylon
	Polyesters: Vinyon*
	Thermoplastic resin: Saran*
	Velon*

* Trade-mark name.

Opposite page: Millions of fibers make up these large soft rovings which later will be spun into yarns—fine or heavier, hard twisted or more loosely spun (American Wool Council, Inc.).

Contribute to the fabric

IN THE BEGINNING

In the evolution of fabrics, spinning fibers into yarns appeared very early as one of man's ancient occupations. Man had to learn how to form the raw fibers into yarn before he could weave, knit, or twist the fibers into fabrics. It is believed that the felting quality of fibers was discovered and used by early man even before he thought to spin the fibers into yarns. Ancient man probably discovered by accident that heat, moisture, and pressure turned the mass of wool fibers into a compact sheet of felt while spinning was more than likely the product of creative thinking about ways and means by which he could change fibers into usable strands of yarn.

Drawings on the walls of the earliest tombs depict prehistoric methods of spinning. Early folklore is sprinkled with the "spindle" and the "distaff." The Greek goddess Athena was the "goddess of the distaff." The rhythmic art of spinning was food for ancient poets. Homer, describing Ulysses following Ajax, says,

"As when some dapper girdled wife near to her bosom holdeth
The spindle whence she draweth out the rove beyond the sliver
So near Ulysses kept and trod the very prints of Ajax."

Theocritus (3 B.C.), the pastoral poet of Syracuse, Sicily, also penned a poem to accompany a carved ivory distaff, a gift to the wife of a friend.

Spinning in the beginning was a very simple but adequate craft. Yarn making has two distinct operations: first, the drawing out

and evening of the fibers, second, the twisting of the fibers into a yarn. The earliest spinner made his yarn by rolling strands of fibers on his leg, with the hands, and so twisting the strands into yarns. The first primitive yarn-making equipment consisted of two implements, a wooden distaff and a wooden spindle. The distaff was a stick or staff on which a mass of carded raw material, called a "roving," was loosely bound. The spindle was a smaller stick, from 9 to 15 inches long. Very early this implement was given a notch at one end and was tapered to a point at the other. Attached part way down was a weight or "wharve," composed of a perforated disk of clay, stone, or wood. This served as a fly wheel or pulley, giving steadiness and momentum to the rotating spindle and twirling the fibers into yarns.

From the roving on the distaff, strands of twisted fibers were attached to the notched end of the spindle. The spindle was set in a turning motion either by rolling against the thigh or by twirling between thumb and finger. Next, the fibers were drawn out into a uniform strand by both hands. When the yarn was of sufficient strength, the spindle was suspended by the yarn until a full stretch had been drawn and twisted. The twisted yarn was then wound on the body of the spindle. The operation was started again and continued until the spindle was full.

The distaff-spindle spinning was not improved upon for centuries. The first changes were attempts to provide mechanical means for rotating the spindle, an automatic method of drawing out the fibers, and devices for working a large group of spindles at the same time. The first invention that took some of the tedious handwork out of spinning and added to its speed was the spinning wheel (a wheel attached to a frame). Such a spinning wheel had long been used in India and was evidently introduced to Europe about the fourteenth century but was not widely used until much later.

In 1530 a wheel smaller in diameter and having a more upright appearance was invented in Nuremberg. This "Saxony wheel" was operated by a treadle, with the operator siting. It changed spinning from an exclusively intermittent to a continuous operation.

The problem of drawing out large masses of parallel fibers and twisting them into uniform strands was partly solved by a few ingenious inventions. Lewis Paul, in 1738, invented the first mechanical spinning device, which, although unsuccessful, was the basis for more successful ones. Richard Arkwright, in 1775, patented the water-twist frame, which consisted of Paul's drawing rollers and the spindle, flyer, and spool from the Saxony wheel. James Hargreaves, in 1775, invented the spinning jenny by the aid of which sixteen or more yarns could be spun simultaneously by one person. Samuel Crompton, in 1779, combined Paul's drawing rollers with the stretching device of Hargreaves. The inventions of these four men are the foundation of all modern systems of spinning.

Western-World Spinning

Yarn making of the ancients in the Western World was accomplished with the same simple devices used by the early peoples of other countries, although there may have been other more elaborate equipment used. There have been found some early Peruvian fabrics made from yarns so fine it seems impossible that they could have been spun by hand.

Early weavers of the Inca period were very advanced. Inca rule was in the form of state-socialism which dated back to earliest times. Among other state controlled industries the spinning and weaving were managed by the state. The weavers received a regular allowance of cotton and llama, or alpaca wool from which they were to spin a certain amount of yarn to be woven into a definite yardage of fabric. This concentration of talent and energy resulted in yarns so fine and even as not to be equaled elsewhere.

Mexico's early fabrics also show a superior knowledge of the fibers used for yarns. The American Indians, instructed by the peoples migrating northward, learned to spin their fibers into yarns that gave both service and beauty to their finished fabrics.

The spinning wheel was a familiar piece in the early colonial home for the colonists were handicapped by the "no export" of spinning machinery from abroad. They later managed to obtain some models. A wooden model of Hargreaves' spinning machine was made in England, cut up in pieces, shipped to France and then to America, and was later reassembled in Philadelphia.

COLORFUL YARNS IN WOOL AND SPUN RAYON FABRICS

It is the combination of the types and colors of the yarns, rather than the weave, that lends attractiveness to these fabrics. Either the fibers or yarns were dyed before weaving the fabrics. They are woven in the plain, twill, or herringbone twill weave.

The two top fabrics are all wool and the others are wool and spun rayon mixtures. The fabric at top left is a colorful, wool donegal, woven in a herringbone twill weave with soft, nubbed yarns. At top right is a wool plaid mackinaw cloth. It has a softly napped surface that subdues the weave.

The remaining fabrics are hard or rough textured, wool and spun rayon fabrics woven with white and colored yarns. The sheer fabric, second from the top on the right, has a wide black check produced by floating yarns on the surface of the fabric.

Other colored illustrations of yarn-dyed fabrics are as follows: filament rayon fabrics facing page 353; cotton and linen fabrics facing page 481.

In 1828 Charles Danforth of Paterson, New Jersey, and John Thorpe, of Providence, Rhode Island, took out separate patents on cap spinning devices. Thorp also was credited with the invention of the traveler, the most important improvement made to the first ring spinning. Since then America has contributed many improvements to spinning equipment, adding power and speed to the machines and quality and newness to the finished yarns.

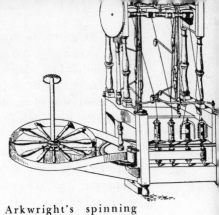

Arkwright's spinning jenny, 1769 (from specifications in the patent office) (New York Public Library).

YARNS

The previous chapters discuss in detail each fiber from its source to the time when it is ready for yarn making. Each filament is developed from the raw materials to the finished filament yarn ready to be made into fabrics. Fabrics may be made of blended fibers or of combinations of fibers in the yarns or in the weaves. In addition, there are many novelty yarns constructed to give special effects in weaving or knitting.

The final personality of a fabric, its appearance, texture, and hand are gained by four factors: the fiber or fibers; the kind of yarn or yarns; the weave or combinations of weaves, or the knit or combination of stitches, or the method of twisting; and the finishes, including dyeing and printing.

The fibers and their contributions to fabrics have already been discussed. This chapter explains how the fibers are blended and combined to give the desired qualities to each fabric, how they are spun into yarns that contribute to the appearance of the fabric, and how the yarns are finally prepared for weaving, knitting, twisting, or felting.

MIXTURES

Every fabric is a mixture. Each wool fabric combines fibers from many fleeces; each cotton fabric contains fibers from numerous cotton plants, often from locations thousands of miles apart; each rayon fabric is made from rayon filaments obtained from many different batches; each silk fabric requires the long continuous strands spun by thousands of silkworms; each linen fabric is woven

257

from flax grown in many fields; each synthetic fiber is a mixture.

It takes highly technical knowledge to skillfully blend and mix these individual fibers and filaments, and yet, to mix the fibers and filaments with each other is even more difficult. A glimpse of the extent of skill needed to blend and combine fibers makes one realize the purposefulness and personality of each fabric.

When planning fabrics the technician does not arbitrarily decide to construct a fabric from certain fibers and yarns. The first consideration is the ultimate use of the fabric, what are its requirements if it is to give satisfaction—satisfaction in heavy duty and long wear; in warmth or coolness; in beauty of color, sheen, sleekness, or rough texture; or in qualities of softness, draping, or stiffness. Each fabric has a purpose and each will have a certain wear-life.

Some fabrics must be closely, firmly woven of tightly twisted yarns; some may be more openly woven of loosely twisted or novelty yarns. Some will have a clear, smooth surface, or a sleek, dull, napped, or rough texture. Some will require dyes that have a high degree of fastness to light, to water, to perspiration. Others will not be subjected to such destructive forces. It is the ultimate use of the fabric that determines its fiber or fibers, its construction, its coloring, and its finishes.

It generally is thought that wools give warmth, cottons and linens give coolness, and rayons and silks give luster. Yet wool may be made into fabrics that give coolness; cottons or spun rayon may be used to construct fabrics that give warmth. Rayons and silks may be made dull as well as lustrous. Cotton and rayon fabrics may look sleek and lustrous. Therefore, the fabric technician has a wide range from which to choose the fiber or fibers that will be suitable and give the appearance required for the fabric. The raw stock is selected carefully to give each yarn and fabric the desired qualities.

First, stock is selected for the spin of the yarn. There must be a certain number of fibers in the diameter of the yarn. For the same diameter, coarser wools take fewer fibers, finer wools take more fibers. Therefore the first basic consideration is that there should be enough fibers for the diameter of the yarn.

Next, the kind of finish to be given to the yarn must be considered. The stock selected must come within these limitations. For example, if the fabric is to be napped, it will require a yarn of a certain diameter, made up of short fibers.

In addition, the strength of the yarn must be studied. For example, a Cheviot is made of coarse, long fibers and the spin is limited to that grade of stock. The fabric is woven, fulled, and the fibers are raised and sheared. Yarns that are to be napped must have strength also.

The most economical way must be planned to secure the best value from the particular kind of fibers used.

Mixture Possibilities

An infinite number of fiber variations are possible in fabrics considering the number of different fibers and their combinations made by blending fibers, and by twisting together and weaving together yarns made of different fibers.

There are three types of yarns: those spun of short fibers; the long, continuous filaments spun by the silkworm or spun through man-made spinnerets; and yarns cut from fabricated materials.

	Short fibers	*Long continuous filaments*
Wool	Glass staple	Silk
Wool specialty	Vinyon* staple	Rayon
Cotton		Nylon
Linen		Vinyon*
Hemp		Glass
Ramie		Saran*
Jute		Velon*
Waste silk		
Rayon staple		
Casein		
Soybean		

The nylon, Velon*, and Saran* filaments have not as yet been cut into short staple fibers but there are possibilities for future production.

* Trade-mark name.

A farm woman shows a visitor how she spins her own yarn as it was spun more than a century ago (Nova Eisnor).

The long continuous filaments of casein have been cut into short staple fibers, as this was found more adaptable to blending with other fibers. Soybean fibers blend with cotton and rayon staple fibers.

Blending the fibers in the raw stock. This consists of mixing together the millions of short fibers before they are spun into yarns. (Naturally, the long continuous filaments are not handled by this method.) More than two fibers, such as wool, cotton, or rayon staple fibers, may be blended to make one yarn. In blending fibers, care must be taken that they are of comparable length and diameter. Blending is started in the breaker and the picker machines. It is continued during carding and combing the fibers, drawing out the slivers and rovings, and spinning the yarns.

Twisting together yarns made of different fibers. Any of the yarns spun from short fibers and any of the filaments may be twisted together. There are seemingly unlimited combinations by this method. For example, wool and cotton may have been blended together and spun into a yarn, which, in turn, may be twisted together with a filament rayon yarn. How yarns are twisted together is explained on the following pages.

Weaving yarns made of different fibers. Again, any of the fibers or filaments may be combined in this manner. Here the possibilities are even wider. The warp may be wool and cotton blended and spun into a yarn, and the filling may be wool and rayon staple fibers blended and spun into a yarn; or the filling may be a filament rayon yarn. Weaving is discussed in Chapter 11.

Fiber and Filament Combinations

Wool. There are three classifications of wool fibers: new wool, reprocessed wool, and reused wool. (These are explained in Chapter 2.) A fabric may be made of all new wool, all reprocessed, or all reused wool; or there may be mixtures of two or all three. Wool yarns are of two types, woolens and worsteds. The latter are made of long combed fibers (see page 39). Woolens and worsteds are blended infrequently in the fibers because of their difference in length; but they may be combined in the yarns or weaves. Wool is most frequently combined with the specialty fibers, rayon, cotton, silk, and casein fibers, but it may be combined with linen.

Cotton. There are two types of cotton yarns: carded, spun from short, or short and long fibers; and combed, spun from long fibers.

Cotton is most frequently combined with wool, rayon, and linen; but it may be combined with hemp, ramie, and jute.

Rayon. There are three filament-type rayons—viscose, cuprammonium, and acetate rayon—and two types of rayon staple fibers—viscose and acetate. Rayon is most frequently combined with wool, cotton, linen, silk, casein fibers, and the specialty fibers.

Linen. This fiber is most frequently combined with cotton and rayon but may be combined with wool, silk, hemp, ramie, and jute.

Silk. There are two types of silk, the continuous filament and the short fibers, called "waste silk." Silk is most frequently combined with wool, cotton, and rayon.

Other fibers. Other fibers, both short fibers and continuous filaments, will undoubtedly be used more for mixed fabrics as experimenting develops suitable fabrics to which the newer fibers will contribute favorable features. Casein and soybean fibers have been blended with other fibers. The polymer fibers have possibilities of combining with the natural and other synthetic fibers.

To simplify the presentation, no emphasis has been given to mixtures of the combinations described above. For example, a yarn blended of cotton and rayon staple fibers may be woven with a two (or more) ply yarn made by twisting cotton and rayon yarns around a rubber core. Because of the infinite variety of fiber blends and yarn combinations, it is impossible today for a person not expert in fabric testing to know the complete fiber content of a fabric. The limitless possibilities for creating fabrics of more than one fiber has brought into being a wide range of fabrics, each with a texture and personality of its own.

MAKING THE YARNS

The long, continuous strands, such as silk, rayon, and nylon filament yarns, are ready for weaving, knitting, and twisting, but it is necessary to spin short fibers into yarns. Over the years, slightly different processes have been developed for making yarns from wool, cotton, linen, and silk. Rayon and the other fibers follow one or another of these processes, depending on the type of yarn to be made.

There are two types of yarn made of each of the four natural fibers, and each is made in a different way. Wool is spun into woolen or worsted yarns. Cotton is spun into carded or combed yarns. Silk is either raw silk, reeled into continuous filaments, or

waste silk (short fibers) spun into spun silk yarns. Linen is either line or tow and is spun into yarns by different processes.

While there is a similarity in manufacturing steps and in machinery, the difference is sufficient to warrant a separate brief explanation of each yarn-making process.

MAKING WOOL YARNS

Carding the wool fibers for woolen yarns. This consists of opening up the fibrous tangled mass, removing any vegetable matter present, blending and mixing the fibers, and straightening them but not necessarily laying them parallel.

The woolen carding machine is in reality

CARDING. Above: Cotton carding machines, showing (below) the cotton entering the machine from the picker lap and (bottom) the card sliver leaving the machine and being deposited in the receiving can. Opposite page, top row: Worsted carding machine showing the sliver leaving the machine to combine with other slivers, and the sliver leaving the last card and being wound on jack spools. Middle row: Woolen carding machine, showing a close-up of the carding web passing from one card to the next. Bottom: Flax tow carding machine showing the can in which the card slivers are deposited.

three machines in one, called the first or "breaker card," the second or "intermediate card," and the last, the "finisher" or "condenser card." The cylinders or rolls on the machines are covered with graduating sizes of wire teeth called "card clothing." The first breaker card has heavy wire, the second has an intermediate size, and the third, or finisher, very fine wires. The action of each of the three cards is the same.

Each card is made up of a "licker-in," a large main cylinder (burr works) covered with wire card clothing, smaller cylinders or rollers called "workers" and "strippers," and a doffer. All rollers are covered with wires that run in a direction opposite to those in the large cylinder.

The wool is carried into the carding machine on a lattice and is fed to the feed rollers. The lickers-in strip it from the rollers. The strippers, in turn, strip it from the lickers-in, and the large carding cylinder takes it from the strippers. The carding takes place between the main cylinder and the workers.

The wool is carried on the teeth of the main cylinder. At the point closest to the worker, the wool is picked off the main cylinder by the teeth of the worker, moving slowly in an opposite direction to that of the main cylinder. The wool is carried around to the stripper (a fast rotating small roll), which pulls it off the worker and carries it back to the main cylinder. The operation repeats itself six to eight times until the fibers are mixed thoroughly and are laid parallel.

The wool leaving the first breaker card passes through combs set in the metallic wires, which distribute it evenly in a wide, filmy web. It is then passed through a rotating funnel and onto the second breaker card. This intermediate card operates the same as the first card, but in place of the large main cylinder there are cylinders with finer wires. Leaving the second card, it passes to the third card, where the wool is further carded with very fine wires.

The wool leaves the last card, again in a wide filmy web, and passing through a condenser, is divided into two or more tapes or ribbons. These ribbons pass through rub-aprons where they are given a false twist. Now in a loose, ropelike form, called a "sliver," they are wound onto jack spools, ready for spinning into yarn.

Carding the wool fibers for worsted yarns. This consists of opening up the mass of wool fibers, blending and mixing them, removing any remaining foreign substance, straightening the wool fibers, and laying them parallel. While fibers for woolen yarns are crisscross, they must lie parallel for the more tightly twisted worsted yarns.

Worsted carding is not nearly as strenuous as is woolen carding, because care must be taken not to break the long fibers that are to be laid parallel later. The worsted carding machine operates much the same as the woolen carding machine except that it has one card in place of three cards. The series of cylinders in the worsted carding machine have different diameters and are covered with wire needles that range from heavy to very fine.

The wool leaves the carding machine in a wide, filmy web, goes through a funnel and is gathered into a narrower, flat strip. It may then be given a slight twist and made into a soft, round ropelike sliver; or it may be delivered to a conveyor belt. Slivers from seven or eight carding machines are collected at the end of the conveyor belt and are combined into one sliver. At this stage the sliver consists of both the long and the short fibers; but since only long fibers are used to make worsted yarns and fabrics, two further steps in manufacturing—gilling and combing—must be completed before the fibers can be spun into yarns.

Gilling. After the groups of slivers from the carding machines are washed, dried, and oiled, they are delivered to a gill box, where they are combined, combed out, straightened, and drawn into a thinner sliver in preparation for combing. For example, sixteen slivers may be combined to form one sliver; then six of these latter slivers may be again combined to form one sliver.

The wool enters the gill box through feed rollers and is drawn out at the other end by additional rollers. The gill box has rows of steel pins, called "fallers." As the wool passes through the gill box, the steel pins come up into the sliver and, traveling in the same direction as the sliver but at a faster rate, comb out and straighten the fibers. The wool then passes through delivery rolls to calender rolls and to a can, ready for combing. The drawing out is done as the wool passes between the feed rolls and the fallers, called "back draft," or between the fallers and the delivery rolls, called "front draft."

Combing for worsted yarns. Combing removes the short fibers, leaving only the long fibers. It further straightens and lays parallel the long fibers and removes remaining foreign substances.

Above: Hand hackling or combing scutched line fiber (Bureau of Plant Industry, Soils and Agricultural Engineering, U.S.D.A.).

There are two systems for obtaining fibers for worsted yarns. The method used depends on the type of worsted yarn to be made. The English or Bradford system is used when the longer, coarser, and straighter wool fibers are desired. In the United States, the Noble comb is usually employed in this method. The French system is used when the shorter, finer, and more crimpy wool fibers are desired. The French comb is employed in this method.

Whichever system is used, the long fibers leave the machines similar in appearance to the carded slivers and are called "comb slivers" or "worsted tops." The short, combed-out fibers are again a tangled fibrous mass, called "noils." These are used to make certain types of woolen fabrics that require short fibers.

English combing. The Noble comb is a large, circular comb. At the top, around the inside, are eight rows of vertical pins, the inside row being the finest and most

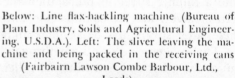

Below: Line flax-hackling machine (Bureau of Plant Industry, Soils and Agricultural Engineering, U.S.D.A.). Left: The sliver leaving the machine and being packed in the receiving cans (Fairbairn Lawson Combe Barbour, Ltd., Leeds).

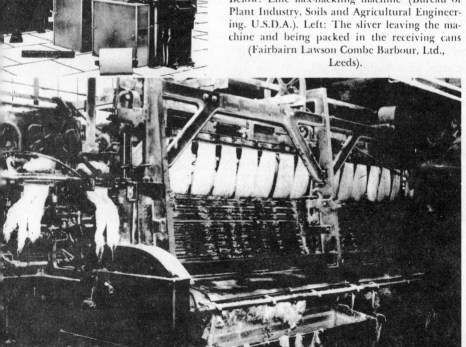

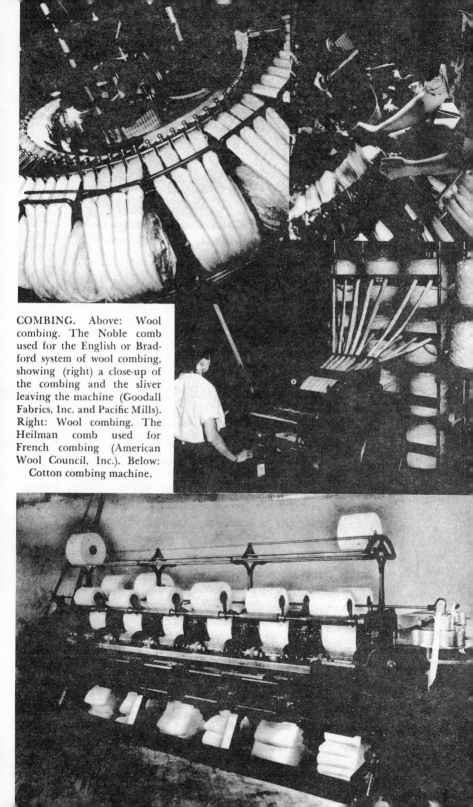

COMBING. Above: Wool combing. The Noble comb used for the English or Bradford system of wool combing, showing (right) a close-up of the combing and the sliver leaving the machine (Goodall Fabrics, Inc. and Pacific Mills). Right: Wool combing. The Heilman comb used for French combing (American Wool Council, Inc.). Below: Cotton combing machine.

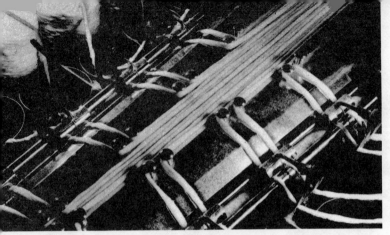

A sliver lap machine, doubling and drafting combed cotton slivers (Pacific Mills).

closely set, and the outside the coarsest and most widely spaced. Inside this circle of pins are two smaller circles of five rows of pins each. Here, the inside rows are the coarsest pins and the outside the finest. Therefore the fine pins of each set almost contact each other.

The slivers are passed through conductors and are laid over the two sets of pins at the point where they converge. Two bristle brushes (called "dabbing brushes") descend and press the wool into the pins of both circles. The two circles of pins rotate, and the fibers are combed and laid parallel as they are drawn through the pins. The long fibers protrude on the outside of the small circles and on the inside of the large circles.

The long fibers are then withdrawn (by drawing off rollers) from the two sides of the comb, combined, and given a slight twist. They are conveyed by delivery rollers to a funnel. Passing through the funnel the twist is removed from the sliver, which is then deposited in a can, ready for further gilling.

Gilling after English combing. The combed wool slivers leaving the Noble comb are uneven and are given two further gillings to insure uniformity, to further blend the fibers, to straighten them, to add water, and to get the fibers into form for drawing out.

These operations are similar to the gilling before combing. The first gill box is like the previously discussed gill box, except that the faller pins are finer. Many slivers are fed to the gill box to be joined as one. Water is added as the wool passes between the front and calender rollers. If the slivers have been dyed, oil

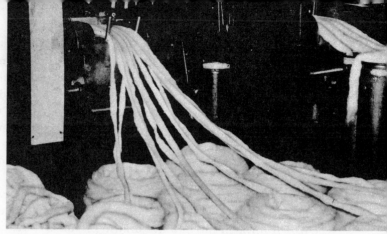

A gilling machine, doubling and drafting combed worsted slivers (Goodall Fabrics, Inc.).

is added. This first gilling is to condition and combine the slivers. The second gilling is a similar operation, except that as the sliver leaves the first roller it passes over a straightening plate and the fibers are evened; then it travels to a folder, which folds the sliver over itself, and winds it on a ball.

French combing. The Heilman comb used in this operation is a straight comb. The wool is passed into the machine in laps or in groups of slivers, generally eight, placed side by side. Drawn off in tufts, the wool passes through feed rolls and a feeding gill into nipper jaws that grasp and hold it while it is being combed.

The comb is circular and has eighteen pin bars. After being combed, the nippers open, and the drawing off rolls, which have moved forward, take the wool from the nippers. In the meantime, a second comb has descended and combs the ends previously held by the nippers while the drawing off rolls hold the combed ends. The tufts of combed wool are laid, one overlapping the other, on a delivery belt. The tufted sliver is then passed through calender rolls and now in a long continuous sliver, is coiled in a can ready for drawing.

A brush removes the short fibers or noils from the comb. Another wire brush and comb take the noils from the first brush and deposit them in a can for later use.

Worsted drawing. The slivers leaving the combing machines must be made into fine rovings that can be spun into yarn. Drawing thins the sliver, straightens and parallels the fibers.

Doubling makes the sliver more uniform. The complete operation produces a roving ready for spinning into yarn. There are three methods used in the United States. The English system is used on long, coarse luster and crossbred wools and produces a hairy yarn; the French system is used on unoiled tops of short, fine wools, and produces soft, smooth yarns; and the cone system is used on oiled tops and produces a fine, uniform yarn.

The *English system* involves five to nine operations. The top slivers pass through two or three gilling operations. Leaving the last gilling machine, the sliver is given a slight twist as it is wound on a bobbin. It is now known as a "slubbing." The slubbing is passed through three to five drawing frames. Here, each is drawn out over sets of rollers traveling at different speeds. The drawing machines double, reduce, twist, and wind the slubbings on spools. This doubling, reducing, and twisting continues on other machines known as finishers, reducers, and rovers. These machines are similar to the first but with faster rotating spindles. The wool is passed through as many machines as are necessary to produce a roving with the desired thinness and twist, ready for spinning.

The *French system* requires from nine to eleven operations in gilling, drawing, finishing, reducing, and roving. The gill box has two sets of fallers with intersecting teeth that comb out the fibers more finely. During gilling no twist is given the slubbing, which is held together by a rubbing motion. As in the English system, the drawing machines double, draw out, make thinner, and twist the slubbing, and wind it on bobbins. These machines differ from other drawing machines in that they have a roller covered with steel pins, called a "porcupine" roller, which moves faster than the back rollers. In this way, the short wool fibers are controlled as the front rollers pull the long fibers.

Upon leaving the last drawing machine, the slubbing passes onto revolving rollers that rub it into a compact roving. It is then wound on bobbins, ready for spinning.

The *cone system* is much like the English system. The difference lies in the method of twisting the slubbing on the bobbins.

DRAWING. Opposite page, top: Cotton drawing frame (Saco-Lowell Shops). Center row, left: Worsted drawing frame (Goodall Fabrics, Inc.). Right: Flax-drawing machine (Fairbairn Lawson Combe Barbour, Ltd., Leeds). Bottom: Close-up of cotton drawing frame (Pacific Mills).

In the English system it is given a hard twist, while in the cone system the twist is less and is also more uniform and accurate, resulting in a fine, uniform roving, softer, and more like a French roving.

Spinning

The woolen rovings from the condenser on the carding machine or the worsted rovings from the drawing and roving machines, are next spun into yarns. There are four methods used for spinning wools. Three of these, flyer, cap, and ring spinning, are used for the worsted English system. Mule spinning, which produces a fine soft yarn, is used for woolen spinning and for the worsted French system. It is being replaced by ring spinning.

Regardless of the method all spinning machines perform three functions: final drawing out or drafting; twisting; and winding the yarn on a package—bobbin or tube. All are continuous operations except mule spinning, which is an intermittent operation.

Mule spinning. Spinning by this method is an intermittent process, but drawing out and twisting are performed simultaneously. This is different from the process in the other three methods, where the roving is drawn out and then is given a twist before it is wound on the bobbin.

The mule spinning machine looks different from the other machines that are upright. The mule machine has several hundred woolen sliver bobbins or worsted roving bobbins mounted on cylinders. In front of them are delivery rolls that carry the strand to the spindles. The spindles are mounted on a long, movable carriage. The carriage moves away from the delivery rolls, drawing out the roving. Simultaneously, a twist is put in by the rapidly revolving spindle, which is being rotated by bands passing from a drum on the carriage.

When the carriage travels about half way on its 60- to 70-inch trip away from the delivery rolls, these stop delivering the roving; and as the carriage slowly travels on, to the end of its run, it draws out the roving to about twice its original length. The carriage then stops, moves in a few inches to allow for the twisting take-up, and the final twisting is given by turning the spindles at a faster speed. Then the carriage stops dead. The turns of yarn at the top of the spindle are backed off, and a wire falls into position to guide the yarn. As the carriage then travels back, the spindles revolve, winding on the yarn. Upon reaching the front rollers, the spinning operation repeats itself—drawing, twisting, and winding.

Ring spinning. This method is used for English spinning of worsted rovings and also, almost exclusively, for cotton spinning. On the ring spinning frame, the rovings are drawn out, twisted, and wound on the bobbin simultaneously and continuously by the rotating bobbin. The roving is mounted on a cylinder on the spinning frame. It passes through drawing rolls that draw it out. The roving then passes through an eye, centered over the spindle, and down to the spinning ring, which is a round track. The spindle is centered inside this track, and the traveler, a semicircular piece of steel, snaps over the spinning ring and slides around it during spinning. The yarn passes from the spinning ring to the traveler, which twists the yarn and guides it on the bobbin.

Cap spinning. This method is used most frequently in the English system of spinning. In cap spinning the spindles are stationary and the bobbins rotate. The rovings are drawn out through rollers in the usual manner, and then pass to the bobbin. A metal tube with a whorl fits over the fixed spindle and rests on a metal plate. The bobbin rotates from 6,000 to 9,000 revolutions per minute by the lifting of the tube. This rotating bobbin twists the yarn, and the tube moving up and down winds the yarn on the bobbin.

Flyer spinning. This method is used less frequently than other methods for English spinning, but it is particularly adaptable to mohair, lustrous yarns, and heavy yarns such as used for carpets. Flyer spinning is also a continuous operation. The machine is much like the English system drawing frame. The roving passes through rolls revolving at different speeds and is thus drawn out. The twisting and winding are done by flyers screwed to the tops of the spindles. The yarn passes to the flyers, which revolve from 2,500 to 3,000 revolutions per minute, twisting and winding the yarn on the bobbins.

MAKING COTTON YARNS

All cottons are carded. From the time the cotton bales are received and opened up in the yarn manufacturing plant, the cotton fibers are being separated from their compressed bunches through the opening, breaker, and picking operations, and through the preliminary cleaning. However, carding is the specific operation for opening up the fibers and eliminating dirt and foreign substances. It also removes some of the very short fibers and somewhat straightens the fibers. The picker "lap" discussed on page 81 is fed into the machine, comes off in a continuous sheet, and is gathered together in the form of a long strand of untwisted fibers called a "card sliver."

While the cotton carding machine is somewhat different from the wool carding machines, it has the same general parts performing the same functions: (*a*) The lap roll rotates, unwinding the sheet of cotton. (*b*) The feed roller carries the cotton to the licker-in. (*c*) The licker-in, with wire teeth uniformly covering its surface, tears away the cotton and delivers it to the large cylinder. Here considerable cleaning takes place. (*d*) The cylinder, moving faster than the licker-in, strips the cotton from the licker-in and carries it upward to the carding flats. (*e*) The flats move slowly in the same direction as the cylinder. There are 110 narrow, flat surfaces (flats) over the large cylinder, each one covered with card clothing, consisting of bent wire teeth. As the cotton passes around the rapidly rotating cylinder, it contacts the wire-covered flats that are moving slowly in the same direction, but the wires are pointing backward. Here carding and cleaning take place as the cotton moves through the almost touching teeth of the cylinder and the flats. (*f*) The doffer, also clothed with wire teeth, takes the cotton from the cylinder. (*g*) The comb combs the cotton from the doffer in a wide filmy sheet. (*h*) The funnel through which the sheet passes delivers the cotton to the receiving can in a soft, round rope called a "card sliver."

Some cottons are combed. Combing is mainly for the purpose of eliminating short fibers, so only the long fibers remain. It also straightens and lays parallel the long fibers and further eliminates foreign substances. As the cotton must be fed to the combing machine in a wide sheet, the card sliver, which is a flat strand, is passed through the first and usually through both of the two following operations before combing. Here they are made into a wide sheet and the fibers are somewhat straightened.

The *sliver lapper* forms the first wide "lap" or strand. Several card slivers are united into one lap. They pass around drafting calender rolls that draw out the strands. Drums in front of the

ROVING. Opposite page, above left and center: Wool roving machines (Goodall Fabrics, Inc. and Mohawk Carpet Mills, Inc.). Above right: Cotton roving frame (Pacific Mills). Below: Linen roving frame (Fairbairn Lawson Combe Barbour, Ltd. Leeds).

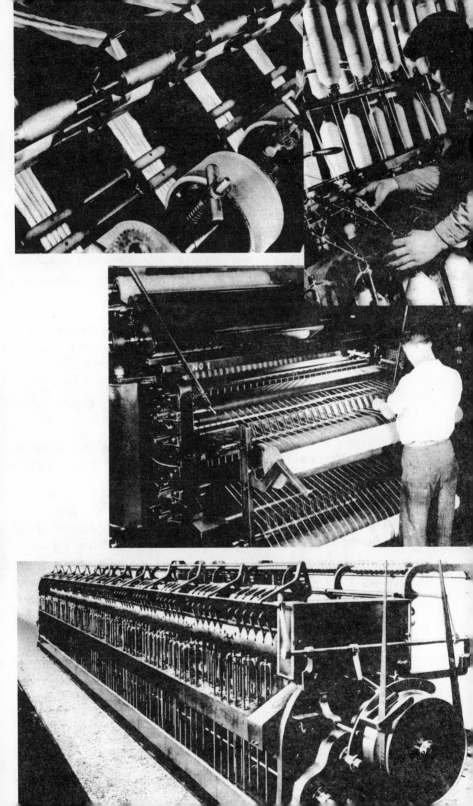

machine wind them into a cylindrical package called a "sliver lap."

The *ribbon lapper* performs the same operation of drawing out and laying the fibers parallel. However, in this machine each lap is drawn separately by an arrangement of drawing rollers and then four or six of these now thin sheets are superimposed one over the other and wound ready for combing.

Combing is the operation that combs out the short fibers and further lays the long fibers parallel. The remaining long fibers are called "combed slivers," and the short fibers, which are removed, are called "noils." Each combing machine has six or eight sections, and each section combs one ribbon lap. The lap is passed into the machine, and a fringe of the comber lap is mechanically held while combs with closely set wires comb out the short fibers, which drop below. However, each time some cotton is drawn back to be combined with uncombed fibers and then passed on to be combed a second time.

The cotton, after being combed, passes through a funnel and forms a "combed sliver." The separate slivers are carried along a table, and all are drawn between drawing rollers into a narrow sheet. Combined into one combed sliver, they are then drawn through a funnel and coiled in a tall container, ready for drawing and roving.

Preparing the Sliver for Spinning

Drawing and roving are necessary before the card slivers or comb slivers can be spun into yarns. *Drawing* further straightens the fibers, laying them parallel as far as possible, and at the same time reducing the size of the strand.

Roving further draws out the sliver and gives it a slight twist. It is then called a "roving." This is the final operation before spinning so the strand must be drawn out sufficiently fine for spinning. Roving processes are repeated until the roving is the desired size. While now it is usual to use only one roving frame for the entire operation, there are four standard frames: slubber, intermediate, fine, and jack. The slubber produces the heaviest, and the jack the finest roving.

Whether one, two, or four frames are used the principle of the machines is the same. Cans of slivers from the carding or combing machines are placed back of the frame. Each sliver is drawn out by passing through a series of rollers traveling at different speeds.

It emerges from the front pair of rollers, is drawn through a flyer and onto the bobbin. The flyer is attached to a spindle which rotates it at a uniform speed. In this way, the flyer gives a twist to the roving as it is wound on the bobbin.

If the sliver passes from the slubber to the other three roving frames, the action is repeated except that usually two rovings are drawn out at one time to form one roving. In the one machine operation, the sliver enters the slubber and is drawn out and twisted into a roving ready for spinning into yarn.

Spinning

This really carries on to a finer degree and completes the drawing operations. The strand is drawn out to the final required size; it is then given the desired amount of twist, and is wound on bobbins, spools, or other suitable packages. In this form it is called "yarn" and can be woven, knitted, twisted, or plaited into fabrics.

Practically all cotton spinning is done by the ring spinning system which is the same as wool spinning by this system, as explained on page 273.

MAKING LINEN YARNS

The flax fibers, freed of straw by scutching, are sorted and graded, ready to be made into linen yarns. The first steps in yarn making are separating and disentangling the fibers and combing them out so they will lie parallel. These processes are called "roughing" and "hackling."

Roughing. This is a hand operation. The "rougher" is a board covered with tin, to which are fastened at wide intervals round steel pins, about 7 inches high.

The scutcher takes a strick (handful) of flax, wraps the top ends around his hand, spreads out the root ends in a whiplike motion, and draws the handful of fibers through the rougher. He repeats this operation again and again if necessary, working up toward his hand. When the root ends are sufficiently roughed, he winds them around his hand and repeats the operation, combing out the top ends. A second roughing is sometimes given on a finer rougher with 5-inch pins set more closely together.

Hackling. In the early days hackling was a hand operation, but now this is done by a hackling machine. In hand operation the hackling boards called "hackles" are similar to the roughing boards except that the needles are smaller and more closely set. The complete roughing and hackling may include four such boards, each with finer teeth. The hand hackling operation is the same as roughing.

Machine hackling accomplishes quickly and effectively the same operations that were performed laboriously by hand for centuries. They are called "vertical sheet hackling machines," as they have a set of endless leather bands (or sheets) that revolve in a vertical direction over a pair of rollers. These sheets are crossed by a double set of steel bars with pins. These bars revolve face to face so that the pins of one intersect with the pins of the other. All hackling machines have several sets of bars, each with finer and finer pins. The first may have four pins per inch and the last forty-five to sixty pins per inch, according to the quality of flax and counts of yarns required. Overhead, between these sheets, is the channel that holds the flax. The channel consists of twelve to twenty-four holders, each 11 or 12 inches long. A holder is made up

SPINNING. Left: Wet spinning frame for line flax. Below: Dry spinning frame for line flax (Fairbairn Lawson Combe Barbour, Ltd., Leeds).

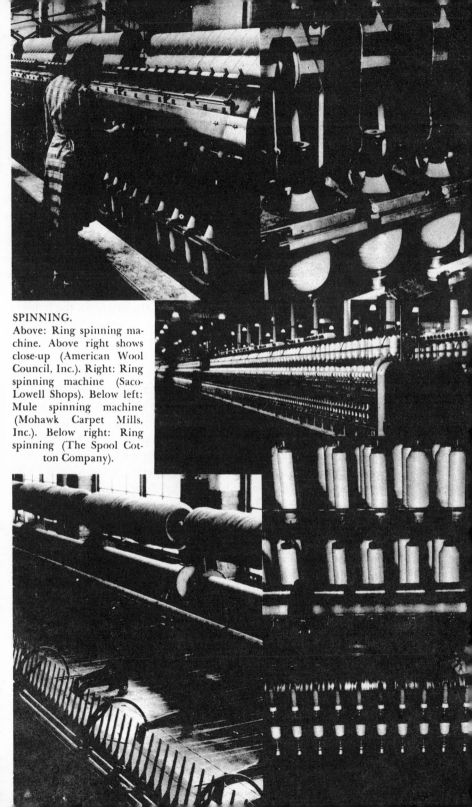

SPINNING.
Above: Ring spinning machine. Above right shows close-up (American Wool Council, Inc.). Right: Ring spinning machine (Saco-Lowell Shops). Below left: Mule spinning machine (Mohawk Carpet Mills, Inc.). Below right: Ring spinning (The Spool Cotton Company).

The long, continuous rayon filaments are ready for fabric construction when they are wound on a suitable package. In the above spooling operation, the rayon filaments are being wound from skeins to spools (American Enka Corporation).

of two heavy iron plates between which the flax is spread and is clamped by screws, with the ends of the flax hanging down.

The flax is clamped into the holders first with the root ends hanging out. The holder channel then falls and rises. As it lowers, the flax is caught between the pins on the bars and is combed. The holder channel then rises and moves this holder to the next set of bars which has finer pins; and the second holder is lowered to be combed in the first bars. This raising and lowering motion continues until the flax has passed through all the sets of pins or hackles.

At the end of the machine, the holders enter the cross channel in front of the machine. They are automatically unscrewed, and the flax is pulled through so that the top ends are in the holders, which are then screwed down. The flax then passes on to the second machine, which combs the top ends. Some machines require four operators to feed the flax to the holders and to change its position in the holders. Automatic machines do the entire job except the initial feeding to the machine, which is done by one operator.

The combed-out short fibers (called "tow") are carried down on the pins of the hackles and are brushed off by a circular brush. A doffer roller takes the tow from the brush. It is removed from the roller by a doffer knife. Tow is carded and spun into yarn used for lower grade fabrics. The combed flax (called "line") is now finely separated into parallel fibers, smooth, glossy, and clean.

Modern spinning mills have a combined transferrer and spreading machine attached to the hackling machine, which converts the single hackled pieces into a continuous sliver or ribbon, ready for drawing and drafting. In some cases, for very fine counts, hand spreaders are used.

Drawing. This process is to prepare the fibers for spinning. They are sorted according to quality and then passed to drawing machines.

The flax sliver from the spreader is fed to the feed end of the drawing frames, which deliver it over feed rollers to the gills or the fallers. The drawing or gill frame has a series of bars with closely placed pins, which pick up the fibers from the feed (or retaining) rollers. The fallers move forward from the feed rollers to the drawing rollers at a speed greater than that of the feed rollers. The drawing rollers draw the fibers out as they move faster than either the feed or the faller rollers. As the fibers are carried from the feed to the drawing rollers, they are combed out through the pins of the fallers.

After the fallers have delivered the sliver to the drawing rollers, two to eight slivers or ribbons may be passed through the machine and drafted together. The slivers are then doubled into one homogenous sliver, which goes to the second, and then to the third, fourth, or any further drafts. Each drawing process reduces the weight and width of the sliver and at the same time, lays parallel the individual fibers.

The number of operations in the drawing frames depend naturally on the fineness of the yarn required and on the quality of the flax used.

Roving. From the drawing frames the slivers are delivered to the roving frame, which is practically the same as the drawing frame except that it has a twisting device. Here the sliver is further drawn out to the required weight for spinning and is given a slight twist to hold it together. The sliver is not doubled in the roving operation. It is delivered from the roving frame in the form of a rove and is wound on bobbins, ready to be spun into yarn.

Spinning. There are three methods of spinning linen, dry, wet, and gill spinning. The first is used for low-count, heavy, and strong yarns; the last two, for fine yarns. The spinning frames for each are similar. They are large double frames with bobbins of rove (from the roving frames) mounted on each side of the frame on rows of stationary pins above the feed rolls. In dry spinning, the rove passes from the bobbin through a guide to a pair of feed rollers and thence to the drawing rollers. Here, the rove is given its final drafting. From the last roller it passes to the flyers and is twisted and wound on spools that are rotating more slowly than the flyers.

In wet spinning, the operation is much the same except that the rove, in passing from the bobbin to the feed rollers, travels through hot water in little troughs. This softens the gummy portion remaining in the fiber. The drafting period for wet spinning is shorter than for dry spinning.

For very special yarns, line and tow yarns are spun on gill spinning machines. In this case the roving frame is eliminated, and an extra drawing takes its place. The sliver from this frame is fed from the combs into the gill spinning frame where the drafting is through gills similar to those of a roving frame.

From the spools or bobbins the spun yarn is next reeled into hanks. If spun by the wet spinning method, the hanks are dried in open air or by artificial heat. They are made into bundles ready for delivery to the weaving plants.

Preparing tow. While line flax leaving the hackling machine is ready for drawing, doubling, and spinning into yarns, the tow must pass through additional processes. Tow, the waste from the hackling and scutching machines, is a mass of tangled fibers.

Carding is the first step necessary for opening up the bunches

Left: Cotton yarns, from the linters in the cotton boll to the finished yarn. Top, right to left: Cotton boll, after ginning, through picker. Below, top to bottom: Card sliver, first drawing, finished drawing, sliver roving, speeder roving, and dyed yarn (Riverside and Dan River Cotton Mills).

Right, top (left and right): Cone and a dye package. Below: Warp bobbin and a filling quill. Right of quill: Large roll of sliver. Bottom: Cheese, a spreader roving, and a slubber roving (Riverside and Dan River Cotton Mills).

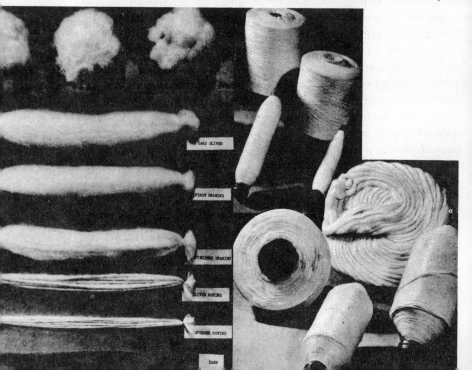

of fibers and, to a certain extent, laying them parallel. The machine is similar to the cotton carding machine. It has one large cylinder covered with metal pins or spikes. The fibers are fed to the machine and are broken up as they pass between the feed rollers. They are carried away from the feed rollers on the teeth of the large cylinder. Rollers, called "workers," contact the cylinder and further break up the fibers, which, as card slivers, are then stripped from the large cylinder by other rollers.

Drawing and doubling. The card slivers are usually drawn out and doubled to further lay parallel the fibers prior to combing.

Combing. This is similar to wool combing. The machine combs out the "neps" or small matted bunches of sliver and also the "strook." The machine takes the sliver from the first drawing after

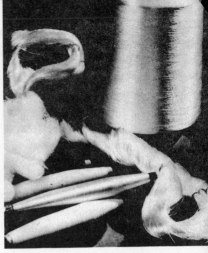

Thousands of yards of rayon wound on a cone (top). Next to it is a skein of filament rayon. The bobbins are wound with dull and with lustrous rayon. Center left are rayon staple fibers (Celanese Corporation of America).

carding and feeds it into the comber, which delivers the sliver into a can. The sliver then goes through a number of tow drawings, required for making the particular rove for the yarn count to be spun.

MAKING RAW SILK YARNS

The bales of raw silk are received, opened, and checked for denier and quality. To soften the gum, the skeins are soaked about twelve hours in a soapy, oily (usually olive oil) bath. They are then dried and placed on reels on the winding frames. The yarns are wound from the reels onto bobbins. For some types of fabrics the silk is wound from the bobbins to cops or beams, ready for fabric making.

Quilling is the term used for winding from bobbins to cones or quills that fit into the shuttle on the loom. Untwisted yarn may be used for the warp of such fabrics as crepe de Chine or radium but, for the most part, silk yarns are further twisted by throwing, as explained elsewhere in this chapter.

The mass of short fibers require many more operations before they can be spun into yarns than do the raw silk filaments, which require twisting only.

Opening the silk. The first step is opening up and beginning to lay parallel the silk fibers. The opening machine has a series of rollers with rows of steel pins and a large revolving drum covered with fine steel teeth. The silk is conveyed to the rollers that hold it. The drum revolves, and the silk is drawn by the steel teeth through the steel pins and is deposited in the drum. When the teeth are filled, the machine is stopped, and the silk is removed in a wide sheet called a "lap."

Filling. This machine is similar to the opening machine except that the drum, instead of being entirely covered with teeth, has coarser steel teeth placed at intervals of 5 to 10 inches. Again, the silk fibers are passed through the steel pins on the rollers and cover the drum, getting hooked on the steel teeth. The machine is stopped, and the silk is cut back of each row of teeth, leaving a fringe of silk hooked on the teeth. The strips of silk are then placed between two boards (book boards) and are pulled off the machine in what is called a "strip." The silk has now been opened, straightened, and cut into lengths, ready for the next step in yarn making.

Dressing. The book boards holding the "strip" of silk are put on the dressing machine, which combs out the short fibers (the noils) leaving only the long fibers (the drafts). There are various types of dressing machines, such as the flat frame used for discharged silks, and the circular frame suitable for schappe silks.

The flat frame holds a number of the book boards taken from the filling machine. The boards are fastened tightly to a frame. This machine has a traveling comb which comes in contact with the strips of silk and combs them.

The circular frame has the book boards fastened to a large drum. There are two small drums covered with fine steel teeth. As the large drum rotates, the fringes of silk pass through the teeth of the fast-moving small drums and the silk fibers are combed.

The teeth of the combs hold the combed-out fibers. These are recombed, and the fibers remaining on the combs are again combed. This process may be repeated several times until the fibers are too short to continue combing. The various combings produce "drafts" of different lengths. The first dressing produces the longest fibers. The last remaining short fibers are called "noils."

Drawing. The "drafts" from the dressing machines are still in the fibrous state and must be put into a continuous form before being spun into yarn.

The drafts are opened up, and small portions are laid end to end on a feeding sheet and are conveyed into a large receiving drum where they are wrapped evenly around the drum. When the desired amount is on the drum, the wide, thin sheet of silk fibers is drawn off. This is called a "lap." It is passed through a second drawing machine, which draws it into a round, soft strand called a "sliver."

A twisting machine, giving yarns their final twist, often combining two or more yarns to form ply yarns (American Enka Corporation).

Eight or more of these slivers are run through a third drawing machine, where they are combined to form one sliver. Eight or more slivers from the third drawing machine are combined on the fourth to form another sliver. This procedure is repeated on two more drawing machines. The doubling and redrawing lays the fibers parallel and makes the sliver even in diameter. From the last drawing machine the sliver is wound on a bobbin.

Roving. This is performed in a drawing machine called a "gill rover." Twenty to forty slivers on bobbins are drawn into one soft, thick yarn called a "slubbing roving." Two or three slubbing rovings are then combined into one. The machine, called a "Dandy roving frame" has back rollers and front rollers, with smaller rollers between them. The slubbing rovings pass over the slowly revolving back rollers to the fast revolving front rollers and are thus drawn out. Wound on bobbins, they are fine rovings, ready for spinning.

Spinning. Spinning spun silk yarns is the same as spinning cot-

Fine rayon yarns, single and ply, tightly or loosely twisted. The fabrics left to right are: 1, taffeta; 2, 4, 5, satin; 3, poplin. (Enlarged 5 times.)

ton and wool; mule, frame, cap, or flyer machines may be used. Here the rovings are further drawn out, twisted into yarns, and wound on spindles or cones, ready for fabric making.

TWISTING THE YARNS

Spinning is the operation that draws out and twists wool, cotton, linen, rayon staple, waste silk, and other short fibers into a long, continuous strand called "yarn." The yarn may be given a slight twist or be very tightly twisted.

Viscose and cellulose acetate rayon and Vinyon filament yarns are given a slight twist, while cuprammonium rayon, silk, nylon, Saran, and glass filaments are not twisted.

Further twisting after spinning is necessary for many yarns, both spun and filament. In addition, some unusual yarns are produced by twisting, which create fabrics with varied textures. The process of further twisting filament yarns such as silk, rayon, and glass, is called "throwing." Twisting operations are either during spinning or are performed on twister machines. These machines merely twist the yarns and do not draw them out (the first operation of spinning).

S and Z twist The yarns may be given a left-hand or so-called "S twist" or a right-hand or so-called "Z twist." A yarn has an S twist if when held in a vertical position the spirals conform in direction of slope to the central portion of the letter S and run from lower right to upper left. It has a Z twist if the spirals conform in direction of slope to the central portion of the letter Z and run from lower left to upper right. A twist is expressed by so many turns per inch or per meter.

Warp yarns, since they require more strength for weaving, usually are given more twist than the filling yarn used in the same fabric, and are given a Z twist. Filling yarns usually are softer yarns with less twist and are given an S twist. Frequently, fabrics

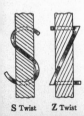

S Twist   Z Twist

Fine yarns, single or ply, tightly or loosely twisted. The fabrics, left to right: cotton organdy; 2, 4, rayon satin; 3, rayon crepe; 5, silk and metal. (Enlarged 5 times.)

such as serge that have a twill running from left to right in a wool will run from right to left in a cotton fabric. This is because cotton yarns are more frequently given an S twist and the twill from left to right requires a Z twist.

After yarns are twisted, they are frequently dyed to distinguish the twists for weaving into fabric. S-twist yarns are dyed yellow, orange, red, or pink, and Z-twist yarns are dyed blue, green, violet, and gray. These dyes are easily washed out and in no way affect future dyeing.

The silk filament consists of two parallel filaments (fibroin) cemented together and covered by glue (sericin). In reeling, four or more or these long, continuous filaments are united and given a slight twist, to form a long, continuous silk yarn of comparatively uniform diameter. Further twisting or throwing is necessary for making a higher twist yarn. The twisting of other filament yarns, as rayon, is also called "throwing."

Grenadine has two or more filament strands twisted together into a single yarn. Then two of these single twisted yarns are doubled and twisted together again in the opposite direction to form a ply yarn. The twists per inch are dependent on the number of strands used. Grenadine is hard twisted. The twist varies from twenty turns of the single and eighteen turns of the double to as high as sixty turns of each. The following are the approxi-

Yarns, top to bottom: The first five are slub or flake yarns and the last two are nub yarns. The second shows elongated slub only. (Enlarged 5 times.)

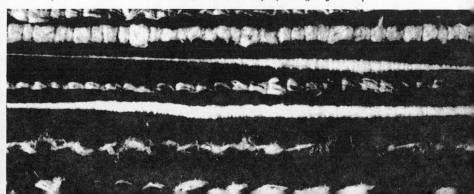

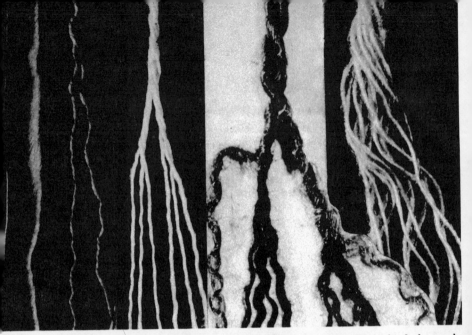

Yarns, left to right: Fine, 1-ply; thick, loosely twisted 1-ply; 2-ply; 6-ply; 10-ply; multiple-ply. (Enlarged 5 times.)

mate number of turns; two strands have a minimum of thirty-two turns; three strands, a minimum of thirty turns; and four strands a minimum of twenty-eight turns. As the strands increase, the number of turns decreases.

Organzine is twisted the same as grenadine, but the turns per inch in the singles and ply are from ten to twenty. Usually the singles are twisted sixteen turns and the doubled yarns are twisted fourteen turns in the opposite direction. Generally, organzine is used for warp yarns.

Tram is twisted the same as grenadine, but with a low number of turns per inch. It is rarely twisted more than six turns to the inch, usually two to four turns. It is used, generally, for filling yarns.

A crepe twist yarn, used for weaving crepe fabrics, may have from forty-five to seventy-five twists per inch either in a single yarn or in ply yarns. Yarns for silk crepe are rarely thrown less than fifty turns per inch and are sometimes thrown as many as one hundred turns per inch. Both S and Z twists are used. In weaving, alternate right- and left-hand twisted yarns are used in both warp and filling to give a balanced fabric.

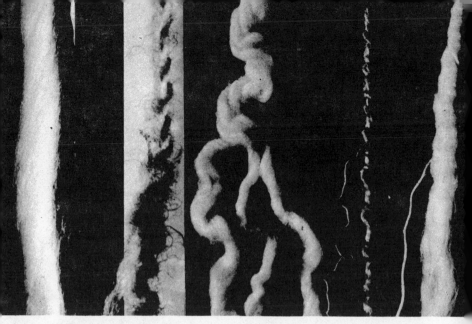

Yarns, left to right: Roving; loosely twisted 2-ply in two colors; thick, loosely twisted 3-ply; thick-and-thin 2-ply; spiral. (Enlarged 5 times.)

Rayon slub yarns may be produced by intermittently increasing and decreasing the amount of spinning solution flowing through the spinneret. This causes the thickening of the filament at intervals.

A ply yarn is made by the twisting together of yarns. For example, a yarn with an S twist may be twisted with a yarn with a Z twist to form a two-ply yarn; or three yarns may be twisted together, an S, a Z, and an S, to form a three-ply yarn. In some cases, one untwisted yarn may be twisted with a twisted yarn. There are any number of combinations. Two thread, three thread, and so forth, means the number of strands that have been twisted together to form the yarn. Three-thread hosiery indicates that three strands were twisted together in the yarn.

A slub yarn is made during spinning. Special attachments on the spinning frame cause variations in feeding the roving, creating thicker places in the yarn. The thick sections are not twisted as much as are the thinner sections, resulting in raised slubs in the yarn.

A flake yarn is similar to a slub yarn but is made on a twister machine. Two yarns are twisted at one time. Between them is a soft, loose roving, used to form the tufts.

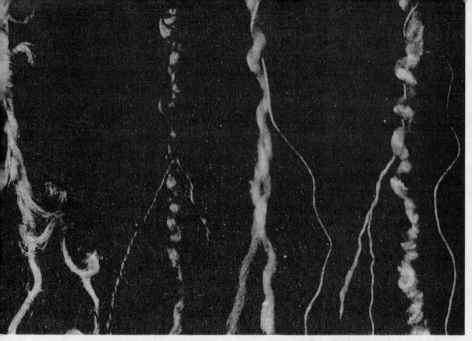

Yarns, left to right: Bouclé; seed, with one heavy and two fine yarns; seed, with one fine and two heavy yarns; ratiné. (Enlarged 5 times.)

A nub or seed yarn is formed with a two-ply, heavier core yarn and one to three finer yarns. Due to the action of the twister machine, in which one roll is stopped, the fine yarn twists around the core yarn. Being fed in excess, it collects and forms a knot or nub. If more fine yarns are used and they are of different colors, they are twisted, first one then the other, around the core yarn and each in turn forms nubs. In this way the woven fabric may have various colored nubs or bunches.

A spiral yarn is made in a twister by twisting together two yarns of different thicknesses, one soft and heavy and the other fine. The heavy yarn is fed faster than the fine yarn and, as a result, winds around the fine yarn in spiral formation.

A ratiné yarn is made on a twister by twisting a spiral yarn a second time with another, usually finer yarn. Since this second twisting is in an opposite direction to the first twisting, the spiral twist tends to unwind, giving the nubby ratiné effect. If the spiral yarn becomes almost straight and is bound at intervals by the other yarn, it produces the so-called "diamond" effect.

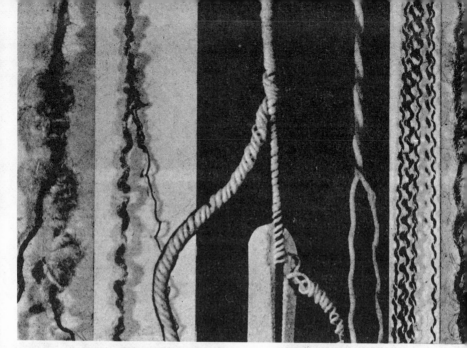

Yarns, left to right: Knob or nub; 2-ply, rubber core covered with two yarns; grandrelle; grandrelle; mock twist. (Enlarged 5 times.)

A snarl yarn is made on a twister by twisting at one time two or more yarns held at different tensions. One may have an S twist and the other a Z twist. The Z-twist yarn twists faster than the S-twist yarn, thereby forming kinks in the yarn.

A bouclé or loop yarn is formed the same as a snarl yarn except that one yarn is made of a stiff fiber, loosely twisted. This yarn is fed faster and forms loops in place of kinks.

A covered core yarn is also of the spiral yarn type. The core yarn may be a single yarn. It may be entirely covered or partly covered. In either case, the covering yarn is fed at a faster speed than the covered or ground yarn and, thus, wraps around the ground yarn. *Rubber* or latex (or a synthetic with elastic properties) is frequently covered with cotton, rayon, or other yarns to make elastic fabrics. The core may be cut strips of rubber or a round extruded filament of rubber. Usually, rubber is covered by being wound first in one direction and then in the other so that the core is completely covered. Special machines are used, since the rubber is fed slowly but is drawn out rapidly in order to stretch the rubber as the yarn is

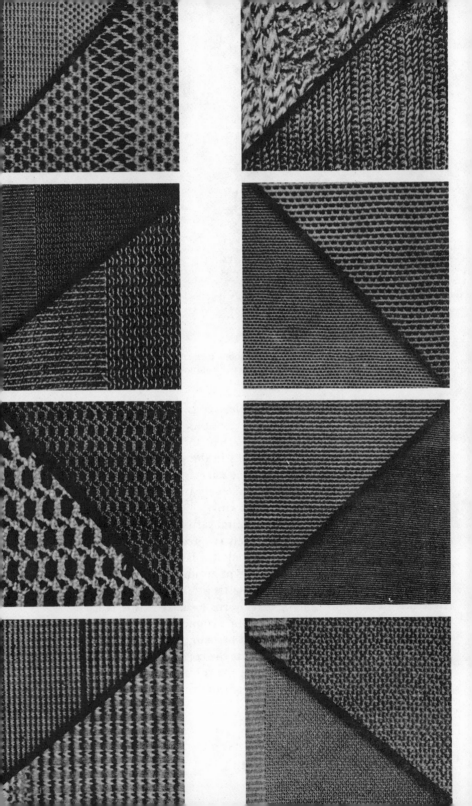

wrapped around it. This allows the fiber yarn to stretch when the rubber is stretched.

Usually an elastic core of rubber (or a synthetic elastic) is wrapped on a twister machine. However, there is one method in which the core is covered during the spinning operation. In this process, a light covering of a sheaf of roving fibers is wrapped around the latex thread.

These elastic core yarns are used for weaving, knitting, and twisting fabrics. By the manner in which the yarns are incorporated into the fabric determines whether the fabric will have a one-way or a two-way stretch.

Metal threads are made with a cotton, rayon, or silk core, around which is wound a flat copper or brass wire (called "lamé") coated with gold or silver.

A mock twist is not a twist but is the effect produced when two rovings, one white and one colored, or both of different colors,

Opposite page: Fabrics made of yarns produced with an elastic core covered with yarn, or yarns, on a twister machine. Top to bottom, left: Knitted, neoprene covered with cotton yarns; leno weave, latex covered with cotton yarns; knitted, neoprene covered with rayon and cotton; knitted, neoprene covered with rayon and cotton. Right: Raschel knitted, latex covered with rayon; bobbinet, latex covered with cotton; rib weave, neoprene covered with cotton; plain weave, neoprene covered with cotton. First fabric has been unraveled to show knitting. A part of each picture shows the fabric construction enlarged .
Right, left and below: Knitted fabric made with fine latex core covered with a soft roving during the spinning process. Center: Elastic lace twisted with a core yarn.

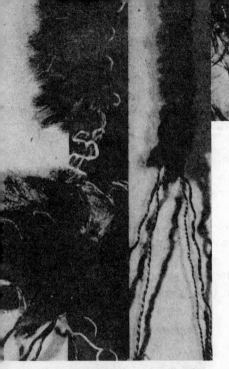

are spun into one yarn. These mottled yarns, called "two-tone," are used for such fabrics as covert.

Grandrelle yarns are two-ply yarns. One may be white and one colored, or they may be of contrasting colors.

Lisle yarns are hard twisted yarns, made of smooth, mercerized and gassed ply yarns, spun from fine-quality, long, combed cotton fibers.

Chenille yarn is not produced by twisting, but by weaving a chenille filling yarn on a regular loom. A so-called "chenille blanket" is first woven and then cut into strips to form the chenille yarn, which is later used as the pile in the woven fabric. Chenille fabrics, including chenille rugs and carpets, are woven with chenille yarns.

Yarns, left: Chenille for rugs (left) and drapery fabrics. Above: Curled yarns form a pile as on astrakhan. (Enlarged 5 times.)

In weaving the chenille blanket, groups of four to eight warp yarns are spaced an equal distance apart. This results in woven strips across the fabric. One half of the distance between the warp groups will be the length of the pile tufts, which are formed by the filling.

The filling weaves, with a plain or leno weave, the first narrow strip; then fills in the empty space and weaves the next narrow strip, and so forth. The filling yards are heavy and softly twisted. After the blanket is entirely woven, it is passed through a machine and a knife cuts the filling yarns half way between the woven strips. The long chenille yarns consist of a narrow woven center with short, projecting filling yarns on both sides.

If a round chenille yarn is desired, it is passed through a twister which twists the yarn so that the strip is entirely covered with the tufts. Or the strips may be moistened, and the tufts pressed into a V shape. When woven, chenille yarn forms a tufted surface texture.

Random yarns are used to produce white or colored flecks in a fabric. The effect is not achieved by twisting the yarn. The necessary fibers are added during carding or in a later yarn-making process. They may be white fibers added to colored fibers to produce a white fleck; or colored fibers added to white stock to produce a colored fleck.

Lace thread for making lace (twisting) is usually made of hard-twisted, fine-quality, combed single yarns, mercerized and gassed for luster and smoothness. Heavier laces are made with carded yarns that may be given a hard or soft twist. Yarns twisted and flattened for use on Nottingham or Levers lace machines are called "brass bobbin yarns."

PREPARING THE YARNS

Yarns leave the spinning machines or the twisters on bobbins, spools, tubes, or cops of many different sizes and shapes, according to the fiber and the type of spinning. Filaments are wound on spools, cones, or cakes. •

Filling yarns (picks). For weaving the yarns are wound on a bobbin that is later placed in the shuttle on the weaving machine. This bobbin drops down into the shuttle, which is thrown back and forth at right angles over and under warp yarns to form the filling of the fabric. If they are not on bobbins, the yarns must be rewound onto bobbins.

Warp yarns (ends). For weaving the yarns are wound on a loom beam, a wide roll, placed at the back of the loom. The yarns are laced through the machine to a large cloth roll in the front.

Warp pile yarns. For weaving most fabrics the yarns are wound on a beam. In rug weaving, the warp yarns may be wound on bobbins that are placed in a frame, as for Brussels or Wilton weaving, or may be wound on long spools placed in long tubes, as for Axminster weaving.

Weft knitting yarns. The yarns are generally wound on large, hollow-center cones.

Warp knitting yarns. The yarns are wound on beams, the same as for weaving.

Twisting yarns. For lace making some yarns are wound on large beams, and other yarns are wound on bobbins.

Winding the warp on the warp beams. Regardless of the fiber or type of yarn a so-called "creel" is used. There are several types. One is a wooden or metallic rack with projecting arms much like a huge hat rack. The creels are of various sizes and one set may have as many as a thousand warp yarns from as many spools, cones, or bobbins on the projecting arms, being rewound to one warp beam.

These yarns are laced through a reel, 62 to 72 inches wide, and from the reel they are laced through a comb that has been adjusted to the exact width the fabric is to be. The sheet of warp yarns then goes over a roll and thence to the warp beam. As the machine operates, the yarns unwind from the packages on the creel and are wound on the beam.

This is one of the more dramatic manufacturing operations, particularly if the yarns have been dyed and hundreds of various colored yarns are traveling in a soft, wavy stream from the yarn packages on the creel to be wound in predetermined sequence to the wide and now colorful beam.

Preparing the Wool Warp Yarn

The wool yarns that come from the spinning or twister machines on bobbins, cones, tubes, or cops of many sizes and kinds do not need to be rewound unless they are to be dyed before weaving.

If the yarns are to be dyed in the skein, they are first wound

WARPING WOOL YARNS
Left: Warp yarns on cones in creel. Right: Warp yarns passing from creel, through reed to warp beam (American Wool Council, Inc.).

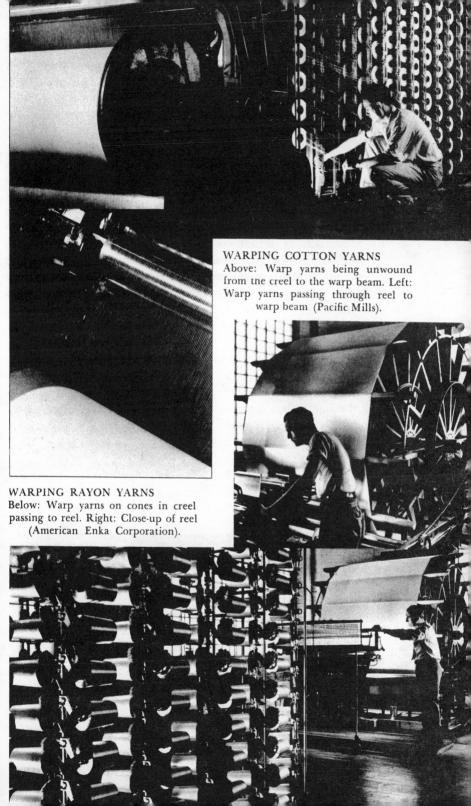

WARPING COTTON YARNS
Above: Warp yarns being unwound from the creel to the warp beam. Left: Warp yarns passing through reel to warp beam (Pacific Mills).

WARPING RAYON YARNS
Below: Warp yarns on cones in creel passing to reel. Right: Close-up of reel (American Enka Corporation).

into skeins. After dyeing, they are wound on cones. If they are wound on jack spools for dyeing, they do not require rewinding, as the jack spools can be placed in the creel and the yarn from them can be directly wound on the warp beam. The same is true of package dyeing as, after dyeing and drying, the packages can be placed on the warping creel.

Warp sizing or slashing. It may be necessary to add sizing to wool warp yarns. This gives them sufficient strength to withstand wear and tear during weaving. At the same time, sizing lays the projecting fibers down, giving the yarns a smooth surface; it sets the twists and adds moisture, all of which contribute to better results in weaving. Even heavier yarns are sized if the warp or the filling is to be of high count.

Animal glue and starches, with the addition of an antiseptic to prevent mildewing, are used for sizing. Regardless of the formula, the mixture must be such that it can be completely removed after the fabric is woven.

A number of wide beams from the warper are placed on beam stands in a machine called a "slasher." This machine has cylinders immersed in the sizing solution. The yarns from the several beams unwind over and under each other and wind, all in one sheet, under the immersed cylinders and are thus given a sizing. The warp goes through rollers that squeeze out the solution and then passes through the drying chamber. As the warp leaves the machine, it is wound on another large beam called the "weaver's beam." It is now ready for weaving.

Preparing the Cotton Warp Yarn

As cotton yarns come from the spinning or twister machines on

Slashing rayon yarns before weaving (American Enka Corporation).

bobbins, they usually are rewound on larger spools or cones. A large package, known as a "cheese," is automatically wound from small bobbins. The cheese packages are placed on the creel and wound onto the warp beam. If cottons are to be yarn dyed, they are collected from the creel through a comb much as for warp beaming, but they are collected in chain form and are wound into a ball. In this form they are dyed. They then are wound onto the warp beam.

Slashing cotton yarns before weaving (Pacific Mills).

Warp sizing or slashing. The sizing of cotton warp is frequently utilized to give a finish to the fabric and is not washed out as is wool sizing after the fabric is woven. Sizing not only permits faster and better weaving but it also adds to the handle or feel as well as to the weight and appearance of the finished cloth. The sizing solution consists of several substances. The main ingredient is starch from vegetables such as potatoes, corn, rice, and wheat. To keep the starch from forming a harsh film, and to give it flexibility and lubrication, a softener is required. This may consist of fats, oils, and waxes of animal, vegetable, and mineral origin.

Just as some dyes require additional chemicals for penetration, so do some sizing solutions require penetrating agents such as soaps, oils, and organic chemicals. Gums are used to toughen the starch and to increase the strength of the yarn. Since the starches, fats, and gums attract bacteria, disinfectants or preservatives may be added to protect the fabric from mildew and bacteria. These frequently consist of chlorides such as chlorides of magnesium and calcium.

The warp slasher is similar to that described for wool fabrics. The cotton warp yarn from the packages mounted on a creel is wound on the beamers. Warp from a series of beamers combines and passes through the size box, is squeezed between squeeze rolls as it leaves the size box and enters the dryer. Here, it winds around drying cylinders and is carried to the warp beam of the loom, ready for weaving.

Preparing the Linen Warp Yarn

If the yarn is in the hank, it is first wound onto spools. If on cross-wound cheeses, these are ready to be placed in the creel to be wound on the warp beam. The yarns are drawn through a guide reed, passing over tension bars and rollers, and onto the beam in front of the creel. The linen yarns are tightly lapped on the beam by means of a pressing roller.

Warp sizing. Various substances such as flour or starch are used for dressing linen. Since linen may become brittle when dressed, a softening agent, such as chloride of magnesium, may be added. The warp-dressing machine also serves as a beaming machine, and the linen yarns pass through the dressing, over and under rollers, before being wound on the warp beam.

Preparing the Silk Warp Yarn

While cotton, wool, and linen require but one operation in winding the yarns, from the spinning frames or twisting machines to the warp beam, it takes two operations for silk. However, silk requires no slashing that necessitates a second winding for cotton, wool, and linen.

The silk yarns from the winder frames or twister bobbins are placed in a creel, such as that used for the other fibers. From the bobbins they are passed by hand through dents (slits) in two or more reeds resembling fine combs. From the reeds they are wound on a horizontal reel from 12 to 60 inches wide, called a "warper." The silk is wound on the warper in sections, and there are as many bobbins in the creel as there are yarns in the section.

After the yarns are threaded through the reed and around the

Slashing wool yarns before weaving (Mohawk Carpet Mills, Inc.).

Quilling, or winding rayon yarns on quills, which are later placed in shuttles (American Enka Corporation).

warper, the large reel slowly rotates, winding the yarn upon itself until the required length, from around 300 to 800 yards, has been wound. The warper is then stopped, the yarns are cut and tied, which completes that section. The operation is repeated until the required number of sections are obtained.

The next operation is to reverse the rotation of the warper and wind the yarns from the warper to the warp beam. The yarns are again guided through the dents in the reed. If for yarn dyed fabrics, the sizing becomes a part of the fabric weight, as in cottons.

The sizing solution consists mainly of gelatin. Other ingredients are similar to those used for cotton. Sulfonated oil from an animal or vegetable base are used for softeners. Other oils give the sizing body and strength. To protect against mildew and bacteria a disinfectant or preservative is included.

Preparing the Rayon Warp Yarn

Both filament rayon and spun rayon yarns are wound on warp beams in much the same way as is cotton. The spools or cones carrying the yarn are mounted on the creel and the yarns are carried to the beamer. In the so-called "silk system" the process is practically the same except that the beam contains all or a complete section of warp.

Warp sizing or slashing. Both filament and spun rayon warp yarns require sizing. If the warp is for gray goods that are to be dyed later, the sizing must be such that it can all be washed out.

Preparing Other Synthetic Warp Yarns

Casein fibers and soybean fibers are prepared the same as the fibers with which they are blended before the yarns are spun. Glass fiber yarns require no additional preparation and are ready to make into fabrics. Nylon yarns are given a sizing with a water soluble material to protect them during subsequent operations. Vinyon* is given either an oil or soap protective finish. Saran* and Velon* are given similar treatments to protect them during fabric construction. The warp yarns are wound on warp beams by one of the above described methods. Other yarns are wound on suitable packages.

YARN NUMBERING

The measurement of the fineness of a yarn is its yarn number. There are two systems used:

The number (units) of standard lengths per standard *weight;* used for wool, cotton, linen, spun rayon, spun silk, glass, and asbestos.

The number (units) of standard weights per standard *length;* used for filament rayon, nylon, silk, and jute.

In the first system, the higher the number, the finer the yarns; in the second system, the higher the number, the coarser the yarns. The following are the measurements of the most important yarns. The names of the units of the yarn numbers are in parentheses after each fiber.

By yarn lengths:

Woolens (cut). The number of 300-yard cuts or hanks per pound.

Woolens (run). The number of 1,600-yard hanks per pound

Worsteds (typp). The number of 1,000-yard lengths per pound.

Worsteds (hank). The number of 560-yard hanks per pound.

* Trade-mark name.

Cellulose acetate yarns are being wound on cones, bobbins, or spools and being given the twists required for various weaving and knitting operations (Celanese Corporation of America).

Cotton and spun silk (hank). The number of 840-yard hanks per pound. Cotton may be as fine as 300, which gives 252,000 yards per pound.

Linen (hank or lea). The number of 300-yard hanks, or leas, contained in 1 pound. (Also it may be the number of 100-yard hanks.)

Glass and asbestos (cut). The number of 100-yard lengths per pound.

Spun rayon (typp). The number of 1,000-yard lengths per pound.

In each of the above length measurements the number of yards can be determined easily for each yarn count. For example, woolens have 300-yard cuts or hanks per pound. Therefore, in 1 pound of No. 40 yarn, there would be 12,000 yards (300 x 40). In 1 pound of No. 10 cotton yarn, there would be 8,400 yards (840 x 10).

Let us find the yarn count of 50,400 yards of spun silk that weigh 1 pound. Spun silk is measured the same as cotton, that is the number of 840-yard hanks per pound. Dividing 50,400 by 840, the answer is 60. Therefore it is No. 60 yarn.

By yarn weights:

Jute (spyndle). The weight of the yarn in pound avoirdupois is called "pounds per spyndle."

Rayon continuous filaments, rayon staple fibers, raw or boiled off silk, and nylon (denier). The number of unit weights of 0.05 grams (a denier) per 450-meter length, or the weight in deniers of a skein 450 meters in length. A denier is equal numerically to the number of grams per 9,000 meters (450 ÷ 0.05). The "new international denier" is equal numerically to the number of grams per 10,000 meters.

If 9,000 meters weigh 40 grams, the fabric is 40 denier. If 9,000 meters weigh 80 grams, it is 80 denier. The 40-denier yarn is twice as fine as the 80-denier yarn, just opposite from the numbering of yarns by length measurement. Considering the 450-meter measurement, if 450 meters of yarn weigh 0.8 gram, the size of the yarn is 0.8 ÷ 0.05, or 16 deniers. If it weighs 2.0 grams, the size of the yarn is 40 deniers (2.0 ÷ 0.05).

After spinning and twisting, the yarns are ready for weaving, knitting, or twisting into fabrics, unless they are to be dyed prior to fabric construction.

SUMMARY

Fabrics may be constructed entirely of one fiber or of two or more fibers by:

Blending the fibers.

Twisting together yarns spun of different fibers or filaments.

Weaving together yarns spun of different fibers or filaments.

PREPARING THE YARNS

Long, continuous filaments: Silk, rayon, nylon, Vinyon*, Saran*, glass. After twisting (often twisted into ply yarns) they are ready for the construction of fabrics.

Short fibers pass through the following processes:

Wool	Cotton and Asbestos	Fiber flax	Waste silk	Rayon staple fiber
Card	Card	Rough	Open	Card
Comb (worsted)	Comb (some)	Hackle	Fill	Comb (some)
Gill (worsted)	Lap (some)	Card (tow)	Dress	
Draw	Draw	Comb (tow)	Draw	Draw
Double	Double	Draw	Double	Double
Rove	Rove	Rove	Rove	Rove
Spin	Spin	Spin	Spin	Spin

To produce the following yarns:

Woolen	Carded	Linen	Spun silk	Spun rayon
Worsted	Combed			

Casein and soybean fibers are prepared the same as the fiber with which they are blended. *Glass* staple fibers are drawn into a strand, twisted, and plied into yarns. *Ramie, hemp, and jute* are handled similarly to flax. *Vinyon** staple fibers are handled similarly to rayon staple fibers.

TWISTING THE YARNS

Yarns are twisted single or are plied with other yarns. Throwing or twisting of silk or rayon yarns include grenadine, organzine, and tram. (Chenille, mock twist, slub, and random yarns are not produced on twister machines.)

Types of twisted yarns

S and Z twist	Ratiné	Grandrelle
Flake	Snarl	Lisle
Nub or seed	Bouclé	Covered core
Spiral	Crepe twist	

* Trade-mark names.

Opposite page: The Jacquard loom, with its overhead cards and hundreds of wires that manipulate the warp yarns, weaving fabrics with simple or intricate designs. This loom is weaving a Jacquard Wilton carpet on a giant Jacquard loom (The Makers of Gulistan Rugs & Carpets).

WEAVING

Creates most of the fabrics

IN THE BEGINNING

"Of all the crafts known to man, weaving is the most ingenious and important," says one writer. Wherever and whenever primitive man began to use his hands, he began to create articles for his convenience. One of the simplest ways to form an article, or a coherent mass, was to pass reeds, fibers, grasses, saplings, and the like, one across the other, over and under, until a definite texture was formed. Thus, weaving was one of the earliest crafts, if not the earliest, practiced by ancient man. From necessity, cords, mats, and baskets were probably some of the first articles woven.

Weaving was in use some 6,000 years before Christ some authorities maintain. Others say it was not practiced earlier than 5000 B.C. But, regardless of the exact date, weaving is known to be one of the handicrafts of the Stone Age, and it may have been practiced even earlier than 6000 B.C. Weaving has been found among the earliest records of man in places as far apart as Peru, Egypt, Mexico, and China, backing the opinion that weaving was a natural outgrowth of primitive man's creative instincts.

Weaving is generally believed to have preceded the knowledge of spinning, for it was carried on with other materials long before man found out that certain fibers had yarn-making qualities. Weaving implements have been found among all excavated ancient paraphernalia that gave testimony of man's earliest way of living.

The first looms were probably made by using a conveniently straight, horizontal branch of a tree, over which the warp yarns were tied. The lower ends were fastened to stones, to hold them taut. Young, pliable saplings and pegs, driven into the ground, may have formed an upright loom. Filling yarns were worked in and out by hand.

The first woven fabrics were very simple and plain. Ornamentation, though, began very early, for as Carlyle remarks, "Decoration is the first spiritual want of man." First decorations represented certain religious symbols. From the very beginning, textile weaving, ornamentation, and decoration have been a continuous record of man's history, experience, and progress, from barbarism to civilization.

Following the progress of weaving since about 3000 B.C. has been comparatively easy. Egypt's method of embalming the dead and burying with them as much of the implements of their daily life as possible, and the peculiar climatic conditions of this section of the earth, which has allowed these early tombs to be well preserved, have given to mankind a definite knowledge of the lives and habits of the early Egyptians. By comparison, it has been less difficult to learn the history of peoples in other places, living in the same era or earlier.

A Navajo woman weaves a rug, using the same crude loom as did her ancestors. The sticks in the warp separate the yarns for weaving in the filling. She pushes the filling in as she weaves. (Frashers Scenic Photos, Pomona, California).

On the walls of the Egyptian tomb of Beni Hassan, 2500 B.C., there is among other paintings one that shows an upright loom and weavers. In the famous Chaldean city of Ur, the reputed home of Abraham, have been found account books, giving the records of weavers working in 2200 B.C. The excavations of the lake dwellers of the Stone Age brought to light bits of woven cloth of flax and wool. Vertical looms have been found not only in Egypt and in India but also in America in the Zuni and Navajo lands.

Mummy cloths have been unearthed that bear witness to the fact that these early peoples practiced the art of weaving so skillfully that it has not been surpassed by modern man's ingenuity or machinery. An example is a mummy cloth of such exquisitely fine linen that it took 540 warp yarns to an inch. Until recently, machinery's finest was 350 warp yarns to an inch.

There is in existence an ornamented piece of linen tapestry dated 1500 B.C. Some colorful linen tapestries were taken out of the tomb of Thothmes IV (1466 B.C.), and young King "Tut," who lived and died about this same time, was buried with many yards of sumptuous fabrics.

The Bible gives many references to woven clothes and weaving. In Job, supposedly the oldest book in the Bible, it says, "swifter than the weaver's shuttle." The priestly garments were "of woven work, all of blue," and "she maketh herself coverings of tapestry, her clothing is silk and purple," are samples of the numerous descriptions of fabrics.

Greek legends are full of weaving tales. Homer, 1000 B.C., tells in his *Odyssey* the famous story of Penelope and her loom. Penelope had remained faithful to her husband Ulysses even after years of his absence. She satisfied the impatience of her many suitors by asking them to wait for her answer until she had finished weaving the cloth she had on her loom. She wove diligently all day and at night raveled out all she had done during the day. She was well paid for her devotion, for she was selected as the patroness of weaving. Ovid, in his *Metamorphoses,* tells of the beautiful Greek maiden Arachne who, daring to compete with the goddess of weaving, Athena, was turned into a spider, doomed to spin forever.

Western-World Weaving

American weaving, like that of all other countries, dates back to the prehistoric. Pre-Inca, as well as other South American civilizations, has been brought to light by excavations. Woven fabrics, found in ruins, prove that these ancient peoples had exceptional weaving ability. Probably they wove with devices far advanced of any other country at that time. The beginnings of these inventions may even predate any mechanical weaving effort of the Old World. But the actual complete equipment and methods of weaving the beautiful ancient Peruvian fabrics have not as yet been found or understood.

Mexico's research into the beginnings of weaving also has given evidence of culture so elaborate and ancient as to equal the highest art of the early times in the Old World. Throughout the length and breadth of the United States there has been found ample proof that weaving was one of the first and most widely practiced crafts of the early North American peoples.

They began, as did the Egyptians, with the plain over and under weaving, but soon the American Indians progressed from this prehistoric simple weaving to a more complex craft. They invented various weaves, such as today's twill and leno. They figured out certain knot-tying embroidery stitches. They created complex geometric designs and wove them into textiles or hand painted them onto cotton fabrics.

These early Americans wove flax, hemp, grasses, and hair from the buffalo, rabbit, and opossum. It was later that they learned about wool. They were familiar with cotton as early as were the peoples of the Old World. Their looms were similar to those found in Egypt except that often, instead of a shuttle, a long cane was used for the filling yarns.

The American Indian weaving ingenuity has colored the pages of American history. In the ancient rock shelters of the prehistoric Bluff-dwellers of the Ozark mountains have been found woven bags, fish nets, feather robes, overshoes, and sandals woven of grass. Ancient earthenware vessels of the Algonquin tribe bear fabric- or cord-marked impressions, showing that the vessels were wrapped in woven material during construction.

The early Basket-makers of 2000 B.C. made twined woven bags and excellent coiled baskets. Civilizations in the arid Southwest that followed these Basket-makers show gradual cultural progress. The fabric art of the cliff-dwellers and the mesa Pueblos consisted of netted and woven articles made of wild vegetable fibers and of fabrics of remarkable beauty. After the early introduction and cultivation of cotton, the fabrics were sometimes interwoven with feathers. The prehistoric Pueblos passed their weaving knowledge on to their descendants, the Pueblos of recorded history. These peoples, in turn, taught the Navajos, who entered the Southwest after the

advent of the Spanish. The Navajos were apt students and soon surpassed their teachers, making finer, more superior woven fabrics.

Today the modern Navajo woman of the Southwest still weaves by hand, much the same as did her ancestors. She sets up the loom with the warp yarns the length and width of the finished size of the blanket. Instead of a shuttle, she prefers to wind her filling yarns on twigs or sticks. She may have carded her wool, which she obtained from the Navajo flock, and she may have dyed it with native dyes used for centuries by her people. She sits on a sheepskin and with deft fingers weaves a blanket, the pattern of which is drawn only in her mind. Even today, she still leaves one untrue spot, or one design break, in her blanket so that the "evil spirit" has a way to escape and thus leave her blanket unharmed. This marking can be readily distinguished on an authentic Navajo blanket.

Weaving Devices

Looms, although probably one of the first complex inventions of man, began as simple affairs. The first hand looms, making plain weave fabrics, were little more than the warp yarns held taut between two drums (one to unroll the warp, the other to roll the finished fabric) ; overhead heddles to raise and lower the warp yarns; and shuttles with filling (weft) yarns to weave back and forth. Foot looms later eliminated the hand manipulation of the heddles.

The drawloom, which was invented in Egypt in 1 B.C., was as important a step in weaving as was the Jacquard loom centuries later. With the drawloom, varied patterns could be formed, because the warp yarns were so controlled by separate strings that any assortment could be raised or lowered at will.

The hand loom had no far-reaching improvement or change until John Kay, in 1733, invented the fly shuttle. Later, in 1760, Robert Kay developed the multiple shuttle boxes. The drop boxes consisted of trays formed in tiers and fitted into ordinary shuttle boxes. Each tray held one shuttle. By a lever, the trays could be raised and lowered as needed, to bring the shuttle containing the desired color into line with the picker. This eliminated changing shuttles by hand and speeded up weaving.

But the most important invention ever applied to the hand loom was the Jacquard device. This device was perfected through the efforts and patents of several men, working from 1725 to 1745.

In 1725, Basile Bouchon substituted an endless band of perforated paper for bunches of looped string, which had been used to select the simples for any shed. In 1728 M. Falcon invented a machine using perforated cards. The machine was attached to the simple cords and was manipulated by a drawboy. Jacques de Vaucanson, in 1745, united in one machine Bouchon's band of paper and Falcon's machine and designed a mechanism for operating it from one center.

Later Joseph Marie Jacquard (1752-1834), while working at the conservatory in Lyons, France, saw the invention of Vaucanson on display there. This new device suggested improvement on his own loom which, although incomplete, had been an attempt at making an automatic net machine. With these new ideas and in an effort to perfect his own loom, he put the finishing touches on what since has been known as the Jacquard. It is applicable to all types of looms, whether weaving, knitting, or twisting. The loom was declared public property in 1806, and Jacquard was rewarded with a pension and a royalty on each machine. He died in 1834, and six years later a statue was erected to him at Lyons.

Step by step, through years of work, it finally became possible to control every part of the loom from one central point. But, as in the growth of other industries, weaving inventions that speeded production and added power fought their way slowly against competition, suppression, and legislation. By hindering their progress some hoped to save the livelihood of thousands of hand workers.

Dr. Edmund Cartwright developed the first power loom in England (1786-1792). It was more or less a crude affair, but it formed the basis of later inventions.

America's Progress

America's fabric mills as we know them today did not exist until about 1800. Our forefathers had a difficult time struggling to clothe themselves. They resorted to all sorts of means to force the early settlers to spin and weave. Bounties were offered to increase the number of sheep and to promote the growth of flax. Laws were passed requiring families to spin and weave. The spinning wheel and hand loom were as

311

familiar in the old fashioned kitchen as were the pots and kettles. The stronger than iron "linsey-woolsey" (cloth of linen warp and woolen filling) was a part of most of the colonial wardrobes.

The early American efforts were retarded by English laws, which attempted to eliminate competition that threatened the homeland industry. It was illegal to export any machinery from England until 1845. It was almost impossible to produce American fabrics that could compete with English cloth either in texture or in price.

But American industry grew slowly and steadily, and step by step, took the work out of the homes. First, neighborhood fulling mills, using nearby streams, were set up, relieving the housewives of the labor of fulling and finishing their woven fabrics.

In 1794 Arthur Schofield invented the first wool carding machine and put it into operation in Byfield, Massachusetts. Carding machines were then added to the fulling mills, and farmers for miles around brought their wool to be carded into rolls ready for the spinning wheel. In 1793 Samuel Slater applied the Arkwright spinning invention to the spinning of cotton and started the first American cotton mill in Pawtucket, R. I. Soon small mills that spun both cotton and wool yarns by waterpower sprang up all over New England. In 1813 Francis C. Lowell made a power loom and established at Waltham, Massachusetts what is believed to have been the first mill in the United States to combine in one establish-

Distinctive original fabrics are still being hand loomed. Dorothy Liebes weaving at a loom in her California studio.

ment the various operations necessary to manufacture the finished fabric from the raw fiber. Thus began the American fabric industry as we know it today. Erastus B. Bigelow (1814-1879), another American, invented looms for suspender webbing, knotted counterpanes, and carpets. Every carpet ever woven was constructed by hand until Bigelow's power loom revolutionized the industry.

Power spinning and power weaving were rapidly applied to wools also, and the household weaving industry disappeared before the greater economy and efficiency of the factory system. Although most of the fundamental inventions were of English origin, America's ingenuity in improving and perfecting these has surpassed all other nations. Some of the most conspicuous American contributions were labor saving, speed, and safety additions to power machinery; mechanisms for expediting processes; and automatic devices for dispensing with intermediate help. America's achievements have been so outstanding that they have changed the entire weaving picture throughout the world.

Yes, progress stops for no man. Today's intricate, complicated, speedy power looms are a far cry from the crude, simple hand loom evolved by our prehistoric ancestors. In addition, time has proved that instead of creating unemployment, the machine age has increased supply and demand until today America's workers, as a whole, have the world's highest standard of living.

WEAVING

The following four terms must be understood before weaving can be explained.

The warp consists of the yarns that run lengthwise in a woven fabric.

The filling (or weft) consists of the yarns that run crosswise in the fabric. They are used to "fill" the warp.

An end is one warp yarn. (It is also used to designate one sliver or one roving.)

A pick is one filling yarn woven between the warp yarns (at right angles to the warp) in one passage of the shuttle through the shed.

THE LOOMS THAT WEAVE THE FABRICS

The principle of the power looms, which weave fabrics so rapidly it is difficult for the eye to discern the operations, is the same as

313

that of the old hand looms. Each part of the machine performs a definite function. Weaving is the interlacing of yarns at right angles. To obtain all the various weaves, the warp and filling yarns interlace in different ways. To better understand how these weaves are produced, the loom that creates the weaves should be first described. The following parts of the loom have to do with the warp:

The warp beam. A wide beam at the back of the loom, containing the warp yarn. This yarn must be drawn through the heddles in the harness, and through the reed to the cloth beam, to form the warp of the fabric.

The whip roll. A roll above the warp beam over which the warp passes as it leaves the warp beam.

The lease rods. Round or oval sticks, under and over which the warp yarn passes on its journey from the whip roll to the heddles. These keep the yarns separated and in order.

The heddle. A slender, flat or twisted rod that has a center eyelet through which the warp yarn is drawn. The heddle has one oblong eyelet at the top and one at the bottom, by which it is fastened to the harness. Lacing the warp yarns through the heddles is called "drawing in."

The harness. An oblong wooden frame with top and bottom steel bars that run through the heddle eyelets, holding the heddles in the harness. There are at least two harnesses in a loom; there may be more, depending on the type of fabric to be woven. If a plain weave, requiring two harnesses (the minimum that can be used) is to be woven, one harness carries the even-number yarns and the other carries the odd-number warp yarns. The harnesses

Rayon warp yarns from a beam back of the loom being joined to yarns that are threaded through the heddles in the harnesses (American Enka Corporation).

move alternately up and down to make the "shed," through which the filling yarn passes, in order to interlace with the warp yarns.

The reed. An oblong metal frame within which flat steel wires are set at close and predetermined intervals. After the warp yarn is drawn through the heddles in the harness, it is passed through the spaces (called "dents") between the wires in the reed. Passing of the warp ends through the reed dents is called "reeding." In weaving, the reed moves forward as each pick is woven and beats or pushes the filling yarn into the cloth.

The shed. The space made as the harnesses, carrying the warp yarns, move up and down. The down harness carries one set of warp yarns and forms the bottom of the shed. The raised harness carries the other set of warp yarns and forms the top of the shed. The shuttle is shot through the shed. Each passage through is called a "pick."

The head motion. Parts of the loom serve to control the actions of the harnesses; that is, the formation of the successive sheds in accordance with a pattern or weave. This method of separation is the most important difference in weave formations. (In the more complicated weaves there are other contributing factors.)

For such weaves as plain and twill and for other simple weaves,

Right: When a warp yarn breaks, a "drop wire," which rides that particular yarn, drops down and forms an electrical contact that instantly stops the loom. At upper left are the harnesses holding the heddles (Goodall Fabrics, Inc.).

Below: Printed warp yarns, laced through a reed to weave a tapestry or velvet rug, are being aligned to conform with the corresponding warp line of checks on the design. Note the elongated yarn pattern that reduces to design size when the yarn loops over wires during weaving to form the pile (Mohawk Carpet Mills, Inc.).

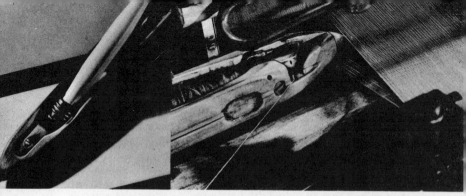

where no more than six harnesses are required, a cam mechanism is sufficient. These are metallic levers, one for each harness. The rotation of the lifting and lowering of the harnesses is such that the fabric is woven in a specific weave. This is more fully explained in the discussion of weaves. With a dobby head the mechanism can control up to twenty-five harnesses. In a Jacquard loom each yarn is separately controlled and can be raised in any predetermined manner.

The following are the parts of the loom that have to do with the filling:

The shuttle bobbin. A hollow cylindrical core on which is wound the filling yarn.

The shuttle. The container for the bobbin, which passes back and forth through the shed, weaving the yarn from the bobbin over and under the warp yarns. One passage of the shuttle weaves a pick.

The following parts of the loom have to do with the fabric:

The breast beam. The part over which the woven cloth passes before reaching the take-up device and the cloth roll.

The take-up device. The train of gears which rotates the take-up rolls delivering the fabric to the cloth roll.

The cloth roll. The roll on which the woven cloth is wound.

Plain and Box Looms

A plain loom provides facilities for one shuttle only, which passes through the shed on the first pick and back on the next pick. A box loom accommodates two or more shuttles and is fitted with a box motion. If a filling pattern or different colored filling yarns are to be woven on different picks, a multiple shuttle arrangement is required. The shuttles are contained in drop or rotating boxes, on one or both sides of the loom, and deliver the shuttle carrying the right filling yarn into place when it is to be woven into the fabric.

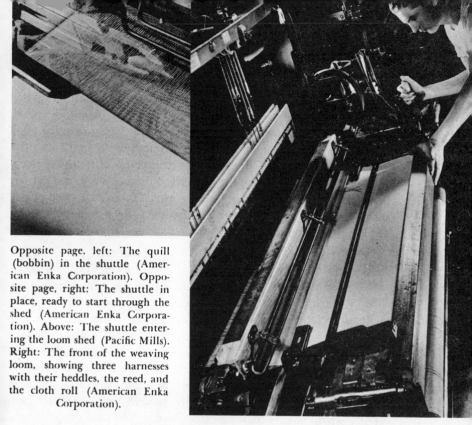

Opposite page, left: The quill (bobbin) in the shuttle (American Enka Corporation). Opposite page, right: The shuttle in place, ready to start through the shed (American Enka Corporation). Above: The shuttle entering the loom shed (Pacific Mills). Right: The front of the weaving loom, showing three harnesses with their heddles, the reed, and the cloth roll (American Enka Corporation).

A mechanism called a "box motion" operates the shuttle boxes.

A loom in motion. As one watches a loom in operation, the harnesses can be seen lowering and rising at a rapid speed. The shuttle, containing the bobbin of filling yarn, is shooting through the open shed; and the forward action of the reed is pushing the filling yarn forward up to the preceding filling yarn.

To weave a fabric, the warp yarns must be unwound from the warp beam (or in some weaving, as rugs, from large spools), the shed formed for each passage of the filling yarn, the filling passed through the warp shed and beaten into the fabric, and, finally, the woven fabric wound on the cloth beam.

This involves five major motions of the loom. (a) *Let-off motion.* This is the motion that controls the rate at which the warp beam turns as it releases the warp yarns which pass through the heddles in the harness. (b) *Head motion.* This controls the dividing of the yarns to form the shed, by the action of the harnesses that separate groups of warp yarns; or by the cards in the Jacquard looms that control the individual warp yarns. Some form the lower part of the shed and others form the upper part. (c) *Picking motion.* This motion hurls through the shed the shuttle containing the bobbin on which is wound the filling yarn. One passage is called one "pick." (d) *Beating up.* This is the forward motion of the reed that beats in the filling yarn so that it lies next to the previous pick. (e) *Take-up motion.* This carries the completed fabric to the cloth roll. Its speed determines the number of picks (or filling yarns) per inch in a fabric. In a more loosely woven fabric the speed of the take-up motion is faster than if there are to be a higher number of filling yarns per inch.

Dobby Attachments. A plain or box loom may have cams or levers that alternately raise and lower the harnesses, or a chain mechanism known as a "dobby head" that controls the harnesses. A loom controlled by cams has two to seven harnesses while one with a dobby attachment may have as many as twenty-five harnesses. The dobby harness control can be used for many types of weaves from a plain weave, woven with two harnesses, to the more complex weaves as well as for small woven patterns that are geometric in form.

The dobby mechanism operates much the same as a Jacquard machine, with perforated cards or by pattern lags into which pegs are inserted. The mechanism causes levers to contact the griff bars, which in turn raise and lower the harnesses as designated by the cards or the pattern lags. Thus the shed is formed and the fabric is woven. The dobby head attachment controls the warp yarns only. It may be coordinated with the shuttle boxes to form both warpwise and fillingwise patterns or distinctive weaves or colored effects.

A loom showing the warp beam, the yarns passing over the whip roll to the harnesses, the reed, the bobbin in the shuttle, and the cloth roll (Celanese Corporation of America).

Leno Device. A leno attachment permits some warp ends to cross adjacent warp ends as picks are inserted. The weave is called "the leno weave" and it produces fabrics with a lacy effect. The device consists of harnesses, called "doup harnesses," and is equipped with heddles quite different from the standard or conventional type. Generally there are two standard harnesses carrying the ground warp yarns, and two or more doup harnesses carrying the doup warp yarns.

The doup mechanism consists of a doup heddle, comprising a U-shaped, thin metal part having one eye at the bend through which the doup end passes. Each leg of the U, the working position of which is inverted, is connected to its own upright. Each one of the two uprights is capable of raising the U part carrying the doup end. The ground warp is passed between the two uprights, above the doup yarns.

In operation, the doup yarns are shifted to one side, then to the other side, of the ground yarns as the heddles raise and lower the doup yarns. When one doup harness is raised, the doup warp yarn is on one side of the ground warp yarn; and when the other is raised, the doup warp yarn is on the other side of the ground warp yarn. In a plain leno weave the ground warp yarns are down and the doup warp yarns are up to form the shed. Leno weaves may be constructed on a plain loom with or without a dobby attachment. A plain weave can be combined with a leno weave if all the doup yarns are down and all the warp yarns are up or vice versa.

Lappet Attachment. By this method a small design is woven into the fabric by means of extra warp yarns. The ground weave may be plain, twill, or satin, using the required number of har-

A velvet loom showing the two warp beams and the two fabrics being rolled face to face on one cloth roll (Cheney Brothers).

nesses. A series of needles are fixed upright in laths, extending the width of the loom in front of the reed. Each needle carries a warp yarn that does not pass through the reed. The needles are guided by a series of pins in front of them and are lifted on each passage of the shuttle and then lowered. After lowering, the needles move sideways the width of the design and are again lifted for the next pick. At the end of each movement the ends are bound by the filling into the ground weave of the fabric.

Swivel Shuttles and Bobbins. These are small shuttles and bobbins for weaving designs. After the regular ground shuttle has gone through its shed, a special shed is formed through which the small shuttles (each bobbin carrying its own color) are passed. Swivel designs can be woven on a box or Jacquard loom.

Double-piece Pile Looms

On these looms two separate fabrics are woven simultaneously, face to face, and are kept at a uniform distance apart. They are joined by warp pile yarns, which are cut on the loom to form a pile on each fabric. The loom is similar to a conventional loom. The ground warp yarns for both fabrics are generally on the same beam. The pile ends are generally on one separate beam, usually located on the upper part of the loom. Rollers deliver the pile yarns in the amount required for the desired length of the pile.

Modern looms are equipped with two shuttles, passing through sheds at different levels simultaneously, each weaving its own piece.

Take-up rollers draw the fabrics apart. A sharp knife speeds across the loom, cutting the pile yarns in the center and producing two separate pile-woven fabrics. The two fabrics then travel to their respective take-up rolls and cloth rolls or are rolled on one cloth roll. These looms are used for such fabrics as velvet and plush.

Cut and uncut pile attachments. A plain, box, or Jacquard loom may have an attachment that forms uncut loops or cut pile on a fabric. There are two sets of warp yarns, one that weaves with the filling to form the ground weave, and the other that weaves the pile. When the pile is being woven the pile warp yarns are at the top and the ground warp at the bottom to form the shed. A wire, the size of which is determined by the size of the required loops, is inserted in the shed. The pile warp lowers and loops around the wire and is bound in by the filling. The wire is withdrawn, the ground weave continues for the required number of picks, and the

ire is again inserted to form another row of pile. If a cut pile is
esired, the wire has a small, sharp knife on one end, which cuts
ie loops as the wire is withdrawn.

Another method of weaving uncut loops is by forming the shed
s discussed above and then weaving several picks of filling. The
ile warp lowers, and loops are formed over the group of filling
arns. After the cloth is woven these filling yarns are withdrawn,
:aving uncut loops.

The Jacquard Loom

For weaving patterns of all types, a more complicated mech-
nism is attached to a regular loom. The hooks of the Jacquard
iachine raise and lower the individual warp yarns to form the
ied through which the filling yarn passes. Since there is one hook
ir each warp yarn, the control of the warp yarns is limited to the
umber of hooks with which the Jacquard machine is equipped.

The patterns are woven by means of Jacquard cards that look
ery much like player piano cards. In fact, the principle is very
iuch the same. The original design is drawn in the form of a
ketch, and then the design is charted on ruled pattern paper,
vhich is used as a guide for punching the holes in the Jacquard

Jacquard loom showing
ie overhead mechanism,
ie cards, the wires lead-
ig to the loom, and the
ack of the loom with the
warp beam.

cards. Each card represents one pick or one filling in the fabric.

The operator perforates the cards, as designated on the design paper, by inserting the unpunched card under a small machine with knobs which he quickly and skillfully "plays" to punch the holes. The weave takes as many cards as are required for one repeat of the pattern. For example, if the filling weaves in fifty times, (meaning fifty picks) before the pattern begins to repeat itself, fifty cards are needed. The cards are constantly traveling, and by the time the pattern is completed the first card is in position to repeat the operation.

The cards are laced together in a continuous chain and mounted on the machine. The Jacquard mechanism consists of an oblong perforated cylinder around which the cards rotate. An arrangement of metal wires coordinated with each Jacquard card raises the predetermined warp yarns to form the pattern.

Across the top of the mechanism, horizontal to the cylinder, are

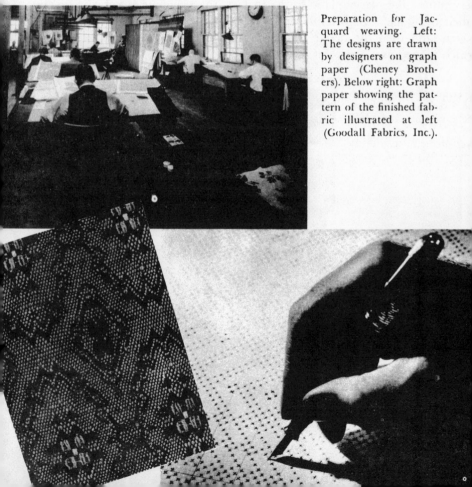

Preparation for Jacquard weaving. Left: The designs are drawn by designers on graph paper (Cheney Brothers). Below right: Graph paper showing the pattern of the finished fabric illustrated at left (Goodall Fabrics, Inc.).

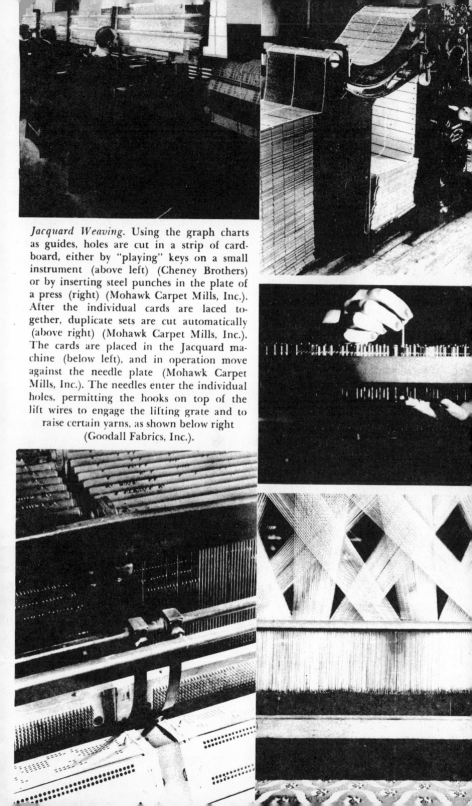

Jacquard Weaving. Using the graph charts as guides, holes are cut in a strip of cardboard, either by "playing" keys on a small instrument (above left) (Cheney Brothers) or by inserting steel punches in the plate of a press (right) (Mohawk Carpet Mills, Inc.). After the individual cards are laced together, duplicate sets are cut automatically (above right) (Mohawk Carpet Mills, Inc.). The cards are placed in the Jacquard machine (below left), and in operation move against the needle plate (Mohawk Carpet Mills, Inc.). The needles enter the individual holes, permitting the hooks on top of the lift wires to engage the lifting grate and to raise certain yarns, as shown below right (Goodall Fabrics, Inc.).

wires, called "needles." There are as many needles as there are hole positions on the card. Lacing through eyes in each of these needle are wires with hooks on each end. The bottoms of the wires are resting on a board. The hook shape of the tops of the wires make it possible to raise them by means of blades or knives operating up and down. Cords (or wires) fasten into the bottom hooks and connect the wires to heddles through the eyes of which pass the warp ends. In operation, one Jacquard card is in position on the cylinder, which presses it against the horizontal needles. Under pressure some needles enter the perforated holes. The needles that strike a blank spot where there is no hole are pushed back by the pressure of the cylinder, and the hooks connected to these needles are pressed back beyond the action of the rising blades. Since the hook are not raised, the cords attached to them and to the heddles, a well as the ends they control, remain *down* to form the lower part of the shed.

The needles that have entered the holes remain stationary or forward. Their hooks contact the knife and are raised. The hooks in turn, through the connecting cords, raised the heddles. The warp yarns laced through these heddles are also *raised,* forming the upper part of the shed through which the filling yarn passes to weave in one pick.

The cylinder, which rotates one quarter of a turn at each movement of the loom, carries the chain of cards, one at a time, against the needles and the particular needles that enter a hole in the card or are pushed back when pressed by the cylinder determine which warp ends form the top of the shed or the bottom.

Rug and Carpet Looms

Rugs and carpets may be woven on standard plain or Jacquard looms, ordinarily larger than those used for weaving other fabrics Wilton or Brussels and Axminster looms for rugs and carpets are different in principle.

Chenille loom. Carpets woven on this loom are the only ones that have the pile formed by the filling. In all others, warp yarn form the pile. Weaving chenille carpets involves two complete weaving operations. First, the chenille blanket is woven. This is cu into strips with protruding pile and is used in this form for the pile filling in the carpet. The weaving of chenille yarns is discussed in the yarn chapter.

Weaving chenilles. Left: Weaving the chenille blanket. Above: Cutting the fabric to form chenille yarn (Mohawk Carpet Mills, Inc.).

In weaving the blanket for rugs, the filling yarns are of various colors that will later form the patterned pile in the carpet. It has as many filling yarns as there are to be tufts in the width of the carpet. The strips of chenille are wound on cops and used as filling. In the second weaving, that of the carpet, five sets of yarns are used: the ground filling yarns; the chenille filling or pile yarns, on bobbins in the shuttles; the chain warp yarns; the stuffer yarns; and the catcher yarns on warp beams.

In weaving, the catcher yarns are raised to form the top of the shed. The chain warp yarns and the stuffer yarns are down. The chenille is shot through and is held by the catcher yarns. The loom is stopped, and the chenille fur is matched with the design of the previous shot, and the fur is combed.

The machine is started and four shots of filling yarn are woven in to secure the chenille yarn. Then the chenille is again woven in and the operation is repeated. A plain loom weaves chenille rugs.

Weaving the chenille rug with the chenille yarn used for the filling (Mohawk Carpet Mills, Inc.).

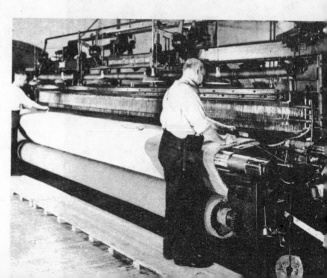

Axminster loom. The distinguishing feature of this loom is the endless chain of cylinders or "spools." In each cylinder is the yarn that makes one cut pile of tufts across the rug.

The colors for the tufts of pile yarns are selected and the fibers or yarns dyed. Each color of yarn is wound on a separate bobbin. The bobbins are placed according to the color arrangement in which they are to be woven in the rug. The yarns are then threaded through a reed and wound onto the long cylinders, which are placed in

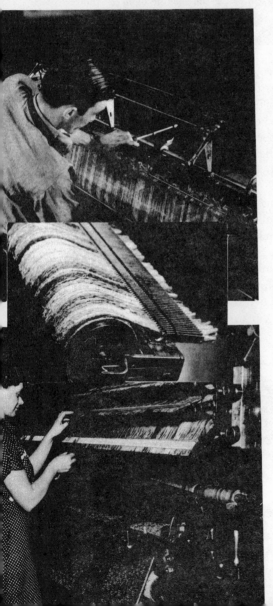

Weaving Axminsters. Top to bottom: The colorist selecting the colors to be used to dye the yarns. Threading the colored yarns from the bobbins through a reed to a cylinder. Threading the loom spool. A close-up showing the yarns wound on a spool and threaded through small metal guiding tubes to assure proper position when being woven in to form the pile (Mohawk Carpet Mills, Inc.). The front of the loom showing rows of cylinders and the pile yarn falling into place for weaving (Alexander Smith and Sons Carpet Company).

their proper position in the loom.

The ground warp is on a large beam, back of the loom, as are the stuffer yarns. In weaving, each row of tufts is bound in by the interlacing of the warp and filling yarns. The tufts of pile yarn are cut, the cylinder moves on, and the next cylinder, carrying its colored yarns, moves into place to deliver its row of tufts.

Wilton or Brussels loom. This is a Jacquard loom with colored pile warp yarns arranged on spools in frames back of the machine (not shown in the illustration). There is a maximum of six frames and each frame usually carries yarns of one color. However, two or more colors may be included in one frame, which makes it possible to use many colors and thus to weave reproductions of Oriental rugs and carpets.

The chain warp yarns are wound on beams back of the machine. Both the chain warp and the pile warp yarns are laced through heddles controlled by the Jacquard mechanism. The pile warp yarns weave over wires to form a pile. For a Wilton rug, a knife on the end of the wire cuts the loops as the wire is withdrawn.

Above: The Axminster loom (Mohawk Carpet Mills, Inc.).

Right: A Wilton or Brussels loom (Bigelow Sanford Carpet Co., Inc.).

327

A velvet or tapestry carpet loom (Mohawk Carpet Mills, Inc.).

Velvet or tapestry loom. These are large, plain, cam-driven looms with a single shuttle. The pile warp yarns and the ground warp and stuffer yarns are wound on beams. The pile warp weaves over pile wires. For velvet carpets and rugs, the pile loops are cut by a knife on the end of the wire as it is withdrawn.

The velvet loom in the figure below is weaving a patent-back rug. The warp yarns are on creels back of the loom. The fabrics are woven face to face, and the connecting pile warp is cut, forming a pile on each fabric. Patent-back frisé rugs are woven singly, the warp pile weaving over a pile wire that is withdrawn, leaving uncut loops.

A velvet carpet loom (Goodall Fabrics, Inc.).

Weaves cannot be classified strictly according to the final appearance of the fabric. Processes and finishes applied to some fabrics after weaving give them surface textures similar to those created by the weave in other fabrics. An example of this is velveteen and velvet, two fabrics woven quite differently but when finished both have a pile surface.

There are many ways in which the weaves of fabrics can be classified, because of the multitude of methods by which warp (or warps) and filling (or fillings) can be interlaced. A textile expert may classify them for the weaver in a way which would be too technical for the layman.

Giving due consideration to the methods of weaving, the constructions of the weaves, the combinations of weaves, and to the appearance of the finished fabric, the following seems to be one logical classification.

1. *Fabrics that are plain woven.* These include fabrics woven in the plain weave, the rib weave, and the basket weave.

2. *Fabrics woven with floating yarns.* These include fabrics woven with one warp and one filling; two warps and one filling; two fillings and one warp; or two or more warps and two or more fillings. Warp, or filling yarns, or both may float.

3. *Fabrics with woven-in pile.* These include fabrics with an extra warp. The pile, cut or uncut, is obtained as the fabric is woven.

4. *Backed fabrics.* These include fabrics with an extra warp for warp backed fabrics; an extra filling for filling backed fabrics; and extra warp and filling for double woven fabrics.

5. *Fabrics with a lacy weave.* These include fabrics woven with one or more warps and fillings, woven so that the filling, after each shot through the shed, is "locked in" by the crossing of the warp yarns.

6. *Fabrics with Jacquard patterns.* These include fabrics woven on a Jacquard machine, flat, raised, and pile (cut or uncut) fabrics.

FABRICS THAT ARE PLAIN WOVEN

The *plain weave* is constructed with but one warp and one filling. The *rib weave* has a number of warp yarns, or a number of filling yarns, weaving as one yarn to form a rib. The *basket weave* has two or more warp yarns and two or more filling yarns. Both the rib and basket weaves weave over and under at right angles to the warp, the same as does the plain weave.

These are the only weaves that are constructed with only two harnesses or if with more than two harnesses, they operate as two would. All other weaves require three or more harnesses operating in different successions. Whether single or groups of warp and filling yarns are utilized, they weave as one warp or one filling.

Plain Weave

This is the simplest of all weaves. Each filling yarn passes alternately over and under one warp yarn, and each warp yarn passes alternately over and under one filling yarn.

The plain weave can be woven with but two harnesses, in which case all the even number warps are on one harness and the odd number warps are on the other. One harness rises, forming a shed through which the shuttle, containing the filling yarn, passes. This harness then lowers and the second harness rises, forming another shed for the returning shuttle. The operation then repeats itself. After each passing of the shuttle through the shed, weaving one pick of the fabric, the reed beats the filling into the cloth.

PLAIN WEAVES.

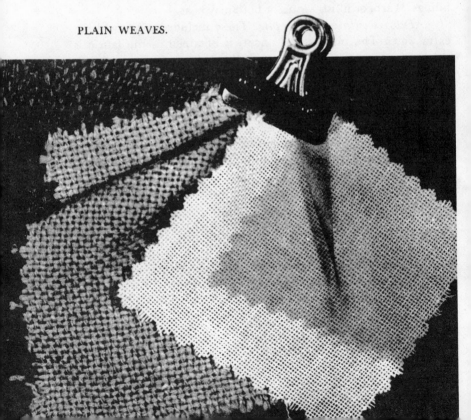

BASKET WEAVES.

Every other pick is the same. The third, fifth, and so forth are the same as the first; and the fourth, sixth, and so forth are the same as the second.

The two-harness plain weave is the simplest to explain. However, it is not the number of harnesses that determines the plain woven fabrics. Rather it is determined by the fact that all warp yarns are equally divided in the shed—there are the same number up as there are down. There may be too many ends in a fabric for two harnesses to carry and two additional harnesses may be required. In that case, two operate simultaneously to form the upper part of the shed and the other two form the lower part.

Variations of the Plain Weave

There are two other weaves made with the warp yarns equally divided and are therefore considered variations of the plain weave. They are the basket and the rib weaves. They also may utilize more than two harnesses, but the explanation is simplified by assuming that only two harnesses are used.

Basket Weave

A basket weave is made up of two or more filling yarns passing over and under two or more warp yarns. Each heddle in the harnesses carries two or more yarns (instead of one as in the plain weave). As the harnesses are raised and lowered to form the shed, two or more successive picks are shot through the same shed, weaving in two or more filling yarns.

In a two-by-two basket weave (utilizing two harnesses) the first heddle in the first harness carries yarns one and two; the next, five and six; the third, nine and ten; and so on. In the second harness, the first heddle carries yarns three and four; the second heddle, seven and eight; the third, eleven and twelve; and so on.

The basket weave may be three by three, or four by four, or many other balanced arrangements. If it is three by three, the first harness carries ends one, two, three and seven, eight, nine, and so forth. The other harness carries four, five, six, and ten, eleven, twelve, and so forth; and three shuttles shoot through the shed each time. There is an unbalanced basket weave, such as four warps to two fillings, made with four yarns laced through each heddle and two shuttles shooting through each shed.

RIB WEAVES.

Because two or more yarns are woven at one time, the fabric is usually looser and less firm than a plain weave fabric. However, fine yarns woven in a close basket weave may produce a firmer and softer fabric than the same yarns woven in a plain weave. There are other types of basket weave, such as a twill basket weave, utilizing two yarns woven as one in a twill weave. This is a fabric woven with a floating yarn and is not classed as a plain woven fabric.

Very attractive fabrics may be woven in the plain basket weave by the use of colored yarns. For instance, colored warp yarns, crossed with one colored and one white filling yarn produce an appearance of a fancy twill weave. Examples of fabrics woven in a simple basket weave are hopsacking, monk's cloth, and some blanket cloth.

Rib Weave

Ribs may be woven lengthwise or crosswise of the fabric. The lengthwise ribs, in the direction of the warp, are formed by the *filling* yarns passing alternately over and under a group of warp yarns. This is called a *filling rib weave*.

The crosswise ribs, in the direction of the filling, are formed by the *warp* yarns passing alternately over and under a group of filling yarns. This is called a *warp rib weave*. In this weave, the shed remains stationary while the required fillings are woven in. It may be two, three, four, or more times that the shuttle passes through the same shed. Just as in the plain weave, the third, fifth, seventh, and so forth, rows are like the first; and the second, fourth, sixth, and so forth, are like the second.

Many fabrics may have ribs running lengthwise with the warp, or crosswise with the filling, which are formed by weaving a thicker warp yarn or a thicker filling yarn at intervals into the fabric. However, this is a plain weave and is not a rib weave. Thicker yarns may be used in any weave to give different surface appearances. A rib weave is a type of weaving and is not determined by the kind of yarns used.

While it may seem that rib weaves are woven with floating yarns, they are not so constructed, as the group of warp yarns weave as one, usually laced through one heddle; and the group of filling yarns weave as one, all in the same shed. These rib weaves are firm and strong and are used effectively for colored

333

stripes. Examples of warp rib weave (in the direction of the filling) are faille and poplin. An example of a filling rib weave (in the direction of the warp) is rep.

FABRICS WOVEN WITH FLOATING YARNS

The classification of fabrics woven with floating yarns includes a wide range of weaves.

Strictly speaking, it could probably be said that in the rib and basket weaves the yarns also float over two or more adjacent yarns. However, since these as well as the plain weave utilize but two harnesses (or if more, they operate as two), and each one, two, or more yarns weave as one, they have been classified as plain woven.

They may have but one warp and one filling. In the *twill weave* the filling yarns float over and under the warp yarns in a regular variation to form diagonal lines. In the *satin weave* the filling yarns float over and under the warp yarns at widely spaced but regular interlacings that do not form distinct diagonal lines. In the *Bedford cord* weave the filling yarns float over and under a different number of warp yarns so they form lengthwise ribs. This weave has but one warp and one filling unless extra warp yarns are used as stuffers for the ribs. In the *huckaback weave* the warp yarns float, producing lengthwise ridges. In the *honeycomb weave* both warp and filling yarns float, producing raised squares or oblongs. In the *bird's-eye weave* the filling yarns float, producing diamond-shaped designs. In the *crepe weave,* used to produce with uncreped yarns fabrics that imitate those woven with creped yarns, there is one warp and one filling. There are many methods of producing this weave. Usually, an adaptation of the satin weave is employed. The floats are short and even and the fabric is so woven that no twill or stripe lines appear. Or it may be a combination of a plain weave alternating on each pick with a satin weave that has floating yarns.

They may have two warps and one filling. The *lappet weave* is an example of extra warp yarns that float on the back and rise to the face of the fabric to form a pattern on the surface. In some fabrics such as those used for draperies, these yarns may not be cut, but may become a part of the back pattern of the cloth. Or they may be cut, to form the face pattern only. Another method used for this weave is to float the extra warp yarns on the face of the fabric, to produce a pattern, and not float them on the back of the fabric.

334

TWILL WEAVES.

They may have two fillings and one warp. The swivel weave is
an example of this. Or an extra filling yarn may float on the back of
the fabric and be raised to the face of the fabric to form the pattern.
The floating yarns may become a part of the fabric or they may be
cut, only a clipped spot remaining on the surface. Rather elaborate
patterns can be woven by using extra filling yarns.

While Jacquard patterns could rightly be classified with this
group, because they are woven with floating yarns, they have been
made a separate classification, because the Jacquard machine can
be used for weaving patterned pile fabrics as well as patterned
flat fabrics.

Twill Weave

The diagonal lines, typical of all twill weaves, are produced by
a series of floats. The type of twill is determined by the intervals
at which the warp and filling are intersected. When there are an
equal amount of warp and filling on the face of the fabric it is a
balanced or equal twill. If there are more filling yarns on the face
the twill is called a *filling face twill*.

335

If there are more warp yarns on the face, the twill is called a *warp face twill*. The twill may be very steep, medium steep, or reclining.

Twill weaves are made with more than two sets of harnesses. The simplest, a 45-degree twill, uses three harnesses. The first harness carries warp yarns one, four, seven, and so forth, the second carries two, five, eight, and so forth, and the third three, six, nine, and so forth. As the first harness is raised, it forms the shed for the first filling yarn to shoot through, then the second harness rises to form the next shed; then the third harness lifts to form the third shed. The first harness starts again, repeating the process. When one harness is up the other two are down.

The first filling passes under one end and over two; under one and over two; and so forth. The second filling passes over one; under one, over two; under one and over two; and so forth. The third filling passes over two; under one and over two, under one and over two; and so forth. The fourth filling yarn starts repeating the weave. The most used is a two up and two down, 45-degree twill. This is a balanced twill. It uses four harnesses of which two are up when two are down. The line that forms the twill is advanced one filling on each row. After four rows of weaving the filling yarn, the operation repeats itself.

TWILL WEAVES.

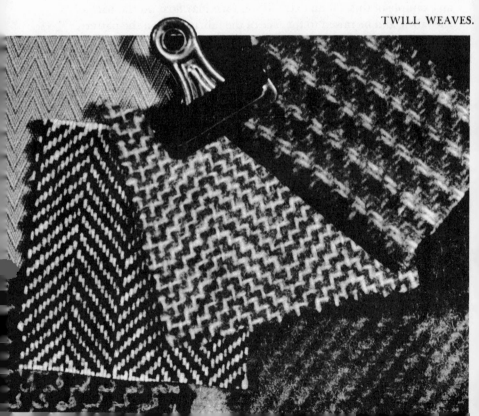

Steep twills are made by advancing more than one filling in each row. For example:

63 degree twill advances the twill two picks for each end,
70 degree twill advances the twill three picks for each end,
75 degree twill advances the twill four picks for each end.

Reclining twills are made when the filling yarn skips warp yarns. For example in:

27 degree twills the filling holds over for two ends,
20 degree twills the filling holds over for three ends,
15 degree twills the filling holds over for four ends.

Variations of the Twill Weave

There are many variations of the twill weave that form fancy patterns in the fabric.

The herringbone twill weave is formed by reversing the twill at intervals. The right-hand twill is woven for a certain number of picks, then the direction is reversed and a left-hand twill is woven for a certain number of picks, then back to the right-hand twill, and so on. At each reversal (turn of the twill) a point is formed.

The broken twill weave is also formed by reversing right-hand and left-hand twills, but instead of being joined by a point there is a space. Also, the number of warp yarns in the right- and left-hand twills may not be the same as they are in the herringbone weave. Broken twills can form many unusual patterns.

The zigzag twill weave is another reverse twill weave. It is similar to the herringbone twill weave in that the right- and left-hand twills are joined; but they are unequal in length. Thus, they form a zigzag or wavy effect.

There are other fancy twill weaves, such as twills that seem to form a rib, twills that form a curve, twills that weave two or more warp or filling yarns as one, or twills that are combinations of two types of twills.

The Satin Weave

In the satin weave, the interlacing of the filling yarns is at widely spaced but regular intervals. If more warp yarns show on the surface, it is called a *warp face* or *satin weave;* if more filling yarns show on the surface, it is called a *filling face* or *sateen weave.*

The regular progression in which the filling interlaces the warp is determined by taking the required number of harnesses and dividing them into two parts. These parts must not be equal, or divisible by the same number, nor can one be the multiple of the other. For example, a five-harness satin weave is repeated on five picks (two and three equal five, but two and three are not equal, nor are they multiples of each other, nor can they be divided by the same number). Therefore, starting the picks with end one, the sequence is one, three, five, two, four. The sequence of yarns in the repeat of the weave for some regular satins are:

Harnesses	Number of Repeats	Sequence of Stitches
5	2 and 3	1, 3, 5, 2, 4
7	2 and 5, 3 and 4	1, 3, 5, 7, 2, 4, 6
8	3 and 5	1, 4, 7, 2, 5, 8, 3, 6

Irregular satin weaves are formed with four and six harnesses, because any two numbers that compose four and six are divisible by a common number. Thus for four harnesses the counters are two and two; and for six harnesses they are two and four, three and three. Two is divisible by two, four is divisible by two, and three is divisible by three. A four-harness weave interlaces at one, three, two, four, and a six-harness, at one, three, five, two, six, four. In the four-harness weave, the first end is over the first pick; the third end

SATIN WEAVES.

is over the second pick, and the second over the third pick, and the fourth over the fourth pick. This produces the scattered interlacing.

Satin weaves are used for many fabrics. Silks, rayons, and wools are woven with the warp satin weave to produce lustrous fabrics. Wool satin weave fabrics are often napped. Cotton fabrics are more frequently woven with the filling face or sateen weave.

Bedford Cord Weave

The cords in this weave run lengthwise in the fabric, in the direction of the warp. In appearance it resembles the filling rib weave (a variation of the plain weave), but is constructed differently. In the Bedford cord weave the filling yarns (every other pick) float under and over certain warp ends to form the rib.

The first pick weaves in a plain, twill, or other weave; the second pick floats under four (or eight or sixteen) warp ends, over two warp ends, again over the group of ends, and so forth. The third pick weaves the same as the first, the fourth pick weaves the same as the second, and so forth.

Stuffing yarns may be used between the floating filling yarns and the face warp. They serve to raise the cord and do not appear on the face of the fabric. They are an extra warp of thick yarns, laced in the loom in the sections that form the rib. Since the stuffer yarns may or may not be used, this weave is not considered a weave with extra warp yarns.

BEDFORD CORD WEAVE.

HONEYCOMB, HUCKABACK, BIRD'S-EYE WEAVES.

Piqué Weave

This is similar to a warp rib weave, as the ribs run in the direction of the filling. However, it is made quite differently, and the ribs are given a raised, corded effect. The construction of the piqué weave is similar to that of the Bedford cord weave. In the latter, the extra filling yarns float under the warp ends to form a lengthwise rib, while in the piqué weave the extra warp ends float under the filling yarns to form a crosswise rib.

The simplest piqué weave has two warps and one filling. The face warp weaves with the filling, forming the face of the fabric. The extra warp ends float under a certain number of filling yarns to form the rib, then float over two picks and again under the designated number of filling yarns. If wider ribs are desired, the extra warp ends float over more filling yarns. There may be an extra filling stuffer yarn, which is not visible on the face or the back of the fabric but lies between the face and back warps and is not woven in. These stuffer yarns may be added on every other pick. Most of the fabrics shown today as piqué are really woven with the Bedford cord weave, with a fine lengthwise rib.

Honeycomb Weave

This weave, also called "waffle weave," is constructed with warp and filling yarns floating unequal distances. Ridges are formed along the longest floats, producing square or oblong-shaped designs on both the face and the back of the fabric. The vertical lines are formed by the floating warp yarns, and horizontal lines are formed by the floating filling yarns.

Huckaback Weave

This weave is constructed with warp yarns floating on the face of the fabric. A typical huckaback weave may have in the repeat of the weave, the first and second warp yarns floating over five filling yarns and, in the following five picks, the fifth and sixth warp yarns floating over five filling yarns. The remainder of the fabric is woven with a plain weave. The floating yarns produce short, intermittent, lengthwise ridges.

Bird's-eye Weave

This weave is constructed with filling yarns floating on the

urface in such a manner that they produce small diamond-shaped
esigns. The back of the fabric shows the warp yarns over which
he filling yarns float.

wivel Weave

This weave constructs patterns by the introduction of extra
lling yarns. In addition to the shuttles carrying the ground filling
arns, there are small shuttles that carry the extra filling yarns.
The ground filling shuttle is passed through the shed; then, a
econd shed is formed for the small shuttle that carries the pattern
arn. Each swivel shuttle, traveling in a circular motion, covers·
nly a certain width of the fabric. The design is made to suit the
wivel attachment, the limitations of which are taken into con-
ideration by the designer. The swivel shuttles may carry bobbins
vith different colored yarns, although usually no more than two
olors are introduced. Dotted Swiss may have the dot woven in
y this method.

The swivel weave can be distinguished from patterns formed
y floating yarns and woven with regular bobbins carrying different
olored yarns. Examining the back of the fabric it may be noted
hat the swivel weave has but one yarn, which travels back and
orth and is cut at the beginning and at the end. When unweaving
t, there is one continuous yarn. A pattern formed with floating

Top row: Both floating filling yarns. Bottom, left to right: Swivel,
lappet, swivel.

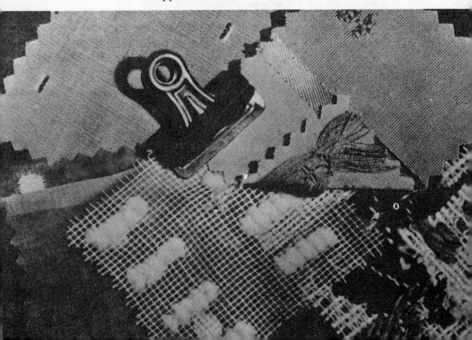

yarns has two or more filling yarns, all of which float on the back of the fabric or are cut.

Crepe Weaves

This is an interesting group of weaves that give the fabric a pebbly, creped surface. True crepe is woven with highly twisted yarns, alternately with a yarn of right-hand twist and then a yarn of left-hand twist, or vice versa. The crepe yarn may be utilized in the filling only, in the warp only, or in both.

A crepe weave simulates true crepe through the weave formation. The fabric is usually soft and spongy, often almost lacy in character. There are numerous ways in which such fabrics are woven. The principle is to have the floating yarns so placed that the fabric shows no twill, no striped effect, or no long floats. The weave may be a variation of the plain or satin weave. It is frequently a combination of these two weaves, as, for example, odd-number warp yarns weaving plain with odd-number filling yarns, and even- number warp yarns weaving a satin weave (with floating yarns) with the even-number filling yarns. Or it may be two weaves, one floating on top of the other.

In this weave construction, the floats are short and of about the same size. A common crepe weave has only two floats, in both warp and filling. There are other methods of producing such mossy, crepy effects, but most of them have the general principle described above.

Corduroy and Velveteen Weaves

Heretofore no special weaves have been set up for these two fabrics. They are classified here as two specific types of weaving

CREPE WEAVES.

because they are made by definite methods and no other weaves are woven in the same way. While both fabrics have a pile surface when finished, both leave the loom as flat fabrics, and the floating filling yarns are then cut to form the pile.

Corduroy weave. The ground filling yarns weave in a plain weave with the warp yarns to form the ground fabric. The pile filling yarns interlace with one, two, or three warp yarns, then float over three or more warp yarns to again interlace with one, two, or three warp yarns followed by floating over warp yarns, and so forth.

After the fabric is woven, the floating filling yarns are cut midway between the interlacings. The fabrics are then finished so that the pile forms cords or ridges. The width and depth of the pile depend on the length of the floating yarns—the shorter they are, the finer the cord. The more rounded cords are formed by uneven lengths of floating yarns.

Velveteen weave. This is similar to the corduroy weave. The ground weave may be plain, twill, or satin. The pile filling does not float uniformly as in corduroy but the floats are scattered. However, they are uniform as required for later cutting.

Velveteen also leaves the loom as a flat fabric. It is then taken to the cutting machine and the floating filling yarns are cut. Since the floats are scattered, a cut pile is formed over the entire surface of the fabric. The length and number of floats determine the depth and closeness of the pile.

LENO WEAVE

This weave does not fit into any of the numerous classifications. It creates a lacy effect and has the appearance of a figure eight. If the filling yarns are drawn out of a fabric made with this weave, the figure eight formations of the warp yarns are very evident.

CORDUROY WEAVE. Top to bottom, left: Flat woven, after the yarns are cut, after scouring. Right: After brushing, after singeing, the finished fabric (The Crompton-Richmond Company).

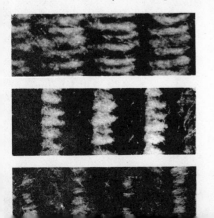
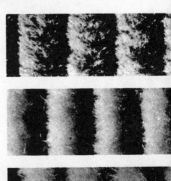

This is a strong durable weave, as the crossing of the warp yarns "lock" in the filling yarns. It is the weave employed for marquisette. Frequently patterns in this weave are woven with the use of the dobby attachment. Leno weave is also combined with other weaves to produce fabrics of unusual surface texture and appearance. (See page 319 for explanation of leno device).

FABRICS WITH WOVEN-IN PILE

All the weaves described so far have had warp and filling yarns lying flat, in longitudinal and transverse parallel lines. The weaves discussed in this section have a third set of yarns that stand upright. For this reason some authorities call them "three-dimensional weaves." They include such fabrics as velvet, frisé, plush terry or turkish toweling, and tapestry, velvet, and Axminster rugs and carpets. The pile, which is formed with pile warp yarns, may be cut or may remain in uncut loops.

There are four methods for weaving these fabrics. One is produced by weaving two fabrics simultaneously, face to face, with pile warp yarns that connect the two. As the fabric is being woven, a knife cuts the connecting yarns which form the pile on both fabrics. Velvet with a cut pile is woven in this manner and the weave, therefore, is frequently called the "velvet weave." Another fabric is produced by weaving-in pile warp yarns over a wire or

LENO WEAVES. All of these fabrics are woven on a loom with a leno device.

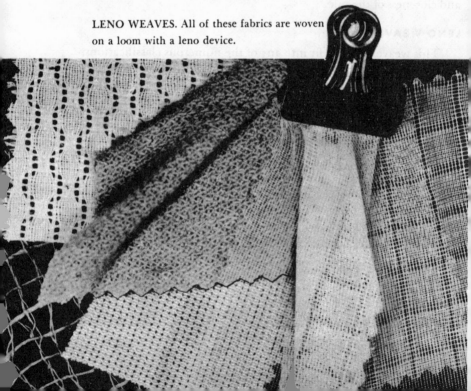

PILE WEAVES. Left to right: Cut velvet, plush, frisé.

over extra filling yarns to form the uncut loops. The wires may
have knives which, when withdrawing, cut the loops, creating a
cut pile. Tapestry and velvet carpets and rugs and frisé, as well as
some velvets, are woven by this method. A third method also has
extra warp yarns that form the uncut pile, but it is constructed
differently from those discussed above. This is called the "terry
weave." As the fabric is woven, certain filling yarns are held back
and are not beaten in by the reed. With the weaving of the next
pick, all are beaten in. The pile warp yarns, weaving over and
under the back filling yarns, are also beaten in, and form loops.
The pile loops may be on the back only, on the face only, or on
both sides. The fourth method fastens in tufts of pile yarns during
the weaving of the warp and filling yarns. This is the way Axminster carpets and rugs are woven and is called "the Axminster weave."

The Pile Woven-in Between Two Fabrics

This fabric is woven on a so-called "velvet loom." Two fabrics
are woven at the same time, face to face. There are two sets of
warp yarns and two sets of filling yarns. The pile warp is on a
third beam. Weaving is simultaneous. Two sheds are formed

345

and one shuttle passes through each shed, weaving separate fabrics.

The ground weave may be plain, satin, twill, or rib. As the two fabrics are woven, the pile warp is woven into each fabric. There are two ways in which the pile warp is interlaced with the ground weaves. One, called the "V pile," weaves with one pick, the other, called the "W pile," weaves with three picks.

After the pile interlaces, a knife shoots across the loom, cutting the pile yarns in the center, producing a cut pile on the surface of both fabrics.

The Pile Woven Over Wires or Yarns

This fabric is woven on a plain or Jacquard loom that usually has a long wire or rod attached, over which the warp pile weaves. There is but one fabric woven, requiring one warp and one filling for the ground weave.

As the fabric is woven, the pile warp is raised to the top of the shed, the wire is inserted and the pile warp drops to the bottom of the shed to weave in with the filling yarns. Usually half of the pile warp is raised at one time. In this way, the pile warp forms the loops. To make a cut pile, a knife attached to the end of the wire cuts the loops as the wire is withdrawn. If both cut and uncut pile is desired, the knife cuts only a portion of the loops.

Another method of weaving uncut pile loops is to use extra filling yarns in place of the wire. The ground weaving is the same. Then a shed is formed and several picks of filling are woven in. The pile warp lowers, forming loops over the filling yarns.

TERRY WEAVE. Left to right: Pile on one side only, pile on one side with stripes of pile on other side, pile on both sides.

CORDUROY AND VELVETEEN WEAVES. Left to right: Corduroy, hollow-cut velveteen, velveteen.

After the fabric is woven, the filling yarns are withdrawn, leaving uncut loops on the face of the fabric.

Weaving Tapestry and Velvet Carpets and Rugs

Tapestry and velvet carpets and rugs are woven the same way, except that tapestry has an uncut pile and velvet has a cut pile. Until recently, tapestry and velvet rugs and carpets were patterned, but now mostly plain colored velvet rugs are woven.

The warp yarns are printed prior to weaving. The complete design is printed on the yarns, which are wound on a large drum. The explanation of this type of printing is given on page 523. After the yarns are printed, they are wound on numbered bobbins which are then arranged according to the pattern and are wound on a warp beam. The original printed pattern is more elongated than the pattern in the finished carpet. This is necessary because, in the weaving, the pile reduces the length of the pattern.

Four sets of yarns are used: the filling; the warp on one beam, laced through two harnesses, thence through the reed to the front roll; the stuffer yarns on one beam, laced through one harness, thence through the reed to the front roll; and the pile yarns on one beam, laced through a harness, the reed to the front rol

As the loom operates, the pile yarns rise to form the top of th upper shed through which the wire is inserted to form the pil loops. Below the wire are one set of warp yarns and the stuffe yarns, forming the bottom of the upper shed and the top of th lower shed. The lower shed through which the filling yarn passe is formed by the set of warp yarns that are down while the othe set is up, simultaneously weaving with the stuffer yarn.

For velvet carpets and rugs the pile is cut. On the end of th wire, over which the pile warp yarns are woven, is a knife that cu the pile as the wire is withdrawn.

The Pile Woven Over Slack Filling Yarns

This is called the "terry weave." The warp is woven in quit differently from the methods already discussed. One warp weave with the filling to form the underweave. The other warp, th pile, is not actually woven into the fabric but is anchored betwee the warp and filling in the underweave.

The loops are usually on both sides of the fabric, but may b on one side only. Some may be in stripes, with pile loops on bot sides in one stripe and on only one side in the next stripe. Som loops are single, others are double or triple, the latter having tw or three pile warps that interlock as one. The pile yarns, bearin little strain in weaving, are usually softer and more loosely twiste than the ground warp yarns.

The ground weave may be a plain or twill. The simplest terr weave is woven on four harnesses, two for the ground warp an two for the pile warp and two sheds are formed. Through one she the pile filling yarn weaves and through the other the groun filling yarn weaves. Two picks of filling are shot through the pil shed and are not beaten in but are held back a short distanc from the cloth. Then, on the third pick, the filling weaves in th ground shed with the ground warp and the reeds beat in th three picks forming loops of the pile warp. The length of the loo is determined by the distance the filling yarns are left back from the fell of the cloth and the length of warp yarn let off when th third pick is beaten in.

Top to bottom, front (left) and back (right): Chenille, Wilton, Brussels velvet, tapestry, Axminster (Mohawk Carpet Mills, Inc.).

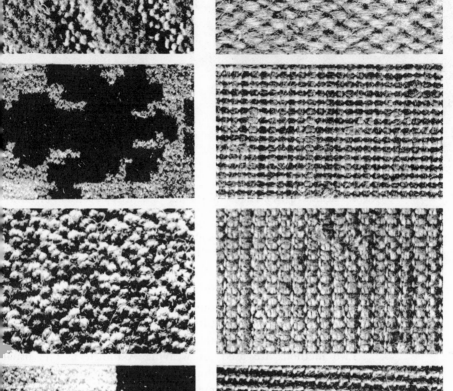

Right: Chenille rug. This shows the chenille yarn (left) prepared by first weaving and cutting a chenille blanket; the heavy chain warp yarns, stuffer yarns, catcher yarns, and the ground catcher yarns. Left: Wilton or Brussels rugs. This shows the filling, the ground warp, the stuffer warp, and the pile warp yarns.

Weaving Axminster Carpets and Rugs

While an Axminster has somewhat the appearance of a Wilton, it is woven by inserting cut tufts of yarn rather than by weaving a loop that is cut to form the pile.

Two methods are used for weaving Axminsters. The usual method is to arrange the colored yarns on long spools. This is accomplished in a setting frame. The operator, guided by the pattern, places on the first spindle the bobbin carrying the color for the first tuft, on the next spindle the bobbin for the color of the second tuft, and so on, until the first row of tufts is completed. The yarns from these bobbins are then passed through a reed and are wound on a 27- to 36-inch spool. Similarly, one spool is made for each row of tufts that form the pattern. The number of spools depends on how many are needed for the complete pattern before it begins to repeat itself. The yarn on the wide spools then is threaded through a series of tubes, one tube for each warp yarn in the width of the fabric.

The tubes, containing the spools, are placed in rows in an endless chain on the loom. As the carpet is woven, one row of the spools falls into place in the warp, and delivers its row of pile yarns that are to form the tufts. As the filling yarn passes through the shed, weaving over and under the warp yarns, it fastens in the tufts. The pile tufts are cut off the yarn and the row of spools passes

350

on, permitting the next row to fall into place. The rows of cylinders carrying the spools of yarn rotate and fall into place whenever the repeat of the pattern calls for each particular row of tufts.

JACQUARD PATTERNED WEAVES

While the forming of patterns has been discussed in the section on weaving with floating yarns, the Jacquard patterns were not included there even though yarns are floated to produce the pattern. The reason for the omission is that the process of weaving and the appearances of the fabrics are quite different. Pile yarns may be woven in during the pattern weaving; the fabric may have one set of warp yarns and one set of filling yarns or more than one set of warp yarns. Jacquard weaving may be a single weave or a combination of weaves.

The fabric may be flat, as damask, or raised, as brocatelle, or

JACQUARD WEAVES.

it may have a cut pile as Wilton rugs or patterned velvets, or an uncut pile, as Brussels rugs; or both cut and uncut pile, as patterned frisé. If a pile weave, the pile yarns (extra warp yarns) may be woven over a wire, withdrawn without cutting for an uncut pile, or with a blade attached that cuts the pile for cut pile. A different surface texture is given the pattern when only part of the loops are cut, thus forming patterns by combining cut and uncut pile.

The discussion of the Jacquard loom (see page 321) explains how the Jacquard patterns are woven.

Weaving Brussels and Wilton Carpets and Rugs

Brussels and Wilton carpets and rugs are woven similarly except that Brussels has an uncut pile and Wilton has a cut pile. They are woven on a Jacquard loom, with a long wire over which the pile yarn loops as it weaves. After weaving the wire is withdrawn; the loops may remain uncut or may be cut.

The weaving involves the use of four sets of yarns: the filling, the ground warp yarns, the stuffer yarns (extra warp yarns to add body and weight), and the pile yarns. Two sheds are formed during the process of weaving. The filling yarns, in bobbins, weave back and forth to form the ground weave. The warp yarns on one beam are laced through two harnesses, which rise and fall to form one shed for the passing of the filling yarn, and then through the reed, in the usual way, to the front roll. The stuffer yarns on a second beam are laced through one harness, through the reed to the front roll, and, being practically in the center, are interwoven with both the pile and the ground weave.

The pile yarns are on bobbins, which are placed on frames or creels. There are as many bobbins as there are pile ends in the width of the carpet. There are from two to six frames, and, usually, all the bobbins on one frame carry yarn of one color. These yarns are carried to the looms, are laced through the harness and reed, and, thence, to the front roll, ready for weaving. The pile yarns are controlled by the Jacquard mechanism, being raised and lowered to form the pattern.

Assuming there are five frames of pile yarns, the carpet will have five pile yarns, two warp yarns, and one stuffer yarn. When the loom is running, the Jacquard mechanism raises one of the pile yarns in each reed. This forms the upper shed through which the wire is introduced to form the pile.

RAYON FABRICS
WOVEN WITH DYED FILAMENTS

This illustration shows how the simplest kind of weave can be used to produce attractive, distinctive fabrics through the combination of colored filaments or yarns, that were dyed before the fabrics were woven. Colored illustrations, elsewhere in this book, show the beauty of printed rayon fabrics.

These fabrics were woven with the plain, rib, or twill weave. On the left, the third fabric from the top had the colored stripes woven in a satin weave.

The stripes, checks, and plaids were produced by the sequence of interweaving colored yarns, or filaments, in both warp and filling. The two bottom fabrics had different colored warp and filling yarns which, when woven together, produced fabrics with changeable colors.

Other illustrations of yarn dyed fabrics are as follows: wool and spun rayon fabrics facing page 257; cotton and linen fabrics facing page 481.

Under this wire, forming the lower part of the upper shed, are the other four pile yarns, the stuffer yarn, and one warp yarn, being woven into the body of the carpet. Simultaneously with the forming of the upper shed, the lower shed has been formed by the raising of one warp harness, leaving one set of warp yarns to form the lower part of the lower shed through which shoots the shuttle carrying the filling yarn. The warp and filling then weave in two or three rows to secure the pile, and then the weaving process repeats itself.

For Wilton carpets and rugs, the machine has a series of knives running parallel to the length of the fabric. As the wire is withdrawn, the knives cut the loops to give a velvety pile texture.

BACKED FABRICS

These fabrics utilize extra warp yarns, extra filling yarns, or both extra warp and filling yarns. Extra warp yarns produce *warp backed fabrics* and extra filling yarns produce *filling backed fabrics*. Reversible fabrics may have two or more sets of warp or two or more sets of filling yarns to one of the other series. Two sets of warp and filling produce *double woven fabrics*. The weave and appearance of backed fabrics, utilizing extra yarns, are quite different from fabrics that are produced by floating extra warp and filling yarns.

The filling or warp backed fabrics are essentially single fabrics made with

BACKED FABRICS. Top to bottom: Double woven, filling backed, warp backed, double fabric.

the extra yarns woven in the face weave, to give the fabric more weight, or more warmth retaining qualities, or a reversible appearance. Double woven fabrics are two separate and distinct fabrics that may be woven in two different ways.

Filling Backed Weave

A filling backed fabric has an extra set of filling yarns. One is the filling for the face of the fabric and the other is the filling for the back of the fabric. The face filling weaves with the warp yarns to form the face of the fabric. This can be any weave but is usually a twill or satin weave, and is woven in the usual way.

The back filling weaves with the warp through a shed that has all of the warp yarns at the top except those few that are to bind in the back filling. These latter form the bottom of the shed. In this way, the back filling is not exposed on the face of the fabric. On the next pick, when the face filling is weaving with the warp to form the face of the fabric, the same warp yarns that were down in the last shed must remain down in this shed to bind the back filling securely.

In weaving, the filling may weave one face pick followed by one back pick; or two or three face picks followed by one back pick; or two face picks followed by two back picks. No stripes, checks, or plaids can be woven into filling backed fabrics as the back filling yarns cover the back of the fabric.

Warp Backed Weave

A warp backed fabric has an extra set of warp yarns. There are usually two warp beams. On one is the warp for the face of the fabric, on the other, the warp for the back of the fabric.

The face warp may weave with the filling in one of many types of weaves, such as twill, fancy twill, satin, rib, or other, to form the face of the fabric.

The back warp weaves at intervals with the filling, to bind in this warp. Stripes or checks can be produced, as the back warp can construct woven-in stripes the same as can the face warp. Warp backed fabrics are frequently called "French back" fabrics.

Double Woven Fabrics

This is not really an individual weave, unless the joining together of two separately woven fabrics can be considered a distinct weave. There are two sets of warps and two sets of fillings that are woven independently but simultaneously on the same loom. They may be fastened together during the weaving process by the yarns of one interweaving, at intervals, with the yarns of the other; or they may be fastened together by an additional yarn. warp or filling. Two separate fabrics, attached at both sides, are not double woven.

The two fabrics are woven at one time, one above the other. All the face warp is above the back filling, so that the two will not weave together. At the intervals, when the two fabrics are to be interwoven, the face warp drops down under the back pick and is interlaced with it. Or the back warp may rise over a face pick and interlace with it, likewise binding the two fabrics together.

Double woven construction is usually used for heavier fabrics. Since there are two separate fabrics, there can be many combinations. One may be plain and the other plaid, making an interesting reversible cloth, or both may be the same, or the face fabric may be of better quality yarns to give warmth, and the back of less expensive yarns to lower the price of the fabric. Some may have a fine, closely woven face fabric and a coarser textured back fabric. Usually heavier yarns are used, but at times this construction is applied to finely woven face fabrics, to give additional warmth.

All of these fabrics are woven on a loom
with a dobby attachment.

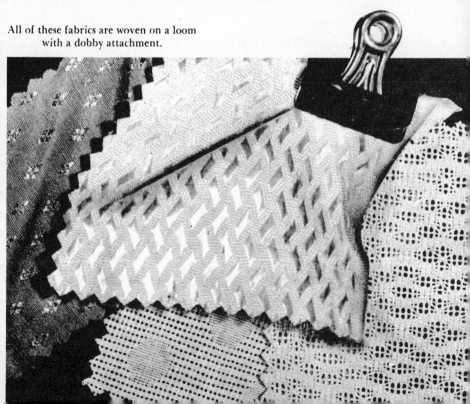

SUMMARY

TO PRODUCE WOVEN FABRICS

Warp: The yarns that run lengthwise in a woven fabric.

Filling (or weft) : The yarns that run crosswise of the fabric and "fill" the fabric.

Weaving: The interlacing of the warp and filling yarns to construct a fabric.

Types of looms: Plain, box, double, pile, Jacquard, Axminster, velvet, and Wilton.

Loom attachments: Dobby, lappet, swivel, leno, and pile.

THE WEAVES

For plain woven fabrics

Plain, rib, and basket weaves.

For fabrics woven with floating yarns

Twill, satin, Bedford cord, piqué, huckaback, honeycomb, bird's-eye, crepe, lappet, swivel, corduroy, and velveteen weaves.

For fabrics with woven-in pile

Velvet, terry, Axminster, and tapestry or velvet weaves.

For backed fabrics

Warp-backed, filling-backed, and double weaves.

For Jacquard patterned fabrics

Jacquard weave.

For leno fabrics

Leno weave.

Opposite page: A circular weft machine knitting an eyelet stitch
(American Enka Corporation).

KNITTING

Provides flexible fabrics

IN THE BEGINNING

Creating fabrics by knitting loops with one continuous yarn was not as recent an idea as some might think. Forming texture by interlacing loops of material is a very ancient art indeed, since the fish nets of prehistoric peoples were probably made by some form of hand knitting. Perhaps even at that early date the craftsmen found it more convenient to draw the loops through one another with the aid of smooth twigs, pointed stones, or polished bones. The simple, primitive method was improved as time went on, new methods of forming loops were added, improved needles were made, different yarns were used. Just how and when all this progress took place is still a matter of guesswork, for it was very late in history, when compared to accounts of weaving, that we find mention of knitting and knitted fabrics.

In the fifteenth century knitted garments were in common use. In Spain, not only was wool knitted, but silk as well, for a fine knitted pair of silk stockings was sent from Spain to Henry VIII. In 1488, England regulated the price of knitted caps, and knitted hosiery was coming into use. By the sixteenth century hand knitting was an accomplished household art in most of the civilized world.

Machine knitting on the other hand had a very definite, entertaining beginning. A young Englishman, William Lee, a curate of St. John's College, met and fell in love with Cecily Yorke. In addition to being a lovely girl, Cecily knew the fine art of stocking knitting and spent much time at her work, so much, in fact, that it is believed the expression "mind your own knitting" was originated by William Lee in his concern over Cecily's knitting. It seems that Cecily, in her devotion to her art, paid more attention to her knitting than she did to William and his wooing. However true this may be, it is known that William did win out and they were secretly married. But not to live happily ever after, for when the college discovered his secret marriage, William was discharged for breaking a college rule which forbade curates to marry.

With William's failure to find other jobs, Cecily was soon forced to go back to her knitting of stockings, to help support the family. One evening, watching his wife's busy knitting needles, William conceived the idea of imitating the movements of hand knitting needles by mechanical means. In 1589, William Lee perfected the first knitting machine, which was a stocking frame in the form of a flat-bed type knitting machine. The needles he designed were of the spring beard type. The yarn was laid by hand across the needles, the bobbins stood on the floor. The jacks, acting in conjunction with the sinkers, divided the yarn in proper proportions for each loop. Lee knew he had in his possession the wherewith-all to revolutionize the stocking industry and asked Queen Elizabeth to give him a patent so he could build many machines. The queen refused because she would have no part in helping to put so many of her hand knitting subjects out of work. Lee was forced to take his machine abroad but died before the world recognized the importance of his discovery.

By 1775 there were many thousands of knitting machines in operation throughout Europe. Numerous improvements had been added. Rib frames and rotary frames had been perfected. In 1775 the first warp knitting machine was patented. Circular machines were invented in 1830. In 1847 Matthew Townsand invented the latch needle. Heretofore all knitting had been done on the spring beard type. This was the first practical alteration of William Lee's original bearded needle of 260 years before. Townsand, an Englishman, moved to America and died in this country. William Cotton, in 1864, patented a new machine, which completely revolutionized the knitting industry, for with it shaped garments, such as hosiery toes and heels, could be made.

Western-World Knitting

Early Americans probably formed textures by making a series of loops, in and out, using a single strand of yarn, rope, or reed.

359

There were many varied types of sandals, baskets, nets, and the like, although most of the fabrics, that have been found, were formed by some type of weaving. Some early fabrics from the Nasca region of South America were decorated with elaborate poly chrome embroidered designs and complex fringes of "needle knit ting." The "needle knitting" on the other hand, might be more closely related to twisting than to knitting. Knitting was an impor tant home industry in colonial times. No home was complete without the familiar knitting needles and no family was properly clothed without warm knitted stockings, mittens, caps, and scarves.

America has also contributed her usual share to improving and advancing the knitting machine. The first knitting machines in the United States were smuggled from England in the early part of the nineteenth century.

In 1850, the Baily Company of New York began to use a power knitting machine for underwear. In 1858, Cooper and Tiffany pat ented a spring needle machine for making ribbed underwear. Here tofore all ribbed underwear had been made on latch needles. With spring beard needles, finer ribbed fabrics could be made.

In 1863, Q. U. Lamb, an American preacher, perfected the first flat-bed knitting machine for knitting wide, flat fabrics. It con tained a new type of frame with two horizontal flat beds. This ma chine could produce fashioned fabric either of the flat type, with selvage, or tubular fabric of any desired width. With the final completion of a fully automatic circular knitting machine in 1870 the United States' knitting industry was well on its way.

In the last fifty years, America's inventive genius has been as prominent in the knitting field as in other phases of our advanced industrialized world. America has made improved flat-bed ma chines, circular machines of countless types, warp machines capable of making endless varieties of texture and design and all the neces sary auxiliary equipment. America has invented special knitting machines for full fashioned hosiery, tie, sweater, double fabric, and pile fabric. Needle grouping devices, design changing appa ratus, yarn selecting equipment, automatic, timesaving, and safety inventions—all too numerous to mention—have not only speeded up production but have facilitated the making of knitted fabrics so different and superior as to vie in beauty, serviceability, suit ability, and popularity with any other fabric. Yes, today's knitted fabrics prove—America has tended to her knitting.

KNIT FABRICS ARE MADE BY INTER-LOOPING YARNS

Knitting is the art of constructing fabric with needles by interlooping yarn to form a succession of connected loops. The essential element of knitting is the *loop*. A loop is a very small length of yarn, taken at some point distant from the end, and drawn through and around some object, usually another loop. As a result, there are two loops, which together are called a "stitch." The loop that has just been drawn through and that is still coiled around the needle is known as the "needle loop." The loop that is around the object or the previous loop through which the needle loop was just drawn is called the "sinker loop." The needle loop plus the sinker loop equal one knitted stitch.

The needle, holding the needle loop it has just formed, again picks up a portion of the yarn and draws it through the loop, thus, the needle loop becomes a sinker loop, and the yarn portion becomes a needle loop.

By such repetition, a chain of loops, each drawn through another, are formed and a fabric results. Since weft knitting is made with one yarn traveling across and back (or round and round), loops are formed that lie in a plane across the fabric and, also in the same plane, up and down the fabric.

The loops lying side by side in a line across the fabric are called "courses." The loops succeeding one another in a line lengthwise of the fabric are called "wales."

Top to bottom: Spring beard needle (left) and latch needle (actual sizes) (American Enka Corporation). Courses (1) and wales (2) in weft-knitted fabric. Warp stitches (1-2-3-4) in warp-knitted fabric (American Viscose Corporation).

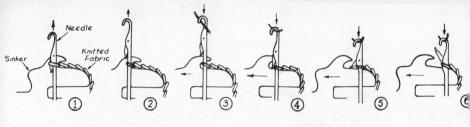

Twelve steps in the basic construction of one knitted stitch (latch needle) (Scott and Williams, Inc.).

The words "course" and "wale" are terms used to grade weft knitted fabric and to explain different types of weft knitted fabrics. For example, the closeness of texture is partially determined by the number of courses and wales per inch of the finished fabric (as in weaving the texture is determined partially by the number of filling and warp yarns per inch). Another example, a design may run "coursewise" or "walewise."

Since warp knitting is made with many yarns, making a series of chains running generally up and down the fabric, the warp knits are measured by the number of stitches or needles per inch for example, thirty stitches to an inch.

Knitting is accomplished in two ways, by hand or by machine.

Hand Knitting

Hand knitting is produced on one circular, or two or more thin, straight, hookless needles, by hand manipulation of the yarn. Any number of types of stitches that are formed by one continuous yarn can be constructed by hand, although the yarn used must be coarse enough to be manipulated by hand.

Since hand knitting is essentially a home industry (as is most other handwork), the majority of knitted fabrics are constructed by a variety of knitting machines.

Machine Knitting

The distinctive feature of the knitting machine is the knitting needle. It is the needles, their placement and manipulation, that loop the yarn, forming a knitted fabric. In machine knitting of every kind, there must be a needle for every loop, and therein lies the difference between machine and hand knitting. The machine needles are actuated or moved about by cams, or some other similar device. There are two types of knitting needles, the "spring beard needle" and the "latch needle."

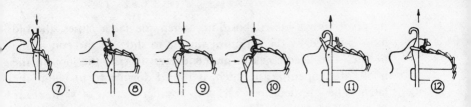

There are many varieties of spring beard needles, but generally
: modern type consists of a long, thin shank with a butt at one
d, bent at a right angle, and a long, flexible hook at the other
d. The shank is formed with a groove, or eye, into which the tip
the beard or hook is buried as the hook is closed.

The spring beard needle forms loops in the following way. The
w yarn, which is picked up and kinked around the needle shank
low the open needle beard, slides by the open beard toward the
ok. At the same time, the previously formed loop, which rests
ll below the tip of the beard, slides upward as the beard is closed
:r the new yarn by a presser. The previously formed loop is
ced up by action of the sinkers, over the closed beard. As the
ter loop continues to move up and off the needle, the pressure
emoved and the beard springs open. The loop in the needle then
ses down to well below the beard, and the open hook picks up
re yarn, and the next looping operation begins.

Spring beard needles are used on machines when closely knitted
rics are to be made with fine yarns. These needles are most
cient when a large number of needles per inch are required. All
any type of knitting machine can be made to use spring beard
:dles, although some machines are seldom fitted with them.

tch Needles

Latch needles for all machines have the same parts, but they
y vary in shape for different machines. The latch needle has a
nk with a butt formed at right angles on one end and at the
er a hook and needle latch, which swings freely on a pin or
et lengthwise of the needle. The latch has no movement side-
e. The latch needle operates in this way: in forming a new loop,
previously formed loop—at the time the hook picks up the
v yarn—is on the needle shank somewhat below the open latch.
the needle moves downward with the new yarn in its hook, the
viously formed loop contacts the open latch and swings it on

its pivot into a closed position. When the latch closes, the old loop is forced by the action of sinkers to slide off the end of the needle. The needle at this moment starts to rise; the loop in the hook, after opening the latch, is forced downward to below the latch. The emptied hook picks up another section of yarn, ready to form a new loop. The constant moving up and down of the latch needle forms continuous loops or a knitted fabric.

Latch needles are used more often than spring beards for fabrics made of coarser yarns, such as ribbed hosiery, heavy sweaters, and cotton underwear.

The gauge or cut of a knitted fabric is determined by the number of needles used to knit a given width of fabric. Knitted fabrics other than hosiery are determined by "cut," which means so many needles per inch. For example, 14 cut means that fourteen needles are used for 1 inch of fabric. Hosiery is determined by "gauge," which means the number of needles used for 1½ inches. For example, 14 cut is the same as 21 gauge $(14 + 7 \ (\frac{1}{2} \text{ of } 14) = 21)$.

THE STITCHES

Regardless of the type of machine upon which a fabric is knitted, all knitted fabrics are divided into two general groups, weft (or filling) and warp fabrics, and all knitting is accomplished by using one or a combination of four basic types of stitches, plain, rib, purl, and warp.

Fabrics that are knitted with one continuous yarn, back and forth across (or around and around) the fabric, are called (for

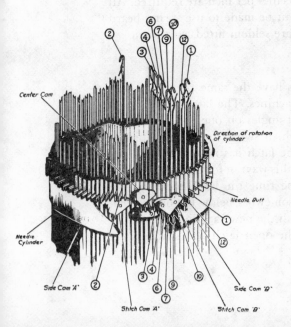

Circular knitting machine with latch needles. Note that the needles, with hook ends 1-12 and their corresponding needle butts 1-12, as they rotate in the cylinder are actuated by the stationary cams.

Center Cam

Direction of rotation of cylinder

Needle Butt

Needle Cylinder

Side Cam 'A'

Stitch Cam 'A'

Side Cam 'B'

Stitch Cam 'B'

the lack of a better name) "weft," or "filling knitted" fabrics. Materials that are knitted with many yarns traveling in a more or less vertical direction—yarns fed from a warp beam—are called "warp knitted" fabrics.

Weft knitted fabrics are constructed by using, singly or in combination, three of the four basic stitches, plain, rib, and purl. Warp knitted fabrics are made by using one type of stitch, which, however, has many variations—the warp stitch. The following is the explanation of the basic stitches.

Plain Stitch

In knitting, plain stitches are made by one yarn, fed to one set of needles, forming loops that are always drawn through in one direction to the *same side of the fabric.* All the *lengthwise* stitches in plain knitted fabrics are visible on one side, termed the "face." These stitches give a lengthwise chain effect. This side of the fabric has a very smooth appearance.

On the reverse side or back, all the stitches are visible in a *crosswise,* ridged effect. The ridges running across the back are coarse looking, giving this side of the fabric a rough appearance. Fabrics knitted with the plain stitch always have a different front and back. Plain knits are also termed "flat knits," or "jersey knits."

If a plain knit is knitted with fine cotton yarn similar to that used for underwear, it may be called "balbriggan." If the plain stitch is made with heavy yarns, as for sweaters, it may be called "shaker knit." The plain stitch is illustrated on the next page.

Circular knitting machine with spring needles. Needles 1-12 are shown in different knitting positions (Scott and Williams, Inc.).

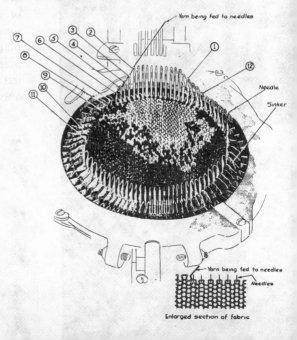

Yarn being fed to needles

Needle

Sinker

Yarn being fed to needles

Needles

Enlarged section of fabric

Warp fabrics that are termed "jersey," as "tricot jersey," are incorrectly named. "Jersey" means weft knitted in the plain stitch only. The plain stitch, due to its construction, will run when a stitch is broken, and the fabric is very elastic; while warp fabrics which are formed by crisscrossing many yarns, will run only with great strain or will not run at all, and have much less elasticity than do jersey knits.

Some plain knits, made by different stitch tensions or certain stitch tying arrangements, which form a patterned surface, may be called "mesh."

Plain knitted fabrics have a great deal of elasticity. The stretch of plain knit is crosswise or coursewise, with little vertical or walewise stretch.

The plain stitch can be knitted on a flat-bed machine, with one bed of needles for selvaged or shaped material. Full fashioned hosiery is an example of plain stitches knitted on flat-bed machines. Plain stitch is also knitted on circular machines, using one set of needles. Some links-

and-links machines can knit the plain stitch but due to the cost of this type of machine, it is mostly used for only purl or links-and-links stitches.

Rib Stitch

Rib stitches are made by one yarn, fed to two sets of needles knitting at the same time, forming loops which are drawn to *both sides of the fabric.* Both sides have visible up and down stitches and visible crosswise stitches, for one set of needles is making stitches by drawing the loops through to the face while the other set is drawing the loops through to the back. This is accomplished in some systematic, alternating order. One set draws a stitch, or stitches, through to the face; then the other set of needles draws the next stitch, or stitches, through to the back. This alternate reversing of stitches makes a *lengthwise ribbed surface on both sides* of the fabric. If the arrangement is one stitch drawn to the face and one stitch to the back and repeated, or in other words, if one rib is composed of one stitch with one reverse stitch in-between, it is a one-by-one rib, or what is called a "plain rib." If each rib consists of two stitches with two stitches in-between, it is a two-by-two rib or a "Swiss rib." By different, systematic arrangements, an endless variety of ribbed effects are possible.

To knit a ribbed stitch the machine must have at least two sets of needles working and two sets of cams, exactly alike. The needles, near their hook ends, must cross each other, and the yarn must be fed simultaneously to both sets of needles. The cams must push the needles of both plates up at the same time and down at the same time. Either latch or spring beard needles are used to

Basic stitches. Opposite page, top to bottom, face (left) and back: Plain stitch, one-by-one stitch, two-by-two stitch, purl stitch. Below: Warp stitch (tricot).
Right: Plain stitch, shaped knitted fabrics (hosiery) made on weft flat-bed machines with needle arrangement devices.

knit the rib stitch on both the plain, flat-bed machines or the circular machines. Many links-and-links machines can be adjusted to knit ribbed fabrics. This type of rib is called "links-and-links rib." Some machines can knit a combination of rib and other stitches.

Ribbed fabrics are more elastic than those made by other stitches. The ribs on both sides of the fabric have a tendency to draw together like pleats. Ribbed fabrics can be knit with a selvage or in tubular form. Rib stitches, unless protected by some special stitch arrangement, will run.

Purl Stitch

Purl stitches are made by one yarn, fed to one set of needles with hooks on both ends, forming loops which are drawn first to *one side of the fabric and then to the other*. The loop is always drawn through in the direction reverse of the preceding loop, making ridges running *crosswise on both sides of the fabric*. The yarn travels across the fabric. Fabric knitted with the purl stitch is the same on both sides and has the appearance of the back of the plain stitch fabric. The purl stitch fabric is quite elastic crosswise and very elastic lengthwise.

The purl stitch is knit only on links-and-links machines. Some of them are circular machines and some are of the flat-bed type.

The needles are either latch or spring beard. With needle grouping devices many patterns can be produced readily. Some links-and-links machines are equipped to make fabrics with mixed

Fabrics knitted with plain stitch on circular machine.

Fabrics knitted with rib stitch on circular machine.

stitches of purl, plain, or rib. A fabric, knitted with a purl stitch, will "run" if a stitch is broken.

Warp Stitch

Warp stitches are made by *many warp yarns* (at least one for each working needle), fed to one bed of needles, forming loops *one above the other* in a general vertical direction. Since each needle needs a separate yarn, the yarns are first prepared on warp beams, the same as for weaving, and then the many ends are fed to the needles. The warp stitch, in simplified form, is a chain stitch; each needle making one separate chain, and the chains tied together by the systematic *zigzag* of the yarns from one needle to the other. The yarn does not form each loop adjoining the previously formed one in the same course but produces it in another course and another wale.

Warp-knitted fabrics, unlike weft fabrics, have little elasticity and the stretch is usually in both directions. The peculiar knitting construction makes warp fabric practically runproof.

Warp knitting is constructed on special warp knitting machines. Either latch or spring beard needles may be used. Warp fabrics are made almost entirely with a flat selvage.

There are many different types of warp machines, knitting innumerable varieties of designs and textures. Some are so closely

369

Fabrics knitted with warp stitch on warp machines. Counterclockwise, starting upper right: Sheer Milanese; satin two-bar tricot; striped two-bar tricot; double Simplex; printed one-bar tricot; checked three-bar tricot; openwork Raschel.

knitted as to resemble tightly woven fabric, others are so loosely knitted as to vie with lace.

The best known warp knits are the following:

One-bar tricot stitch makes lengthwise rows of loops on the surface and small looplike forms crosswise on the back. It resembles the plain stitch in weft knitting. Plain tricot has some elasticity and can drop a stitch in but one direction.

Two-bar tricot stitch makes a fine-lined surface and a distinct cross-ribbed back. It is more closely knitted than plain tricot and is runproof. There is little elasticity in two-bar tricot.

Milanese stitch makes a fine, twill rib, running diagonally in the fabric. The back and face are nearly alike. It is more elastic than tricot but only in the direction of the ribs.

Raschel stitches can be made in so many different combinations of textures, in such a large variety of patterns, that there is no distinct Raschel stitch or pattern.

Stitch Variations

The four basic stitches can be given many variations by combining two or more of them or by manipulations of either yarn or needles. Stitches may be dropped at intervals, yarns may be floated from one place to another, stitches may be knitted together, some yarns may be held firmer or looser than others, some needles may be given more loops, some needles may be left out, all of which add variety to the finished fabric. Some stitch arrangements are in such common use as to have acquired definite names. The best known are the following:

Tucking is formed by needles which, still holding one loop, take on one or more additional loops and then knit them off as one stitch. Selected needles do not cast off their loops for a number of courses, thereby producing a multiple series of loops at the pattern positions. *Tuck stitch* is used to make fancy patterns of the basic stitches. Tuck stitch can be made on any type of knitting machine, although the latch needle is most suitable to it. Tucking is used in combination with basic stitches and with other design variations. Some fabrics made with tuck stitches in definite positions have become well known by definite names.

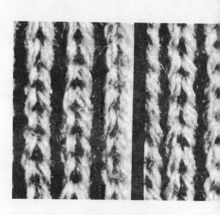

The cardigan fabrics are a combination of the rib and tuck stitches. *Half cardigan* is made by setting one bed of

Stitch variations. Top to bottom, face (left) and back: Rack stitch, half cardigan, full cardigan. Right: Double fabric. Left: Links-and-links.

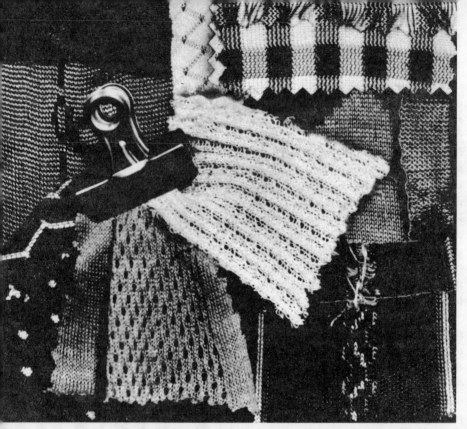

Fabrics knitted with stitch variations. Counterclockwise: Two-bar tricot eyelet; plain-knit float stitch; three-bar tricot, cut presser design; plain stitch, eyelet; plain-knit wrap stitch; plain-knit reverse plated; three-bar tricot float stitch. Center: Plain-knit tuck stitch.

needles to form a tuck stitch on every other needle and setting the other bed to knit the rib stitch continually. *Full cardigan* is made by adjusting both beds to tuck every other stitch but with alternate needles. Thus, tucking does not occur in both beds at the same time but first in one bed and then in the other. Full cardigan is also called "welt stitch."

Tucking makes a fabric that is thicker than other knitted fabrics from the same yarns.

Plating or plaiting is formed by having two or more yarns delivered at one feeding point in such a way that one or part of the yarns appear on the face of the fabric and the other or others on the back. Plating can be done on any knitting machine and with one or more colors or types of yarns and in all different stitches. If the yarns are controlled so that their positions in the fabric are interchanged, the pattern is called "reverse plating."

Racking is formed by shifting rows of loops to adjacent needles, first to the right and then to the left, making a zigzag series of stitches on the surface of the fabric. Rack stitch is best suited to plain knitting.

Eyeletting is formed by drawing a new loop through more than one loop at a time, thereby forming open spaces or eyelets through the fabric.

Floating is formed by knitting two strands of yarn in the same course, one yarn forming the loop while the other yarn, which does not enter the needle hook, is carried along or floated across between some needles. The same yarn always remains on the surface except where it is occasionally caught. This stitch method is used to form patterns especially when two yarns of different colors are used.

Interlocking is formed by a variation of the rib stitch. It is accomplished by knitting with two sets of needles in each of two needle beds (four sets of needles in two beds). In knitting, rib stitches interlock, forming fabric which has the elasticity of plain rib but which is thicker and more compact. Especially built machines are needed for the interlock stitch.

Wrapping or splicing is formed by feeding an extra yarn occasionally to selected needles, which knit it into a raised pattern on the surface of the fabric. Wrap stitches may be knitted along with plain or ribbed stitches and different colored yarns may be used for weft knitted fabrics.

Cut presser designing is formed on warp fabrics by carrying an extra yarn of different color or texture by means of a cut presser bar. This yarn is held and allowed to be knitted into the fabric at a given place forming a design on the face of the fabric. When not in use, the yarn is carried along the back of the fabric. This is similar to wrapping in weft knitting.

Double knitting is formed in two ways. One way is by adding an extra yarn to the knitting yarn or yarns. This gives warmth retaining quality, firmness, or design to the fabric. Many heavily napped fabrics are constructed with a heavy extra yarn. The other way is to knit two separate fabrics and by needle manipulation tie the two fabrics together at intervals. This type of double

fabric can be knitted with two types of yarns, as wool for one fabric and cotton for the other. To knit double fabrics, machines have to be specially constructed.

There are two general types of knitting machine, flat-bed and circular.

FLAT-BED MACHINES

Flat-bed machines, as the name implies, are those on which needles are placed in a straight line position on flat, oblong beds. There are many varieties of flat-bed machines but they can be grouped under three distinct types: basic flat-bed, links-and-links, and warp.

A basic flat-bed machine carries the needles in two parallel grooved plates or beds. The plates are held on a frame, with the grooves or slots on top, in such a way as to form an acute angle to each other; the needles of one bed cross the needles of the other near their hook end, with the fabric passing down between the two beds. This space between the two beds, called a "throat," is made the correct distance to

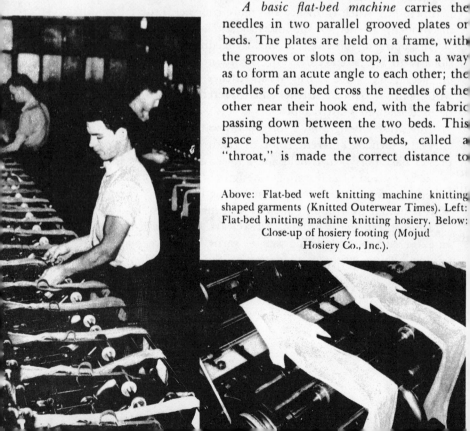

Above: Flat-bed weft knitting machine knitting shaped garments (Knitted Outerwear Times). Left: Flat-bed knitting machine knitting hosiery. Below: Close-up of hosiery footing (Mojud Hosiery Co., Inc.).

it the spacing of the needle slots in the
ds, about ⅜ inch in coarse machines to
inch in fine machines.

The needles are placed so that only the
edle butts project above the surface of
e bed. The protruding butts come in con-
ct with the moving cams, which move the
edles up and down individually so that
eir hook ends may engage with the feed-
g yarn successively. A mechanism is inter-
osed between the needles to assist in form-
g the loops. The carriage, to which the
ms are attached, slides lengthwise of the
edle bed not only moving the cams to the
sired positions, but guiding the necessary
ngth of yarn to the right place for each
edle to catch the yarn. A flat-bed machine
oduces flat fabrics.

Plain stitch material is produced when
ly one set or bed of needles is used. It can
narrowed or widened by addition or sub-
action of needles. The machine knits
bbed fabric if the yarn is fed to both sets
needles simultaneously. Rib fabric can
widened or narrowed in the same man-
er as plain knits.

The flat-bed machine can be constructed
use either type of knitting needle; it can
ake fine or coarse fabrics, and use any
pe, quality, or size of yarn. The needles
n be cammed and arranged to produce a
riety of stitch combinations. These ma-
ines by the use of loop-transferring de-
ices can produce shaped fabrics. Full
shioned hosiery, ties, underwear, and
ther shaped knitted garments are made on

Above: Links-and-links knitted fabrics. Right:
Links-and-links circular machine with two beds
of needles for knitting ribbed hosiery (Scott &
Williams, Inc.).

375

special flat-bed machines. These machines are constructed to manipulate the needles in special ways, but, fundamentally, their action is similar to the basic flat-bed machine.

Flat-bed machines may be equipped with Jacquard attachment for designing, which is simply a means to select and put into operation any desired needle or needles at any predetermined place in the fabric. This is the same function as performed by the Jacquard machine in weaving.

The majority of the flat-bed machines knitting weft stitches are machines with special needle arrangements, allowing the machine to make only shaped fabric. Yardage weft knitted fabrics are as a rule made on the less expensive, speedier circular machines.

A links-and-links machine has two needle plates but only one set of needles. Each needle is constructed with a hook on both ends but no butt with which to operate it (the needle ends may be either latch or spring beard). The cams do not operate directly on the needles but indirectly through what are called "jacks." The jacks are small bits of metal that have a needle notch at one end and a butt at the other. Each fits the slot of the needle. The cam carriages operate above the needle beds.

The links-and-links needle plates are set flat or, in other words, in the same plane with the needle slots directly opposite each other. Each double-hooked needle has two slots and two jacks (one of each in each plate), which cause the needle to slide from one plate, or bed, of the machine to the other.

This machine construction knits the purl stitch. The hook of one end of the needle picks up the yarn and, as the needle travels from one

One-bar tricot knitted fabrics.

bed to the other, the stitch formed falls off the opposite end. As the needle travels back, the next yarn is picked up by the other end of the needle and falls off in the direction opposite to the preceding stitch. The loops are drawn through the fabric, first in one direction and then in the other, which makes the distinctive purl stitch.

Links-and-links machines are equipped with needle-grouping devices so that a variety of patterns can be made. Jacquard attachments also may be used with this machine.

If the links-and-links machine is equipped with two cylinders and one set of double hooked needles and with special yarn selecting devices, it can make a large variety of fancy or patterned ribbed stitches. The fancy ribbed fabric made on a links-and-links machine is termed "links-and-links fabric" and is said to be knitted with the "links-and-links stitch."

Links-and-links machines can be adjusted so that only one end of the double-hooked needles knit, thus making a plain stitch or a ribbed stitch; but, since plain circular or plain flat-bed machines are less expensive, this adjustment is seldom made.

Above: Three-bar tricot knitted fabrics. Below: Tricot machine. (American Enka Corporation).

A *warp knitting machine* (with very few exceptions) is a flat-bed machine constructed to use yarns that have first been prepared on a warp beam the same as warp yarns for weaving. For warp knitting there must be at least as many yarn ends as there are working needles. (Often not all of the needles are knitting. Some are left out to produce a particular design.) Occasionally, for certain textures, two or more yarn ends are knitting together in the same needle, but every yarn must be in a needle.

In preparing the warp yarns, many strands of yarns are wound parallel, side by side, onto a beam or onto several beams. The yarns from the beam are threaded through guides (each beam, if more than one beam is used, has a separate group of guides) and carried to one set of needles.

In warp machines, the warp yarns are fed throughout the fabric in a vertical or at least a more or less vertical direction, making a series of chains, bound together by the transversing yarns. Each yarn does not always knit on the same needle but, on passing to right or left, knits on any one of two or more needles at different times.

Some warp machines have one bed of needles, others have two. Some are constructed to use latch needles, some spring beards. Some needles are placed vertically, others on horizontal beds. Warp machines can have many and various needle control features, capable of making innumerable types of fabrics. Jacquard patterns also can be applied to warp machines. All warp machines have guides that wind the yarn around the needles in some definite way.

Two-bar tricot knitted fabrics.

To add speed to knitting, sometimes the needles of warp machines are cast in lead and all work together. They are bolted onto the needle bar and act as part of the needle bar. The guides are cast in lead also and act as part of the guide bars.

There are many types of knitting machines using warp yarns. Although they have similar features, they also have conspicuous differences. The most common warp machines are described in the following.

A single-bar tricot machine has a needle bar made to operate in a vertical position so that the fabric leaves the machine in direct view of the operator. Plain tricot is knitted with one beam of warp (or one set of warp yarns) and one set of yarn guides, in a single bar.

The needles move up and down, and the guides, each with its yarn, move backward between the needles, traverse in back of the needle, and swing forward between other needles. This movement of the guides measures off and winds a yarn around each needle, which in rising has dropped the old loop onto the shank. The needle descends and the old loop is cast off. As the needle rises again, the new loop is held in position. The guide changes from one needle to another, and a new stitch begins.

Each yarn travels to right or left, knitting on any one of two or more needles at different times. For instance, starting at zero, the guide bar goes forward through the needle at one, and falls to zero at the back of the needle. From zero, the yarn is taken through at one again, and then, rising to two for the next motion, the yarn is passed from one needle to another. This connects the series of chain stitches and produces a one-bar solid fabric on two needles which is called a "basic one-bar tricot."

The knitting, alternately first in one direction and then in

Raschel knitted fabrics.

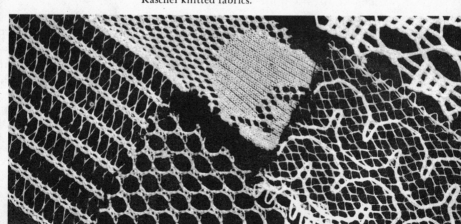

the other, produces a close textured fabric with a striped effect running crosswise. The yarn of one guide is on the face of the fabric in the form of loops, covering the yarns of the other guide

Single-bar tricot machines can be equipped with needle controlling devices and with a Jacquard attachment. Some tricot machines have an extra bar known as "cut presser bar," which presses certain needles into action at the desired time. With a cut presser bar, extra warp yarn can make patterns on the tricot. This is similar to the dobby attachment on the looms, and the wrap stitch attachment on weft machines.

Tricot machines can be adjusted to use either latch or spring beard needles, although most tricot machines have spring beard needles. Tricot machines are very speedy, knitting from 400 to 500 courses a minute.

A two-bar tricot machine is a tricot machine using two guide bars and at least two beams of warp yarns, or two sets of warp yarns; one set is knitted in one direction and the other in the opposite direction, producing a chainlike surface on the face of the fabric. Using two bars makes two-bar tricot a heavier, firmer fabric. The machines can be equipped with extra attachments the same as for single-bar tricot.

A three-bar tricot machine has three guide bars and at least three warp beams. The third bar allows for great variation in design. It is possible to knit one-bar tricot fabrics on either a two-bar or a three-bar machine.

A Simplex warp machine is constructed similarly to a tricot machine, except that it has two sets of needles, at least two sets of warp yarns, and two guide bars, by which two separate fabrics are knitted and laced together simultaneously. In other words, Simplex machines knit double fabrics. It is possible to use two different kinds of yarns on the two sets of warps. Usually, spring beard needles are used on Simplex machines. Simplex machines can make a ribbed fabric instead of a double fabric in the same fashion as weft-ribbed fabrics are made.

A Milanese machine is a huge machine, often as wide as 20 feet, in which the needle arrangement is horizontal. There are two warp beams, two guide bars, and one set of needles. The warp yarns are mounted on spools or bobbin carriages and are placed on bed plates, which are gradually traversed around an endless chain, under and over the machine. The two groups of yarns are shifted, step by step, in the same direction across the machine, with one set of yarn going downward to the left and one set downward to the right. Each warp yarn, as it reaches the end of the machine, moves step by step backwards, across the machine again, by the use of an ingenious mechanism.

Milanese produces a diagonally designed fabric, which has

Below: Simplex machine (Alfred Hofmann, Inc.).
Right: Double fabric knitted on Simplex machine.

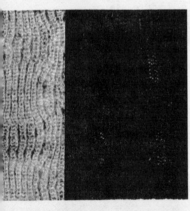

stretch in both directions. Milanese machines can use either latch or spring beard needles, and have additional needle control equipment.

A Raschel machine is a machine which usually uses latch needles. It may have one or two needle beds. If two beds are used, the machine knits the rib stitch. The guides (in horizontal bar) are placed in a vertical position above the needles (in horizontal bed). Raschel machines use from one to six guide bars, which are controlled by means of links of varying heights built into chains in proper sequence for some predetermined design. Each bar of yarn may be controlled by any one of three pattern chains. The latch needles of Raschel machines are usually cast in lead.. The warp beam (or beams) is held above the needles and guides. The yarns are kept in their proper places by being passed from the beam through metal plates, just above the leaded guides. The plates have two lines of holes, a hole for each guide. The yarns go through the holes before passing through the guide proper. The guides acting in conjunction with the needle bar produce the knitting action.

Raschel machine's latch needle construction allows the machine to construct fabrics of heavy yarns. A variety of patterns are pos-

Top to bottom, left: Plain-knitted fabric with extra heavy yarn on top, which has been deeply napped; plain-knitted fabric made by holding some yarns loose to give a rough surface; one-by-one rib, with needle arrangement making a pattern; one-bar tricot with rayon and metal yarns. Right: 1, 2, and 3, plain-knitted fabrics with extra yarns that are napped giving a design on the face of the fabric; 4, and 5, one-bar tricot, knitted with metal yarns for design.

ble as well as a combination of textures om very coarse and heavy to open and icy. Jacquard attachments and arrangements for handling elastic yarns may be art of a Raschel machine. Raschel knits are sually more elastic than other warp knits. aschel machines are more versatile but ower than the tricot machines.

IRCULAR KNITTING MACHINES

Circular knitting machines, or circular ody machines, as the name implies, have ne needles arranged in circular fashion on ound beds. Since the needle bed is round, ne fabric knitted is always tubular.

The needles are mounted to slide indiidually in a vertical cylinder, with an dditional mechanism functioning between ne needles to assist in forming the loops. mall cams act on the needle butts, to cause nem to move up and down individually nd successively. Sometimes the cylinder is ationary, and the yarns revolve, but more ften the yarns are stationary and the cylnder revolves because, if the yarns are ationary, more yarn can be wound on the ools and more fabric can be made without rethreading the machine.

The needle action is the same as on flated machines. For a plain stitch, there is ne bed of needles which is bent into a verical cylinder with the needle slots on the utside. If a ribbed fabric is to be made, here needs be, also, a horizontal bed which

Above: Plain, circular sweater knits with extra yarn carried on the back to reinforce and to prevent running. Right: Plain circular machine with Jacquard attachment for women's hose (Scott & Williams, Inc.).

383

is made into a circular plate or dial with the needles in a circle at right angles to the needles in the vertical bed. All needles are individually and successively movable. Some times one, or both sets are mounted in conical cylinders.

Circular machines vary in size from those making small circular stockings with a width of ¼ inch, to machines constructing fabric a wide as 40 inches in diameter. Some machine use latch needles, others spring beards. Some roll the finished fabric above the machine others down below the needle beds.

Practically every type of knitting machine has been constructed in circular form as well as on a flat bed. Besides circular machines that knit plain and ribbed fabrics, there are links and-links circular machines, and some circular warp machines. But, because the circular warp machines require tedious handling and are slower to operate, the warp machines in the United States are mainly the flat-bed type.

Some circular rib machines are equipped with double-edged needles, which can be cammed to knit a variety of rib widths on the same piece of fabric. Both hooks of the needle are used only when the machine is changing from one type of rib to another. The rest of the time, only one end of the needle knits. Circular links-and-links machines, like the flat-bed links-and-links can be equipped with needle arranging devices to knit the "links-and-links" stitch. This type is used particularly for fancy ribbed hosiery.

Circular knitting adds speed to knitting

Circular underwear knits. Top to bottom: Plain stitch plain stitch, mesh, one-by-one rib, two-by-two rib tuck stitch, double fabric knitted in plain stitch, re verse of double knit.

384

l many types of needle-selecting and knit-
g devices can be applied to a circular ma-
ne as well as to a flat-bed. The Jacquard
achment is used with circular machines
. Tubular fabrics can be cut and sewn,
l garments that need to be tubular, as
iery and some other shaped garments,
only be made on circular machines.

Whether flat-bed or circular, knitting
chines are, for the sake of economy and
ed, cammed for one type of knitting us-
one type of needle and one gauge;
ns of different fibers may be used as long
hey are in the same weight range. Often
chines may be equipped with two or
re extra cylinders of different gauge and
nged back and forth as the need war-
ts. Any machine may be equipped with
quard or any of a large variety of other
dle-selecting devices. Any and all types
yarns may be knitted with proper ma-
ne adjustment. The range of possible
ric types is tremendous—from the sheer,
n net to the heavy, double napped
tings.

Weft-knitted fabrics, made on either cir-
ar or flat-bed machines, are constructed
give elasticity—a necessary requirement
such items as hosiery, underwear, sweat-
, gloves. Stitch tying, extra tie yarns, Jac-
ard arrangements, or strengthening the
in or rib stitch to prevent running, de-
ase the elasticity—the distinctive feature
the weft-knitted fabric.

Warp knits, due to their construction,
particularly suited for garments that re-
re less give and more firmness. Warp
ts and weft knits are different in suit-
lity, and each has a distinct place to fill
the family of fabrics.

Above: Plain-stitch circular
machine for men's hose:
Note yarn for wrap stitch
at top of machine (Scott
& Williams, Inc.).

SUMMARY

TO PRODUCE KNITTED FABRIC

Knitting: The art of constructing fabric with needles by inter looping yarn to form a succession of connected loops.

Loop: A small length of yarn taken at some point distant from the end, and drawn through and around some object, usually another loop.

Stitch: The loop on the needle (the needle loop) plus the previous loop (the sinker loop) equal one stitch.

Course: The term used for the loops lying side by side across the weft-knitted fabric.

Wale: The term used for the loops lying one above the other lengthwise of the weft-knitted fabric.

Needles: Spring beard, latch.

BASIC MACHINES

Flat-bed
Basic flat-bed.
Links-and-links.
Warp: Tricot, Milanese, Simplex, Raschel.

Circular
Plain circular.
Rib circular.
Links-and-links.

BASIC STITCHES

Weft stitches: Plain, rib, purl.
Warp stitch.

BASIC STITCH VARIATIONS

Tuck	Plate
Eyelet	Interlock
Rack	Double-knit
Wrap	Cut presser design
Float	

Opposite page: A hand-made Chantilly, French bobbin lace, nineteenth century (The Metropolitan Museum of Art).

TWISTING

Makes the lace fabrics

IN THE BEGINNING

Lace has a very vague origin, beginning at least a few thousand years before Christ. Probably some imaginative as well as thrifty woman conceived the idea that it might be easier and more attractive to fill up a torn or worn out section of a piece of fabric by twisting and looping the thread into a pretty design.

This practice producing satisfactory results, might later have led some ingenious women to ask the question that if the method were good for old holes why not for new ones? So it became popular to cut holes or pull threads in order to fill them in with designed stitches. Thus it seems that lace, the aristocrat, began its career as a lowly, ordinary "patch."

Buratto, which is really a weave rather than a lace, began in Babylon and Egypt as early as the twenty-fourth Dynasty. But is was not until the beginning of the first century after Christ, the Coptic Period, that what might now be called "lace" made its appearance. This lace was found in the form of plaited thread headdresses worn by the native Egyptians. The curious twisting of thread, which suggests bobbin work, was called "Egyptian plaiting" although it may not have been original with the Egyptians.

Early forms of lace have been found in the Roman and Egyptian tombs. When the ancient Roman settlement of Claterna was unearthed in Italy, there were found a series of small cylindrical objects, which in size and shape are very similar to modern lace bobbins.

It is hard to tell which type of lace was made first, needle point or bobbin. The earliest type, called "needle point" was the work of nuns, done for church decoration. This type consisted of drawing threads from pieces of linen and filling the spaces with fancy stitches, using the exact number of threads withdrawn because church equipment must be perfect. The early pieces in museums show that this first needle point was done with needle and thread on cloth, similar to Italian hemstitching.

Twisting bobbin lace on a pillow (Aldine, 1872. New York Public Library).

Needle point without the aid of cloth could not have begun until after the invention of brass pins, and the first mention of pins was in a wardrobe account of 1347: 12,000 pins for the trousseau of Joanna, daughter of Edward III.

In an Harleian manuscript, dated about 1471, directions are given for making "lace Bascom, lace indented, lace bordered, lace covert, a brode lace, a round lace, a thynne lace, an open lace, lace for hattys, ..." The manuscript is illustrated with a woman making lace. The processes seemed to be a type of hand-weaving, with the fingers serving to carry the thread, which was dropped occasionally to form an openwork braid.

The sixteenth century witnessed the evolution of needle point from embroidered and cut linen to actual needle-point lace. In 1542, the Sumptuary Edict prohibited the extravagant use of silk and gold thread in lace. This marked the beginning of elaborate white work, which discarded cloth and made lace entirely of thread.

Italy established a lace technique that was not only beautiful but also meticulous and accurate. This country supplied the world during the greater part of the seventeenth century, and laid the foundation for lacemaking in other countries. Handmade needle point and bobbinet flourished in Italy during this time, and continued to flourish not only there but in other European countries for several centuries.

Machines for making lace have an uncertain origin. It is generally thought that a man named Hammond, a Nottingham worker, was one of the first to construct a machine capable of making an openwork mesh. In 1768, improvements were made on the

389

stocking frame, whereby lacy materials could be made. In 1771, Robert Frost invented a machine for making coarse, square net, which was used for wigmaking; and later an hexagonal-mesh machine appeared. In 1795 an improvement was made on the warp machine by which warp edgings could be twisted. But it took the invention of Joseph Marie Jacquard (about 1800) to mark a new era in the world of machine-made fabrics. Jacquard, while working in a Lyons (France) factory, constructed an improved loom.

In 1801 or 1802, the French government offered a reward of 10,000 francs to the person who would invent an automatic machine for making net. Jacquard exhibited his invention at the industrial exhibition at Paris. His model was brought to the notice of Napoleon I, who asked Jacquard:

"Are you the man who pretends to do what God Almighty cannot—tie a knot in a stretched string?"

His model, which was then incomplete, was awarded a prize and he himself received an appointment in the conservatoire. While there he was able to study Vaucanson's invention. Armed with this new idea, he improved his loom and in so doing perfected the attachment since known as the Jacquard. It was not only the most outstanding invention the lace industry had seen but it also marked a milestone in all loom history. The Jacquard could be applied to all types of looms.

Another outstanding figure in the era was John Heathcoat, who invented the bobbinet machine in 1809. By improving and modifying the arrangement of the bobbins and carriages in the Heathcoat machine and applying the Jacquard cards, John Levers, in 1813, perfected a machine capable of making all types of lace. The Levers machine is the foundation of all modern lace machines. The Nottingham lace-curtain machine is a modification of the Levers machine. Like all other machines, lace machines had to fight their way through the prejudices of hand workers. But the result of the struggle was a tremendous lace industry that developed into an English monopoly prohibiting other countries for many years from participating in the benefits of the inventions.

Western-World Lace

Early America produced lace as did other ancient countries. The Incas in Peru in about the ninth century made a type of *buratto* or a sort of woven net on a loom or frame, in which the warp yarns were twisted at each passage of the weft. In South America there was also a type of net worked by needles over a cardboard pattern or spool, termed "nanduti."

In the United States the hand-making of lace was never very extensive. Ipswich, Massachusetts, seems to have been the only place where much lacemaking was carried on. An account of 1786 says that the Ipswich workers were at this time producing 42,000 yards of silk lace annually. This was bobbin lace, made on round pillows by use of slender wooden sticks without bulbed ends. There was bobbin lacemaking in the Hudson Valley but it never developed into an industry.

A church veil. Handmade, modern Flemish lace, bobbin made (Paul M. Stokvis).

Medway, Massachusetts, was the seat of the first machine-lace factory in 1818. A second was founded at Ipswich, in 1824. However, machine lacemaking in the United States developed slowly, because no lace machines were allowed to be exported from England. The only ones in the country had been brought over piecemeal and assembled by people who had worked on them before coming to America.

Machine lacemaking as an industry in the United States did not get started until 1909. It was not the desire for beauty that brought machine lacemaking to this country then, but the need for protection against germ-carrying insects. During the Spanish-American War in 1898, many American boys died from malaria, because our government could not supply them with the protection of mosquito netting. As a result, for eighteen months in 1909 and 1910, the duty on lace machines was lifted and lace machines from England were installed in America. Today, there are fifty-three lacemaking concerns in the United States, producing annually thousands of yards of lace and net for all purposes of modern living.

LACES ARE MADE BY TWISTING YARNS

Lace is an openwork fabric formed by looping, interlacing, or twisting yarns either by hand or by machine. Regardless by what method a lace is twisted, it receives its name from its appearance—from the pattern by which it is made.

It is the type of réseau, the kind of toile, the presence or absence of cordonnet, whether the scallop is shallow or deep, whether the texture is fine or coarse—one, two, or all—that determine the pattern of lace.

Lace has been characterized by such words as original, real, artificial, imitation, reproduction, lacelike, or lacetype. What is meant by a lace being original or real and by imitations or reproductions? If by original and real is meant that lace is not lace unless made as it was in the beginning, then only darned holes or fancy drawn work can correctly be called "lace," and no cotton lace can be real lace either, for in the beginning lace was made of linen thread. If it means the first lace made without the use of cloth background, then only needle point is original lace, and bobbin lace is imitation.

If only lace made entirely by human hands is really lace, then all laces are reproductions, for even needles, bobbins, and crochet hooks are mechanical devices. If handmade lace, in contrast to power machinery lace, is the only real lace, then hand appliquéd Carrickmacross, handmade princess lace, and other lovely types of appliqué on machine-made nets are only imitations.

If a lace is real only when it is made in the town where the pattern originated, as Binche and Alençon, then the lace ceases to exist with that town's lacemakers. Even the English word "lace" lacks distinguishing derivation. It is taken from the French word *"lacis"* which, when correctly used, denotes only "darned netting."

So just where on the pages of the history of our culture can a finger be placed and it be said, "Here is the original lace. That only is real lace?" Where, among the many varied and complicated constructions of modern lace, can one say that real lace ends and imitations begin?

Lace methods have progressed with civilization, and lace patterns, handed down through the ages, are being made by hand methods, by machines, or by both. New patterns appear continually, some developed by hand-lace artists, others by designers for the lace machines. The name of a lace explains its pattern, not its quality or its method of construction. Value is determined not by how lace is made but by the quality of workmanship and suitability.

Corner lace tea cloth, 64x64, completely made by hand. The pattern is both needle point and bobbin. The floral designs that are surrounded by fine mesh are Flanders pattern and were made on pillows with bobbins. The fine mesh surrounding the Flanders, as well as the centers of the flowers, and all of the rest of the cloth is point de Venise made by needles. The lace was twisted with very fine linen thread (No. 1500) by some 100 different lace makers, working for over a year (1935-1936). At that time it was valued, $7,500. The cloth was designed and constructed by Manufacture Lava of Malines, Belgium, under the supervision of Paul M. Stokvis (Francis G. Mayer, photographer).

Lace is divided into two general types, handmade lace and machine-made lace.

HANDMADE LACE

Handmade lace is a work of art and, since no form of art ever really dies, there will always be handmade lace. It is created by three different processes:

With a needle: needle point. This type developed from the earlier drawn work, or linen cutwork, in which certain yarns in the fabric were removed and the open spaces filled with lace stitches. If the yarns were drawn out, the work was termed "drawn work"; if the spaces were made by cutting out pieces of fabric, the work was called "cutwork."

Needle point, today, is made by first drawing a pattern for the type of lace desired. The art work is then accurately computed into number and kind of stitches necessary to make the given pattern. This technical pattern is then divided into sections of about 2 inches by 2 inches (no needle-point pattern can be any longer than the front finger of the worker), and the outline is drawn on a black or dark piece of paper. The paper is placed on a double layer of cloth, and the design is outlined by closely spaced stitches holding a heavy thread along the design.

The worker, with needle and thread, then makes the piece of needle-point lace by actually embroidering the pattern on the two-by-two piece of paper. The paper is torn away, by cutting the threads between the two layers of cloth, and the innumerable squares of lace are assembled on another paper pattern, also double lined with cloth, and the background of net or of bars is worked.

Thread for needle-point lace may be of various weights and types. Some needle point has been constructed of thread as fine as No. 4,000. Contrary to some belief, needle lacemaking is not necessarily hard on the eyes. In one Belgian city, employing 2,500 lacemakers, only about one hundred need to wear glasses. The workers sew on dark blue, black, or green paper with ivory thread, and the work is changed from fine to coarse every few hours. Also, modern needle point is not made entirely by aged workers, stitching away at home as pictures of lacemaking suggest. Today's needle point and bobbin are made largely by skillful, young technicians, working for business firms under competent supervision. Most handmade needle point as well as pillow is of foreign make.

Needle point is made with one thread in a needle. The characteristic design, or "toile," has a ground of net or of bars, which

Handmade needle-point lace. Left to right, top: French Point d'Argentan; Belgium Brussels Point de Gaze. Bottom: French Alcençon; Italian Venetian rosepoint (The Metropolitan Museum of Art).

may or may not have picots. The decorating pattern, or "à jour," is closely textured or clothlike.

With bobbins, pins, and pillow or cushion—bobbin lace or pillow lace. Bobbin lace is made with a pattern placed on a pillow and with threads that have been wound on long spools or bobbins. The design is pricked on a strong, stiff paper, and the paper is placed on a pillow. The filled bobbins are hung in pairs at the beginning of the pattern. After the ends of the threads are pinned on the beginning of the pattern, the lace is made by crossing the bobbins, in and out, over and under, in predetermined order. As the threads are crossed, pins are placed in the holes on the paper to hold the looped threads in correct position until, with further crossing, twisting, and braiding, the design and ground are made and are tied sufficiently to allow the pins to be removed to different positions and the pattern continued.

Bobbin lace is made with many threads. The pattern is more difficult to execute and requires more planning than needle-point designs; but the lacemaking is easier and goes faster. The pattern can be memorized and the work becomes almost automatic.

Bobbin lace is, as a rule, finer and sheerer than needle point. The à jour can be large and open or small and close, the mesh

Handmade bobbin lace. Left to right, top: Belgium Brussels, bobbin sprays on machine net; Flemish Binche; Italian Milanese. Bottom: French Valenciennes; Italian Genoese (The Metropolitan Museum of Art).

small or large. Bobbin designs can be very similar to needle point for it is possible to use similar types of thread and draw designs that result in similar finished lace, although there are some patterns that are definitely bobbin lace and some that are always made by needles.

With a crochet hook: crochet lace. Crochet lace is made with a long needle having a hook on one end. With the hook, the worker is able to loop and twine the thread or yarn in and out, to make a given texture. Crochet laces can be fine or coarse according to type of thread used, although the crochet hook is such that it would be impossible to crochet lace as fine as could be made with a needle or with bobbins.

There are laces that are combinations of the types of handmade laces described above. Sometimes the cordonnet, or cord outlining the pattern, is included as the lace is made and sometimes it is sewed onto the lace after it is made. Lace also may have motifs sewed on later. This is called "appliqué." Some lace is made by employing tape as a ground and needlework to twist a design in and around the looped tape.

MACHINE-MADE LACE

Machine-made laces are of many different kinds. They are made on machines, the distinctive feature of which is their bobbin construction. These machines can twist a great many of the handmade lace patterns and in addition make many simple and elaborate laces that cannot be reproduced by hand. Machines can handle yarns too fine for the human hand or eye to follow, or make very coarse textures achieved by complicated yarn manipulations.

All machine-made lace begins with the designer. He must be more than an artist with originality and drawing ability. He must have technical understanding of lace machines and lacemaking. He draws a design on paper bearing in mind its interpretation in loops, twists, and knots of yarn.

The artist's drawing is then made ready for reproduction on the machine by the draftsman. This person must be a real technician, for he must mathematically and exactly adapt the artist's picture to the potentialities of the machine. It is he who, by understanding and experimenting with the lace machine, is able to develop new lace types and new effects. He must know each part of the machine, the movement of each yarn, and the properties and uses of all types of fibers and yarns in order that design, yarns, and machine together may produce the desired effect.

The draftsman, using the original drawing as a model, charts the design on very narrow, ruled drafting paper, each square of which represents a certain yarn motion. Or he may enlarge the drawing and with this, transfer the design in the form of key numbers to a chart, each number representing a yarn motion. Either form is the key to the pattern, which is to be punched on the cardboard pattern cards of the Jacquard.

These cards (made of treated cardboard in various sizes, such as 3 by 8⅜ inches or 2½ by 36 inches) are perforated by a punch machine, which is operated by "playing" keys by hand on either side of the machine.

When all the cards that complete one pattern have been punched, the finished group is laced together into two packs by a lacing machine. One pack is for the cylinder of the front

Corded fabriclike all-over lace made on Levers machine.

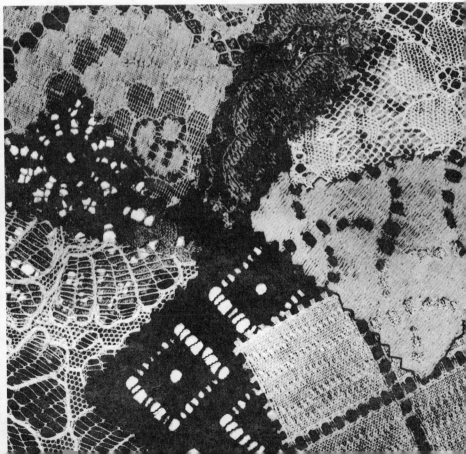

motion of the Jacquard and the other for the cylinder of the back motion. The Jacquard principle, in some respect similar to that used for weaving, is adapted to the special requirements of the lace loom. With the pattern completed the lace machine is now ready to be threaded.

Levers Lace Machine

The following explanation is of the Levers lace machine. There are several types of lace machines built with some definite differences but in general they are only modifications of the original Levers machine.

The lace machine is an enormous, complicated piece of machinery, weighing as much as 33,000 pounds and equipped to handle at one time as many as 10,000 yarns and more across its width of from 144 to 260 inches. Each machine is adjusted to a definite gauge, and for reasons of economy, is seldom changed.

Lace machines twist with three different types of yarns, bobbin yarns, warp yarns, and beam yarns. The bobbin yarns travel in a more or less vertical direction except for occasional twisting to tie in warp and beam yarns. The warp and beam yarns also twist in a general vertical direction except when they are moved to right or left to form the pattern.

The *bobbin yarns* are termed thus because they are manipulated by bobbins, the unique feature of lace machines. A brass bobbin consists of two fine, round disks, grooved on the inner surface and riveted together so that a narrow space is left between them. Each bobbin has a square opening in the center, which fits over a square shaft for the purpose of filling the bobbins. The circle of rivets acts as a base upon which the yarn is wound between the two discs.

The brass bobbin winding machine holds and, by revolving the square shaft, fills 120 or more bobbins at one time. The amount of yardage, depending upon the size of the yarn, averages 80 to 120 yards.

The filled bobbins must fit accurately into a given space on the machine, such as twenty bobbins to an inch for a ten-point machine. Since the bobbins are too thick when filled, they are put into a pressing machine and are pressed by hydraulic power sometimes as high as 20 tons per square inch. Held by a flat plate and a nut, the bobbins are prevented from expanding when the

Levers lace machine (North American Lace Company).
Right: Bobbin and carriage used on lace machines
(Native Laces & Textiles, Inc.).

pressure is removed. The frames of the bobbins are then heated with steam in an oven. When cooled, they can be unclamped, as the yarn will stay the size it was pressed.

The bobbins are next threaded and inserted in the carriage. A carriage is a thin steel holder, triangular in shape, with a curved, smooth bottom. The bobbin fits into a circular hole in the center, which serves as a rail upon which the bobbin revolves and is controlled in tension by a spring. On either side is a notch or "nib." The full number of bobbins are held and moved by two comb bars directly opposite each other. They are equipped with catch bars, which engage with the nibs of the carriage and alternately swing the carriage backward and forward. Thus, the bobbin yarn has a double movement. It is rotated at great speed within the carriage and then moved back and forth by the carriage itself, not only winding the yarn but twisting it at the same time.

The *warp yarns* are wound on a large beam the same way as warp for weaving. The *beam yarns* are wound on smaller beams.

To make the simplest patterns for which only one take-up of yarn is required, or in other words, for which the same length of warp yarn is measured off for each twist, all the warp yarns can be arranged on the same beam. This, of course, gives the same tension to all yarns. But for most laces, the various sets of yarns require many different lengths of take-up to make a given design. Consequently, it is necessary to have yarns that can be individually measured off in the amount needed, without affecting the adjacent yarns. Therefore, beam yarns are wound on separate small beams, which can be adjusted to work independently of all other beams. Most laces require a large warp beam and many small beams, the number depending on the design.

Both the large warp beam and the smaller beams are placed in the lower part of the machine and are threaded up in a vertical direction, reaching contact with the bobbin at the top of the machine. Inside and beneath the working parts of the machine is a perforated guide called a "sley." It is 12 inches wide and runs from end to end. It is made of light metal and is similar to window screening. This perforated metal is rubbed with soot and lampblack to fill the holes and is then painted over with varnish to give it hardness and durability. The holes conforming to the selected lace pattern are then reopened or "set in" the sley.

The warp beams are placed below and behind the sley and pass directly to the sley, while the numerous smaller beams are behind the warp beams and are first laced through looped wire eyelets and then upward through the sley. This "setting up" of a machine is done by hand and it takes two men about two weeks to place correctly these thousands of yarn ends in readiness for lace twisting.

With the help of the Jacquard mechanism, which divides, groups, and shifts the warp and beam yarns, the bobbin yarns are able to swing between and around the warp and beam yarns, making toile, réseau, or picot, as the pattern demands. Yarns can be so manipulated, arranged, and rearranged as to make practically countless types of lace, ranging from heavy cloth-textured, thickly figured lace to the finest open-meshed veiling, in patterns from the narrowest edging to as wide as the machine itself.

Lace Finishing

After the lace is taken from the machine, it is inspected and mended. It is then bleached by boiling, scouring, and beating with big wooden paddles in tubs of soapsuds. After dipping in hydrochloric lime and later in acid, it is again washed.

Lace is next dyed with the same dyes and in a similar manner as are woven fabrics.

Bleaching and dyeing leave the lace limp, so it is necessary to give it a dressing. It is dipped in a weak solution of soda ash and sized with thin starch to which bluing has been added.

The wet lace is then passed through a tentering machine (similar to that for woven fabrics), which dries and "dresses" it. Some lace is dressed the same width it was made, some much wider. Veiling is stretched to four times the width it is on the machine, losing length in proportion.

Nottingham lace made on Nottingham lace machine.

Instead of using the tentering machine, some lace is dressed on a stationary dressing machine, similar to household curtain stretchers, and is fanned dry by rotating fans above the frame.

Some lace is given more and specialized finishing. Ciré lace is treated with a shellac that makes it stiff and shiny, Repoussé lace is heavily starched and then embossed by ironing, which raises the pattern.

Nottingham lace machine (Wilkes-Barre Lace Mfg. Co.).

Lace is completed by drawing out the dividing threads and allowing the widths to fall apart. If the pattern or lace type requires it, the loose yarns, which pass or "float" from one detached pattern to another, are then clipped. This may be done by machine or by hand. The surplus material on the uneven edge may have to be cut away. After the lace is drawn, clipped, and scalloped, it is again examined, mended, and calendered, to give a glossy appearance. After being wound or jennied, or wrapped on cards, and hydraulically pressed if needed, the lace is now ready for the market.

Other Lace and Auxiliary Machines

There are many other lace-making machines or machines employed to finish the lace, of which a few, because of their widespread use, are worthy of note.

A Nottingham lace machine is a huge machine, on the whole similar to the Levers but with some different features. The machine always works with three yarns, called warp, spool, and bobbin. The warp yarns are wound on a large warp beam, which fits horizontally in the lower part of the machine and is threaded vertically through the machine. There are as many warp yarns as bobbin yarns.

The spool yarns are each wound onto a small spool, about

8 inches long and about 1 inch in diameter. The spools are placed in the lower rear of the machine, and the yarns are fed up through the machine vertically and parallel to the warp yarns. There are as many spools of yarn as there are bobbins used for the given design. The spool yarns, which have individual tension, are used to make the part of the design that requires different lengths of yarn.

The bobbin yarns are wound on bobbins exactly like those in the Levers machine except a trifle larger. The bobbins are so situated in the machine as to allow the yarns to run vertically and parallel to the other two yarns. Like the Lever bobbins, these bobbins run back and forth on a track on the machine. By manipulating the three yarns (with the help of the Jacquard) in a basic V motion, a mosaic tile effect is accomplished. This effect is a characteristic of Nottingham lace.

To obtain different effects, two warp beams can be used instead of one; two sets of spools or a double spool board can be used instead of one; the warp yarns, if the design requires it, can be wound onto spools to allow for varied tensions; or there can be a combination of the above variations.

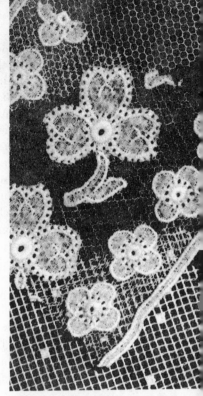

Top to bottom: Bobbinet filet, net appliqué, Nottingham filet.

An eight-gauge machine (eight bobbins to the inch), 360 inches wide, makes lace using 8,640 yarns at one time. Like other machines, the more bobbins per inch, the finer the lace. American machines are from six gauge to sixteen gauge. In Europe there are some machines as fine as twenty gauge.

There are some 500 Nottingham lace machines in the United States, making all types of lace curtains, tablecloths, and similar lace fabrics.

A bobbinet machine makes only net. As a result of its construction, the net made is much stronger and more even than that made on a Levers machine. The bobbins on the net machines have not only the two movements of revolving and moving back and forth but have a third movement. The net-machine bobbins, by means of teeth or gears at the bottom of the carriages, also travel all the way around the machine. By not staying in the same place, the bobbin action adds strength and evenness to the net. Net machines can make a variety of meshes within the net by leaving out bobbins and can work with an adapted Jacquard attachment.

A schiffli machine is not a lace machine but an embroidery machine. It is a big machine about 15 yards long, equipped with needles and small shuttles and working similar to an ordinary sewing machine. It must have a ground fabric upon which to work.

It is a double-decker machine, embroidering two rolls of fabric with the same design at the same time. The top mechanism is exactly like the bottom and both are controlled by the same power and the same set of Jacquard cards, placed at the end of the long machine. The design is drawn and the cards are punched in practically the same manner as are lace designs made for the Levers machine. The fabric to be embroidered is rolled on two long beams, and each is so placed as to roll up away from the needles after it has been embroidered. As much as 10 yards in width may be embroidered at one time.

Below: Schiffli embroidery machine. Insert: Shuttle and bobbin, of schiffli machine (Saentis, Inc.).

The schiffli machine embroiders with two sets of yarns. One set is wound on small spools, about 4 inches long. There are as many spools as there are needles required for the pattern. A machine may use 1,000 needles at one time, 500 for the lower part and 500 for the upper. The spools are placed in a more or less horizontal fashion in the front of the machine and are fed to the needles in a stationary horizontal bed, directly behind them. The needles, very similar to ordinary sewing machine needles, are cammed to go in and out of the fabric as it is moved by the Jacquard action, from right to left and from left to right, allowing the needles to make the design. It is the fabric that moves and not the needles.

The other set of yarns is wound on tiny bobbins which are placed in small 2-inch long shuttles. The shuttles are placed in a bed in a vertical position, just below the needles and behind the fabric. They move up and down and contact the needle yarns as they go through the fabric, thus fastening the pattern yarns securely. There are as many shuttle yarns as there are needles used for the design.

Above: Point de Venise construction. Top to bottom: Treated cotton fabric, embroidered treated fabric, lace after burning out fabric. Below: Machine-made point de Venise.

Fabrics embroidered on schiffli machine.

There is also, beneath the needle bed, a set of boring needles that may be brought into action if the design requires eyelet work. The boring needles are heavier, with sharpened ends, and are used to cut holes in the fabric, which are later embroidered into eyelets by the needles.

After the whole piece of fabric is embroidered, a machine cuts it into correct sections. There are machines that cut the fabric away from the scallops, divide the motifs, or in other words arrange the embroidered work ready for use on garments.

If net or other fine fabric is embroidered, it is usually backed by some strong fabric to give it firmness. This can be removed by means of heat later.

Besides being capable of embroidering all types of designs, from fine outline to heavy clothlike insignia on sheer silk or on heavy cotton or wool, the schiffli machine also makes many interesting laces. This group of laces are termed "burned-out laces." To make this type of lace, the ground must be either of a different fiber or so treated chemically that it will react to heat, lye, or another chemical without harm to the yarns that embroidered the design.

Most of this type of lace today is embroidered on a very thin, wild silk fabric, named "Etz," or on chemically treated cotton cheesecloth, called "Londat." After the design is embroidered, the fabric is burned away, leaving only the lace, which is then bleached, divided, and made ready for use.

The schiffli can make many lace patterns, the most popular of which are Cluny, rose point, Irish crochet, and point de Venise.

A braiding machine is a circular machine using needles that braid or plait the yarns into tubular fabrics. Some machines make fancy braids, others simple braids as shoe laces. One circular machine, known as "Barmens machine," is capable of making certain lace patterns, the most common of which are Cluny and point de Paris. The lace or braid is plaited in circular form and the dividing threads are pulled out later.

A Corneley machine has needles that make lock sitches. It is used mostly to put cordonnet on lace, particularly on Alençon.

A Bonnaz machine is a small machine that makes a chain outline stitch. It works like a sewing machine. It is used to make dots, figures, and edges on veilings, as well as for decorating other types of fabrics.

LACE TERMS

Lace is described by many features, each of which is given a name. The most common parts of a lace are:

Réseau, the background of the lace as distinguished from the design. The effect is irregular as compared to the regularity of mesh. Réseau is also more open than mesh.

Mesh, the net part made by needles or by bobbins in hand-made lace, or by machines.

Ground, the inside part of the design giving more richness to the pattern.

Toile, the solid design of the lace.

À jour, the openwork design which makes the pattern.

Picot, loops used to finish the edge of a lace or to decorate parts of the pattern.

Cordonnet, a heavy thread outlining the design.

LACE STYLES

Lace is made in a variety of widths, each a special type. Today's lace styles are seven in number.

Edging. Narrow lace with a straight back, a scalloped, picot, or some other form of indented front, following a definite pattern. Widths may be from the narrowest, slightly more than a picot, to many inches.

Insertion. Narrow lace with both edges straight to be used inserted between two pieces of fabric. Widths may be from $\frac{1}{4}$-inch to 10 or 12 inches or more.

Galloon. Lace with both edges scalloped or indented in some way. Widths vary from fractions of an inch to many inches.

Beading. Lace or insertion or galloon type, with slits through which ribbon or some form of tape may be laced.

Flouncing. Lace with plain edge and indented front. Widths are from about 12 to 54 inches.

All-over lace. Wide lace, varying in width. It is made to be cut up into desired shape.

Single motifs. Lace made in complete or single design, such as emblems or doilies.

Below: Single motif. Above: Top to bottom: All-over lace, wide edging, narrow edging, beading, insertion, galloon.

408

LACE PATTERNS

Most lace patterns have names, many of them received from the towns or villages in which they were first made or where they first became popular. Some of the types were originally needle point, others bobbin laces. Most of the patterns are now made also by machine. There are many lace patterns that are made only by machines. Since lace is a twisted fabric, no fabric made by a weaving loom or a knitting machine can correctly be termed lace.

Some patterns are constructed by both machine and handwork, as hand embroidery on bobbinet or hand-run cordonnet on machine-made lace. Also, there are many laces that are not one distinct pattern but a combination of two or more patterns. Although the names of many laces may be familiar, some are seldom made or seen at the present time.

The distinguishing characteristics of the most popular lace patterns in use today are outlined in the following:

Alençon. Needle point, or made on Levers machine. The ground is usually very fine, hexagonal mesh, which may be dotted. The toile is heavy, usually of floral design, and the à jour is conspicuous. The edge has little scalloping but may have picots. Alençon's distinguishing feature is the cordonnet which outlines the toile. The cordonnet may be put on by hand, put on by machine and clipped by hand, or left without clipping.

Breton. Hand embroidered on net or embroided on net on a schiffli machine. The long embroidery stitches are conspicuous on the net. The net may be plain bobbinet or a novelty net

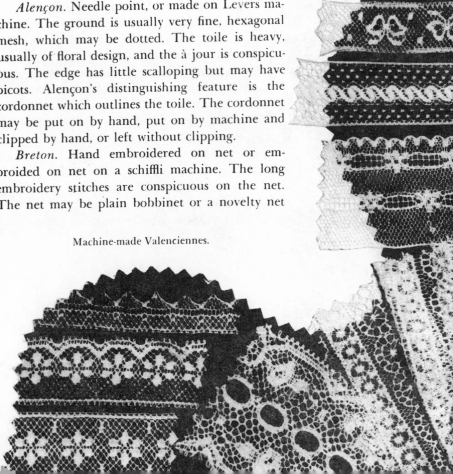

Machine-made Valenciennes.

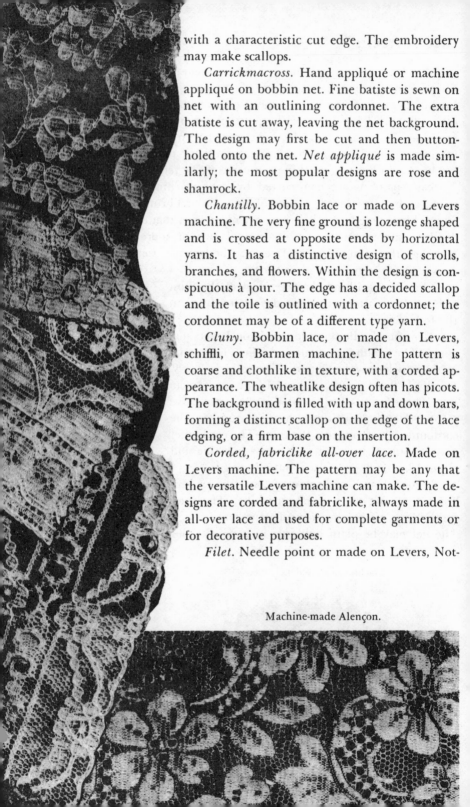

with a characteristic cut edge. The embroidery may make scallops.

Carrickmacross. Hand appliqué or machine appliqué on bobbin net. Fine batiste is sewn on net with an outlining cordonnet. The extra batiste is cut away, leaving the net background. The design may first be cut and then buttonholed onto the net. *Net appliqué* is made similarly; the most popular designs are rose and shamrock.

Chantilly. Bobbin lace or made on Levers machine. The very fine ground is lozenge shaped and is crossed at opposite ends by horizontal yarns. It has a distinctive design of scrolls, branches, and flowers. Within the design is conspicuous à jour. The edge has a decided scallop and the toile is outlined with a cordonnet; the cordonnet may be of a different type yarn.

Cluny. Bobbin lace, or made on Levers, schiffli, or Barmen machine. The pattern is coarse and clothlike in texture, with a corded appearance. The wheatlike design often has picots. The background is filled with up and down bars, forming a distinct scallop on the edge of the lace edging, or a firm base on the insertion.

Corded, fabriclike all-over lace. Made on Levers machine. The pattern may be any that the versatile Levers machine can make. The designs are corded and fabriclike, always made in all-over lace and used for complete garments or for decorative purposes.

Filet. Needle point or made on Levers, Not-

Machine-made Alençon.

tingham, or bobbinet machine. Filet originated in the net used by fishermen. If handmade, the ground is formed by knotted square mesh and the pattern is darned in in geometric outlines. If machine made, the mesh is square with filled-in squares forming the design, and the scalloped edge finished in buttonhole stitch.

Irish crochet. Crochet or made on Levers or schiffli machine. The ground is made of square mesh with or without picots. The design is usually roses or leaves. The edge is made firm by the irregular buttonhole stitch. Irish crochet is usually made of coarse, durable thread.

Maline and tulle. Made on bobbinet or Levers machine. Both have a diamond shaped mesh, made of fine yarns. Maline has a more open net than tulle. Both are a type of net.

Mechlin. Bobbin lace or made on Levers machine. The réseau is very fine hexagonal mesh. The à jour is very open and sheer, and the toile is dainty, both making a floral design, with a heavier yarn forming the cordonnet. There is little or no scallop, but the edge is usually finished with picot.

Milan. Bobbin lace or made on Levers machine. The chief characteristic of Milan is the tape or tape effect making the design and an openwork mesh forming the réseau. Often the lace is made with machine tape forming the pattern and hand buttonhole stitches filling in the background. The edges are usually made in tape design. In the machine-made Milan the tape design is twisted along with the background.

Nottingham. Made on Nottingham machine. The pattern is mosaiclike with a V design. The texture, whether sheer and open or closely woven and fabriclike, always shows a V twist. This lace is always made in wide widths or in finished pieces such as curtains and tablecloths.

Point d' esprit. Made on bobbinet or Levers machine. It is characterized by a solid dot on a net background.

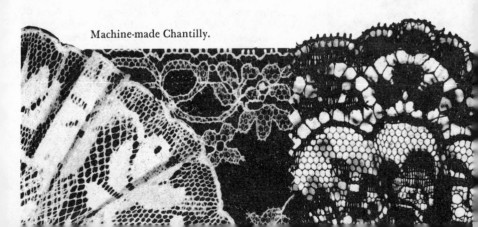

Machine-made Chantilly.

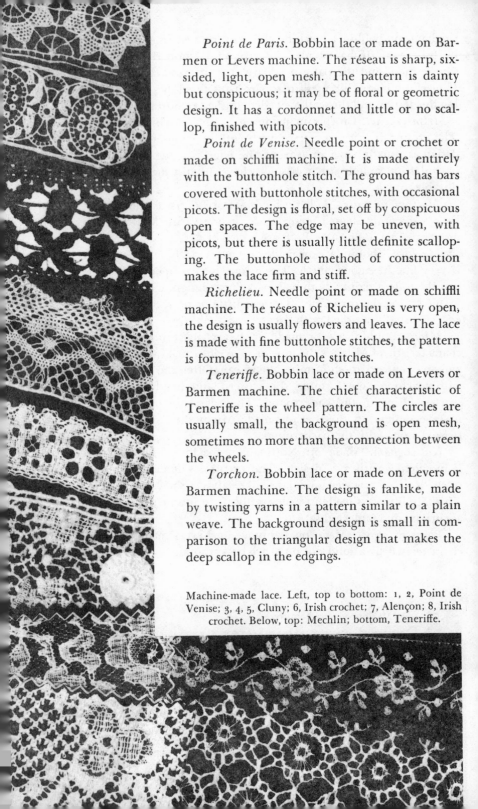

Point de Paris. Bobbin lace or made on Barmen or Levers machine. The réseau is sharp, six-sided, light, open mesh. The pattern is dainty but conspicuous; it may be of floral or geometric design. It has a cordonnet and little or no scallop, finished with picots.

Point de Venise. Needle point or crochet or made on schiffli machine. It is made entirely with the buttonhole stitch. The ground has bars covered with buttonhole stitches, with occasional picots. The design is floral, set off by conspicuous open spaces. The edge may be uneven, with picots, but there is usually little definite scalloping. The buttonhole method of construction makes the lace firm and stiff.

Richelieu. Needle point or made on schiffli machine. The réseau of Richelieu is very open, the design is usually flowers and leaves. The lace is made with fine buttonhole stitches, the pattern is formed by buttonhole stitches.

Teneriffe. Bobbin lace or made on Levers or Barmen machine. The chief characteristic of Teneriffe is the wheel pattern. The circles are usually small, the background is open mesh, sometimes no more than the connection between the wheels.

Torchon. Bobbin lace or made on Levers or Barmen machine. The design is fanlike, made by twisting yarns in a pattern similar to a plain weave. The background design is small in comparison to the triangular design that makes the deep scallop in the edgings.

Machine-made lace. Left, top to bottom: 1, 2, Point de Venise; 3, 4, 5, Cluny; 6, Irish crochet; 7, Alençon; 8, Irish crochet. Below, top: Mechlin; bottom, Teneriffe.

Valenciennes. Bobbin lace or made on Barmen or Levers machine. The background may be small, round or diamond shaped mesh. The toile is sheer and clothlike. The patterns are usually floral, the mesh containing oval or round spots. The edge is finished with picots and with little or no scallop.

NETTING

Netting is knotted fabric. Net-making antedates man's historic past. Relics of netting have been found in most prehistoric ruins. Netting was one of the arts carried on by prehistoric man in every part of the world. The ancient Egyptian hieroglyphs depict netting needles and reels. Hand needles were used entirely until about 1840. In that year the first machine that could make fabric by tying knots was patented.

Today some netting is still made by hand netting needles. The needles have various lengths, are pointed and have a long eye in which is a pointed tongue. The other end is double pointed. Twine is wrapped on the needle, using the tongue and the forked bottom as anchors. With this, knots are tied by hand in the twine making a netting.

The modern netting machine is a special machine, differing greatly from other fabric-making machines. It makes fabric by tying a knot known as the "fisherman's knot." This machine handles heavy twine instead of yarn, knotting a fabric that, regardless of its use, is termed "fish net-

Machine-made lace. Right, top to bottom: 1, 2, Torchon; 3, Breton; 4, 5, point d'esprit; 6, 7, point de Paris. Below: Narrow and wide Teneriffe.

Above: Lace elastic. Below: Handmade lace. Left: Irish Carrickmacross, applied on machine net. Top to bottom, right: Irish crochet; Spanish network filet, sixteenth century; Irish machine net Limerick (tambour work), similar to Breton (The Metropolitan Museum of Art).

ting." It may be of a variety of meshes and may be used not only for fishing, but for bags, furniture, screens, sport items, and the like.

There are several types of netting machines but most are similar to the pound netting machine. This machine uses two sets of twine. One is wound on spools and is carried from the creels at the back of the machine up and over the top of the machine, and down over a tension roller, to the long needles (average length of which is 4 to 5 inches).

The other twine is wound on large, round bobbins. The long needles with a hook on the end are placed on a bed in such a way as to work on a slight angle to vertical position. The round bobbins are in a horizontal bed, directly in back and beneath the needles. There is one needle for each mesh across the width and a bobbin and a spool for each needle. Netting machines can make a variety of meshes, ranging from ¾ inch to 4 or 6 inches, using any quality twine.

Both the bobbin and the spool twines enter simultaneously into the knotting, the

Below: Hand netting needle (actual size). Right: ish netting made on knotting machine.

machine tying a complete row of knots at one time in each revolution. In the netting the twines are alternately from the spool and the bobbin, because each knot, as tied, has bobbin twine and spool twine in it. Alternate rows of knots are reversed and inverted in position to the adjacent rows. Netting is so made that strain tends to tighten and not loosen the knots.

After the netting comes off the machine, it is inspected and, when necessary, stretched on a stretching machine to draw the knots tighter. Often edges or knots are given certain handwork known as "making up." Netting twine may be of cotton, linen, or other bast fibers, constructed similarly to yarns for other types of fabric, for twine is but multiply yarn.

For quality netting, the fibers must be strong and the twine must be sturdy and firm enough to endure the pull of tightening knots and the force of heavy weights. For example, netting must withstand the sudden strain encountered during such activity as commercial fishing or merchandise loading of ships.

SUMMARY

TO PRODUCE LACE

Warp: The yarns that run crosswise in a lace fabric.

Filling: The yarns that run lengthwise in a lace fabric.

Twisting: The twisting together of warp and filling yarns to construct a fabric.

Types of machines: Levers, bobbinet, Nottingham, schiffli.

LACE TYPES

Handmade

Needle point	*Bobbin*		*Crochet*
Alençon	Chantilly	Torchon	Irish crochet
Filet	Cluny	Teneriffe	Point de Venice
Point de Venice	Milan	Valenciennes	
Richelieu	Mechlin	Point de Paris	

Machine-made

Levers	*Bobbinet*	*Schiffli*
Alençon	Net appliqué	Cluny
Chantilly	Filet	Irish crochet
Cluny	Maline	Richelieu
Maline	Tulle	Point de Venice
Teneriffe	Point d'esprit	Breton
Milan	Carrickmacross	
Tulle	Commercial nets	*Braiding*
Point de Paris	Breton	Cluny
Valenciennes		Point de Paris
Filet	*Nottingham*	Torchon
Mechlin	Nottingham lace	Teneriffe
Corded all-over	Filet	Valenciennes
Carrickmacross	Net	
Net appliqué		
Commercial nets		

LACE STYLES

Edging, insertion, galloon, beading, flouncing, all-over single motifs.

LACE TERMS

Réseau, toile, à jour, picot, cordonnet.

Opposite page: A thick rug cushion that has been felted with 100 per cen
cattle hair. The design was pressed into the cushion after it was felted
(Clinton Carpet Company).

Produces fibrous fabrics

Felting constructs a fabric by pressing together and interlocking masses of loose fibers. Heat, moisture, and pressure are required
to produce felt. Wool fabrics that are woven and then fulled to
shrink them, often called "woven felts," are not felts.

It is the ability of the wool and hair fibers to intermingle—as
moisture and pressure cause the fiber to bend towards its root—
that permits felting. Felts are also made of cotton, spun rayon,
casein and soybean fibers, and other synthetic fibers.

The fibers to be felted into a fabric are mixed and carded. If
for yarns, the fibers leave the carding machines in a rounded soft
roving. For felt, they leave the carding machine on what is known
as a "former," in a wide soft web. The web of fibers passes over a
canvas or lattice apron and is laid in layers called "bats," until
the desired thickness is obtained.

The bats are placed in a large press, one bat on top of the other
with a moistened fabric between them. Steam is applied, and the
cover of the press rotates and presses down the bats. In this manner
heat, moisture, and pressure produce felting of the fibers and condition them for further felting.

The fibers are then soaped and entered into some type of fulling
or felting machine, which presses the soaped pieces together.

The felt is washed to remove the soap, rinsed, dyed, and dried.
Some are given a starch bath for a firmer body. Others, such as
those intended for hat felts, are napped, brushed, sheared, and
pressed. Napped felts are dyed after napping. Shoe felts are dyed,
dried, sandpapered, and then sheared to give a smooth surface
texture.

Another method of producing felts is by punching the bat into a burlap cloth. Punching needles cause the fibers to adhere to the cloth. The fabric is then fulled, washed, rinsed, dyed, and pressed. This is a much less expensive method.

Felting Rug Cushions

A wide use of felting is in the making of rug cushions. The following is a brief description of one method used.

The hair is removed from the cattle hides, washed, and sterilized. After the bales of hair reach the manufacturing plant, the fibers from several bales are blended, as one hide has many qualities of hair, just as different breeds of sheep produce wool fibers of different qualities. After the fibers are mixed in a mixer machine, foreign matter is removed, and the fibers are passed to a burr picker machine. Here they are further cleaned and combed. Three wide layers of fiber, one on top of the other, leave the machine simultaneously. Over these three bats a cover of burlap is laid, and on top of the burlap are laid three additional bats of hair. The cushion is now ready for compressing and felting. The cushion is moistened and passed into a heated "jiggler" chamber. The heat converts the water into steam, which penetrates the cushion. This is then subjected to a pressure of several tons and is compressed to about one twelfth of its former thickness. During this compressing, the cushion is resting on a steel bed and is shaken or jiggled to further interlock the hair fibers.

After felting, the cushion is dried and is ready for cutting and marking for shipment.

Left: Hair from the cow hide may be used as fibers for making felted fabrics. Right: Layers of hair batting that later will be compressed into a thick, felted fabric (Clinton Carpet Company).

SUMMARY OF FABRIC CONSTRUCTION

Fabrics are constructed by the methods listed below. Technically, the word "textiles" applies to woven fabrics only and is but one section of the fabric industry. However, as used in the trade, it includes the fabrics constructed by the other methods.

Weaving: The construction of fabrics by interlacing two sets of yarns, at right angles to each other.

Knitting: The construction of fabrics, with needles, by interlooping one continuous yarn to form a succession of connected loops.

Twisting: The construction of openwork fabrics by twisting together two sets of threads. One set of threads pass between and round a second set of threads.

Knotting: The construction of fabrics by tying together two sets of yarn.

Braiding: The construction of flat or tubular fabrics by interlacing two sets of yarns at oblique angles.

Felting: The construction of fabrics by pressing together and interlocking masses of fibers by means of heat, moisture, and pressure.

Opposite page: Fabric chemists are constantly exploring, experimenting, improving, and standardizing chemical compounds to produce more beautiful, serviceable, and suitable fabrics (American Wool Council, Inc.).

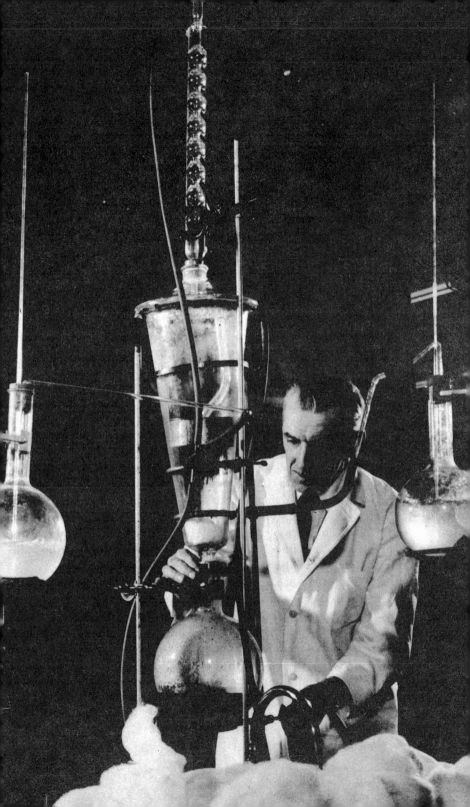

CHEMISTRY

A part of today's fabrics

This chapter would not be considered adequate by someone familiar with the wide ramifications of fabric chemistry. It merely endeavors a brief explanation of the chemical terms used throughout the book, in parts dealing with the making of rayons and other synthetics; with bleaching, finishing, dyeing, and printing fabrics; or with additional functional finishes. It contains subject matter that is elementary and much that is incomplete, because full discussion of the particular subject would be too technical in a book such as this. The chapter is included because it was thought that, while it is not possible or essential for the layman to understand the complex formulas and technical methods and procedures, it is worth knowing whence come the raw materials that have such a great part in making fabrics beautiful and colorful.

Even though chemistry is a science many hundred years old, only in this century have chemicals been used extensively to widen the range of fabrics, to give fabrics made of both natural and synthetic fibers many new qualities, and to contribute improvements to the existing finishing processes.

The study of chemistry is vast. It involves everything on the earth, in the air, and in the water. Some few of the elements were discovered centuries ago, long before the birth of Christ, some are recent, spectacular discoveries. Over the centuries man, little by little, has gained new knowledge of the elements and their compounds and has learned and is learning how to use them.

The chemist in the fabric industry uses complex formulas in

making compounds, usually mixed with other compounds, forming one chemical after another until he arrives at the one that answers the purpose for the finish, the dye, or the synthetic fiber. It requires years of intensive study to become a fabric chemist. In addition to the multitude of chemicals with which he must become acquainted he must learn the reaction of each one to the various fibers which are also chemical compounds. Animal and plant fibers react quite differently to the same chemical. Even wool and silk, both animal fibers, often require different chemical compounds for dyeing and finishing.

Today, numerous synthetic fibers are being introduced. They are chemical compounds within themselves, and, as a result, require a whole new field of study even for the experienced chemist.

Chemistry is the science that studies the composition of matter, what it is, how it changes, what makes it change, and the conditions that control these changes. It studies its actions and reactions, its energy, and its possible uses.

MATTER

Matter is anything that occupies space and has weight. It cannot be created or destroyed but it can be changed. Matter, such as water, air, or coal, is called a "substance." There are three types of matter: solid, liquid, and gas. Each is made up of elements and compounds.

In a solid (such as coal) the molecules are very close together, in a liquid (such as water) they are more widely separated, and in a gas (such as air) they are the farthest apart. The distinction between solids, liquids, and gases is essentially in the difference in the speed with which the molecules move. In gases they move with great speed. In liquids they are more closely spaced, and move slightly, so that a liquid takes the shape of its container. In a solid they are so closely placed that they cannot move freely, and the solid takes a definite shape.

Properties. Physical properties of a substance include its appearance, its color, its odor, and taste. Chemical properties include the lack of action or the speed of action with other substances; basic or acidic properties, and whether or not a substance burns or supports combustion.

Changes. All matter contains energy that can neither be created nor destroyed but that can produce physical and chemical changes.

A physical change is brought about when a substance is changed outwardly but does not lose its original chemical composition and characteristics. A glass bowl is broken and thus is changed physically, but the glass remains glass. Viscose and cuprammonium rayon are called "regenerated cellulose rayons" because the cellulose, while changed physically (by chemical means), remains cellulose.

A chemical change is brought about when the chemical composition of a substance is changed. It loses the original characteristics by which it is recognized and is changed into one or more new substances with new physical and chemical properties and with new energy. Such processes as burning, rusting, souring, or decaying are chemical changes. Mixing with other chemicals frequently results in a physical change in the substance only, while others produce a chemical change.

Kinds of Matter

There are three types of matter: elements, compounds, and mixtures.

An element is a substance that cannot be broken down into simpler substances by ordinary chemical methods. Each element has its own particular kind of atom. Elements are found in the air, water, earth, and vegetation. There are now ninety-six recognized elements, and many of these are not important for our discussion. The elements, and their symbols, most frequently used for fabrics are:

Ca —Calcium	N —Nitrogen
C —Carbon	O —Oxygen
Cl —Chlorine	K —Potassium
Cu—Copper	Si —Silicon
F —Fluorine	Na—Sodium
H —Hydrogen	S —Sulfur
Fe —Iron	Sn —Tin
Pb—Lead	

Elements are divided into two classes, metallic and nonmetallic. In the above list the metallic elements are calcium, copper, iron, lead, potassium, sodium, and tin. Calcium, potassium, and sodium are alkaline metallic elements. Heat and electricity will pass through metallic elements but light will not. These elements are malleable. They are solids (at ordinary temperatures) with a

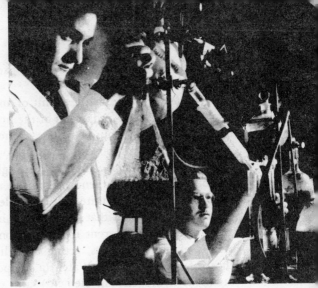

Experimenting and testing is constantly going on to produce better dyes and finishes and to create new fibers (American Wool Council, Inc.).

metallic luster. With the exception of copper, most of them are grayish white or silver in color. Chemically, metallic elements combine with nonmetals and form salts. They form positive ions and lend electrons.

The nonmetallic elements are carbon, sulfur, silicon, chlorine, fluorine, hydrogen, nitrogen, and oxygen. The last five are gaseous nonmetallic elements. The properties of the nonmetallic elements are opposite from the metallic properties. They are poor conductors of heat and electricity. They are nonlustrous solids and gases (at ordinary temperatures) and are not malleable. Chemically, nonmetallic elements combine with metals, form acids and negative ions, and borrow electrons.

A compound is a substance formed when two or more of the elements unite chemically. In other words, a compound is a substance that by chemical means can be broken down into two or more simple substances. Each element loses its original characteristics and the compound acquires its own characteristics that differentiates it from other compounds. Water is a compound made up of two elements, hydrogen and oxygen, and water always contains the same elements in the same proportion.

A mixture is a substance formed by two or more elements or compounds that are not chemically united. These elements or compounds retain their individual properties and can be separated by mechanical means. Air is an example of a mixture.

An atom is the smallest unit or particle of an element that takes part in chemical change.

A molecule is the smallest unit of an element or a compound that contains all the properties of the original substance. A molecule of an element consists of one or more atoms of the element. A molecule of a compound contains groups of atoms.

The atomic theory. In 1803 an English school teacher, John Dalton, developed a theory regarding atoms. Briefly stated, all matter is made up of finite particles called "atoms." Each element is composed of atoms that are identical in size, weight, and shape, and are entirely different from the atoms of any other element. Atoms of most elements, acting as individual units, unite with those of other elements. Water (a compound) is made up of two atoms of hydrogen (an element) and one atom of oxygen (an element) which form a molecule. It is noted as H_2O. Hydrogen chloride (a compound) is composed of one atom of hydrogen and one atom of chlorine, which forms a molecule. It is noted as HCl.

A radical is a group of atoms in some compounds which act as a unit in many chemical reactions.

A catalyst is a substance that changes the rate of a chemical reaction while not being permanently changed itself. *An enzyme* is a catalyst that is a complex organic substance and that changes one organic compound to another. Alcohol may be made by the action of the enzymes on sugar (dextrose) changing it to carbon dioxide and alcohol.

The electron theory. It was the twentieth century scientist who made the great contribution of the electron theory regarding the atom. This theory has added immeasurably to the understanding of atomic structure. Dyes and finishes have been improved as the electron theory more fully explains the action and affinity of chemicals, the processes of reduction and oxidation, and how chemicals are combined. According to the electron theory, an atom is a small particle of matter made up of a nucleus (contain-

Opposite page: This is an electronic picture of the formation of sodium chloride from its elements, sodium and chlorine, and its dissociation in solution (Visualized Chemistry, by William Lemkin, Ph.D. Published by Oxford Book Company).

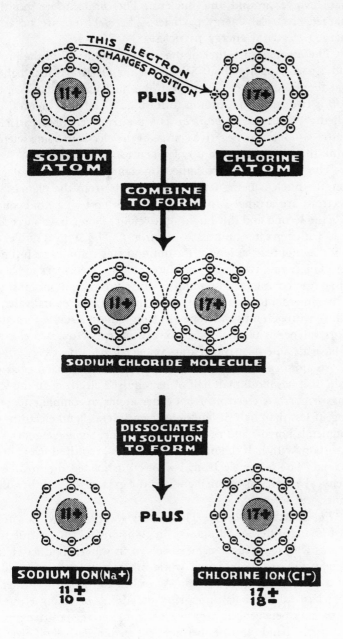

THIS ELECTRON CHANGES POSITION

PLUS

SODIUM ATOM

CHLORINE ATOM

COMBINE TO FORM

SODIUM CHLORIDE MOLECULE

DISSOCIATES IN SOLUTION TO FORM

PLUS

SODIUM ION (Na+)
11 +
10 −

CHLORINE ION (Cl−)
17 +
18 −

ing protons, neutrons, and positrons) and one or more electrons that revolve around the nucleus. The *protons* are positively charged electrical energy particles. The *electrons* are negatively charged electrical energy particles.

The *nucleus,* or center of the atom, contains all of the protons and a part (usually about a half) of the electrons. It is believed that there are no free electrons in the nucleus but that each electron is tied up with a proton to form an electrically neutral mass called a *"neutron."* A *positron* is a positively charged particle of matter equal to the negative charge of the electron. Some scientists think the proton is made up of a neutron and a positron.

The atom is often explained by comparison to our solar system, the nucleus being the sun and the electrons the planets. The electrons are arranged in rings, the innermost ring holds one or two, the second and third hold up to eight. (Some have succeeding rings with a greater number of electrons.) The atom is stable, that is, it does not react as long as the outmost ring contains its full number of electrons, two, eight, or more, and it always seeks to have its outer ring complete. The number of electrons an atom must gain or lose in order to have its outer ring complete is called its "valence." Thus, hydrogen must lose one electron to have a stable formation, so its valence is positive one. Oxygen must borrow two electrons to have a stable formation, so its valence is negative two.

An atom completes its outer ring with as little change as possible. If it has more than half of its required electrons in the outer rings it borrows electrons from other atoms to complete its outer ring. If less than half, it lends electrons. Metals lend electrons and nonmetals borrow and thus readily combine. Some elements, such as carbon, with half the number of electrons required may borrow or lend. The fewer electrons borrowed or loaned the greater the chemical activity of the element and the greater the stability of the compound formed.

The ionization theory. This theory was developed as a result of studying different substances in solutions to answer questions such as why some substances, dissolved in water, conduct electric current and are decomposed while others are unaffected and do not conduct an electric current. According to this theory, when an atom has stabilized its outer ring by adding or losing electrons, the balance between the negative electrons and the positive protons is disturbed and the atom becomes charged, negatively if electrons were added, positively if electrons were lost. The charged particle is called an "ion." Ions may also be groups of atoms such as the sulfate ion. Acids, bases, and salts, when dissolved in water, break up into ions. Water solutions of compounds are electrically neutral as they have an equal number of positive and negative ions. Neutral also means neither acid or basic. Many water solutions of compounds are not neutral in this respect.

Metals lend electrons so metallic ions have positive charges; nonmetals borrow electrons and nonmetallic ions have negative charges. As an electric current passes through a solution the positive ions called *cations* move to the negative pole and borrow electrons. The negative ions called *anions* move to the positive pole and lend electrons. This is called *electrolysis*.

The atom has an incomplete outer ring and is chemically active; while the ion has a completed outer ring and is inactive.

Alpha and beta are used to designate the electric rays given off by certain atoms. An alpha ray consists of ions and is positively charged; a beta ray consists of electrons and is negatively charged. It has about one hundred times the penetration of the alpha.

MIXTURES

In mixtures, particles of one substance are mixed with another. Mixtures are not compounds unless they are of definite proportions and undergo a reaction to form new molecules. For example, hydrogen and oxygen may be mixed but will not form the compound water unless present as two parts of hydrogen to one part of oxygen and then are reacted to form water.

Particles may be mixed as dry powders, in a solution, in the colloidal state, or in suspension. In a true solution only single atoms or molecules are mixed. In a colloidal mixture the particles are usually groups of molecules, hence they are larger than the units in a solution. In a suspension the particles are larger units.

A solution is a mixture made up of two parts: the *solute,* the substance that is dissolved, and the *solvent,* the substance that dissolves another substance. The solvent is usually a liquid such as water or alcohol but may be a solid or a gas. The solute may be a liquid, gas, or solid. The solute becomes distributed throughout the solvent but separates by evaporation of the liquid—a physical change only. A dilute solution has a small proportionate amount of solute; a concentrated solution has a large amount; an unsaturated solution has less solute than it can dissolve; a saturated solution has all of the solute the solvent can normally dissolve; a supersaturated solution has more of the solute than it can dissolve.

The solubility (ability to dissolve) is a physical property; and, of course, some solutes dissolve readily while others do not. Temperature is usually important, and agitation (as stirring) frequently increases rate of going into solution. High temperature usually increases the solubility of solids but decreases that of gases. In other words, solids are generally more soluble in heated liquid, and gases are more soluble in cold liquid.

Solubility of compounds. In general, the following are the solubility characteristics of the more important compounds used in fabric chemistry. *Soluble in water:* Most ammonium, potassium, and sodium compounds; all acetates, chlorates, chlorides, and nitrates; all sulfates except calcium sulfate which is only slightly soluble. *Insoluble in water:* Most carbonates, hydroxides, and oxides except those of potassium, sodium, and ammonium (mentioned above as being soluble). Calcium hydroxide, an exception, is slightly soluble.

A suspension is a mixture of a solid and a liquid, such as clay and water. The solid can usually be separated from the liquid by filtering or, if left standing, the solid will sink out of the liquid. When the suspended solid is large, the liquid is cloudy and darkened.

An emulsion is a suspension of one liquid in another in which it will not dissolve, such as oil in water. The stability of emulsions varies and agents are often added to make an emulsion more permanent.

The colloidal state exists when very small particles are scattered throughout another substance, usually a liquid. The particles held in colloidal suspension are called "the dispersed phase," and the medium in which they are suspended is called "the dispersing medium." Either may be a liquid, gas, or solid. The particles in colloidal suspension may be suspended in, but are not combined with, a solid, liquid, or gas. The particles appear to be electrically charged and remain in suspension more or less indefinitely. A passage of electric current will neutralize the colloidal particles causing them to coagulate and drop out of the suspension. Colloids play an important part in fabrics. They are used in making rayon, rubber, glue, starch, plastics, soap, finishes, and dyes.

ACIDS, BASES, AND SALTS

Since there are hundreds of thousands of compounds it is necessary to separate them according to their properties. Most compounds can be classified into four important groups: acids, bases, salts, and oxides. This last group is discussed under oxygen (page 433).

Acids

An acid is a compound of hydrogen ions and an acid radical whose water solution contains hydrogen ions as the only positive ions. An acid turns blue litmus red. Many hydrogen compounds are not acids as only those that form positive hydrogen ions are acids. The hydrogen of all acids can be replaced by a metal. Acids react with bases (metallic hydroxides) to form salt and water. Acids may be formed by the action of sulfuric acid with the salt of an acid or by the action of acid anhydrate and water. Acids have a sharp, sour taste. Some are weak and others are strong. The strength of an acid is dependent on the degree of ionization. Nitric acid is a highly ionized, strong, and active acid. Citric acid is a slightly ionized weak acid. Fruit juices are examples of weak acids.

The standard for measuring the acidic strength of a solution of acid is called "pH." A pH of seven indicates a neutral solution; a pH greater than seven indicates a basic solution, and one of less than seven, an acidic solution. The smaller the pH, the greater the acidic strength of the solution.

Bases

A base is a compound which forms a hydroxyl (hydrogen and oxygen radical) ion, as the negative ion when put into solution. A base turns red litmus blue. Bases combine with acids and non-metallic oxides to form salt and water. Not all hydrogen and oxygen radicals are bases, as many do not form ions. Bases are formed by (a) metals, such as calcium, potassium, or sodium, which combine with water and replace the hydrogen ions in water; (b) the salt of the desired base combined with another base; and (c) water and a metallic oxide. Water soluble bases are called "alkalies." Examples of bases are sodium and potassium hydroxide, which are strong bases (frequently called "alkalies"); and calcium and

431

ammonium hydroxides, which are weak bases. Most hydroxides are practically insoluble in water.

Salts

A salt is a compound containing positive ions other than hydrogen ions, and negative ions other than hydroxyl ions. There are many more salts than acids or bases. There are acid, basic, normal, and double salts. Salts may be formed by (*a*) an acid and an oxide of a metal; (*b*) a base and a nonmetallic oxide, and (*c*) an acid and a base. Most salts are crystalline solids, many of which are soluble in water. The water solutions of normal salts, acid salts, and basic salts (if soluble) conduct electricity.

In the newer concept, *acids* may be molecules, positive ions, or negative ions. It is a substance which lends protons. *Bases* are molecules and negative ions and will combine with a proton. Thus acids and salts combine to form bases. *Salts* are made up of positive and negative ions and are good conductors of electric current.

The prefixes and suffixes of acids and their corresponding salts explain their type. The following is the nomenclature for acids and bases and an example of each.

Binary	*Acids*	*Salts*
(Composed of two elements—hydrogen and a nonmetal)	hydro-ic (as *hydro*chlor*ic* acid)	-ide (as sodium chlor*ide*)
Ternary		
(Composed of three elements—hydrogen, a nonmetal, and usually oxygen)	-ic (as chlor*ic* acid)	-ate (as sodium chlor*ate*)
Ternary with less oxygen	-ous (as chlor*ous* acid)	-ite (as sodium chlor*ite*)
Ternary with least oxygen	hypo-ous (*hypo*chlor*ous* acid)	hypo-ite (sodium *hypo*chlor*ite*)

THE ELEMENTS AND THEIR COMPOUNDS

In the following brief discussion of the elements and their compounds it will be noted that the nonmetallic group—the compounds of oxygen, hydrogen, nitrogen, fluorine, chlorine, carbon, and sulfur—are expressed as nonmetallic ions, such as oxides, hydroxides, nitrates, fluorides, chlorides, carbonates, and sulfides; and the compounds of the metallic group—sodium, potassium, calcium, silicon, and copper—are expressed as metallic ions, such as sodium sulfate. Knowledge of the simple compounds will help in understanding the chemicals discussed throughout this book.

OXYGEN

Oxygen, one of the world's most important elements, is a tasteless, odorless, colorless nonmetallic gas, but when compressed becomes a blue liquid. It constitutes about one quarter of the weight of the air just over the earth's surface, eight ninths of the weight of all water, and it is combined with other elements in about one half of the surface of the earth. It occurs in a free state and in chemical compounds. Animals and plants require oxygen for life. Oxygen may be obtained by heating certain metallic oxides or certain oxygen-containing salts, or by the action of water on sodium peroxide. It is slightly soluble in water.

Oxygen Compounds and Their Uses

The most outstanding property of oxygen is its ability to readily combine with other elements and to support combustion. Oxygen forms oxides.

One of the greatest uses of chemicals is in dyeing. Here wool yarns are being dyed in huge steaming vats filled with dyestuffs (Cyril Johnson Woolen Co.).

An oxide is a compound of oxygen and another element. The oxides are formed by heating the metals in the air (which contains oxygen), or by decomposing the salts with oxygen. *Oxidation* occurs when the substance is combined with oxygen.

Deoxidizing is the removal of oxygen from its compound. Any substance that removes oxygen is called a "deoxidizing" or "reducing agent."

Sulfur dioxide, a colorless gas, results when sulfur or a sulfur compound is burned in air or oxygen.

An anhydride is an oxide that combines with water to form either acids or bases. The oxides of nonmetals are acid anhydrides. These compounds plus water form acids. For example, nitrogen and oxygen are a nitric anhydride (or anhydrous). When this is combined with water, it becomes hydrated nitric acid or merely nitric acid. Two other acids so formed are:

Sulfur + oxygen + water form sulfuric acid.
Carbon + oxygen + water form carbonic acid.

Oxides of metals are basic anhydrides. Upon the addition of water they tend to form bases. For example, calcium, oxygen, and water form calcium hydroxide, a base.

Uses. The following are a few of the specific uses of oxides. Copper oxide in ammonia is used for making cuprammonium rayon; sulfur dioxide for bleaching wool; hydrogen peroxide for bleaching wool and cotton, and for an aftertreatment for vat and sulfur dyes.

HYDROGEN

Hydrogen, another important element, is a colorless, tasteless, and odorless nonmetallic gas. It occurs in free state in gases and in the sun's atmosphere, it is one ninth of the weight of water, it is in all acids and in many carbon and organic compounds. Hydrogen is readily obtained from an acid such as hydrochloric acid or sulfuric acid, when replaced by a metal, such as zinc. Hydrogen is the lightest known gas. It is slightly soluble in water.

Hydrogen Compounds and Their Uses

Hydrogen combines readily with other elements, particularly with the nonmetallic elements, such as sulfur and chlorine, and especially with oxygen for which it has such a strong attraction that it can remove the oxygen from certain of its compounds. It thus acts as a deoxidizing or reducing agent. Hydrogen is used in dyes and for bleaching animal fibers.

Hydrogen peroxide, an oily liquid, is a compound of the two elements, hydrogen and oxygen. Water has two atoms of hydrogen and one atom of oxygen, while hydrogen peroxide has two atoms of hydrogen and two atoms of oxygen. Hydrogen peroxide is unstable and tends to give off one of its atoms of oxygen, leaving water. This newly released, nascent oxygen is used to oxidize or bleach the color in wool and cotton. It is also used as an after-treatment for vat and sulfur dyes.

NITROGEN

Nitrogen, a gaseous nonmetallic element, important to life, is the most abundant substance found in air. It is slightly lighter than air, odorless, tasteless, and colorless. It is less soluble in water than is oxygen. At one time, potassium and sodium nitrate were the chief sources of nitrogen, but now it is commercially obtained by liquifying air. It does not readily combine with other elements. When it does, the compound can be easily decomposed and is used in consequence to manufacture explosives. Ammonia is one of the important compounds of nitrogen.

Nitrogen Compounds and Their Uses

Nitrogen and its compounds are used for fabrics as reducing and cleaning agents, as solvents, in the making of dyes and rayons, and for many other uses.

Oxides of nitrogen are formed from a compound of nitrogen and oxygen with a different number of atoms of each. They include:

Nitrous oxide, made by heating ammonium nitrate; *nitric oxide,* made by the action of dilute nitric acid on an inactive metal; *nitrogen dioxide,* made by nitric oxide uniting with oxygen. *Nitric acid* is the combination of a nitrate and sulfuric acid.

The nitrates are an important group of nitrogen compounds. They are formed by the combination of nitric acid with metals, oxides, hydroxides, or salts. Sodium nitrate, potassium nitrate, (saltpeter), and calcium nitrate are the most important.

Ammonia, a colorless gas, is a nitrogen compound formed by a combination of nitrogen and hydrogen. It may be prepared chemically from soft coal, as ammonia is one of the gases given off when coal is distilled. It is also manufactured by the catalyzing combination of hydrogen and nitrogen of the air under pressure. Ammonia is extremely soluble in water. When the ammonium radical replaces the hydrogen in an acid it forms a salt. Ammonia and

sulfuric acid form *ammonium sulfate*
nitric acid form *ammonium nitrate*
carbonic acid form *ammonium carbonate*
hydrochloric acid form *ammonium chloride*.

When ammonia forms a compound with a nonmetallic element such as sulfur it becomes ammonium sulfide. When ammonium carbonate is combined with acetic acid it forms *ammonium acetate*.

An *amine* is formed by hydrocarbon radicals replacing hydrogen in a compound of the ammonium type.

Uses. The following are a few of the specific uses of nitrogen compounds. Copper ammonia is used in the making of cuprammonium rayon, ammonium acetate for retarding acid and chrome dyes, and ammonium sulfate in dyes for mordanting. Sodium nitrite is used for diazotizing certain dyestuffs.

SULFUR

Sulfur, a nonmetallic element, is found extensively in nature, both free and combined with other elements. It is present in ores of metals as a sulfur compound and also as a compound with alkaline metallic elements, such as calcium and sodium. Some 75 per cent of the world's supply of sulfur is produced in the United States, in Louisiana and Texas. It is a pale-yellow, crystalline solid, tasteless and odorless.

Sulfur is similar to oxygen in chemical activity. Each has a minus two valence and each needs to borrow two electrons to complete its outer ring. Sulfur combines with metals to form sulfides as oxygen combines to form oxides. It also combines with nonmetals, such as with chlorine to form sulfur chloride.

Sulfur Compounds and Their Uses

Sulfur and its compounds are widely used in the fabric industry in making coal-tar products, dyes, mordants, special finishes that render fabrics resistant to insects and mildew, bleaching agents for organic materials like silk, as dehydrating agents, for vulcanizing rubber, for making rayon, and for many other uses.

Sulfur dioxide, a colorless gas with a strong odor, is found in gases and is commercially prepared by burning sulfur. The sulfur combines with oxygen and forms sulfuric (or sulfurous) anhydride, which when combined with water forms *sulfuric (or sulfurous) acid* (sulfur, hydrogen, and oxygen) . When the hydrogen in the acid is replaced with a metal, sulfates of the metal are formed.

Uses. Sulfur readily combines with nitrogen and with most metallic elements. The following are typical of the many specific uses of sulfur compounds as applied to fabrics: Ammonium sulfate as a mordant in dyes. Copper sulfate as after treatment for sulfur dyes. Sodium sulfate in the making of viscose rayon, as a retarding agent for acid and for mordant dyes; as an exhausting agent for sulfur, vat, diazotized, chrome, and direct dyes. Carbon disulfide in the making of viscose rayon. Sodium sulfide for reducing sulfur dyes. Bisulfite of soda in wool shrinking and bleaching; bisulfate of soda as an exhausting agent for acid dyes; and sulfite for bleaching cotton linters. Alum (a double sulfate of a trivalent and univalent metal) is used as a retardant for basic dyes.

Sulfuric acid is used as an exhausting agent for acid and mordant dyes, in the making of viscose rayon, and as a catalyst in the making of cellulose acetate rayon. Sulfurous acid is used in the bleaching of wool.

THE HALOGEN OR SALT-FORMING ELEMENTS

There are four halogen elements, two of which—chlorine and fluorine—are used in the fabric industry. These nonmetallic elements combine with metals and form compounds that resemble salt.

CHLORINE

Chlorine is a nonmetallic, yellowish green, heavy, gaseous element, the second most active nonmetal. It is found extensively in nature, usually in combination with sodium, an alkaline metallic element. It is never in a free state. It is found in salt beds in deposits of solid salt and in salt water. Chlorine may be manufactured by the electrolysis of sodium chloride (common table salt) , but is generally prepared by treating hydrochloric acid with an oxidizing agent. Chlorine is poisonous, with a disagreeable

odor. It is slightly soluble in water and quite soluble in organic solvents.

Chlorine Compounds and Their Uses

The greatest use of chlorine for fabrics is for bleaching cotton and wood pulp. It is also used in dyeing, as a disinfectant, and as a water softener.

Hydrogen chloride, a colorless gas, is a compound of hydrogen and chlorine. It can be oxidized to form water and chlorine. *Hydrochloric acid* is a colorless solution of hydrogen chloride and water. When a metal such as magnesium, aluminum, or zinc is combined with hydrochloric acid, the metal replaces the hydrogen in the acid to form a *chloride,* a salt.

Uses. The following are a few of the specific uses of chlorine and its compounds. Chlorine is widely used to bleach cotton. Sodium chloride is used as an exhaust for direct, sulfur, and vat dyes; calcium hypochlorite to shrink wool; vinyl chloride as a resistant to mildew; and toluene sulfochloride forms an ester of cellulose (discussed later) and immunizes cotton. Hydrochloric acid is important as a solvent, particularly when combined with nitric acid. It is also used in shrinking wool and as a diazotizing agent in dyeing.

FLUORINE

Fluorine is a nonmetallic, pale yellow, gaseous element, similar to chlorine. It is found in small quantities in vegetable and animal substances, and is most abundant in a mineral in the form of calcium fluoride. It is highly soluble in water and very poisonous. It is the most active of all the elements. Fluorine is prepared by the electrolysis of sodium or potassium hydrogen fluoride.

Fluorine Compounds and Their Uses

Fluorine has the strongest tendency of all elements to form chemical compounds. *Hydrogen fluoride,* a colorless gas, is a compound of hydrogen and fluorine. This combined with water forms *hydrofluoric acid.* Fluorine readily forms compounds with carbon and sulfur. It also combines with alkaline metallic elements.

Uses. Silicon fluoride is used for finishes that make fabrics resistant to insects.

THE ALKALI ELEMENTS

Sodium and potassium are in the alkaline metallic group while calcium is in the alkaline earth metallic group. None of these three elements is ever found in a free state and each is more important in compounds than as a free element.

SODIUM

Sodium is an alkaline metallic element found extensively in nature but always combined in a compound. It is found in rock salt beds, salt lakes, and oceans. Common table salt and soda are compounds of sodium. It is an important element in the earth's crust and is a very soft metal, with a soft metallic or silvery luster.

Sodium Compounds and Their Uses

Sodium compounds have a wide variety of uses in fabric manufacturing, such as for dyes, as deoxidizing or reducing agents (replacing hydrogen in water), as bleaching agents, in the manufacture of viscose rayon, in mercerizing of cotton, and for many other uses.

When sodium replaces one of the hydrogens in water, it forms *sodium hydroxide* (a compound of sodium, hydrogen, and oxygen). It is a white solid very soluble in water and commonly called "caustic soda." When sodium replaces the hydrogen in an acid it forms a *salt*. Sodium and

> sulfuric acid form *sodium sulfate* (Glauber's salt)
> nitric acid form *sodium nitrate*
> carbonic acid form *sodium carbonate* (soda ash)
> hyrochloric acid form *sodium chloride*

Sodium combines directly with the nonmetallic elements, such as chlorine and sulfur, to form sodium salts.

When sodium replaces one of the hydrogens in carbonic acid, *sodium bicarbonate* (baking soda) is formed. When sodium carbonate is combined with acetic acid, an organic compound, it forms *sodium acetate*.

Uses. The following are a few of the many specific uses of sodium compounds. Sodium sulfate is used as a retarding agent for acid and mordant dyes; as an exhausting agent for sulfur, vat, diazotized and developed, and direct dyes; in the making of viscose rayon. Sodium sulfide is used to reduce sulfur dyes; and sodium hydrosulfite to reduce vat dyes. Sodium carbonate is used to retard direct dyes, to reduce sulfur dyes and as a water softener. Sodium chloride is also a reducing agent for direct, vat, and sulfur dyes. Sodium nitrite is a diazotizing agent for dyes. Sodium hydroxide (caustic soda) is noted for its use in the mercerizing of cotton and in the making of viscose rayon. It is also used to make vat dyes soluble and to develop azoic and diazotized and developed dyes. Sodium dichromate is used as a mordant for dyes, and sodium perborate is used for aftertreatment for vat and sulfur dyes and for bleaching wool.

CALCIUM

Calcium is an alkaline earth metallic element found only in compounds. It is a silvery white, hard metal. Limestone is a form of calcium carbonate, a white solid. Marble is pure calcium carbonate. Calcium may be prepared by the electrolysis of calcium chloride.

Calcium Compounds and Their Uses

Calcium compounds are used for water softeners and bleaching agents. When calcium replaces one of the hydrogens in water it forms *calcium hydroxide*. When calcium replaces the hydrogen in an acid it forms a *salt*. Calcium and

> sulfuric acid form *calcium sulfate*
> nitric acid form *calcium nitrate*
> carbonic acid form *calcium carbonate*
> hydrochloric acid form *calcium chloride*

POTASSIUM

Potassium is an alkaline metallic element found as a compound in soil which contains decomposed granite rocks and in salt lakes. It is essential food for plant life. It is similar to sodium but softer, silvery white in color, with a blue tinge. Potassium may be prepared by the electrolysis of its hydroxide or chloride.

Potassium Compounds and Their Uses

Potassium is used for dyes and also as a deoxidizing (reducing) agent.

440

When potassium replaces one of the hydrogens in water it forms *potassium hydroxide*. When potassium replaces the hydrogen in an acid it forms a *salt*. Potassium and

> sulfuric acid form *potassium sulfate*
> nitric acid form *potassium nitrate* (saltpeter)
> carbonic acid form *potassium carbonate*
> hydrochloric acid form *potassium chloride*

When potassium forms a compound with sulfur it becomes *potassium sulfide*. When potassium is combined with carbon, hydrogen, and oxygen it becomes *potassium tartrate,* a salt.

Uses. The following are a few of the typical uses for fabrics of potassium compounds. Potassium bichromate is used as a mordant in dyes and as an aftertreatment for vat dyes. Potassium antimony tartrate is also used as a mordant.

CARBON

Carbon, an odorless and tasteless solid, is found in living things, animal and plant; in the air, water, and earth. Cellulose, a substance in plants, is a compound of carbon, hydrogen, and oxygen. Proteins (in plant and animal cells) are carbon, hydrogen, oxygen, and nitrogen. Carbon in the air is in the form of carbon dioxide. It is a free element in the earth, in coal and graphite, and a compound with metallic elements such as calcium carbonate. Diamonds are pure carbon in a crystallized form. Carbon is insoluble in ordinary solvents and will not react with acids, bases, or ordinary solvents. It unites readily with oxygen and is used extensively in combination with hydrogen, although a high temperature is necessary to unite them.

Inorganic Carbon Compounds and Their Uses

The inorganic compounds of carbon include carbon monoxide and carbon dioxide. Carbon will also combine with sulfur, calcium, silicon, and other metals to form carbon disulfide, calcium carbide, silicon carbide, and other compounds.

Carbon dioxide is essential to all life. A very small proportion of the air and water contains carbon dioxide. Plants utilize solar energy to form starch by uniting carbon dioxide and water, and oxygen is a waste product. If it were not for plant life, all the oxygen in the air would change to carbon dioxide. When carbon

441

burns, carbon dioxide is formed. This gas does not burn nor can fuels be burned in it. When soft coal is burned in air, carbon dioxide gas is separated from the nitrogen and other gases.

A carbonate is carbonic acid (carbon, hydrogen, and oxygen) which has the hydrogen replaced with metals. An example is calcium carbonate (calcium, carbon, and oxygen).

Uses. The following are a few of the specific uses of inorganic carbon compounds. Sodium carbonate is used to retard direct dyes and reduce sulfur dyes. Carbon disulfide, an inflammable liquid, is used to dissolve rubber, in the manufacture of rayon, and as an insecticide.

Organic Compounds and Their Uses

There are a few hundred thousand carbon compounds. In fact they are so numerous that organic chemistry is the study of nothing but carbon compounds. There are from 225,000 to 500,000 organic compounds compared to some 26,000 compounds of all other elements. It was the discovery in 1828 that *urea* (an organic compound found in the body) could be produced synthetically from the inorganic substance ammonium cyanate (an acid composed of ammonia, nitrogen, and hydrogen) that led to the laboratory synthesis of these thousands of organic compounds. In fact, the discovery of vitamins and hormones is a result of this first synthetic organic compound.

Carbon with its valence of four is not an active element. The reason for its tremendous use is that the carbon atoms have the unique feature of forming chains or rings of two to sixty atoms combining with each other.

Today, organic compounds have the widest use of all chemicals in the manufacture of fabrics. Many of the formulas are highly complex. They are used in some form in practically all dyestuffs, as oxidizing and reducing agents, in soaps, as solvents, for rayons and other synthetic fibers, for practically all the special finishes, as well as for a vast number of other uses. Some compounds coat the fibers while others impregnate the fibers.

Organic compounds are divided into two general groups: hydrocarbons, which are made up of the two elements, hydrogen and carbon; and compounds derived from hydrocarbons, which have, in addition to hydrogen and carbon, oxygen and other elements such as nitrogen and sulfur.

Source. Numerous organic compounds are obtained by distillation, which is the operation of driving off gas or vapor from liquids or solids (as by heat) and condensing the product. Many are synthetically made but a great majority are from natural sources. Crude petroleum by distillation produces gasoline, kerosene, naphtha, and paraffin. Coal by destructive distillation produces coal tar and carbolic acid. Bituminous coal is used. It contains carbon, hydrogen, oxygen, sulfur, and mineral ash. Coal tar by distillation produces anthracene, benzene, naphthalene, and toluene. Wood by destructive distillation produces acetic acid, acetone, charcoal, and wood alcohol. Fruits and grains by fermentation produce acetone, acetic acid, and ethyl alcohol. Grains, fruits, and vegetables by separation produce starch and sugar. Cotton and wood pulp by treatment produce cellulose. Animal tissue, seeds, and nuts by extraction produce fats and oils.

Hydrocarbons. These are organic compounds containing two elements, hydrogen and carbon. There are two classes: (*a*) those that form chain compounds with the carbon atoms, combining into long molecules in chain form; and (*b*) those that form ring compounds, with the molecules in form of rings. They include the following series, of which the first three are chain and the last three are hexagonal ring formation.

The *methane or paraffin* series: methane, the first compound in the series is an odorless, colorless gas, present in natural gas. The *acetylene* series: acetylene is a colorless gas formed by the action of water in calcium carbide. The *ethylene* series: ethylene is a colorless gas, formed by heating ethyl alcohol with sulfuric acid. The *benzene* series: benzene is a colorless liquid, secured from fractional distillation of coal tar. *Phthalates,* used in alkyd resins are obtained by oxidation of various benzene derivatives. Toluene (or toluol), a colorless liquid, is in the benzene series. It is used for dyes. The *naphthalene* series: naphthalene is a white, insoluble solid used in the making of dyes and in moth-resistant finishes. The *anthracene* series: important as a source of alizarine.

The following compounds contain oxygen in addition to carbon and hydrogen.

Alcohols. These organic compounds are obtained from a hydrocarbon by substituting hydrogen and oxygen (hydroxyl) groups for one or more of the hydrogen atoms. *Ethyl alcohol,* commonly called "alcohol," an inflammable, colorless liquid, is formed by

the fermentation of sugar, starch, or cellulose. It is useful as a solvent and is used in dyeing. *Methyl alcohol,* commonly called "wood alcohol," a colorless, odorless liquid, is formed by combining hydrogen and carbon monoxide or by the destructive distillation of wood. Both of these alcohols react with acids to form esters. *Butyl alcohol* is obtained from the fermentation of starch. It is widely used as a solvent. *Glycerine,* a thick liquid, is a byproduct of soap or petroleum.

Phenol (carbolic acid) is a white crystalline solid, secured by the fractional distillation of coal tar. Phenol is synthetically made from benzene. It is used in making synthetic resins, one of which gives a crease-resistant finish to fabrics.

Ethers are alkyl oxides, similar to inorganic oxides. That is, two carbon and hydrogen radicals are combined to one oxygen atom. *Ethyl ether* is a light, colorless liquid made by the reaction of ethyl alcohol and sulfuric acid.

Aldehydes. These organic compounds are obtained from the hydrocarbon by substituting an oxygen atom for two hydrogen atoms. *Formaldehyde,* the simplest aldehyde, is a colorless gas. It is formed by oxidation of methyl alcohol. It has many uses, such as a part of a special finish given to fabrics and a part of a chemical wool shrinkage process.

Ketones are obtained from the hydrocarbon by substituting an oxygen atom for two hydrogen atoms. *Acetone,* a colorless liquid which is an excellent solvent, is prepared by heating calcium acetate, by fermentation of starch, or by destructive distillation of wood. Acetone is an important chemical compound in the manufacture of cellulose acetate rayon.

Acids. These organic compounds are obtained from hydrocarbon by forming a certain carbon, oxygen, and hydrogen group. They are formed by oxidizing the corresponding alcohol compounds. Organic acids are used extensively in dyes. They include the following: *Acetic acid,* a colorless liquid, is produced from ethyl alcohol. *Glacial acetic acid* is the name given to the strongest acetic acid. It is used in the making of cellulose acetate rayon. *Formic acid,* a colorless liquid used in dyeing, is produced from methyl alcohol, which first forms formaldehyde which is oxidized to produce the acid. *Oxalic acid,* a white crystalline solid, is produced by the reaction of an oxalate and sulfuric acid or heating sawdust with caustic soda. It is a reducing agent and used for

444

bleaching such fibers as flax. *Tartaric acid,* a transparent crystalline solid, is from an acid found in fruits. It is used in dyeing.

A tartrate is a salt or ester of tartaric acid. *Tartar emetic,* a white crystalline salt, is an antimonyl potassium tartrate, which is a tartrate combined with the two elements, antimony and potassium. It is a basic salt used as a mordant.

Esters. These are organic compounds formed when an alcohol reacts with an acid. Just as a base and an acid unite and produce salts (alkalies) and water, so alcohol and acid unite and produce esters and water. *Methyl acetate* is an ester obtained from acetic acid. *Ethyl acetate* is obtained from acetic acid and ethanol. *Ethyl nitrite* is an ester made with an inorganic acid. Animal and vegetable oils and animal fats are esters of glycerine and organic acids.

Carbohydrates. These are organic compounds containing hydrogen, oxygen, and carbon, in which there are twice as many hydrogen atoms as oxygen atoms. Plants obtain carbon dioxide from the air and from water and convert it into starch and later into sugar. The sugar is then reconverted into starch or becomes the cellulose of the plant. Thus, carbohydrates include *starch, sugar,* and *cellulose.* Rayon is made from cellulose of wood and cotton linters.

COPPER

Copper is a soft, heavy, reddish metallic element. It is found in a free state in copper ore and in compounds in other ores. Copper is next to silver as a conductor of electricity and heat.

Copper Compounds and Their Uses

Copper has a more limited use for fabrics than most of the other elements. When heated it combines directly with oxygen, sulfur, chlorine, and fluorine. When copper forms a compound with oxygen it becomes *copper oxide.* When copper replaces the hydrogen in an acid, a salt is formed.

Copper oxide in ammonia is used in the making of cuprammonium rayon. Copper is also used in rendering fabrics resistant to mildew.

SILICON

Silicon is a nonmetallic element, crystalline in form. Next to oxygen, it is the most abundant element in nature but is never

445

found in a free state. It forms some 27 per cent of the earth's crust and is found in many salt compounds. It belongs to the same family as carbon. Silicon dioxide (silica) is a common mineral and includes quartz and sand. Silicon can be prepared by reducing silicon dioxide with carbon.

Silicon Compounds and Their Uses

Sodium silicate (formed by boiling sand with sodium hydroxide) is used for sizing and as a fire-resistant finish. Silico-fluoride is used in making fabrics moth-resistant. Other silicon compounds are being used as water-repellent finishes. Silicon is also a part of glass fibers.

BORON

Boron, a nonmetallic element, is a brown powder. Its compounds resemble those of carbon and silicon. In nature it is found, particularly in volcanic regions, as borax (sodium tetraborate) and borates. It never occurs in a free state. Boron as an element has no important use.

Boron Compound and Its Use

Borax, a white crystalline salt, is the most important compound of boron. It is fairly soluble in water and has a strong alkaline reaction. It is used as a watersoftener, as part of high-gloss starches, for stiffening fabrics, for dyeing and fireproofing.

Other Metallic Elements

A few other metallic elements are used, but less frequently, for fabric. *Iron* is found in a free state and more in compounds of carbonates, oxides, sulfides, and silicates. *Tin,* a lustrous white metal, is found usually in a compound with oxygen. *Stannous chloride* (tin and chlorine) is used as a mordant in dyeing and as a reducing agent. *Lead,* a heavy, soft metal, is found in ores, some of which also contain silver and gold. Both tin and lead are used for weighting silk.

THE CHEMISTRY OF FIBERS

The original animal, vegetable, and mineral fibers have a more or less definite chemical analysis and can be broken down into elements. In addition a whole new series of synthetics are being developed by building up elements into complex compounds. The chemical compounds of all fibers, natural and synthetic, have certain properties that contribute special characteristics to the yarn and the finished fabric. They also have certain exact reactions to acids and alkalies, reactions which form the basis for all cleaning, dyeing, and finishing processes.

There is one great chemical difference between animal and vegetable fibers. The basis of all vegetable fibers is cellulose and the basis of all animal fibers is protein in some form, such as gelatine or some albuminoid body. The animal fiber does not contain any cellulose. Pure cellulose is composed of carbon, hydrogen, and oxygen, while albuminoids are carbon, hydrogen, oxygen, and nitrogen. Vegetable fibers possess great chemical inertness, having very little affinity for other bodies and can scarcely be acted upon by any reagents except mineral acids, which are harmful to cellulose. Alkaline solutions are usually used for vegetable fibers, while acid solutions are better for animal fibers. Gelatine has a higher specific gravity than cellulose, hence animal fibers sink in water while vegetable fibers float. Since the ultimate vegetable cells are larger than animal cells, there are more animal cells to the square inch than vegetable cells, and therefore the tenacity of animal fibers is greater than that of vegetable fibers or, in other words, the animal fibers as a rule are stronger than vegetable fibers. The following are the major elements that make up the fabric fibers. There are other elements in most of them but their percentage is very small.

Wool and silk are natural animal or protein fibers composed of different parts of the following: carbon, hydrogen, nitrogen, oxygen, and sulfur. Wool is composed chiefly of keratin, which is an amino acid. It has excellent response to most acids. If care is not taken, chlorine will make wool harsh. All alkalies are destructive to wool. The lanoceric acid in wool is an aid in dyeing, as it can precipitate mordants and dyestuffs from the solution fixing the coloring matter on the fiber. Wool will shrink with the application of heat and pressure.

The two filaments of silk that make the silk fiber are fibrons, which are held together by sericin. Strong organic acids and alkalies will damage the fiber but weak solutions will have little effect on the fiber. Silk has less resistance to acids and more resistance to alkalies than does wool.

447

Cotton, flax, hemp, jute, and ramie fibers are natural vegetable fibers composed of different parts of the following: carbon, hydrogen, and oxygen. Cotton is almost pure alpha cellulose while the bast fibers—linen, hemp, jute, and ramie—are liquified cellulose. Cellulose fibers are little affected by alkalies and organic acids, but they can be destroyed by mineral acids. Flax and the other bast fibers are more easily affected by alkalies and acids than is cotton. Cellulose fibers are little affected by heat and some can be boiled with no harmful results.

Rayon is a synthetic fiber made from the cellulose of wood and cotton.

Viscose and cuprammonium rayon are regenerated cellulose and react much the same to acids and alkalies as do other vegetable fibers, but extreme heat and boiling water will destroy the fibers. *Acetate rayon* is made of cellulose esters and it reacts quite differently to chemicals than do other cellulose fibers, due to the chemical change of the cellulose during the manufacturing process. Strong organic acids will destroy acetate rayon, while alkalies and boiling water will deluster it. Viscose and cuprammonium rayon have more affinity for dyes than does cotton. Acetate rayons require special dyestuffs, different from those required for other vegetable fibers.

Asbestos and glass fiber are mineral fibers. Asbestos, a natural fiber, is composed of silicon and magnesium. Glass fiber, a synthetic mineral fiber, is made of silicon and carbon (in the form of limestone, which is a carbonate).

The following are also synthetic fibers. *Casein fiber* is a regenerated animal protein fiber containing the same elements as the natural animal protein fibers. *Soybean fiber* is a regenerated vegetable protein fiber containing the same elements as the natural vegetable fibers.

The remaining synthetic fibers are made from chemicals. *Nylon* is composed of carbon, hydrogen, oxygen, and nitrogen. Its complex compounds are adipic acid and hexamethylene diamine. *Vinyon** is composed of carbon, hydrogen, oxygen, and chlorine. Its complex compounds are vinyl chloride and vinyl acetate. *Saran** is composed of carbon, hydrogen and chlorine. Its complex compound is vinylidene chloride. *Velon** is also made from the vinylidene chloride compound.

* Trade mark.

448

PRINTED FABRICS
ADD COLORFUL INTEREST

This is a group of printed fabrics made of cotton, filament rayon, spun rayon, and silk. All dyeing and printing involve a knowledge of chemistry. The colorist must understand the chemical compounds of the dyestuffs and know how they react with each of the fibers which are also chemical compounds. Starting at upper left, counterclockwise, they include the following fabrics and printing methods.

The soft, lustrous silk satin was printed in a roller printing machine by the blotch printing method.

The next is an acetate, knitted tricot. It was direct and overprinted by hand screening.

The fine glazed chintz was printed in a roller printing machine by direct printing on a white ground.

This light weight, spun rayon crash was direct (or application) printed on a white ground, using roller printing.

In the center is a glazed, twill weave, cotton fabric. It was roller printed by the direct method.

This plain weave, cotton longcloth was first dyed blue and the color in the pattern was an application overprint.

The glazed cotton cambric was printed in a roller printing machine by the application method on a white ground.

The spun rayon crash was roller printed by the direct method. The gold is a blotch print. The red is a pigment color.

The last fabric is a boldly printed, cotton piqué which was roller printed by the application method.

Other colored illustrations of fabrics, printed by various methods, are the frontispiece and the one facing page 513.

The synthetic resins are usually made from organic compounds and are used to make plastic solids and synthetic fibers. They are also applied to fabrics in the form of additional functional finishes, rendering them water-repellent, resistant to spots and stains, fire, and mildew. It may give them a durable crispness or glaze. The effect depends on the compound, its formula, and its application. Some are soluble, others are insoluble. Some have the ability to become insoluble and infusible upon being treated.

Physically, resins may be divided into two groups: those that remain soluble, including (a) the alcohol soluble or spirit soluble resins, and (b) the benzene soluble and oil soluble resins; those that are initially soluble but become insoluble by oxidation or when heat is applied. These include formaldehyde, phenol, and certain alkyd resins. These resins are used to impregnate the fiber or to form a film or coat on the fiber.

Chemically, resins may also be divided into two groups: those in which the molecules unite through condensation or the elimination of water (the resins of urea formaldehyde and alkyd are of this type); those in which the molecules unite through polymerization, that is, one molecule attaches itself to another in a long chain of larger molecules. These include the acrylic resins. Polymerization is discussed in Chapter 8. It is by polymerization that nylon and other such fibers are formed.

The urea formaldehyde resins are widely used to make plastics for tableware, piano keys, doors, wall panelling, table tops, toys, and novelties. These resins are used for finishes that render the fabric crease-resistant or give it a glaze. It is this same resin that is used for impregnating fibers. This lightweight, clear, soluble resin can easily be made insoluble with the application of heat. The water is eliminated, causing the molecules to unite and harden on the fiber.

The phenol formaldehyde resins are used to make such plastics as Bakelite. The fully polymerized resin is insoluble in all the usual solvents except nitric acid and caustic alkali. It has high tensile strength, and is used widely for electrical insulation due to its property to withstand high temperatures. These resins are used for finishes that render the fabric water-repellent or waterproof.

449

The alkyd resins containing phthalates are used in great quantities by the automotive industry for a protective coating material. The resins have a high gloss. They are a compound of glycerol and ethylene glycol. They all stem from the action of glycerine and phthalic acid, and because they are esters, they have unlimited combinations of fatty acids and alcohols producing resins. They have a tendency to coat the fiber and are usually applied from an emulsion. Since they cannot be easily dyed, they are applied after dyeing. They can be used to bind the color. Alkyd resins give crispness to the fabric, render it resistant to slippage and abrasion, and aid the fabric in retaining its original physical appearance.

The acrylic resins have crystal clarity. They are methacrylic acid polymerized and are light, tough, and impervious to moisture. They are used extensively for lenses, dials, and are the familiar plastic known as "Lucite." They are the carbon, hydrogen, and oxygen acid series of resins and are used as a water-repellent finish for fabrics.

The vinyl resins include an extensive group, such as vinyl acetate, polyvinyl acetate, polyvinyl chloride, polyvinyl butyrol, copolymers of vinyl chloride, and vinylidene chloride. They are tough and adhesive and can be easily colored.

Vinyl is a univalent radical. That is, it has the combining power of one atom of weight. Acetate is a salt or ester of acetic acid. A polyvinyl chloride means a number of radicals combined with chlorine. Polymers of vinyl chloride (used with vinyl acetate to make Vinylite*) are different in that the resin is formed by polymerization of the compound. The same is true of vinylidene chloride, which is the resin used to make Saran* (see Chapter 8) by polymerization.

The above resins are used to give the following properties to fabrics: crush-and crease-resistant, water-repellent and waterproof; fullness and luster; nonslipping qualities, greater strength, and some shrinkage control.

The coumarone resins are obtained from certain fractions distilled from coal tar naphtha. They are polymerized with sulfuric acid, and are used for special finishes. They will dissolve in benzene but not in alcohol.

The polyamide resinoids are protein-like amides. They are made from the four elements, carbon, hydrogen, nitrogen, and

* Trade mark.

oxygen, and form intermediate chemicals of adipic acid and hexamethylene diamine. These are the resins used for nylon, the production of which is fully explained in Chapter 8.

The melamine resins are comparatively new. A melamine is a cyclic compound, containing carbon and nitrogen in alternate position. Combined with formaldehyde, it produces heat-resistant materials with high resistance qualities. Alpha-cellulose-filled melamine makes tableware for United States airlines and buttons for the United States Army. It is durable, light, and strong. Melamines are used for finishes that make the fabric crease-resistant or give it a glazed surface. It is also used in a chemical wool shrinkage process.

THE CELLULOSE DERIVATIVES

In addition to the use of cellulose for making rayon, it also forms compounds that are used for finishing fabrics. The action of certain salts, such as zinc chloride and similar compounds, on cellulose produce vulcanized fibers. The action of strong alkalies on cellulose produces mercerized cotton. The action of regenerating cellulose from its esters is the process for making cuprammonium and viscose rayon.

A cellulose ester is a cellulose compound consisting of carbon, hydrogen, and oxygen. Nitric acid is combined with cellulose. The hydrogen of the acid and the hydrogen and oxygen of the cellulose are eliminated and are replaced by the nitrogen and oxygen groups, forming nitrates. If they form one group they are called mononitrates; two groups, dinitrates; three groups, trinitrates. Dinitrate was the basis of the nitrocellulose rayon made by Chardonnet.

Cellulose reacting with acetic anhydride forms acetate esters which, in the presence of sulfuric acid or some other catalyst, form a triacetate that is insoluble in acetone. Upon further heating and partial hydrolysis, the triacetate changes and becomes soluble in acetone. It is by this chemical change that cellulose acetate rayon is made.

A cellulose ether results when cellulose, acting as a trihydric alcohol, is condensed with the residue of another alcohol and the water is eliminated. One or more of the hydrogen-oxygen residues may react forming any number of groups that give alcoholic properties to the hydrogen-oxygen group.

SUMMARY

Matter: Anything that occupies space and has weight. There are three types, solid, liquid and gas.

Substances: Kinds of matter, as water, air, and coal.

Element: A substance that cannot be broken down into simpler substances by ordinary chemical methods. The elements most frequently used for fabrics, and their symbols are:

Metallic Elements	Nonmetallic elements
Cu —Copper	C —Carbon
Fe —Iron	Si —Silicon
Pb —Lead	S —Sulfur
Sn —Tin	
	Gaseous, nonmetallic
Alkaline Metallic	Cl —Chlorine
Ca —Calcium	F —Fluorine
K —Potassium	H —Hydrogen
Na—Sodium	N —Nitrogen
	O —Oxygen

A compound is a substance formed when two or more of the elements unite chemically.

An atom is the smallest unit or particle of an element that takes part in chemical action.

A molecule is the smallest unit of an element or a compound that contains all the properties of the original substance.

Resins Used for Fabrics:

Urea formaldehyde	Vinyl
Phenol	Coumarone
Alkyd	Melamine
Acrylic	Polyamide resinoids

ELEMENTS IN FIBERS

Animal or protein fibers—wool, silk, casein fiber (synthetic) : Carbon, hydrogen, nitrogen, oxygen, and sulfur.

Vegetable fibers—cotton, rayon (synthetic), flax, hemp, jute, ramie, and soybean (synthetic) : Carbon, hydrogen, and oxygen.

Mineral fibers—asbestos: Silicon and magnesium. Fiber glass (synthetic) : Silicon and carbon.

Synthetic fibers—nylon: Carbon, hydrogen, oxygen, and nitrogen. Vinyon*: Carbon, hydrogen, oxygen, and chlorine. Saran*: Carbon, hydrogen, and chlorine.

* Trade name.

452

Opposite page: Long lengths of wool fabric are being scoured preparatory to dyeing and finishing (Goodall Fabrics, Inc.).

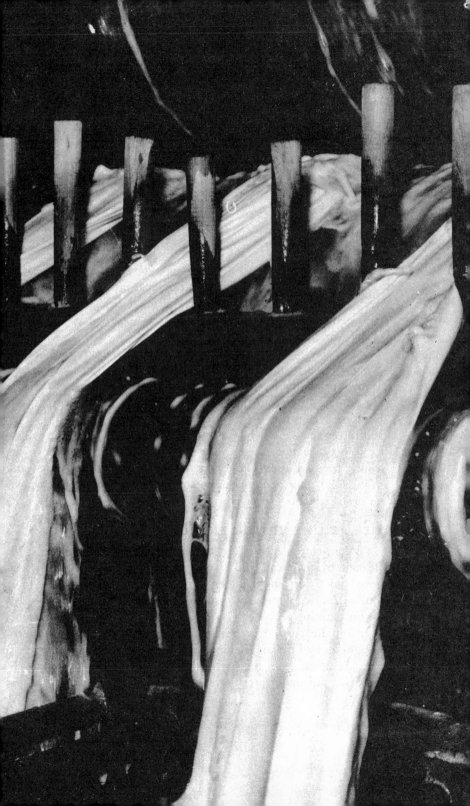

Prepare the fabric for finishing

After the fabric has been constructed it must be prepared for dyeing, printing, and finishing. In addition, fabrics are often given treatments that affect the final coloring, absorbency, and luster of the fabric.

Most wools are scoured, and cottons and linens are kier boiled. Wool is often treated to remove the vegetable matter by charring. Many fabrics are bleached. This may be for the purpose of producing white fabrics or to prepare the fabric for later coloring. Silk must be degummed before dyeing. Cotton may be mercerized to add luster and absorbency and to produce clearer, brighter colors, or it may be treated to resist cotton dyes and to acquire an affinity for acid dyes. These preparatory treatments are important to the final color and appearance of the fabric. Synthetics require no pretreatments except possibly bleaching.

KIER BOILING COTTON FABRICS

Before cotton fabrics can be bleached and dyed they must be cleaned by chemicals and by boiling. Huge tanks, about 9 feet deep and 6 feet in diameter, called "kiers," hold some 2 to 5 tons of fabric. Water and usually some form of alkali are used for the cleaning process.

After the fabric has been placed in the machine, it is closed airtight and all the air is eliminated. The scouring solution sprays over the fabric from the top of the machine. The boiling, which is carried out under pressure, may be only for two or three hours or as long as twelve hours, depending on the fabric, the strength of the scouring solution, and the results desired.

MERCERIZING COTTON YARNS OR FABRICS

This is applied almost exclusively to cottons. It gives the cotton fabrics greater strength and luster, and it increases the fiber's power to absorb dyes so that clearer and brighter colors result. Either the cotton yarns or the cotton fabrics may be mercerized. The fabric can be mercerized before or after bleaching, and occasionally a dyed fabric is mercerized. The cotton yarn or fabric is saturated with a strong caustic soda (sodium hydroxide) solution which causes the cotton fibers to swell. As the fiber swells, the yarn shrinks in length. Instead of the normal, flat, spiral form, the fiber is now round, and smooth, more reflective to light, resulting in a high luster.

Fabric mercerization is in one long continuous machine. The fabric, saturated with sodium hydroxide, goes through two squeeze rolls onto the tentering machine, which holds it under tension, to its correct width (as the mercerizing solution tends to shrink the fabric). As the fabric leaves the tenter frame it passes between rotating rolls, through boxes of warm and cold water and dilute sulfuric acid, to wash out most of the caustic soda. Then it is washed again, rinsed, and dried.

Yarn mercerization is either in the warp (in rope form) or in skeins. The result is similar to that achieved by mercerizing the fabric. The yarns go through a series of boxes, where they are scoured, saturated with caustic soda solution, washed, soured, rinsed, and dried.

Continuous fabric washer, which washes and rinses from 13 to 35 yards per minute (Riggs and Lombard, Incorporated).

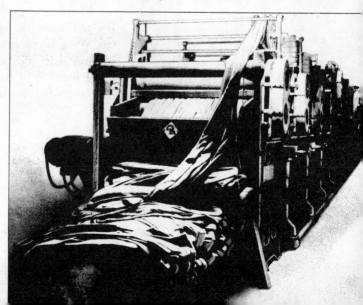

IMMUNIZING COTTON FABRICS

This is a process applied specifically to cotton and is used to give it a resistance to certain cotton dyestuffs, making possible a wide range of effects. Cotton is changed into alkali cellulose and then is treated with paratoluene sulfochloride to form an ester of cellulose. (Cellulose acetate is also an ester).

Immunized cotton has practically the same strength and resistance to finishes and chemicals as untreated cotton. It takes the same dyes as do cellulose acetates, which are different from dyes used for cotton, wool, and regenerated cellulosic rayons. Therefore, it is possible to dye a fabric woven with immunized and ordinary cotton first with direct, developed, or vat dyes, and then dye the immunized cotton with an acetate dye. These produce a paler color than when used on cellulose acetate. The process creates a wide range of color combinations on decorative fabrics.

SCOURING WOOL FABRICS

As the yarns have been spun, the warp sized to give it body and strength for weaving, the fabric woven, and the cloth inspected and mended, the yarns and fabric have passed through many processes, from one machine to another. Naturally they have accumulated oils, emulsions, sizing, and grease and oil spots, all of which must be removed prior to the dyeing and finishing processes.

There are three types of scouring machines in which (*a*) the fabric, sewn in one continuous length, is scoured in rope form; (*b*) the fabric is scoured in full width; and (*c*) the fabric is tumbled and scoured in bulk.

Dry-fulling used for dusting and crushing the fabric after carbonizing. The fabric passes between the rolls in the machine (Riggs and Lombard, Incorporated).

The scouring solution consists of water, soap, and weak alkali. As in the scouring of the raw wool, the alkali combines with some of the oils, and the soap breaks up the oils into minute particles that are carried off by the water.

In a typical rope washer the fabric, sewed into a continuous rope, goes from the scouring solution through two rolls of a wringer, which squeeze out the solution, and then again passes into the scouring solution, to be rewet while more fabric is going through the squeeze rolls. This process is continued for one or more hours, according to the weight of the fabric. The machine has several sets of rolls and as many pieces of fabric are scoured at once as there are sets of rolls. Each rope of fabric is kept in place by guide rings so that it will not come in contact with other ropes being scoured.

In the open width washer the fabric, full width, is run through the scouring solution and then through two long rolls of a wringer, which squeeze out the solution. The fabric is returned to the liquid to float and then again it goes through the squeeze rollers.

In the laundry wheel machine the fabric is placed in the scouring solution. There, a perforated cylinder tumbles the fabric in much the same way as do certain types of home washing machines.

Carbonizing unit for wool fabrics, including the wetting-out tank, the acid soaking tank, the dryer, and the carbonizer (James Hunter Machine Co.).

After the fabrics have been scoured the allotted time, the scouring liquid is drawn out, clear water is admitted into the machine, and the fabric is thoroughly rinsed to remove all traces of the soap and alkali.

CARBONIZING WOOL FABRIC

If the wool has not been carbonized in the raw stock, it is usually necessary to destroy chemically the vegetable matter after the fabric has been woven. On page 38 is a discussion of carbonizing wool before weaving. Carbonizing after weaving is practically the same. The wool fabric is steeped in an acid solution; it passes through heated chambers in which the acid and heat char the vegetable matter; the charred material is dusted out of the fabric; and the fabric is washed to remove all the acid.

This process may take place right after the fabric has been scoured, or the fabric may have been fulled or even dyed first. However, it usually is not dyed first, because the vegetable matter does not take the wool dye and when removed it may leave a spotty discoloration in the fabric.

If the fabric is not in a wet state, it is first wetted and the moisture is squeezed out so that the fabric will absorb the acid bath evenly. It is then delivered to the large acid vat, which contains six to eight rolls at the top and the bottom. The fabric passes around these rolls, through the acid bath. Leaving the acid vat, it passes through two squeeze rolls that press out the excess acid and then possibly over a vacuum extractor that removes more of the acid.

Now that the fabric is thoroughly saturated, it is delivered directly to a large carbonizing oven, which dries and then chars the vegetable matter. The oven has several chambers, each of which has top and bottom rollers around which the fabric passes. The first few chambers dry the fabric, the last few char the vegetable matter.

The next step is to remove the charred material from the fabric. The fabric, leaving the oven, is passed to a dry "fulling well." This is also called a "riffler," as the cloth is literally "riffled" or shaken to remove the charred particles. The cloth is beaten back and forth, and twisted and wrung to dispose of the carbon.

The fabric, passing from the riffler in rope form, runs through a machine that opens it up. It is rinsed and then washed in an alkali solution of sodium carbonate (soda ash) to neutralize the acid. A final rinsing is then given to the fabric to remove all traces of the alkali.

DEGUMMING SILK

The sericin (gum) must be removed before silk can be dyed. The gum may be boiled off after doubling and twisting, while the silk is in skein form, or after the fabric is woven. In either case the same method is used.

The silk skeins or fabrics are boiled for about an hour in olive-oil soap and water, and then thoroughly rinsed and dried. Usually the material is given a second boiling with half the quantity of soap, rinsed, hydroextracted, and dried. After degumming, the silk is delicately soft, pearly white, and lustrous. The filaments in the yarn are now separated. Perfect scouring removes from 20 to 27 per cent of the silk's weight, leaving a filmy fabric, pleasant to the touch but none too usable. For certain fabrics only a part of the gum is removed. Such silks are called "souples."

The liquor from the boiling-off process, containing the sericin, is saved and may be added to the dyeing bath. This gum solution returns to the fabric some of the natural weight it lost, but most silk fabrics are weighted with other substances as discussed in Chapter 19.

BLEACHING

This process is necessary for all fibers in order to obtain white fabrics and also as a preliminary treatment for most fibers that are later to be dyed or printed. For many years, fabrics were bleached by exposing them to the sun. Now they are chemically bleached, with the exception of some sun-bleaching of linen. Regardless of the chemicals used, it is usually oxidation that ultimately causes the fibers to turn white. Bleaching may be done in either raw stock, yarn, or in the fabric.

Bleaching cotton. The cotton fiber contains certain impurities such as oils, waxes, pectose, proteins, coloring matter, and insoluble salts. These materials must be removed from most fabrics. Cotton is usually bleached after the fabric is woven. Bleaching involves two steps. The first is kier boiling the fabric, which is a high temperature treatment with dilute alkali. This dissolves the pectose and emulsifies the oils and waxes. The second step is the bleaching (chemicking), which destroys or dissolves the coloring matter, producing a white fabric.

Cotton may be bleached in a cold solution of calcium or sodium hypochlorite by first wetting completely and allowing the wet cloth to lay "wet" until the proper "white" is obtained.

In another method, sodium chlorite alone, or mixed with sodium hypochlorite, is used. The solution is maintained at room temperature. Sodium carbonate and sodium bicarbonate are used to maintain the pH of the solution. Bleaching, if done with care, does not harm cotton fabrics.

There is also a continuous peroxide bleaching process used for cotton and synthetic combinations. The fabric is bleached in rope or open width form. It is first singed, desized, and may be mercerized, or given other treatments prior to bleaching. The fabric is then dipped in a solution of caustic soda or alkaline peroxide. When the fabric is saturated uniformly, it is squeezed to eliminate excess chemicals and is passed to the heater tube, where it is heated from the saturation temperature to the operating temperature.

The fabric is then passed to the J box where it remains for an hour at operating temperature. It is then pulled out and is washed free from alkali and passed into the alkaline peroxide solution. After saturation the fabric is squeezed and again passed through the heater tube, stored in the J box, washed, and is ready for dyeing or finishing. In this method 90 to 100 yards may be bleached per minute, the total processing taking about two hours.

Bleaching linen. The same bleaching methods as used for cotton may be employed for linen. Because of the coloring matter and the intercellular substances, linen is more difficult to bleach than is cotton. Grass bleaching of linen, which is the oldest method, is still considered by some the best for the finest linens.

Bleaching other bast fibers. Ramie and hemp are bleached like linen, which may be the same as bleaching cotton. Jute is extremely difficult to bleach but can be bleached with sodium hypochlorite.

Bleaching Wool. Wool is usually bleached in the yarn or the fabric state. It does not bleach as white as do cotton and linen.

The material is first scoured and then bleached while in a wet state. Wool may be bleached by exposing to gaseous sulfur dioxide.

This is known as "stoving." The hanks are hung on rods and are placed in closed chambers. The fabric is passed through slits and over rollers in the stove. This does not produce a permanent bleach and can be removed by soaping.

Another method is to steep the wool in a strong solution of bisulfite of soda. Whether this or sulfurous acid is used, the color of the wool slowly returns after it is washed.

A more permanent result is obtained by the use of hydrogen peroxide rendered alkaline with ammonia or silicate of soda. The material is saturated and steeped for about twelve hours. After bleaching, the wool is washed, soured with dilute acetic acid, washed and rinsed to remove the bleaching agent. The fabric is then dried.

Bleaching rayon. The wood pulp and cotton linters, that produce cellulose for rayon, are bleached. (See Chapter 7.) Rayon yarns, in skeins, and rayon fabrics may be bleached also. The bleaching agents are such compounds as hydrogen or sodium peroxide, sodium perborate, chlorine, and sodium hypochlorite. This last compound is used for cellulose acetate rayon. If acetate is blended with wool, hydrogen peroxide is used.

Other agents are often added, such as wetting agents, neutralizing chemicals, and stabilizing agents. After bleaching, the fabrics are rinsed.

Bleaching silk. Only wild silks, which are dark in color, may need to be bleached. Much of the original color of silk is in the sericin or silk gum, which is removed during degumming. However, the fiber is usually not pure white and stains and spots are often evident. Silk may be bleached in the yarn or fabric state. It may be placed in a closed machine and exposed to sulfurous acid fumes. The best method is the use of hydrogen peroxide, with or without the addition of ammonia. The silk is entered, the temperature of the liquor is raised to 90° C. After five or six hours, the material is washed, soured, washed again, soaped, rinsed, and dried.

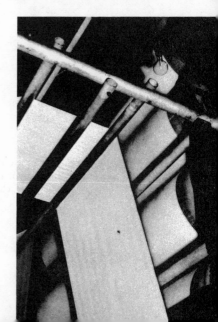

After the fabric is thoroughly washed, boiled, or scoured, it is dried by passing around heated rolls in a machine (Bernside Mills).

SUMMARY
TO PREPARE FABRICS FOR DYEING AND FINISHING

TO PREPARE WOOL

Scour

Carbonize

Bleach—if desired

TO PREPARE COTTON

Kier boil

Bleach—if desired

Mercerize—if desired

Immunize—for special effect

TO PREPARE LINEN

Kier boil

Bleach—if desired

TO PREPARE SILK

Degum

Bleach (spun silk usually)—if desired

TO PREPARE RAYON

Bleach—if desired

Opposite page: Fabrics are dyed with the dyestuffs most suitable to the fiber and to the requirements of the fabric for fastness to water, light, and certain chemicals (Pacific Mills).

DYEING

Gives color and life to the fabric

IN THE BEGINNING

"Let there be light; and there was light." Thus with the birth of light began the ever-present sensation of broken light or what is known to man as "color." Man has always loved color. Color is, has been, and always will be a primary universal need, for it answers man's search for beauty and for social and spiritual importance. He has responded to the lure of the abundant color in nature from the earliest time and he has sought to introduce the colors of nature into his life, his garments, his home, his religion, his festivities. He used color for his kings, for his flags, for burying his dead.

To search out and reproduce color that could be used for his own personal life, early man found many interesting and devious ways. There are two methods of coloring fabrics. One is to cover the surface with small particles of opaque material, called "pigments," which reflect the desired color. The second is to allow the object to absorb a colored solution, called a "dye," which gives the desired color to the material being dyed.

Ancient man probably discovered coloring agents and methods of applying them before he learned to weave fabrics. From the dawn of history he has used colored pigments. Men of the Stone Age decorated their caves with earth colors. Pottery, baskets, ceremonial equipment, body adornment, and the like were decorated with colored designs. Some early tribes dyed their bodies before they wove textiles. It was a logical step then for early man to transfer his knowledge of colors and designs to the fabrics he created.

Saffron Gatherer of Knossos. An ancient fresco painting showing a child gathering crocus to be used for saffron yellow dyes.

Designing

The first designs were probably religious symbols. Man attempted to associate his life with nature as he saw it and with the Divine as he felt it. He observed waves on the water, rays from the sun, winding vines, craggly rocks, lightning, and rain.

The fundamental method of expressing these sights and thoughts, or designing, was developed many thousands of years ago. There can be only a certain number of effective designs in a given space and it has been proved that man has discovered them in every part of the world without intercommunication. For instance, the swastika is a universal design. It was not communicated from one place to another or from one person to another, for it seems that even small children, when left to decorate a square, will spontaneously draw a swastika.

There is much evidence to ancient man's artistic accomplishments. Pictured on the walls of the tomb of Beni Hassan, 2100 B.C., were many people whose clothes were trimmed with stars, chevrons, frets, and other conventional designs. In the tomb of Thothmes IV (1466 B.C.) was found natural colored linen decorated with lotus flowers, birds, the tree of life, and other symbols popular in the land of the Nile. Fabrics found in the burial grounds of the pre-Incas show many symbolic designs: bird heads standing for the "bird men," the puma god (half man and half beast), the llama, and the cat. There were many geometrics, scrolls, and curves, but Peruvian designs are conspicuous for the absence of flowers, leaves, and such designs taken from nature. The American Indian drew squares, triangles, and parallel lines; he conventionalized the birds and animals and his gods. In like manner, each section of the ancient world produced designs that portrayed the native surroundings and interests. Designs and de-

465

signing have continued to change, gone backward or forward with the retreat or advancement of civilization—for "as a people are so they do."

Dyeing

The first coloring was probably staining of material with colored juices of vegetation such as flowers, fruits, or leaves. The Bible speaks of garments dyed in the blood of grapes. Any such discovery would lead to a search for other parts of nature from which dyes could be obtained. Not later than 2000 B.C. the secret of mordanting was discovered, probably in India. This led to a further advance of dyeing, for many natural coloring matters could not be applied to fabrics without a mordant.

Old-World Dyes

The following were the most outstanding dyestuffs of the ancient peoples of the Old World.

Kermes was obtained from a wingless female insect found on the leaves of the kermes oak. This scarlet dyestuff is the most ancient on record. It was known in the time of Moses. It is mentioned in the Scripture by its Hebrew name "tola," meaning scarlet. Kermes became obsolete with the discovery of the Western World and its brighter scarlet dyestuffs.

Indigo is the blue dyestuff obtained from the plants of the genus *Indigofera*. The coloring matter is present in the leaves and is easily extractable with water. Indigo remained the most important of all dyestuffs from very ancient times to the middle of the nineteenth century. The method of preparation of indigo was given in ancient Sanskrit literature in India. There has been found in Thebes a garment dyed in indigo, which has been dated 3000 B.C. The word "blue" in Exodus probably refers to indigo. Today the dyestuff from the plant is used very little and almost entirely in Bengal.

Tyrian purple is obtained from types of shellfish or sea snail (*Purpura* and *Murex*) found in the Mediterranean. The dye secretion is contained in a small cyst adjacent to the head of the animal and this puslike matter, when spread on fabrics in the presence of sunlight, develops a purple-red color.

Tyrian purple is the most celebrated of all dyestuffs. To the city of Tyre is given the credit for its discovery, supposedly about

3500 B.C. Whether or not this is true, it is known that the Phoenicians became wealthy through their monopoly and trade in Tyrian purple. The shells of these shellfish have been found near to ancient dyeworks in Athens and Pompeii. The terms "royal purple" and the Scripture's "clothed in purple" refer not only to the dyestuff but to its expensive character. It is said that it took some 12,000 shellfish to make 1.4 grains of color. The more popular the dye became, the harder it was to find the shellfish, until to own a purple dyed garment one really had to be royalty. Tyrian purple is no longer in use although these shellfish are again plentiful. To the modern eye Tyrian purple would appear a bit dull and drab, but to the eyes of the ancients it was synonymous with the best that life had to offer.

Madder is the dyestuff taken from the root of the herb called by the same name. This dyestuff was also known to the ancients. Cloth dyed in madder has been found on Egyptian mummies. Madder is made from the ground root of the plant. It yields shades of great beauty and fastness in a considerable range of colors— yellow, red, brown—through variation in the manner of application. It is interesting to note that madder colored red the bones of animals that fed upon it, also the claws and beaks of birds.

Alizarine, the red coloring principle of madder, was made synthetically in 1868, and madder ceased to be used commercially.

Saffron is a flowering plant of the common crocus family. The yellow dye was taken from the stigmas and from part of the style. The chief center of saffron growing was the town of Corycus in Cilicia in about 2500 B.C. The earliest known fresco painting of a human figure, unearthed on the Island of Crete, and dated 1900 B.C., is the "Saffron Gatherer of Knossos" a blue painted child gathering crocus. Saffron yellow was mentioned by Homer and by Hippocrates. Today saffron as a dye has been replaced by synthetics.

Weld is a herbaceous plant cultivated in France, Germany, and Austria. The yellow dye is obtained from the dried tops and seeds of the plant. It is the oldest European dyestuff known. Today, it has almost disappeared from the market.

467

Archil is a purple dye obtained from various species of lichens. The most valuable lichens were gathered in Angola in Africa. The use of archil is very ancient. Pliny referred to it as serving to re-enforce the shade of Tyrian purple. The coloring properties are developed by special treatment. The weeds are washed and ground with water to a thick paste, then mixed and fermented with ammonia (such as stale urine) in an iron container. Lichen dyes were formerly very popular but now are used only by some primitive tribes.

Lac is the resinous incrustation formed by certain insects. They exude a resinous secretion over their bodies, forming a cocoon from which the females never escape. After impregnation, each female develops into an organism consisting mainly of a large, smooth, shining crimson sac (ovary) and a beak. The red fluid in the ovary forms the lac dye and the resinous material is made into "shellac." Lac is an extremely ancient dyestuff and was used in the East many centuries before it was known in Europe. This dye was at one time very important and the shellac only an unimportant byproduct. Now the situation is reversed. Today, the crimson lac is seldom employed as a dye.

Mineral dyes were also discovered by ancient man. He learned that certain minerals made fast dyes and that others caused vegetable dyes to become more brilliant. He found out that minerals dyed by depositing insoluble pigment in the fibers of the dyed fabric. Thus, he could weight his fabric as he dyed it, and this was, for example, the function of Prussian blue on silk and wool. Since mineral dyes were not soluble in water, they were rubbed into the fabric.

Early man found out that certain rocks, pounded to powder, also could be used for dyes. He made color from old iron or rust; he burned materials and obtained a fine black powder, carbon, which he used for pigment. He used certain muds in his dyeing, soaking his fabrics in streams containing minerals in order to aid the dyeing processes.

Western-World Dyes

The ancients of the Western World were as clever as those of the Eastern World in hunting out and learning to apply vivid, lasting colors. In many respects, they were even more outstanding than their contemporaries on the other side of the globe. In the

East, the development of dye sources and treatments was the result of exchange of discoveries and methods between highly civilized peoples of various countries. The peoples of the Americas found the same colors and arrived at similar heights of the craft quite by themselves.

Most of the natural dyes these Westerners perfected were so much more brilliant, more easily used, and faster than the ones common in the Old World that upon the discovery of the Americas a large number of the Old-World dyes went out of existence and a whole new series of colors became popular in Europe and the East. These New-World dyes continued to be used exclusively until the dramatic entrance of synthetics.

There were countless ways by which the early peoples of the Americas found and applied colors to their fabrics. Some are incidental, others are not as yet authentically explained, but all are interesting and entertaining. Indians, like most primitive peoples, had a great love of color. In the magic world colors were personalities breathed out of the mouths of the gods. As if they had some knowledge of the eternal source from which come the vibrations of colors, the aboriginal Americans gave to colors the attributes of persons and interwove them in their folklore as naturally as they intertwined reeds into baskets. Fabrics found in the tombs of the ancients of Peru were gay with vivid colors—achieved by dyes that have withstood the wear and tear of centuries. There were purples, greens, browns, blues, much yellow, and a great deal of red—all in harmonious combinations that compete favorably with the best that modern artists have to offer. These early Peruvian dyes were probably vegetable and insect dyes used with advanced mordanting knowledge. Exactly from what all these dyes were made and how they were mixed and applied are questions still unanswered. Mexico is also unearthing similar, very ancient, unsolved coloring mysteries.

Although the first records of the North American Indians are not as ancient as Mexico's or Peru's, they are very colorful. They show many ways by which the aborigines dyed their yarns and their fabrics.

469

The Tlingit Indians of the Northwest made black, purple, red, and yellow in varying shades. To dye black, they soaked the material in black mud and water of sulfur springs, or in mud boiled with salt water to which had been added hemlock bark. They also dyed black by soaking fresh-stripped hemlock bark in a bath of strong urine. The yarns were plunged into this solution and boiled. Purple was made by mashing and boiling in water large blue huckleberries. To obtain red, they steeped the material in children's urine that had been left standing in a vessel dug out of the trunk of the alder. Pieces of alder were added to deepen the shade. Yellow was made by boiling wolf moss, a tree lichen, in water, making a yellow solution. Greenish shades of blue were produced by boiling hemlock bark with oxide of copper, scraped from old pieces of metal or from rock.

The early Indians of California, Oregon, and Washington made yellow from the root of the Oregon grape. They dyed yarns orange or red-brown by chewing white alder bark and then drawing the material to be dyed through their mouths until the desired color was reached. They mixed chewed salmon eggs with a vermilion to give a peculiar red color.

The Cherokee extracted a purple dye by rubbing the petals of the purple iris over a rough surface and using the paste to rub onto the fabrics. The Hopi Indians obtained pink to purple from the amaranth and blue from the sunflower.

The Navajo has acknowledged superiority over other North American Indians not only in weaving but also in dyeing. They were undoubtedly the most original in the application of color. No native tribe has carried the art of weaving and dyeing to such perfection or has shown less European influence. One of their most ingenious dyes was black. They first boiled branches of the sumac, then they crushed yellow ocher and roasted it until it was brown. This brown powder was then mixed with piñon gum and again roasted until the mass became a fine black powder which, when mixed with the sumac brew, made an excellent permanent black dye. Red was produced from the alder, and yellow in two ways: either by steeping the tops of a yellow flower and mordanting with alum; or by crushing a large fleshy root to a soft paste, adding almogen (native alum) and rubbing the paste onto wool by hand. Navajos originally made blue but later secured indigo from Mexico and added urine to give a fast, durable color.

When the Spaniards landed on the Western Hemisphere, they found many dyestuffs in use by the natives. The following are the most outstanding.

Cochineal was a natural dyestuff made from the female insect which fed upon some species of cacti, especially nopal. This cactus is native to Peru, Central America, and Mexico. The females of the insect, which are found in the proportion of 150 to 200 to one male, were brushed from the branches of the cactus into bags and then killed by dipping in hot water or exposing to sun, steam, or oven heat. The dried insects have the appearance of irregular grains and it takes about 70,000 of them to make 1 pound. They were crushed and made into dye by boiling. This dyestuff was scarlet and with different mordants became crimson, orange, and other tints. This dye is little used today. Cochineal was familiar to the ancient inhabitants of Mexico, Central America, and Peru. After its introduction into Europe, kermes ceased to be used, for the dyeing qualities of cochineal were so superior that 1 pound of cochineal produced the same effect as 10 pounds of kermes. In 1630, Drebbel, a Dutchman, discovered how to produce a new brilliant scarlet on wool by using tin and cochineal.

Logwood was, and to a certain extent still is, an important dyestuff. It is obtained from the logwood tree, which is native to tropical America. This is a large and rapidly growing tree, having a peculiar ribbed appearance. The wood is hard and dense and white in color when first cut, but it changes to a brown-red hue after exposure to air. The freshly cut wood is put into water and the dye is secured by the evaporation of the solution. Logwood makes black dye and with mordants produces many compound shades. Logwood dye, when used for silk, leaves the fiber opaque. It was used by the early people in America, and is still used, to make inexpensive black dyes for wool and cotton.

Fustic or yellow wood is the wood of a large tree of the mulberry family and was used to produce yellow-browns and olives. *Quercitron bark*, the inner bark of an oak tree, gave a bright yellow dye. *Brazil wood*, a small shrub of tropical America, gave a dye that with a mordant, developed into brick-red, pink, and purple.

Synthetics Began

The year 1856 might be called a "purple-letter" year in the history of dyeing, for that date marks the beginning of the end of natural dyes and the birth of the important era of synthetic dyes. To one W. H. Perkin of England is given the honor of ushering in this new field. In an attempt to prepare quinine from aniline he stumbled by accident onto the first synthetic dye, aniline purple or mauve. The next few years saw a fine beginning. Verguin in France obtained magenta from aniline; Nicholson found a blue (the first solid acid dye for wool); and aniline yellow and the first vast group of azo colors were discovered.

In 1868, Germany started its important discoveries of synthetics. Graebe and Liebermann found that alizarine (madder) could be prepared from anthracene, a hydrocarbon compound of coal tar, and the synthesis of the first natural coloring matter was effected. By the end of 1885, Germany had perfected other hydrocarbons and their derivatives: methylene blue (first basic blue soluble in water); malachite green (first green of real dyeing value), and tartrazine and congo red (first colors to have direct affinity for cotton) from organic compounds. With these discoveries, Germany's leadership in the dyestuffs field was definitely established.

By 1913, Germany produced over 80 per cent of all dyestuff used and held control over much of the other 20 per cent, since other countries manufactured dyestuffs from "intermediates," which were of German origin. During this time, German manufacturers had taken out patents in the United States, not only on the dyestuffs they were making but on key processes, with the intention of blocking the development of a synthetic organic chemical industry in America.

The United States began to make synthetics only a few years after Europe. There is a record, dated 1864, which states that magenta was made by a Thomas Holliday of Brooklyn, N. Y. Work and experimentation continued until by the twentieth century American chemists and manufacturers realized that America might be faced with a chemical famine. Starting from the beginning, chemists tried to piece together the complicated jigsaw of the coal-tar dyestuff puzzle. This was difficult, since all the information was held secret by German patents. By trial and error, however, American intelligence and ingenuity found some purer materials and some better ways of production. In 1914, the United States made 104 different dyes, almost wholly from German "intermediates."

Very early in World War I it became evident that ownership of important American patents by the enemy interfered with the production of our war supplies. As a result, the government seized all enemy owned patents. The Chemical Foundation, Inc., was formed for the purpose of administering the chemical patents. They were to license their use for a small uniform fee to any "bona fide" American manufacturer who wanted to make dyes.

Since then, America has progressed. By 1937 the United States had 1,000 distinct types of dyestuffs. Organizations, both industrial and governmental, have been formed to further interest and research in organic chemistry. Dyes and dyeing processes have been developed not only to add new and distinctive colors to the old, familiar fibers and fabrics but also to take care of the many strange new fibers that are entering our fabric world in startling numbers.

DYEING

The successful dyeing and printing of materials demand a knowledge of the chemical and physical properties of both the dyestuffs and the fibers to be dyed. Dyes and fibers are both chemical compounds, and it is the reaction between the two that determines the kind of dye to be used and its application. New dyes are being developed constantly, particularly since the introduction of new fibers and of the blending and combining of various fibers into one fabric.

Dyestuffs have a more or less marked acidic or basic character. Animal fibers are proteins or derivatives of amino acids and they have both acidic and basic characteristics. A basic dye may combine with the acidic portion of the animal fiber to form a salt; or an acid dye may combine with the basic portion of the animal fiber, also forming a salt. Therefore, it is supposed that in dyeing wool a chemical change occurs and not just a mechanical absorption of the dyestuff.

Vegetable fibers are cellulosic. Cotton is almost pure alpha cellulose, while the bast fibers—linen, hemp, jute, and ramie—contain lignin as an impurity, which gives them a greater affinity for basic dyes. Cotton does not have this affinity.

There is an affinity between the cellulosic fibers and the direct colors and a strong affinity for the leuco form of vat dyes. There is a difference in opinion regarding the theory of dyeing vegetable

473

fibers. Some think it is a mechanical operation, others believe it is colloidal; some advance electrical theories, and some chemical theories. However, it is generally agreed that most dyeing of vegetable fibers involves a physical rather than a chemical change.

Viscose and cuprammonium rayon, being regenerated cellulose, react to dyes much the same as does cotton. In general, viscose rayon has more affinity for the dyes than does cotton, and cuprammonium rayon has still greater affinity.

Cellulose acetate rayon is made of cellulose esters and is quite different from the regenerated cellulose rayons, viscose and cuprammonium, and does not take the same dyes. Cellulose acetate rayon does not appreciably swell when wet as do the regenerated cellulose fibers and hence cannot be dyed with the same dyestuffs in the same way. As a result, a new group of dyes was developed for acetates.

Nylon, Vinyon*, and Saran* require special dyeing. Cellulosic types of dyes have been used but have some disadvantages in application and fastness. Nylon can be dyed with some of the same dyes as wool and acetate rayon, giving satisfactory shades and fastness.

Synthetic protein fibers, such as casein and soybean, are dyed very similarly to the natural wool, protein fiber. However, fastness and method of application may vary somewhat.

In general, vegetable fibers are dyed in neutral or alkaline solutions while animal fibers are dyed in neutral or acid solutions.

Terms Used in Dyeing

In addition to the dyestuffs, other chemical substances are frequently required to color fabrics successfully. Therefore to comprehend the application of various types of dyes, the following terms must be understood first.

Exhausting agents are chemicals added to the dye bath to help force the dye from the bath onto the textile fiber. They are usually salts but, as in the case of acid dyes, may be acids.

Retarding agents are chemicals that slow down the process of dyeing or decrease the amount of exhaustion. An example of a retarding agent is glue, which is used in the dyeing of vat colors.

Mordants are chemicals that fix soluble dyestuffs during dyeing. Some fibers will not take the dye but will take chemicals that will fix the dye that is applied later. The fiber is treated in the mordant

---

* Trade-mark names.

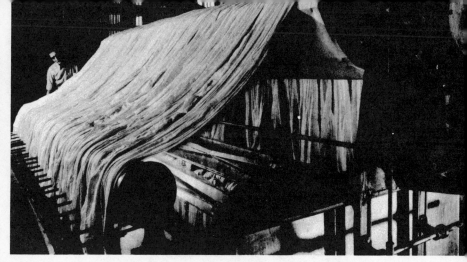

Long ropes of rayon fabric being dyed in a dye beck
(American Enka Corporation).

solution and the mordant is fixed on the fiber. The mordanted fiber is then dyed in the dye bath and the color is fixed to the mordant.

Leveling agents are compounds which produce more level or uniform dyeing of a fabric. This is usually accomplished by slowing down the dyeing process. Thus, most leveling agents are retarding agents.

Reduction is the term used to describe a special type of chemical reaction. Usually, as in the case of organic compounds, such as dyes and dye intermediates, reduction involves a loss of oxygen atoms or a gain of hydrogen atoms by the compound that is reduced.

Oxidation is the opposite of reduction. Thus, in most organic compounds, an oxidation reaction involves a gain of oxygen atoms or a loss of hydrogen atoms by the compound that is oxidized.

Reducing agents and *oxidizing agents* are chemical compounds that bring about reduction and oxidation, respectively. In dyeing with vat dyes, which are insoluble, it is necessary to solubilize the dye before it can be used. This is accomplished by reducing the dye to the leuco form with a reducing agent such as sodium hydrosulfite. The dye can then be oxidized on the fabric with the help of an oxidizing agent, such as potassium dichromate or even oxygen of the air.

Diazotization is the chemical reaction that takes place when

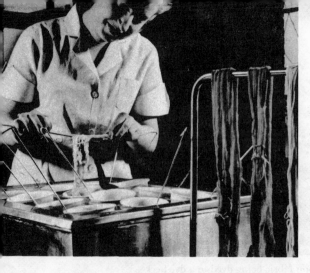

Dye mixtures are carefully tested before the materials are dyed (American Wool Council, Inc.).

amines are reacted with nitrous acid in the presence of an excess of an inorganic acid. Usually sodium nitrite and hydrochloric acid are used in the diazotization of an amine.

Developing chemicals or *coupling compounds* are compounds that react with the diazonium compound, formed by diazotization, to form an azo dyestuff.

THE DYES

Most dyes used for coloring fabrics are synthetic dyestuffs manufactured from coal-tar products. They are complex compounds made from the element carbon combined with other elements, such as hydrogen, nitrogen, oxygen, and sulfur.

Practically all dyestuffs are obtained from one of five hydrocarbons: anthracene, benzene, naphthalene, toluene, and xylene. The most available source of these hydrocarbons is the coking of coal. The hydrocarbons used for dyestuffs are separated from other compounds present by boiling, distillation, and washing with acids and alkalies. Benzene, toluene, and xylene are colorless, volatile liquids, and anthracene and naphthalene are white solids. Certain of the hydrogen atoms in these hydrocarbons are replaced by groups of atoms from other elements to form the dyestuffs.

Some dyes are used more successfully on animal fibers, some on vegetable fibers, some on rayon and other synthetic fibers.

Some dyes are applied directly to the fibers, others not being soluble in water require reducing agents, and still others necessi-

476

tate mordanting the fibers before the dyestuffs can be applied.

Some dyes go into the fiber and produce a chemical change by combination with the fiber; others enter the fiber and are deposited or adsorbed by the fiber to be held by physical forces.

Dyes are classified in two ways: by their chemical structure and by their dyeing properties and affinity for particular fibers. The chemical classification is determined by the atomic groupings in the dyestuff; and such a classification is used only in the chemistry and manufacturing of dyestuffs. The more common classifying by application is the one used here and groups the dyes into the following types:

Acid	Mordant and chrome
Azoic	Pigment
Basic	Sulfur
Direct	Vat
Diazotized and developed	Acetate

In the following summarization of the classes of dyestuffs, only a few representative chemicals are mentioned. Many other chemical compounds, acids, bases, and salts may be used in place of those given. For example, if an acid is listed as a retarding agent then that or usually another acid may be used. If a salt is listed as an exhausting agent then that or often another salt is used. There are many methods for applying dyestuffs to fibers, yarns, and fabrics. Therefore, it should be clearly understood that while an effort has been made to have the information factually correct, the following is not all-inclusive.

ACID DYES

Dye protein materials—silk, wool, nylon, and casein—directly from an acid bath. Some will dye from a neutral bath.

Dye vegetable materials—with a mordant.

Acid dyes have a wide color range and will produce bright shades. Their fastness to water, perspiration, sun, and alkalies varies from poor to excellent, depending on the type of the acid dyestuffs and the method of application. Some are fugitive to light, while others are very fast.

The name "acid dyes" is derived from the fact that they are best applied in an acid bath. Acid dyes are salts, usually sodium salts of color acids. A chemical combination (probably salt formation) occurs between the acid groups in the dye and the amino groups of the protein molecules. They are soluble in water and require no mordant in dyeing protein fibers. In acid dyes the chromophore (coloring group) is in the color acid.

Chemicals used. To exhaust the dyestuff: sulfuric acid or bisulfate of soda. *For slower rate of exhaustion:* Glauber's salt or ammonium acetate.

The process. There are several methods by which acid dyes are applied, depending on the type of fabric to be dyed and the results desired. The dyestuffs are combined in different ways with the chemical compounds used for exhaustion. The fabric is entered into the dye bath and is boiled for about one and a half hours, with the material constantly moving about. After dyeing is completed, the fabric is rinsed. Some dyes produce more uniform shades, others deeper shades, while still others produce a better degree of all-round fastness.

BASIC DYES

Dye animal materials—wool, silk, and feathers—directly from a neutral or slightly alkaline bath.

Dye vegetable materials—linen, hemp, jute, cocoanut fiber—directly, without a mordant.

Dye vegetable materials—cotton and regenerated cellulose rayon—with an acid mordant.

Basic dyes produce brilliant shades but have a poor degree of fastness to light and washing. Therefore, basic dyes are used where bright shades are required that will not be exposed to sun or water. Frequently they are used to top or redye, to obtain bright colors on fabrics that have been dyed with sulfur or direct dyes. Since this is a slower and probably less effective method of dyeing, it is not used as frequently as are the other dyeing methods.

Dyeing skeins of silk with dyestuffs that have an affinity for the animal fiber (Cheney Brothers).

A basic dye is a salt of an organic color base and an acid, for example, a hydrochloride. Some color bases are soluble and others are insoluble. An acid usually makes the insoluble bases soluble.

Basic dyes dye animal materials, wool, silk, casein, and feathers, directly from a neutral or slightly alkaline bath. A chemical combination possibly occurs between the carboxy groups of the protein and the basic color group of the dye, to form a colored salt on the fiber. Basic dyes have a strong affinity for animal fibers.

Basic dyes dye some vegetable materials—cotton and viscose and cuprammonium rayons—by first applying a mordant of tannic acid and then applying a metallic salt, which produce insoluble compounds that will combine with the dye base. Basic dyes can be applied to the vegetable fibers, linen, jute, hemp, and cocoanut fiber without a mordant.

Chemicals used. The vegetable fiber mordant: tannic acid and a fixing bath containing tartar emetic (potassium antimony tartrate). *To retard the action:* alum and acetic acid or other acid salts and acids.

The process. To mordant the vegetable material, the fabric is treated in a tannic acid bath, cooled, entered into a cold fixing bath containing tartar emetic, and then rinsed. After this, it is ready for dyeing.

The dyestuff is mixed with acid and warm water. The dye bath consists of part of the dyestuff, acetic acid, and alum; and needs no other exhausting agent. Tannic acid treated vegetable fibers and untreated animal fibers, have a great affinity for these dyes. The material is entered into the dye bath, which is then heated. When the dyeing is completed the material is rinsed and dried.

DIRECT DYES

Dye vegetable materials—cotton, linen, regenerated cellulose rayon—from a neutral or alkaline solution. No mordant is required.

Dye animal materials—wool, casein, and silk—from a neutral or an acid bath. This is true only of a few direct dyes.

Direct dyes have a wide color range but do not produce the brilliant colors of acid dyes. They are inexpensive to apply and give even, level colors but have a varied degree of fastness. They have a poor fastness to washing as they have a tendency to run. Some are sensitive to acids and alkalies. According to the type of the dyestuff and its application, they may have from a poor to excellent degree of fastness to sunlight. They are used primarily for fabrics that require infrequent washing. The direct dyestuffs are often used for viscose or cuprammonium rayons. Special classes of direct dyes have been developed for these rayons to give excellent, level shades. Acetate rayons have no affinity for direct dyes.

Direct dyes, like acid dyestuffs, are sodium salts of color acids that have a molecular structure that produces a specific affinity for cellulosic materials. It is supposed that in the application of direct dyes the dye molecule is absorbed as a whole.

Chemicals used. To exhaust the dyestuff: sodium chloride or calcined Glauber's salt. *To retard the action:* use less exhausting chemical.

The process. The cold or lukewarm dye bath may consist of a combination of sodium chloride, sodium carbonate, and sodium sulfate. The concentration of the dye bath is important because the affinity for direct dyes varies with the concentration of the dye and the salts that are present. The dyestuff is added and the fabric is entered. The dye bath is raised to a low boil, and the fabric is left in the bath for about an hour until the material absorbs the dyestuffs. This is usually an alkaline dye bath. The method may vary with different dyestuffs.

It is thought that the affinity of materials for direct dyestuffs is caused by the long chain molecules of the dyestuffs that enter the pores in the cellulose fibers and are precipitated in the fiber during the dyeing by the heat and salt.

MORDANT AND CHROME DYES

Dye animal materials—wool and silk—with a mordant.
Dye cellulose materials—cotton and linen—with a mordant.

Mordant dyes are excellent for animal fibers but only in a few cases are they suitable for vegetable fibers. The one important use of a mordant dye on cotton is the old, well-known Turkey or alizarine red. It is dyed on a calcium aluminum mordant. These dyes are used mostly to color silk, raw wool, wool slubbings, wool yarns, and fabrics. They produce a wide range of colors, although not as bright as those from acid dyestuffs. The colors have a high

COTTON AND LINEN FABRICS, DYED AND PRINTED

These are familiar cotton fabrics that are a part of our great utilitarian group. Colors that have a high degree of fastness to laundering and sunlight, have added much to their usefulness and wearability, as well as to their attractiveness.

In the left row, top to bottom, are included the following fabrics. The plaid gingham, the checked madras, the heavier, striped Cheviot, and the broad striped chambray were all woven with colored yarns and appear the same on both sides of the fabric. The bottom fabric in this row is a poplin with printed stripes.

The following are the fabrics in the right row, top to bottom. The oxford cloth with green and white stripes was woven with two fine warp yarns and one coarse filling yarn which is the size of the two warp yarns. The fine gingham was woven with dyed yarns. The third fabric is a loosely constructed linen, woven in a twill weave with dyed yarns. The last two fabrics are seersucker. Both were woven with colored yarns. The crinkle was obtained by certain yarns being held at a different tension than others. The last fabric has a changeable appearance because of the interweaving of different colored warp and filling yarns.

Other colored illustrations of yarn dyed fabrics are as follows: filament rayon fabrics facing page 353; wool and spun rayon fabrics facing page 257.

degree of fastness to sunlight. Their degree of fastness to alkalies and washing is much better than that of acid dyes.

Mordant and chrome dyes are colored compounds, capable of forming insoluble compounds with metallic salts, particularly with those of aluminum, chromium, copper, and iron. When the metal salt is added, it combines with the wool and also with the color, fixing the color in the fiber. Silk may be mordanted the same as wool.

Most of the synthetic mordant dyeing colors contain acid groups and are thus classified as acid mordant colors. These are mainly used on or with a chromium mordant and thus are referred to as chrome colors.

One of the newest ways of applying the chrome-mordant dyes is what is known as the "metallized dyes." In this method the chrome is built into the chrome dye molecule and the metallized dye can then be applied as if it were an acid dye, except that greater quantities of sulfuric acid are necessary to obtain the best fastness and best working properties.

Chemicals used. The mordant: sodium or potassium bichromate is used usually for the three methods of application. Metachrome adds ammonium acetate or ammonium sulfate. Bottom chrome adds tartar. *To exhaust the dyestuff:* usually acetic acid, formic acid, or sulfuric acid. *To retard the action:* Glauber's salt.

The process. The mordant dyes are applied by three processes. In the top-chrome or after-chrome method the material is first dyed as in the acid dyeing process described above, and the chrome mordant is applied later. The dye bath consists of the dyestuff, Glauber's salt, acetic acid, formic acid, or sulfuric acid. The shade is developed by adding bichrome or chromate of soda and boiling. In the metachrome method the mordant and dyestuffs are com-

Yarn dyed cotton fabrics. Left to right: Candy stripe, cluster stripe, satin stripe, blazer stripe.

bined in a single bath. The mordant is usually sodium or potassium bichromate combined with ammonium acetate or sulfate. This bath frees the acid, permitting the material to absorb the dyestuff and mordant slowly and gradually. Acetic or formic acid may be added to improve the exhaust. In the bottom chrome method the chrome mordant is first applied and then the material is dyed. Metallic salts are precipitated on the fibers in the form of chromium oxide. The mordant is usually obtained by boiling the wool in sodium bichromate (chromate of soda) and tartar. The material is then rinsed and dyed. The dye bath consists of the dyestuff and acetic acid.

DIAZOTIZED AND DEVELOPED DYES

Dye vegetable materials—cotton, all rayons, linen.
Dye animal material—silk.

These are usually direct dyes and are applied in the same way. However, after dyeing, the fabric is diazotized and developed to produce more satisfactory colors that have a greater degree of fastness to water and washing and to acid cross-dyeing. They are frequently used for discharge printing. In addition to direct dyes, any class of dyes that contain a free amino group can be converted on the fiber in the same way.

Chemicals used. These are usually the same as used for direct dyes. *To exhaust the dyestuff:* Glauber's salt or common salt. For rayon, sulfonated oil is added. *The diazotizing chemicals:* sodium nitrite and hydrochloric acid (and/or sulfuric acid). *The developing chemicals:* beta naphthol and caustic soda.

The process. In one method the dye bath consists of the dyestuff and Glauber's salt. The fabric is entered and boiled a short time before the salt is added, after which the boiling continues for about thirty minutes. For rayons, the dye bath is charged with sulfonated oil.

After rinsing, the dyed fabric is diazotized in a bath of sodium nitrite and hydrochloric acid (or sulfuric acid). The solution is cold and the fabric remains in it from fifteen to thirty seconds.

After repeated rinsing, the color is developed directly in the fiber with a developer, such as beta naphthol with caustic soda to dissolve it, actually producing a new dyestuff in the fiber. By using various developers, a wide range of colors is produced. After the color has been developed, the fabric is washed, rinsed, and dried.

This method can be used in continuous machines. Some dyes should be dried after padding them on the fabric and before they are diazotized and developed.

For dyeing acetate rayon, the dyes are first dissolved in a solution containing such compounds as hydrochloric acid, formic acid, or soda ash. The fabric is then dyed, with the addition of Glauber's salt and acetic acid to the dyestuff's solution. After rinsing, the dye is diazotized with sodium nitrite and hydrochloric acid. The material is again rinsed and immersed in the developing solution which may be mildly acid with acetic acid, the dyestuff being absorbed by the fiber. However, more recent methods of dyeing acetate rayon make use of special acetate dyes.

AZOIC DYES

Dye vegetable material—cotton and regenerated cellulose rayon.

Azoic dyes are also called "naphthol" dyes. They produce an all-round degree of fastness, excelled only by vat dyes. Some of them have a high degree of fastness to washing, bleaching, sunlight, acids, alkalies, chlorine, and cross-dyeing. They are frequently used to dye yarns that will be combined with white, as they will withstand boiling water and chlorine. Because they are formed by precipitation, they have a tendency to crock. The azo colors are insoluble colored precipitates, made by adding a solution of a diazo compound to an alkaline solution of a naphthol. Since the solution must be kept cool, they are often called "ice colors."

Chemicals used. Developing chemicals: naphthol, caustic soda, and alcohol. *Diazotizing chemicals:* sodium nitrite and hydrochloric acid.

The process. This process is usually carried out in two steps. In the first, the cloth is prepared with the naphthol (developer) and the fabric is dried. In the second step, the color or the intermediate, containing the free amine group, is diazotized. This diazotized product is unstable unless kept cold. The prepared cloth is then passed through this diazotized product, whereupon it couples to the naphthol on the fabric and a color is formed. This method is the reverse of that described above as diazotized and developed. Para red is an excellent example of an ice color.

In order to eliminate the undesirable use of ice and the necessity for drying cloth prepared with beta naphthol, new naphthols, and stabilized, fast color salts have been developed. It still is necessary to prepare the cloth first, then apply the color salt, whereupon the desired color is obtained. Passing a dyed or printed fabric, so obtained, through a steamer improves the color fastness.

Finally, it was found possible to combine both the naphthol and the stabilized diazo compound if the material were kept alkaline. This product is usually applied to the fabric by printing and the color is developed by acid ageing.

SULFUR DYES

Dye vegetable materials—cotton, linen, and regenerated cellulose rayon.

Sulfur dyes produce a wide range of rather dull colors. They have a remarkable degree of fastness to water and are used for fabrics that require frequent and hard washing. However, they have no fastness to chlorine, which is used in bleaching and in commercial laundries. Sulfur dyes are extensively used to dye fabrics for automobiles. They have a fair degree of fastness to perspiration, acids, alkalies, and light.

Sulfur dyes differ from all others in their chemical composition and method of application. They are made by the reaction of sulfur with organic compounds. They are insoluble in water and therefore must be put into solution with sodium sulfide and alkali. Since sulfur dyestuffs must be applied to fabrics in an alkaline solution, they are not used for wool or silk, for alkali is destructive to animal fibers.

Chemicals used. Reducing agents: sodium sulfide and sodium carbonate (soda ash) to reduce to leuco form. *To exhaust the dyestuff:* sodium chloride or sodium sulfate (Glauber's salt). *After treatment:* bichromate of soda and acetic acid, or hydrogen peroxide, or sodium perborate (copper sulfate may be used as an after treatment).

The process. Since the sulfur dyestuff itself is insoluble, the color is heated for a few minutes in a solution of sodium sulfide and soda ash. This reduces or dissolves the dyestuffs, probably forming leuco (soluble) compounds. This is diluted with water and the material is dyed in this bath for one to two hours. Exhaustion is obtained with salt. After the fabric is dyed, it is given an oxidizing aftertreatment to develop the dye (to make it insoluble) and to give a fast and brighter shade. The fabric is washed, rinsed, and dried. The color is oxidized in a bath such as bichromate of soda and acetic acid. The dyed material is then given a final washing with soda to remove the acid.

It is supposed that the dyeing properties of sulfur dyes are due to the chain molecules, which are reduced in the alkaline solution and are taken up by the fiber, after which they are oxidized to become regenerated insoluble dyes.

VAT DYES

Dye animal materials—wool and silk (limited in use).
Dye vegetable materials—cotton, linen, and all rayons.
Dye synthetic materials—nylon (limited).

Vat dyes are considered to have the most satisfactory degree of fastness of all dyestuffs, especially on cellulose. They have a wide range of shades, being second only to the number of direct dyestuffs. They produce the clearest, brightest colors with the exception of the little used basic dyestuffs. Just as with all other dyes, on the kind and quality of the dyestuff and its method of application depend the degree of fastness to water, sunlight, and alkalies. However, as a class, vat dyes have an excellent degree of all-round fastness. Many can be washed with chlorine and boiled with alkali.

Indigo is probably one of the oldest dyestuffs known to man as well as the oldest vat dye. Today, two other groups are very important. One includes the thioindigo vat colors, used for animal and vegetable fibers. Sodium hydrosulfite is usually used as a reducing agent for both types of fibers. The other group includes the anthraquinone vat colors, used for dyeing and printing cottons, linens, and rayons. These require more caustic soda in their application and so are not usually used for wool.

Vat dyes are supplied in paste or powder form. They are insoluble in water and must be reduced in order to become soluble. They are applied to the fiber in the form of their alkali-soluble leuco compounds. In the leuco state they possess attraction for the fiber. They can then readily be reoxidized by air or by an oxidizing agent.

Chemicals used. Reducing agents: sodium hydrosulfite. *To make soluble the dyestuff:* sodium hydroxide (caustic soda). *To exhaust the dyestuff:* sodium chloride or Glauber's salt may be added. *To retard the exhaustion:* glue. *Aftertreatment:* potassium bichromate and acetic acid, sodium perborate or hydrogen peroxide, or simply atmospheric oxygen.

The process. The vat dye in paste or powder form is made into a paste with water, sulfonated oil, and caustic soda. The dye is then reduced by adding sodium hydrosulfite and heating. The reduction is usually accompanied by a change in shade, as most leuco compounds have a different color than the parent dyestuff. The reduction is usually quite rapid, although this depends upon the temperature of the dye bath. The reduced color has a strong affinity for cellulose fibers.

The dye bath contains sodium hydroxide (caustic soda) and sodium hydrosulfite. The reduced vat dye is mixed with this bath and the fabric is entered. For some fabrics, sodium chloride or Glauber's salt is added. Glue may be used to retard the dyeing and to protect the wool fiber. Vat dyes may be applied under varying conditions of dye bath composition and temperature, which is usually from 120° to 180° F. Cottons may be dyed in a few minutes, but it takes from thirty minutes to two hours to dye wool with vat dyes, depending upon the procedure used.

After the fabric is dyed, it is rinsed and given an oxidizing aftertreatment. This is by exposure to air, or with potassium bichromate and acetic acid, or with sodium perborate or hydrogen peroxide.

ACETATE DYES

Dispersed dyes for acetate rayon include certain types of insoluble compounds of the azo dyes or anthraquinone vat dyes that are maintained in a colloidal suspension by means of sulfonated castor oil or soaps. They are dispersed during dyeing and in insoluble form are absorbed by the fiber. The dye solution is mildly alkaline. These dyes have no affinity for any other cellulose or animal fiber and can be used with other dyes to produce two-tone effects. For example, acetate rayon can be dyed with acetate dyes while cotton is being dyed with direct dyes. Acetate dyes may be considered as pigments, but the pigment classification does not completely cover acetate dyes.

Dye vegetable material—cotton.

Dye synthetic materials—rayon, nylon.

Dye mineral material—glass fiber.

Pigments differ from dyes because they are insoluble in the media used in their application and usually have little affinity for fibers. There are two general classes of pigments, natural and synthetic pigments. Although there are several thousand different pigments, only a comparatively small number have proved satisfactory for use in the pigment dyeing of fabrics. The so-called "mineral dyeing process" consists essentially of forming a synthetic pigment on or in the fabric fibers. However, most of the other methods for pigment dyeing consist of bonding pigments to fabric fibers mechanically by means of synthetic resins or other film-forming materials.

The satisfactory results of pigment dyes are dependent on two things: the property of the pigment and the methods used. Some have an excellent degree of fastness to washing, drying, light, acids, and alkalies.

Crocking and resistance to rubbing is one of the problems in pigment dyeing. Some pigments with some binders can be washed. The same pigments with other binders can be dry cleaned only. The binders insure the fastness to washing and will modify the crocking to some extent. One of the several advantages of pigment dyeing is the uniformity of shade that can be obtained from lot to lot and from the beginning to the end of a run of goods, making it easier to match shades.

Dyeing. The pigment, either mineral or synthetic, is first prepared in an extremely fine state. It is then necessary to incorporate this finely divided pigment into the resin-water system to give a uniform, stable emulsion that will not "break." Furthermore, the emulsion must not permit the pigment to settle out. The purpose of the pigment is to color or dye the fabric. The purpose of the resin is to bind the pigment to the fibers or fabrics. The pigments, resin, solvent, water, and emulsifying agent form the dyeing composition. This mixture is usually applied by padding and then, bonds the pigment to the cloth with a permanent binder.

There are four systems now in use for pigment dyeing. These

are: water-in-oil emulsion system; oil-in-water emulsion system; solvent dispersion system; aqueous dispersion system.

Water-in-oil emulsions are prepared by dispersing pigments in a solution of synthetic resins in an organic solvent, and then stirring in water with the aid of a high-speed mixer. The high-speed mixer breaks up the water into small globules or droplets. The synthetic resins bond the pigment firmly to the fiber. The process is carried out by running the cloth through the pigmented emulsion on a padder, and then drying and curing.

Oil-in-water emulsions are prepared by dispersing pigment colors in a solution of synthetic resins and then stirring in an organic solvent. With this type of emulsion it is the oil or organic solvent that is in the form of the small globules or droplets. Dyeing is carried out in a similar manner to that described above.

Solvent dispersions are prepared by dispersing the pigments in an organic solvent solution of synthetic resins. The dispersion is applied on a padder, after which the fabric is dried and cured.

Aqueous dispersions are prepared by dispersing the pigment colors in a solution of water soluble binders, such as starches and gums with resins, urea formaldehyde, or a specially treated starch. A specific method is using a solution made by dissolving hydroxy cellulose ether in a solution of caustic soda. The fabric is padded through the dye liquor, next through an acid bath, and then washed, neutralized, and dried.

COLOR PRIMARIES *

The subject of color primaries is often a point of discussion. The physicist usually calls his primaries red, blue-violet, and green, while the colorist, color photographer, or artist considers his primaries to be red, blue, and yellow. For many reasons, because of the complexity of color, it is not practical to use only three basic colors and with these three match exactly every desired color. However, for general discussion, it is possible to choose certain color ranges and state that these are the primaries of a certain color system. That is, with three colors only, it is possible to obtain mixtures that closely resemble all colors. Thus, in color photography and color printing, three colors and black or three colors alone are used for color reproduction.

In dyeing and most other coloring work, a dye or pigment is

* Written by Dr. G. L. Royer, Calco Chemical Division, American Cyanamid Company.

RELATIONSHIP OF COLOR PRIMARIES

The primaries of both the additive and subtractive color mixtures are shown in this figure and their color composition is demonstrated by indicating the portions of the spectrum which they embrace. The colors of the additive system are used by scientists when mixing colored sources of light, while the artist in mixing pigments or dyes uses the colors of the subtractive system Figure 1.

EFFECT OF MIXING
SUBTRACTIVE PRIMARIES

When the three subtractive primaries are mixed with each other, they give three binary colorants which are similar to the colors of the additive primaries. This can be seen in this figure where circular areas of three subtractive primaries, bluish red, yellow, and greenish blue overlap to produce the orange-red, green, and violet-blue. The mixture of all three produces black which can be seen in the center where all three of the circles of primary colors overlap. Figure 2. (Calco Chemical Division, American Cyanamid Company, Bound Brook, New Jersey).

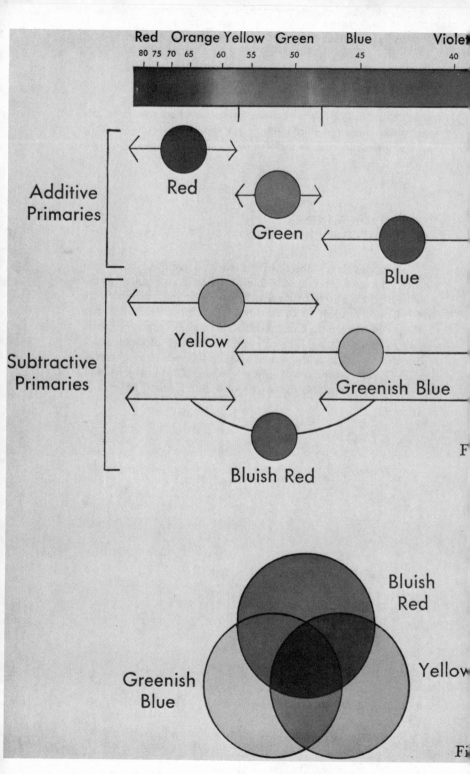

Red Orange Yellow Green Blue Violet
80 75 70 65   60   55   50   45   40

Additive
Primaries

Red

Green

Blue

Subtractive
Primaries

Yellow

Greenish Blue

Bluish Red

F

Bluish
Red

Greenish
Blue

Yellow

Fi

chosen that most closely matches the desired shade, and other dyes or pigments are added to make the match more exact. No attempt is made to match a shade by the mixture of only three primaries. Other than from the point of view of the difficulty of exact color matching with three primaries, there are other factors which must be considered in all coloring problems. Some of these factors are dyeing or pigment working properties, fastness to light, fastness to washing, and fastness to rubbing off. All these factors must be considered in any result, in addition to color. Therefore, it can be seen why it would be difficult to use only three primary dyes or pigments to match any sample. As mentioned before, in color printing and particularly color photography, three color primaries are used for color reproduction and while the color reproduction is not exact, it does approach the original. In dyeing and other coloring work, similar general principles relative to color mixture and matching exist, so that a general discussion of color primaries and color blending is desirable.

When white light is passed through a prism, it is spread out into its spectrum or component parts. The spectrum can be divided into the following well-known order of the colors of the rainbow: red, orange, yellow, green, blue, and violet. Each of these is defined as covering definite spectral ranges (see Fig. 1). If all of these colors are passed back through another prism, white light will again be obtained. Thus, both by analysis and by synthesis, it can be seen that white light is composed of all of the colors. The spectrum can be divided in an arbitrary way to give combinations that can be called primaries. It can be divided into three parts, the violet and blue, the green and yellow, and the orange and red, which can, in general, be described by the terms blue, green, and red respectively. If these three portions or colors are added together, they will form white. These are, therefore, known as the primaries of the additive system of color mixture. The additive system is used in the scientific field, where light from different sources is added together. In this system all the colors of the spectrum must be put together to produce white light. No light gives darkness.

The artist or colorist divides the spectrum differently to obtain his primaries, because he uses the subtractive system of color mixture. He does not deal with different colored sources of light but with white light that falls upon or illuminates his subject, which is colored with dyes or pigments. The pigments or dyes remove or absorb portions from the white light and reflect the remainder, which is seen by the observer. This reflected light, which remains after the pigment or dye has subtracted certain portions of the white light, is the color that the colorist or artist sees. This system is often called the "subtractive" system. Thus, if green light is removed or subtracted by the dye or pigment from white light, the result is a bluish red that is a combination of portions of the blue and red parts of the spectrum. This colorant is usually called the "red primary" by the artist or colorist and by the color photographer is often called "magenta." When the pigment or dye absorbs the red part of the spectrum, the blue and green portions are reflected, and this colorant is a greenish blue. This is the blue primary of the artist or colorist and is distinctly a green-blue and not the violet-blue of the additive system. When the blue portion of the spectrum is removed from white light, the red, orange, yellow, and green are reflected. The colorant that accomplishes this appears yellow and is the third primary of the subtractive system.

Thus, it can be seen that each of the primaries of the subtractive system reflects light that is similar to a mixture of two of the primaries of the additive system. In subtractive color mixture, a mixture of any two of the primaries reflects light similar to a primary of the additive system. A mixture of all three subtractive primaries substantially reflects no light and is black. No subtraction gives white. Thus, the subtractive system is somewhat opposite to the additive. The subtractive primaries remove or absorb one of the additive primaries from white light, thus leaving the other two additive primaries. Consequently, the subtractive primary, red or magenta, is often called "minus green," the blue or greenish blue is called "minus red" and the yellow "minus blue." That is, the light reflected from a subtractive primary is the same as white light with one of the additive primaries subtracted or removed. The additive system is used with the mixture of light and, therefore, it is of specialized interest. The subtractive system is used with mixed colorants and, therefore, this system is of major interest to fabric colorists.

When the three primary colorants of the subtractive system, a bluish red, a yellow, and a greenish blue, are mixed, the results shown in Fig. 2 are obtained. Where two primaries overlap, binary colors are formed. Where all three overlap, black results. These binary colorants reflect light that is similar to the additive primaries, orange-red, green, and violet-blue. It is obvious that the confusion that often occurs in the naming of the primary colors is due to a confusion of the two systems of color mixing and to an indiscriminate description of the two colors red and blue. In the additive system the red is an orange-red and the blue a violet-blue while in the subtractive system the red is a bluish red (or magenta) and the blue a greenish blue. In the majority of the color mixing problems the subtractive system is used and so the primaries are a bluish red or magenta, a greenish blue, and a yellow.

If these six colors, the three primaries and the three binaries, are arranged around a center, a color wheel is formed. (See Fig. 3.) This is one of the well-known methods of color arrangement, and it can be used in a general way, to understand the mixing of colors. When opposite colors are mixed, blacks or grays tend to form. If colors once removed are mixed the color between is approximately obtained. Since no colorants are pure color and since no three perfect primaries exist, it is not possible to set up a color wheel of colorants that can be mixed to give all hues. However, general practical conclusions can be established from a color wheel even if an exact color mixture is not obtained. For example, a mixture of yellow and red gives an orange-red, but this will not be as pure an orange-red as can be obtained by the choice of a specific colorant having properties nearer to a pure orange-red.

Other colors or portions of the spectrum could be chosen and designed to be primaries of a specific color system. They in turn would work together to give various color combinations. However, the systems which have been discussed are those in most general use and the primaries chosen are those usually employed in the discussion of practical color blending.

THE BLENDING OR MODIFICATION OF COLORS *

The blending or modification of colors can be discussed from two viewpoints. Each view is highly important in certain industrial, art, and mercantile fields. From one approach, color modification involves the skillful use of materials that have been dyed or otherwise given color, in combinations that are pleasing or useful for a purpose. The second approach involves the mixture of col-

* Written by William H. Peacock, Calco Chemical Division, American Cyanamid Company.

orants, such as dyes or pigments in various media, to impart color to materials.

When colored materials are viewed in juxtaposition, each one reflects light to the eye, and the optical mechanisms tend to combine or add all these lights together to get additive blends. However, many factors modify the effects obtained. The uniformity or completeness of the blend will vary with the relative sizes of the colored areas viewed and the distance of the objects from the viewer. For example, a signal flag composed of large red and blue checks will tend to appear purple when viewed at a considerable distance. Similarly, a red and blue small-check gingham will appear purple at a lesser distance, while a pointillé-like blend of red and blue, separately dyed fibers, similar to oxfords and heather mixtures, will appear purple at a relatively short distance from the viewer. Motion of the separately colored areas also tends to give an additive blend, as, for example, when one semicircle of a circle area is colored red and the other half is colored blue, the whole circle will appear purple if rapidly rotated.

Even when the colored areas are too large, or too close to the viewer for their reflected lights to be blended in the eye, each tends to appear somewhat different in color than when viewed alone. For example, with colors approximately equal in saturation or strength and value:

Reds viewed with oranges and yellows appear bluer.

Reds viewed with greens appear brighter and slightly bluer or yellow, depending on their hues.

Reds viewed with blues and purples appear yellower.

Oranges viewed with reds or purples appear yellower.

Oranges viewed with yellows or greens appear redder.

Oranges viewed with blues appear brighter and yellower or redder, depending on their hues.

Yellows viewed with reds and oranges appear greener.

Yellows viewed with greens and blues appear redder.

Yellows viewed with purples appear brighter and greener.

Greens viewed with reds appear brighter and bluer or yellower, depending on their hues.

Greens viewed with oranges and yellows appear bluer.

Greens viewed with blues and purples appear yellower.

Blues viewed with reds or purples appear greener.

Blues viewed with oranges appear brighter and redder or

they appear greener, depending on their hues.

Blues viewed with yellows or greens appear redder.

Purples viewed with reds or oranges appear bluer.

Purples viewed with yellows appear brighter and redder or bluer, depending on their hues.

Purples viewed with greens and blues appear redder.

All chromatic colors appear brighter viewed against a black than against a white background.

Grays assume a cast towards the complementary of the color with which they are viewed.

Browns also vary in appearance when viewed with differently derived browns or other colors. The effects are complex, but, in general, reddish browns, yellowish browns, and greenish browns will tend to be affected much as the major hues from which they are derived, although to a less noticeable degree.

The second approach to color blends and modifications is from the viewpoint of the dyer, stainer, printer, coater, or painter, that is, the application of coloring agents. These persons blend color by blending colorants. Here, a knowledge of the spectral transmission or reflectance characteristics is essential for correct work, although undoubtedly the majority of colorists usually depend upon experience and judgment acquired through years of trial and error applications. Colorists who are experienced in the use and interpretation of spectrophotometric data can make some colorant blends on a mathematical basis.

In general, it has been found that most dark-color effects can be obtained by making blends of magenta, yellow, and blue colorants so balanced that a mixture of equal portions of them yields a neutral black or gray. A complete discussion of such blends would be lengthy. It may be helpful, however, to discuss the four general ways in which such a use of three so-called primary colors may be blended to obtain color modifications. These are:

1. By admixture of one hue with another:
   a. By the addition of a small quantity of the second colorant, to obtain a "cast."
   b. By larger additions, to obtain intermediate hues, approaching binary (two-color) effects.

2. By admixture of all three primaries, which is the equivalent of adding black, since an equal portion of each primary yields black:

    a. An excess of one primary plus smaller additions of the other two tends to yield shades, in much the same effect as the addition of black.

    b. An excess of two primaries, which is the same as adding black to a binary color.

3. By similar mixtures of the three primaries used in blends in relatively weaker concentrations, to yield colors tending toward pastels, and similar to tones made by the addition of gray or by the addition of both black and white. This effect is frequently obtained with pigments by adding a white colorant to the three primaries, or by adding gray to one or two primaries.

4. By using one colorant alone in varying amounts to produce tints. With pigments this is accomplished by the addition of a white colorant.

These methods of blending colorants for modifying color effects may be summarized as follows:

Admixture of one hue with other hues to produce casts, binary and multicolor effects.

Admixture of a hue with black to produce shades.

Admixture of a hue with gray to produce tones.

Admixture of a hue with white to produce tints.

Combination mixtures of these four types of blends.

Illustrative of the first method is the procedure where each hue may be changed by the addition of red, yellow, or blue modifiers. This practice is an important feature of color matching. By means of adding one colorant to another the hues can be changed as follows:

Red can be made bluer or yellower.

Orange can be made redder or yellower.

Yellow can be made redder or greener.

Green can be made yellower or bluer.

Blue can be made greener or redder.

Purple can be made redder or bluer.

Each hue also can be changed by the addition of more than one colorant to it. For example, a blue and a yellow might both be added to a red. The result will change not only the hue but the brightness and saturation also, because in almost every case a three-color or triad mixture of coloring agent introduces grayish effects that dull or "tone down" the hue.

A second method of modifying a color is to admix it with black. Such additions produce "shades." These are less bright than the

494

COLOR WHEEL OF HUES

The three subtractive primaries and the three binaries produced by combination or mixing can be arranged so as to produce a color wheel of hues. The opposite colors of this wheel are complementary colors and when mixed, blacks or grays tend to form. If the hues are mixed with each other, a color intermediate is formed. Addition of white and black to any hue gives tones, tints, and shades. Figure 3. (Calco Chemical Division, American Cyanamid Company, Bound Brook, New Jersey).

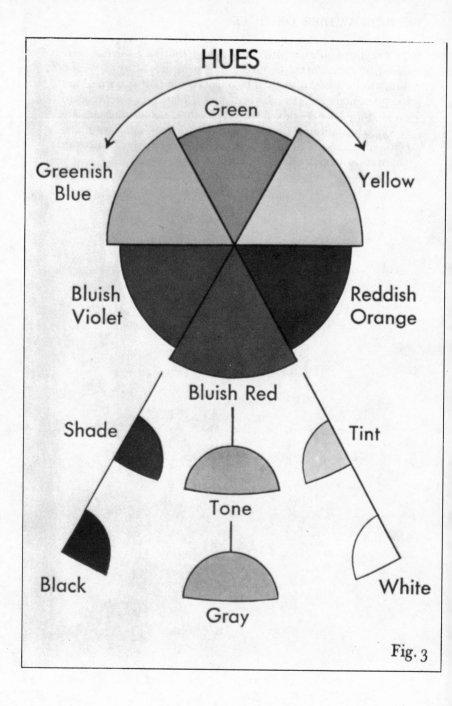

Fig. 3

original color. Because almost every black coloring agent is not a "true" or neutral black, it will be found that the hue is usually changed as well as the brightness and the saturation. This is particularly apparent when a black is added to a yellow, because the yellow is almost invariably changed to a green. Note too, that this use of the term "shade" to denote a hue mixed with a black, should be carefully differentiated from the present prevalent use of this term in the fabric industry, as being synonymous with color— "on shade," "off shade," means "on color," and "off color."

5. By these mixtures with black:
Red yields colors such as maroon, mahogany, and chocolate.
Orange yields colors such as golden brown to seal brown.
Yellow yields colors such as moss green and drab.
Green yields colors such as olive and evergreen.
Blue yields colors such as navy, midnight, and blue black.
Violet yields colors such as prune and eggplant.

A third method of modifying hue is the addition of gray. As grays may be considered merely a mixture of black and white, some writers omit this classification. The addition of both black and white or a gray to a color tends to reduce its brightness and saturation, to yield "tones." By this means:
Red yields such colors as garnet and pigeon gray.
Orange yields such colors as cocoa, beaver, and beige.
Yellow yields such colors as chartreuse, old gold, and khaki.
Green yields such colors as blue spruce and palmetto.
Blue yields such colors as Copenhagen and West Point gray.
Violet yields such colors as plum and purplish gray.

A fourth method of modifying a color is to admix it with white. When the coloring agents are used in the form of pigments, this can be done by a physical mixture of the chromatic pigment with a white pigment to produce a "tint." When soluble colorants (dyes) are applied, the same effect is produced by using less dye. Usually the addition of white tends to brighten a color. By this means:
Red yields tints such as strawberry, scarlet, and pinks.
Orange yields tints such as pastel orange to peach.
Yellow yields tints such as pastel yellow to off-white.
Green yields tints such as emerald and tourmaline.
Blue yields tints such as sky blue and turquoise.
Violet yields tints such as mauve and lavender.

It should be noted that in this discussion it has been assumed that magenta and yellow will blend to give an orange, equivalent in brightness to the two primaries, and that, similarly, the green from a blue and yellow mixture and the purple from a red and blue mixture will be as bright as the original primaries. In practice this does not occur, for the binary color obtained is invariably duller than the color given by either of the two primary colorants alone. Consequently, the blending of two colorant primaries modifies not only the hue but also the brightness. Every coloring agent, therefore, tends to act as though it were a binary or even a three-color blend, or as though it contained some portion of black. The experienced colorist frequently will add black or gray colorants rather than a black made by using equal portions of magenta, yellow, and blue, and will obtain visually similar color effects.

These orientations of color are a practical aid to color matchers. Actually, much of a colorist's success is based upon his recognition of the reddish, yellowish, bluish, and grayish elements present in both his materials and the individual colorants, and on his skill in balancing these and other factors to obtain the desired color effects.

Raw stock dyeing. Top to bottom: Raw stock being lowered into the dyeing machine (North Star Woolen Mill Co.). A stainless steel, stock-dyeing machine showing the heavy cover being lowered and clamped to the machine (Riggs and Lombard, Incorporated). A large quantity of loose fibers being dyed in a huge vat (Mohawk Carpet Mills, Inc.).

496

DYEING METHODS

There are various methods of applying dyes. The raw fibers may be dyed, which is called "raw stock dyeing." The long, combed wool fibers may be dyed, "top" or "slubbing dyeing." The yarn may be dyed, "yarn dyeing." The fabric may be dyed, "piece dyeing." In deciding which method to use, consideration is given to the fiber, the ultimate use of the fabric, the type of dye being used, the result expected, the time required, and the cost.

Different types of machines are used for each method as well as more than one type for one method. For example, fabrics may be dyed in rope form or in open, full width. Yarns may be dyed in hanks or in package form. The following section describes each method, the form in which the material is dyed, and the machines used.

Raw-stock Dyeing

Fibers of wool, cotton, linen, casein, rayon staple, and waste silk can be dyed in the raw stock. There are two types of machines used for raw stock dyeing, regardless of the kind of fibers dyed. One is the rotating cylinder machine, which has rotating cylinders that circulate the raw stock through the dye bath. The other is a circulating machine in which the dye bath circulates and the stock remains stationary.

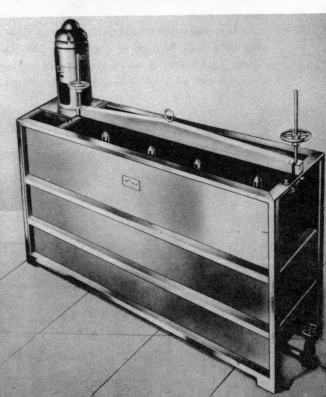

Top dyeing. A machine for dyeing wool tops or slubbings. Inset shows unloading arrangement (Riggs and Lombard, Incorporated).

CHAIN HOIST

CLAMPING BAR

CAP

LIFTING PLATE

TUB

UNLOADING ARRANGEMENT

The rotating cylinder machine is less frequently used because of the possible matting and twisting of the fibers. In dyeing rayons it is used principally for dark sulfur dyes. This machine is a large tank with an inner cylinder that has compartments in which the raw stock is placed. The cylinder is packed outside the tank and is then placed in the tank that contains the dye bath. The cylinder revolves, forcing the dye liquor through the stock.

The circulating machine is a large tank with a removable false bottom. The stock is placed in the tank and the dye liquor, pumped through pipes to the tank, circulates through the stock.

In both machines, after dyeing is completed, the dye bath is drawn off, the stock is rinsed, taken out of the machines, and dried.

Top or Slubbing Dyeing

Tops or slubbings are the continuous, practically untwisted slivers of wool fibers that result from combing. The short fibers or noils are combed out and the remaining long fibers are used for making worsted yarns and fabrics. The tops may be in hanks, rolled into soft balls, or wound on spools.

Hanks or skeins of the sliver may be hung on rods that are placed in a machine containing the dye bath. These rods or carriers revolve, constantly turning the tops or slubbings through the dye bath.

Soft balls of the slivers are placed, one on top of the other, on long rods. Several of these balls are then transferred to and compressed into a short, perforated cylinder. The cylinder is then placed in a dye tank and the dye circulates through the tops.

Silk skein dyeing. In this machine the dye liquor remains stationary and the skeins circulate through the bath (Cheney Brothers).

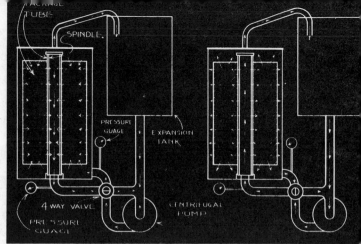

Yarn dyeing in packages. A yarn dyeing apparatus showing the flow of the dye from outside in, and from inside out (Calco Chemical Division, American Cyanamid Company).

Another method is to dye one ball alone rather than several compressed balls.

Perforated dye spools, on which the sliver is wound under tension, are frequently used. The top and bottom of the wide perforated cylinder are closed, forcing the dye bath through the perforations and the material. The spools, fastened securely in place, are stationary and the dye bath circulates. The top dyeing machine will hold up to eighteen spools but any lesser number may be dyed at one time.

Yarn Dyeing

Yarn dyeing is used for yarns made of any fibers, with the exception of glass. It is applied most extensively to cottons and rayons. While it does not permit as complete or as even

Yarn dyeing in skeins. Skeins being submerged in heated dye liquor that circulates through the yarns (Mohawk Carpet Mills, Inc.).

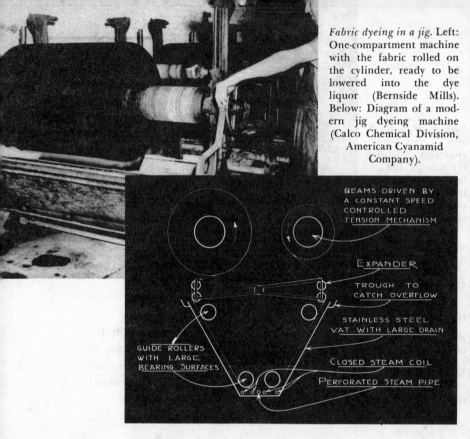

Fabric dyeing in a jig. Left: One-compartment machine with the fabric rolled on the cylinder, ready to be lowered into the dye liquor (Bernside Mills). Below: Diagram of a modern jig dyeing machine (Calco Chemical Division, American Cyanamid Company).

BEAMS DRIVEN BY A CONSTANT SPEED CONTROLLED TENSION MECHANISM

EXPANDER

TROUGH TO CATCH OVERFLOW

STAINLESS STEEL VAT WITH LARGE DRAIN

CLOSED STEAM COIL

PERFORATED STEAM PIPE

GUIDE ROLLERS WITH LARGE BEARING SURFACES

penetration as raw-stock dyeing, it is a more rapid process.

One now practically obsolete method was to place layers of hanks at right angles to each other in the tank and circulate the dye bath through them. The following are better methods.

The hanks are hung on spools or rods and are placed in the dyeing machine. They may be dyed in various ways. The spools may rotate, causing the hanks to circulate through the dye bath. The hanks may hang stationary and the dye bath be forced through the yarn. The spools may rotate and also move up and down, circulating the hanks through the dye bath as well as moving them up and down. The dye bath may be forced through perforated spools on which the yarn is wound, flowing through the yarn into the dye box. This latter method is used extensively for rayon.

The yarn-on-package system is used for wool, cotton, rayon, and

Fabric dyeing in a jig. Six-cylinder machine (each batch of cloth involves two cylinders) showing the squeeze rolls through which the fabric passes as it leaves the machine (Riggs & Lombard, Incorporated).

silk. This may be called the "spindle system." Wool yarns may be in the form of packages or cheeses and dyed on perforated spools. Cotton and rayon yarns may be dyed on perforated spools or tubes, or on tubes made of metal springs.

The yarn is wound on the perforated tubes or spools and is placed over perforated spindles and fastened down. The spindles are placed in the dyeing tank; the top is then fastened down, and the dye bath is circulated through the tubes and through the yarn. After the yarn is dyed, the dye bath is drawn off, the yarns are soaped, rinsed, and dried by circulating air through the package.

Dyeing warp yarns. Cotton and rayon warp yarns may be dyed. They may be wrapped on perforated beams before dyeing, or be dyed in skein form.

The dyeing on perforated beams is much the same as for package dyeing of yarns; in fact the same machines may be used. Several hundred yards of warp yarn are wound on a large perforated beam, which is placed vertically and fastened stationary in the dye machine. The dye circulates through the yarn.

The beam may be placed horizontally in a container shaped like a half cylinder, which holds the dye bath, and the beam rotates through the dye bath.

Another method is to collect a number of strands of warp yarn into a long chain and then double and redouble the strand to produce thick bundles. These strands are then passed through a dyeing machine that has compartments. Between each compartment are squeeze rolls. The strands are guided through the

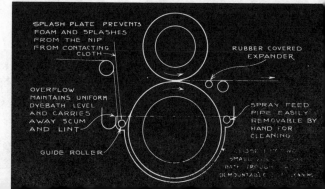

Fabric dyeing in a padder. Modern, universal padder, adapted to varied types of dyeing (Calco Chemical Division, American Cyanamid Company).

dye bath by immersed guide rolls. The squeeze rolls aid in the penetration and leveling of the dye by squeezing out the dye liquor.

Piece or Fabric Dyeing

Fabrics made of any fiber may be dyed in the piece. Cotton fabrics are usually held under tension and pass around rollers in the dye bath and then through squeeze rollers. This is accomplished on different types of machines. Wool and rayon fabrics are dyed in a loose condition and are not held taut.

As a general rule, piece dyed fabrics have not had as great a degree of fastness as fabrics woven from yarns that have been dyed in raw stock form or in skeins, as the dyestuffs do not as readily impregnate the fiber in fabric form. However, present-day equipment has reduced this difference. Piece dyeing is a quicker and more economical procedure and, if correctly applied with reliable dyestuffs, it provides colors that are completely satisfactory for the purposes for which the fabrics will be used.

Jig or jigger dyeing is used for cotton and rayon, when it takes a considerable time for the dyes to penetrate. The length of time needed for dyeing depends on the type of fiber or fabric being dyed and on the type of dyestuffs being used.

The machine consists of a V-shaped box, holding 75 to 150 gallons of the dye bath. It has guide rolls to immerse the fabric in the dye bath. There are two beams, one on each side. They may be above the box or be inside the box and immersed in the dye bath. The fabric to be dyed is wound on one beam. It leaves the beam, travels around the guide rolls, into and out of the dye

bath, and winds onto the beam on the opposite side of the box. The machine may be reversed, and the fabric run through the dye bath as many times as are necessary. Each passage through this machine is called an "end."

Pad dyeing for cotton and rayon is a high-speed operation and is therefore usually limited to dyeing of light shades. It does not produce a high degree of fastness but it is an economical way of producing even shades with satisfactory fastness.

A padder is used for finishing (such as applying sizing) as well as for dyeing, as it is really a means of saturating cloth and then wringing out the excess liquor. The machine consists of a tank with two rollers made usually of rubber. The lower roll may rest in the liquor, which is in a pad-box immediately underneath, and pad the dye or finish on the fabric. The other roller presses against the pad roller and squeezes the excess liquor out of the fabric. Feed pipes continuously carry additional dyestuffs into the tank. The fabric, usually dry, is run into the tank in open width. Guide rollers carry it into the dye bath, around the lower roller, and up between the two rollers, where the pressure of the rollers squeezes out the dye liquor and helps to level and penetrate the dyestuffs.

Pad-jig dyeing combines pad dyeing with jig dyeing. The dye is applied in the padding machine where the dye in pigment form is padded on the fabric. The dye is then reduced and the fabric is dyed by the jig method. This method is used for applying vat dyes. Often the padder is the first part of the continuous operation described below.

Dyeing fabrics in a kettle. Opposite page: Ropes of fabric circulating in a dye kettle (Goodall Fabrics, Inc.). Right: A stainless steel kettle showing the knurled-hump, drum reel that helps eliminate slippage and entangling of the fabrics (Riggs & Lombard, Incorporated).

Continuous dyeing for cotton and rayon may include in one continuous operation, dyeing, drying, and finishing. It is a fast and economical method, particularly efficient for large batches of fabrics that are dyed the same color. If properly applied, it gives good uniformity and a satisfactory degree of fastness.

The fabric in full or open width enters the dye bath, passes through two pad rolls, is then carried above the machine on so-called "sky rolls" to allow time for penetration. It then passes through the various compartments filled with developing bath, soap solution, and rinse water. As the fabric goes from one compartment to another, it passes through squeeze rolls that remove excess liquor. Leaving the last compartment, the fabric passes to a drier with numerous cans (or cylinders) that are steam-heated. The fabric is dried as it rotates around the cylinders.

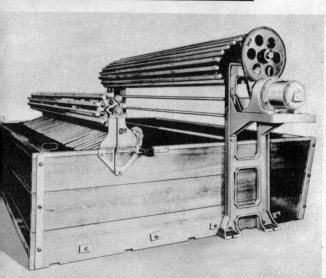

Dyeing fabric in a dye beck. Top to bottom: A pile fabric being dyed in open width in a dye beck (Cheney Brothers). Wool being dyed in rope form in a dye beck (American Wool Council, Inc.). A modern, dye beck machine used for silk and rayon fabrics (Riggs & Lombard, Incorporated).

The fabric may then be finished with starch or softener, after which it is usually dried on a tenter frame. This tenter frame evens the fabric to a uniform width and removes creases and wrinkles.

While it is possible to dye and finish in one operation, it is not usually done. However, a continuous dyeing machine is a real timesaver.

Winch or kettle dyeing is the principal method used for wools and rayons. It is rarely used for other fibers. This is the method of dyeing fabrics in a loose state rather than held taut or under tension as in the three methods just discussed. The fabric is dyed in rope form. The machines may be called "winch," "dye vat," "dye kettle," or "dye beck."

The machine is a long rectangular box, which contains the dye bath. Lengthwise across the top of the box is mounted a slat, reel, or winch. One reel may carry a number of pieces of fabric. The two ends of each piece are sewn together. As the pieces are usually 30 to 50 yards long, and the distance around the machine is 5 to 10 yards, the fabric is in a "loose" condition most of the time and taut only from the surface of the liquor to the top of the reel.

Cross-dyeing

A fabric may be cross-dyed if it contains yarns made from different kinds of fibers. Since acetate rayons are usually dyed with colors that have no effect on animal or vegetable fibers (including viscose and cuprammonium rayons), the fabric usually contains acetate rayon as warp or filling and one of the other fibers as filling or warp, respectively. One fiber may be dyed first, the dye bath dropped and a second dyebath used for the other fiber; or both fibers may be dyed in one dye bath, each taking its own dye; or one may remain white and the other dyed.

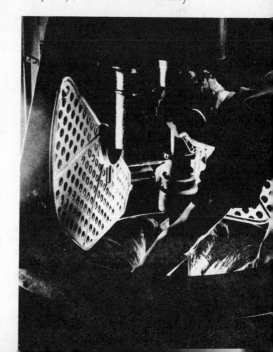

Dyed yarn is being placed in centrifugal extractor where it will be whirled at a high rate of speed to take out surplus moisture (Mohawk Carpet Company).

SUMMARY

DYES	HOW APPLIED	Wool	Cotton	Linen	Silk	Viscose and cuprammonium rayon	Cellulose acetate rayon	Casein	Nylon	Vinyon*	Saran*	Asbestos	Fiber glass	Feathers
							APPLIED TO:							
Acid	Direct	×			×			×	×					×
Azoic	With mordant		×	×		×								
Basic	Direct	×			×	×		×						×
Direct	With mordant		×	×		×								
Diazotized and developed			×	×		×								
Chrome	With mordant	×			×			×	×					×
Resin-bonded pigment			×	×		×	×		×	×	×	×	×	
Sulfur			×	×		×								
Vat		×	×	×	×	×	×	×	×					
Acetate							×		×	×				

*Trade-mark names

Opposite page: A roller printing machine printing a fabric. The raised roller (carrying the part of the design that prints one color) will be lowered into the dye box before printing resumes. Note the lower rollers carrying other parts of the design that will print other colors (American Enka Corporation).

PRINTING

Decorates the fabric

IN THE BEGINNING

The application of color to form designs originated so long ago that it is hard to tell exactly where its beginnings were. It may have been Egypt, it may have been India, or it may even have been Mexico or Peru. Since cotton was so well known to the ancients of India and since the cotton fiber is so very receptive to dyes, it is probable that these early peoples invented the art of printing. Egypt has given numerous evidences of very ancient printing, and Mexico is unearthing many interesting prehistoric relics.

Although true cotton fabric of about 3000 B.C. was found in a tomb at Mohenjo Daro in India, it is not known whether these ancient fabrics were printed or not. On the other hand, on the walls of the tomb of Beni Hassan, of Egypt (2100 B.C.), were drawn figures clothed in costumes containing stripes, spots, zigzags, and the like, and it is thought that these were probably stamped onto the fabric. Near Thebes a piece of fabric has been found, a wall or tent hanging of 1594 B.C., which was patterned in red and blue. An Egyptian painting of Hathor and King Meneptha I, 1320 B.C., shows their costumes stamped with complicated, orderly patterns. The earliest sample of printing in existence in the East is a piece of batik (the process of waxing to prevent certain sections of cloth from dyeing), found in a temple of Java and dated 1200 B.C. The word "batik" is a Javanese word meaning "wax painting."

Ancient India not only knew batik and block printing but also used stencil and tie and dyeing methods. This method consists of tying up certain parts of the fabric so tightly that the tied

508

parts do not take up the color when the fabric is dipped. It is necessary to re-tie for each color. Sometimes the yarns instead of the fabric are tie dyed.

Mummy cloths of later Egyptian periods were printed with borders of blue and tan and some bore texts of ancient documents. Joseph's cotton coat of many colors is said to be an example of printing.

Printing had become quite well known by 500 B.C. Herodotus, (484-425 B.C.) in writing of the garments of the peoples of the Caucasus, says that the pictures of various animals were dyed into them so as to be irremovable by washing.

Greek writings of 445 B.C. mention the popularity of gay colored Indian prints worn by the women of the Mediterranean. Another Greek traveler to India in 300 B.C. wrote, "They [the Indians] wear flowered garments made of the finest muslins." Pliny, in A.D. 61, described in detail the Egyptian method of printing with color. Egypt used small blocks for printing and the process was a very tedious affair, for each small design had to be printed in small sections, with a separate application of each color in each design.

Some bits of Egypto-Roman fabric of about A.D. 200, which is in existence today, show bold, printed designs made with wooden blocks. A fragment of distinctly printed cotton was found in the grave of St. Caesarius who was a bishop at Arles about A.D. 542.

By the end of the twelfth century, pattern printing on textiles had become a developed industry in many parts of Europe. Some authorities believe that in some sections, especially in the Rhenish towns, printing on fabrics preceded printing on paper, and that it became very elaborate with patterns in gold and silver on very fine silk and linen. With the increasing use of decorative weaving and embroidery, printing began to decline in the fourteenth century and was not revived for about three centuries.

Western-World Printing

In the Western World the ancients of Peru and Mexico had perfected methods of printing also. Excavations in Peru prove that these peoples practiced tie and dye, and batik methods. They printed with blocks as well as with rollers. Small terra-cotta rollers

for printing figures on fabrics have been found in many ruins. Some of these rollers were cut for printing narrow bands on tunics. These ancient Americans had also perfected a finish of some sort that gave a high glaze to fabrics.

The primitive methods were used with little change until 1783, when a revolution took place in the printing industry. A Scotsman by the name of Bell invented the important new method—roller printing. With this quicker printing method and with the perfecting of peg or surface printing, a new industrial field was organized.

The early American colonists, who highly prized printed fabrics, were able to obtain only the very few that reached America from other countries. It was well in the eighteenth century before the colonists did any printing on their own woven fabrics. An advertisement of 1761: "stamps linen china blue or deep blue, or any other colour that Gentlemen and Ladies fancies." A linen factory in Boston in 1762 advertised that it was equipped to print checks and stripes. The first calico printworks was built near Philadelphia in 1772, mainly through the efforts of Benjamin Franklin who greatly appreciated gay printed fabrics. Mrs. Washington is said to have visited this factory and to have bought some printed calicos for herself and her home.

As in all other fields of the fabric industry, American inventiveness and ingenuity has transformed printing from the meager efforts of our colonial ancestors to an immense, thriving industry, producing fabric patterns that are not only unique, colorful, and artistic but outstandingly American as well.

Speed and power have also been added. Roller printing has now reached such a stage of perfection and speed that sixteen rollers can print as many colors on fabric so fast that 100 yards of material can leave the machine in one minute. New ways of printing with screens, cameras, and air brush have been devised. Old block methods have been speeded up. New methods of handling yarn and fabric, before and after printing, have been perfected to add beauty and permanence to the finished fabric.

Western-World designs are becoming more distinctively American. Designers have created and executed patterns that typically

WIDE VERSATILITY IN PRINTING

The various printing methods produce surface textures which are widely diversified in appearance. Some of these fabrics have printed designs and some have crinkles that appear woven in. Starting at upper left, counterclockwise, the fabrics include the following.

This lawn has electrocoated dots. The bold, colorful print was produced in a roller printing machine, by direct printing on a white ground. In parts, color was printed on color

The design on the pink satin was produced by flock printing.

The bird's-eye piqué had the color printed on a white ground by screen printing.

The batik printed on a plain weave cotton fabric was a hand operation.

The printed velvet had the color application produced by hand-painting the fabric.

The color on the center fabric was produced by a combination of printing methods. The crinkled white stripes are the result of plissé printing. The fabric was first dyed blue, then the pattern color discharged, and the color in the design printed on in a roller printing machine.

The color on the blue and gold cotton fabric was produced in a roller printing machine. However, here some of the discharged pattern was permitted to remain white.

The crisp, colorful organdy was given its appearance by a patented acid-mercerization process. The colors were applied by machine application methods.

This open weave, crisp, cotton fabric had the color and white printed on by the lacquer printing method.

The rayon faille was first dyed blue and then the metallic design was applied by overprinting.

Other colored illustrations of fabrics, printed by various methods, are the frontispiece and the one facing page 449.

portray American interests. The American Indian, colorful American scenery, historical America, and modern America have found a place on the gayly printed bolts of American fabrics.

With the crumbling of many recognized centers of art in the Old World, will American designing become definite and reach a point of lasting distinction? Only time will tell.

PRINTING COLOR

Printing is one method of applying dyes and pigments to produce color. Dyeing, the other method, results in an all-over, uniform coverage, while printing applies color in a predetermined pattern or design. There are still other ways of decorating fabrics, such as embroidering or appliquéing.

While the majority of today's fabrics are printed by the roller printing method with machines, there are many fine screen and block printed fabrics—results of applications that are entirely made by hand.

Printing Pastes

Dyestuffs are in liquid solutions for dyeing fibers, yarns, and fabrics, while they are in paste form for printing fabrics. This is necessary, especially for roller printing, so that the color adheres to the engraved portion of the roll only. Printing pastes are made of dyestuffs, thickeners—such as starch, flour, gums, dextrine or albumen—water, and the necessary chemicals to fix the dyestuff on the fabric.

The same dyes are used for printing as for dyeing and for the same reasons; vat dyes because of their wide range of colors and all-round excellent degree of fastness; azoic dyes (while not as fast to light) because of their serviceable fastness and inexpensiveness; basic dyes because of their wide range of colors and their suitability to the discharge method of printing; mordant and chrome dyes because of their full range of colors and their high degree of fastness on animal fibers; acid dyes because of their bright shades; direct dyes because of their wide range of colors and economical application; diazotized and developed dyes because of their higher degree of fastness to water than direct dyes; and acetate dyes because of their degree of affinity for acetate rayon.

Printing may be considered as localized dyeing. Whereas dyeing causes the fabric to be colored one shade only and without any design, printing permits the fabric to be decorated in many colors and in any preconceived design. If dyes need a mordant this must be put on the cloth before printing.

Pigments

Pigment colors have been used for many centuries for printing fabrics. However, it is only within recent years that methods have been developed for bonding the pigments firmly to the fibers, thereby permitting the production of printed designs which will withstand repeated washing or dry cleaning.

At one time, blood albumin was employed almost universally as the fixing agent for pigments, but the results were never satisfactory from the standpoint of fastness to washing. During the 1930's the so-called lacquer printing process came into vogue. In this process, the fabrics are printed with a dispersion of pigments in a solution of nitrocellulose or ethyl cellulose. One drawback to this method is that the printed parts of the fabric are usually stiff and harsh.

There is a recent process by which the pigments are bonded firmly to the fibers by means of synthetic resins. The printing paste is a pigmented emulsion usually of the water-in-oil type. It is different in form from the pastes previously discussed for printing with either pigments or ordinary dyestuffs.

Opposite page. Top to bottom: *Cross section photomicrographs.* A cotton fabric which has been printed with resin bonded pigments. The pigment can be seen deposited upon the fibers. Both the filling and warp yarns can be seen. This particular type of application was designed so that the pigment would be distributed throughout the fabric.

A cross section of a cotton fabric also printed with resin bonded pigments. In this example, the application was designed to have most of the pigment remaining on the top surface of the cloth. This is in contrast to illustration above where it was distributed throughout the cloth.

A cross section of an unpigmented rayon fabric which has been printed with resin bonded pigments. In this application the resin is deposited throughout the fabric with some concentration on the surface.

A vat printed fabric prepared immediately after the printing step. In this case the insoluble vat color can be seen deposited along the threads of the cotton.

The completed vat printed fabric. The insoluble pigment shown in illustration above has been reduced to a soluble form and has penetrated inside the individual cotton fibers. It is then oxidized to an insoluble form within the fiber. The size of the vat color prepared in this way is very fine and cannot be seen by an ordinary microscopical examination (Calco Chemical Division, American Cyanamid Company).

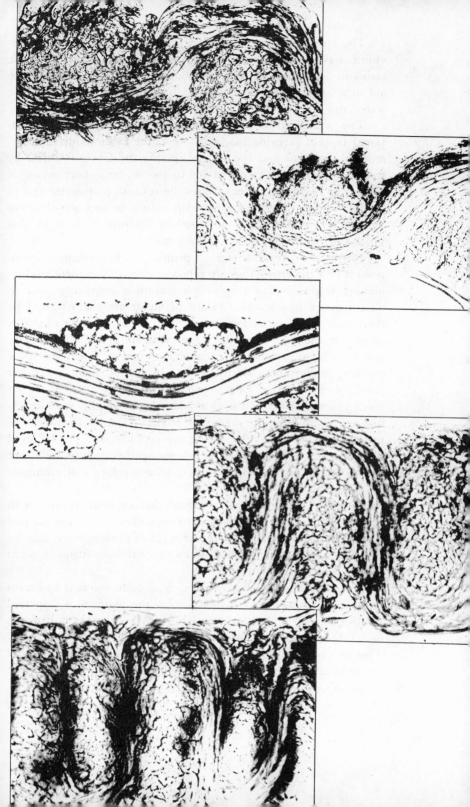

The emulsion is prepared by dispersing pigments, which previously have been ground extremely fine, in a solution of synthetic resins in an organic solvent, and then stirring in water with the aid of a high-speed mixer. The high-speed mixer breaks up the water into small globules or droplets.

The pigmented emulsion is applied to the cloth on the conventional type of printing machine. Next, the fabric is dried to remove the solvent and the water. Finally, the fabric is cured by treating it at a temperature of 300 to 400° F. for a short period of time. The curing treatment causes the resin to polymerize and to become insoluble in alkaline soap solutions and dry-cleaning agents. Polymerization is discussed in Chapter 8. It is by this process that some synthetic fibers are made.

One important advantage of printing with pigmented emulsions is that the engravings are reproduced with complete fidelity, making it possible to employ photographic engravings and to obtain sharp, fine marks. Certain pigments are characterized by their high degree of fastness to light and their brilliancy of shade. At present the range of shades is somewhat limited. This, however, is not a serious drawback, as pigment colors can be employed for printing in conjunction with aniline black, vat, azoic, and other classes of coal-tar dyes. Fabrics printed with pigment colors do not bleed upon washing, but dark shades usually exhibit some crocking when subjected to the usual laboratory tests. Considerable progress has been made in solving this problem, and it is believed that use of some of the newly developed synthetic resins and other improvements in manufacture of pigment colors will eliminate crocking.

Among the various types of fabrics that are being printed with pigmented emulsions are cotton shirtings, dress goods, pajama fabrics, draperies, and curtains. Pigment colors are employed also for printing fabrics made from all types of continuous-filament rayon and spun rayon yarns.

A second process for printing with pigment colors is by means of aqueous dispersions. In this process, the pigment is dispersed in a solution of a modified starch. Printing is carried out on the usual type of printing machine, after which the fabric is dried. This process does not give maximum fastness to laundering and crocking and therefore, is confined almost entirely to low-priced cotton fabrics.

Roller printing is a machine method, used for almost all fabric printing in the United States. The designs are engraved on copper printing rollers. As the fabric passes through the machine, it comes in contact with the printing rollers and the colored designs are printed on the fabric. There are various methods of roller printing, each giving a desired result. However, the operation of the machine is the same for all. The difference lies in the handling of the fabric, in the pastes used, and in the sequence in which the color is applied.

Engraving the Rollers

The first step in roller printing is the engraving of the copper rollers. It starts, as does all fabric, with the designer who plans the design which is then worked out on graph paper from a drawing and painting of the design with its colors. A separate roller is required for each part of the design represented by a separate color.

The rollers may be engraved entirely by hand, by hand and then by machine, by pantograph, or by photoengraving. For any of these methods the design is engraved in the metal. The depth of the etching is dependent on the kind of fabric to be printed. To engrave the entire copper roller by hand is a slow and expensive method, and machine methods usually are used. The pressure and the pantograph methods are both methods of engraving rollers by machine.

The pressure method transfers the pattern to the roller by machine. The pattern, which is called a "repeat pattern," may be hand engraved on soft steel. This is then hardened and called the "die." The die is placed in a machine where it is rotated against a revolving soft steel roll under pressure. Thus the design on the die is transferred in reverse to the soft steel. This is called the "mill." The mill is then hardened and is again rotated, this time against the copper roller, under pressure. This transfers the design to the copper roller. By repeating this process, the entire roller is engraved with the one design it is to cover.

The pantograph method is the other machine method of engraving rollers. One repeat pattern, that has been drawn or carved, is transferred from the design to the copper roller, which is covered with a thin film of wax or varnish. The pantograph machine used for this has a needle which, as it traces the drawn design, also transfers the tracings to the treated copper roller. This is repeated until the entire pattern for one color is completed. One roller is engraved for each color.

The roller is then given an acid bath that dissolves out, or etches, the exposed parts while the film of varnish protects the remaining portions.

The photoengraving method is another means of engraving rollers. The design to be reproduced is drawn on tracing cloth or tracing paper, or photographed on photographic film. A bichromate and gelatin solution is applied to the copper roller which is then allowed to dry. This operation must be done in a dark room, illuminated only with a photographic safety light, inasmuch as the bichromate gelatin is light sensitive.

Using a modification of the method employed for photographic contact printing on paper, the design is transferred onto the surface of the bichromate-gelatin sensitized roller. Strong light is projected on the film, placed around the roller, and penetrates the light spaces in the film, causing the gelatin to harden and become insoluble where exposed to light. The roller is then washed to remove the soluble gelatin from the unexposed or protected parts.

The roller is now given a bath in iron chloride which "eats out" the unexposed portions. The hardened solutions are then washed off, leaving the portions beneath in their original state, and the roller is ready for printing the color and design on the fabric. As with all roller printing, one roller prints one color, and the portion of the design that carries that color.

The Roller-Printing Machine

This is another fascinating machine to watch in operation. The yards of unprinted cloth, white or dyed, are fed into the machine, pass around a cylinder, contact the engraved, color-

Opposite page. Roller printing. Upper left: Cutting the design on a zinc plate. Upper right: Transferring the same design from the zinc plate to the copper roller by the pantograph operation. Center left: The copper roller being etched (Cheney Brothers). Lower left: Pouring the color paste in the color box. Copper rollers are in the roller printing machine (Pacific Mills). Below right: The roller printing machine showing a fabric being printed (Cheney Brothers).

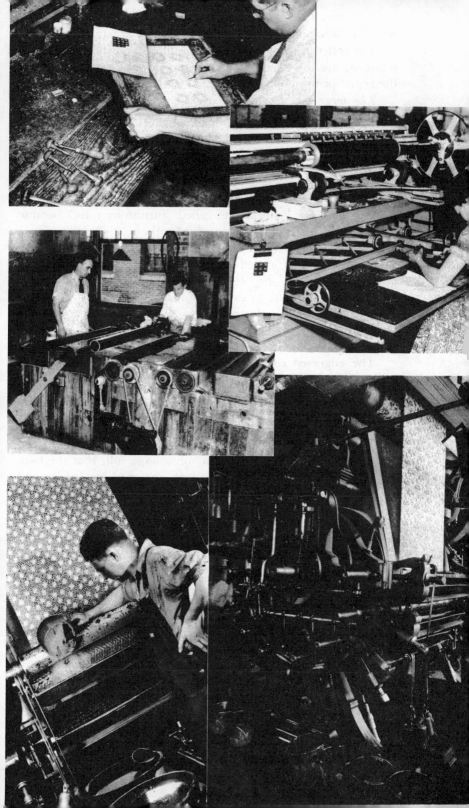

coated rollers, and emerge a colorful, patterned fabric.

The roller-printing method is not difficult to understand. The following are the essential parts of the machine that function in the actual printing of the fabric:

The large metal cylinder covered with woolen felt or other fabric, around which the unprinted fabric travels as it contacts the printing rollers.

The engraved copper rollers, each of which carry the etched-in design for one color in the pattern.

The feed rollers which rotate in the color box and feed the dye paste to the engraved copper rollers. These are small rolls covered with soft material and called "furnishing rolls," because they furnish the color to the engraved rolls.

The color box, which holds the print paste.

The two knives, called "doctors," which rest against the engraved roller. One, the "cleaning doctor," scrapes off the color from the smooth part of the roller before the fabric contacts it. The other, the "lint doctor," scrapes from the roller any lint left by the fabric or any previous color picked up by the roller from the cloth.

The engraved copper rollers are placed in the machine in sequence of the desired application of colors. A machine may have as many as sixteen copper rollers, carrying sixteen different colors, but fewer are generally used. The color paste to be printed with the design on each roller is poured into the color box under the roller. If there are to be four colors, then four color rollers are used. The fabric, on a large roll, is at the back of the machine. It is drawn through the machine, under and around the large cylinder.

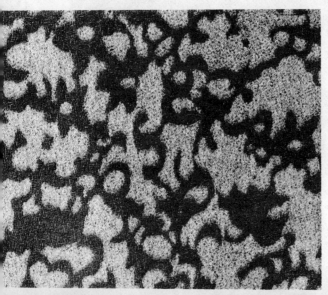

An acid printed fabric. Acids are used the same as color pastes. The acid "eats out" the fibers according to the design etched on the roller.

In operation, the large cylinder rotates, carrying the cloth. The machine is geared so that the rollers rotate at the same speed as does the cylinder. As it turns, each copper roller, which has been fed the color paste by the furnishing roll, prints its portion of the pattern. The two doctor blades clean the roller before and after each printing, leaving the roller ready for the repeat print of the pattern. The fabric, when fully printed, is carried from the cylinder through a series of drying chambers heated by steam, and the color paste is dried. The goods are then steamed or aged. If vat dyes are used, they are oxidized at this point. Later soaping and washing may be necessary to remove the gum and unused chemicals.

It is possible to secure more colors than there are rollers, as the same roller may contain different depths of engraving. Also overprinting of a color will produce a different color. A fabric may be printed on both sides, by first printing on one side, then turning the fabric, and running it through the printing machine a second time.

PRINTING METHODS

There are three distinct printing methods: direct printing—color printing on a white or pastel ground; discharge printing—white or colored pattern on a colored ground; resist printing—white or colored pattern on a colored ground. These are used for all types of printing, whether by machine or by hand.

Direct printing prints the color directly on the fabric. This method is widely used. Direct printing is sometimes called *"application printing,"* or application is considered one method of direct printing. Usually, the designs are small, covering a minor portion of the fabric. *Blotch printing* is another form of direct printing.

Printed velvet
fabrics.

In direct printing the color is printed on white fabric generally but may be printed on dyed fabric, usually a pale color. When printed on a dyed fabric, the dyed color is not destroyed. Fabrics for summer wear are more frequently printed by this method.

In application printing the colored pattern and at times a differently colored ground are printed on a white fabric. The design is formed by the colored pattern and not by the ground. In blotch printing the entire surface is printed except the portion that forms the design. In other words, the design is formed by the ground rather than by the pattern. It receives its name from a blotch roller that has the ground etched in the roller instead of the design. This method usually colors a major portion of the fabric, and the design is frequently white.

Discharge printing results in a white or colored pattern on a colored ground. The entire fabric is first dyed and then it is run through the roller printing machine in the usual way. The rollers are engraved with the patterns, the same as for direct printing. A discharge white or a color is printed on the previously dyed fabric.

In a discharge white print paste no colors are used, but only the chemicals necessary to destroy the ground shade. The dyed color is discharged when the fabric is steamed after printing. Sometimes pigments are added to the print paste to make whiter the portion where the color has been discharged. This method leaves a white pattern on a colored ground.

In color discharge printing, chemicals are used for the de-

Direct printing on spun rayon appears like woven-in stripes.

struction of the dyed ground. The color that is to be printed in (usually vat colors) resists the action of these discharge chemicals, with the result that the dyed ground shade is destroyed in the printed areas, leaving only the newly printed color. The newly printed color is then developed by steaming. In this process, the result will be a colored pattern on a colored ground. Discharge printing may also be applied by the screen method.

Resist printing also results in a white or colored pattern on a colored ground, but the method is different from that of discharge printing. Resist printing may be chemical or mechanical. The copper rollers are engraved with the pattern, the same as for printing color.

For chemical resist printing, the chemical is printed on the fabric (natural color, white, or dyed) for the purpose of destroying a color applied subsequently, either by printing or by dyeing. A substance that destroys the dyestuffs chemically prevents the developing of the printed color.

For mechanical resist printing, a gum, wax, or some such substance is printed on the fabric for the purpose of preventing the fixation of a color applied subsequently either by printing or by dyeing. A substance that mechanically protects the fiber does not permit the printed design to be dyed.

The undyed fabric may be printed in a second design with a printing paste. The fabric is then piece dyed, and the portion of the pattern covered with the resist printing substance remains white or light colored. This type of printing is not used widely. Resist printing may also be applied by the screen method.

PLISSÉ PRINTING

Plissé printing, also called "crepe printing," results in a permanent crinkled or puckered effect given to rayon or cotton fabrics. It is applied to white, piece dyed, yarn dyed, or printed fabrics. This may be accomplished by one of two methods. In one method, as the fabric passes through the roller printing machine, the design, stripe or otherwise, is imprinted with print paste that contains strong caustic soda. The fabric is first dried, then subjected to a moist steaming operation. The portion printed with the strong caustic paste shrinks, while the remainder of the fabric retains its original size. The shrinkage results in a crinkled effect on the unprinted portions.

Warp printed fabrics. Left: Taffeta. Right: Cretonne.

By another method the design on the fabric is printed with paste containing resisting gums. The fabric is then given a caustic bath and the unprinted portions shrink. When the gums are washed out, the printed sections have the puckered or crinkled appearance. This effect of the caustic soda is much the same as in mercerization, except that in plissé, the fabric is permitted to shrink, while in mercerization the fabric is kept in a stretched condition.

WARP PRINTING

As the name indicates, this is the printing of warp yarns after they have been wound on the warp beam and before they are woven into fabrics. They are printed the same as fabrics. The warp yarns pass around the rollers, each of which prints its pattern and color on the yarns. When these yarns are woven with plain-colored or white filling yarns, the pattern on the fabric takes on a softly subdued outline and color. Both sides of the fabric show the printed design. Cretonnes are woven with warp-printed yarns.

VIGOREAUX PRINTING

This is a method of printing stripes on wool slubbings or tops. In place of the rollers on the printing machine carrying the design, there are cylinders with raised bars running lengthwise. The width of the bars and their frequency determine the width and number of stripes.

As the slubbing is run through the printing machine, the colors are printed in crosswise stripes on the wool. The wool is then steamed, washed, and dried, ready for combing and spinning into yarn.

This method of printing is less frequently used, but it does give the finished fabric a soft, even color, because subsequently the slubbing is drawn out and spun into yarn, and the printed stripes are broken up and appear on the yarn as small spots of color more or less evenly intermingled with the yarn that has not been colored.

DRUM PRINTING

This is a method of printing warp yarns for the pile of tapestry and velvet rugs. The large drum or wheel is equal in circumference to the length of the rug to be woven or to one pattern repeat. It may be from 4 to 18 feet in diameter. A single strand of yarn is wound in one layer around the drum. This provides yarn for a single, lengthwise series of pile loops for a designated number of rugs, to be woven exactly alike. For example, a 9-by-12

Drum printing. The drum printer stops the drum as designated on his design paper and sends the color carrier forward and back underneath the drum and thus prints the color on the yarn that is wound around the drum (Mohawk Carpet Mills, Inc.).

rug is 9 feet wide. Assuming, there are to be ten warp yarns per inch, the 9 foot width would require 10 x 12 x 9 or 1080 warp yarns; and there would need to be 1080 drums of warp printed. If each drum were fully wound, there would of course be enough yarn to weave some 800 rugs of this one group of warp yarns.

The dye paste is in traveling buckets at the base of the large drum. Each bucket holds a different color of print paste. As the carriage conveys the bucket across the platform, a rubber roller takes the color paste from the bucket and, passing beneath and across the face of the drum, presses the color paste onto the yarns. This one color is applied where designated in the pattern, in strips equal in width to one tuft around the entire drum.

As the wheel continues to revolve, each color is applied according to the previously drawn color chart, until all colors are printed on the yarn. To prevent bleeding of colors, they are applied from light to dark. Upon completion, the yarns are taken from the drums and, in skein form, are given a wet steaming to set the color. They are then washed, dried, and wound onto spools, ready for warp beaming. The spools are arranged according to color for the pattern and wound onto beams, ready for weaving the pile into the rugs.

BLOCK PRINTING

Block printing is a hand operation. The color is printed directly on the fabric by the use of square blocks covered with color paste. A separate block is used for each section of the design represented by a separate color.

This form of printing can be used for any size patterns, since the designs are not controlled as is necessary in roller printing. The dye being applied directly, in this manner, produces clear, rich, translucent patterns. The colors may be very bright, strong, or pastel. Because of the method by which each block prints one color, the outline of the pattern is usually uneven. Block printing can be recognized by dots in the fabric, caused by the pitch pins that guide the blocks. However, imitation block prints also show these dots on the fabric to simulate hand block-printing characteristics.

For hand block printing the fabric is securely and tautly clamped to a table about 3 feet wide, 3 feet high, and varying in length from 3 to 12 yards. The table is covered with felt lin-

Block printing showing the printing of the first block (Goodall Fabrics, Inc.).

ing or a woolen blanket. A cotton cloth is placed over the blanket to protect it.

The blocks used for printing are wood and may be covered with metal or linoleum. They vary in area and may be square or oblong but are usually about 18 inches square and are always 2½ or 3½ inches thick. The design carrying only one color is applied to the block in various ways. The design may be cut in relief in the block of wood or it may be burnt into the wood. If lines too fine to be carved in wood are desired, the design is built up on the flat surface of the block in strips, using rods or pins of copper or brass. In this method, the outline is cut out and the holes are drilled to attach the pins or strips, which are hammered into the design.

The block carrying its one-color design is pressed on the printing paste and then applied to its proper place on the fabric and tapped with a mallet. The entire fabric is printed with this one color first. After this color is dried, a second block is used for the second color; each color being dried before the next is applied. The correct placing of the block is assured by means of four pins at each corner of the blocks. Care is taken that the pitch pins at the top and one side of the block fit into the pin impressions made by the previous block, to ensure the continuity of the pattern. The color is applied to the raised parts of the design. This is opposite to roller printing, in which the depressions are filled with color paste and pressed on the fabric.

SCREEN PRINTING

Screen printing is also a hand operation. The fabric is fastened to long tables
and the screen is laid over the cloth and
the color printed on the fabric. A separate screen is used for each color in the
pattern.

Screen printing permits large, bold
patterns that have a rich, almost transparent clarity and a feeling of depth to
the colors. The design does not have
the perfect outlines achieved by machine
printing but the slight imperfections give
it value, evidencing that it is hand made.

The long printing table has along
one side a guide rail, to which adjustable
stop clamps are fastened at intervals. On
one side of the screen frame is a gauge.
As the screen is placed in position on the
fabric, the gauge is brought up close to
the clamps, which are placed so that they
correspond to one repeat of the pattern.
In this way each screen, carrying its own
color and design, will fall in proper position on the fabric in printing.

The prepared fabric, cleaned and
usually bleached, is run the full length
of the long printing table and its sides
are pinned to the table in a straight line.
The screen, on wood frames of about
40 by 30 by 6 inches, is made of silk,
nylon, or Vinyon treated with gelatine.
The part of the design that will be
printed in one color is traced on the
screen's fabric and then the entire screen

Screen printing. Top to bottom: The first, second, third and final colors being applied to the
fabric. Two men operate the screen (Goodall
Fabrics, Inc.).

except the sketched design is painted with an alkali resisting paint. The gelatine is washed out of the unpainted portion, and the screen is ready for use in printing. Each screen carries the design for the printing of one color. There may be as many as fifteen colors in one fabric, but usually there are less than half this number.

One screen is laid in place on the fabric in the first position, adjusted according to the spot clamps which were placed in the correct position before the actual printing started. The color paste is poured on the screen, and it is stroked and pressed into the fabric by a wooden squeegee that is wedge shaped and almost as long as the stencil. When the color has been applied to the fabric, the screen is lifted and placed in its next position. After a short period the second screen, used to apply the second color, is placed over the first position. Continuing, the entire length of the fabric is printed. Upon completion, the fabric is unpinned and by pulley raised above the table, where it remains to dry. The fabric is then run through an ager to set the dye.

Imagination and creativeness have a wide range in screen printing, for direct, resist, and discharge printing can be combined in many ways in one design to produce unusual color combinations.

STENCIL PRINTING

Stenciling is less frequently used than other forms of printing except in the rug industry. Rugs containing various fibers such as kraft fiber and sisal, are printed in large, colorful patterns with stencils.

Screen printing. Above: Printing one portion of the design that carries one color. Below: Lifting the screen after the color has been applied. This is a one man screen operation (American Enka Corporation).

Stencil plates are sheets of paper or thin metal with perforations forming the pattern, or with the pattern entirely cut out, or with the background cut out, leaving the pattern. The stencil is laid over the material, and the color is applied by brushing over the design.

As a rule the stencil covers the entire rug or at least one half of the rug. One stencil carries the entire design in one color and there are as many stencils as there are colors in the rug. The stenciled design is on one side of the rug only.

SPRAY PRINTING

For this method of printing, dyestuffs are more liquid than in the usual printing pastes. The fabric is mounted in a spray machine. A screen, much the same as that used for screen printing, carries the design for one color. It is placed over the fabric and the color is sprayed through the design in the screen onto the fabric by a spray gun similar to that used for painting automobiles. Heavier fabrics, such as velvets and draperies, may be spray printed.

Spray printing is also used for solid coatings of fabrics such as applying waterproof finishes. In such cases it is a finishing as well as a printing operation.

BATIK PRINTING

This is similar to resist printing. The fabric is entirely coated with wax, by hand, except those portions that are to form the design. The fabric is then treated with the dye paste or solution in cold surroundings so as not to melt the wax. In the process, the wax tends to crack and break, permitting the color to seep in fine lines onto the waxed portions of the fabric. One color may be applied, then the wax removed, and the fabric given another wax coating for another part of the design. In this way, several colors may be applied.

Left: Applying the color through a stencil. Right: Removing the stencil from the rug (Deltox Rug Company).

FLOCK PRINTING

Another method of decorating fabrics is to first print an adhesive on the fabric by the stencil or roller-printing method, and then to attach fibers onto the fabric by means of the adhesive.

In the roller-printing method the original design is transferred to the roller by the same methods as are designs for printing color pastes. The fabric is run through the roller-printing machine and the adhesive is printed on the fabric according to the design.

In the stencil-printing method the actual design is punched in small holes or figures in a copper sheet which is made into a 50-inch hollow cylinder and placed in a machine, with a doctor blade inside the cylinder. The adhesive is poured into the cylinder. As the fabric passes around it, the blade forces the adhesive out through the holes and onto the fabric.

The fibers. Cotton or rayon staple fibers are used although other fibers could be applied. The fibers are cut very short and, if necessary, they are bleached and dyed.

The process. After the fabric passes around the roller or copper cylinder, and while the adhesive is still wet, the fabric goes into the tunnel where the flock is freely circulated and attached to the adhesive. The fabric is then slack dried by hanging, in long strips, in a drying chamber for 24 hours. After drying, the fabric travels under a vacuum brush that removes the loose fibers that are not attached to the adhesive.

A fabric may be printed with a color paste as well as with adhesive as it passes through the roller-printing machine. One roller carries the adhesive design and another the color-paste design. In this way the fabric can be decorated with color printing as well as with a flock design. The printing pastes used are

Left to right: Metallic stencil printing, flock applied by the roller-printing method, flock applied by the stencil method.

quick drying and are dried before the fabric reaches the flocking tunnel.

Fabrics with the stencil designs can be distinguished by the characteristic small dots or bunches of fibers, while those produced by roller printing have a continuous, unbroken line of fibers.

Flock designs are produced on many types of rayon and cotton fabrics from the sheer organdy to heavy poplin. The fibers are firmly anchored and the fabrics, so decorated, can be dry cleaned, washed, and ironed. The fibers produce a velvetlike, textured, and often colorful design that adds immeasurably to the attractiveness and suitability of the fabric.

STENCIL LACQUER PRINTING

The stencil is prepared the same as discussed previously. In place of adhesive, the cylinder contains lacquer which is forced through the design holes and deposited on the fabric. After printing, the fabric is dried. The metallic designs are composed of clear lacquer with metallic powder added.

ELECTROCOATING

This is a method of depositing a pile, all over a fabric or in design, by an electrical process. This law of electrostatics is that equally charged substances repel each other and unequally charged ones attract each other.

Cotton, rayon, or wool fibers can be attached in this manner. The fabric most used is white, printed, or dyed voile, on which

cotton or rayon fibers are deposited and fastened. There are three major steps in the process: preparing the fibers, applying the adhesive to the fabric, electrocoating the fibers onto the fabric.

The fibers. The fibers are cut to an even length, bleached or dyed, and dried. The dyeing must be carefully matched as the fabrics are not treated after electrocoating.

The adhesives. The fibers are "shot"

Flock applied to fabric by stencil method. The fabric is also printed with a color paste.

530

into the adhesive, which must be of a consistency that will permit penetration of the fibers, and must harden so that the fibers will remain fixed to the fabric through wearing, washing, or dry cleaning. The adhesives consist of solutions or dispersions of such materials as rubber, resins, oils, or combinations of these.

If the fabric is to be an all over pile fabric, the entire fabric is coated with the adhesive. If in design, the adhesive is applied according to the pattern with a stencil or by some other means.

The electrocoating. The electrocoating machine has two electrodes, between which is a strong electrostatic field. The fibers are fed to a conveyor belt. One electrode is under this belt. The other electrode is over the fabric. The conveyor belt moves through the machine, carrying the fibers. The fabric with its adhesive side down also moves through the machine.

As the fabric and the fibers (on the conveyor belt) reach the space between the two electrodes, the fibers are polarized and charged and are of the same polarity as the bottom electrode. They are thus repelled by the bottom electrode

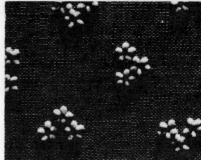
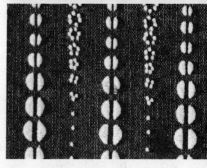

Fibers attached to fabrics by electrocoating.

and hurled into the adhesive coating on the fabric. There they are retained, standing on end, uniformly spaced from each other. The fabric is then fed into a dryer which dries and cures the adhesive.

It is found that electrocoated fabrics can be washed, ironed, and dry cleaned and that the fiber coating will last as long as does the fabric. Such fabrics are used extensively for dresses, blouses, curtains, and so forth. The designs have a soft, raised, velvety appearance and are often in lovely colors. The all-pile fabric is for such uses as upholstery, draperies, and linings for the side walls of automobiles. It is also adaptable to making rugs.

531

SUMMARY

Printing applies color in a predetermined pattern or design.

KINDS OF PRINTING

Roller printing

   Machine operation using engraved rollers. *Plissé* and *warp* printing are also roller printing.

Block printing

   Hand operation using blocks.

Screen printing

   Hand operation using screens.

Stencil printing

   Hand operation using paper or metal stencils.

Drum printing

   Mechanical operation using a rubber roller.

Spray printing

   Hand and machine operation using a spray gun.

PRINTING METHODS

Direct printing

   Color printing on a white or pastel ground. *Application* and *blotch* printing are also direct printing.

Discharge printing

   White or colored pattern on a colored ground. Discharging agent and design color are applied after the fabric is dyed.

Resist printing

   White or colored pattern on a colored ground. Resist agent is printed on the fabric before the fabric is dyed.

Opposite page: One type of tentering machine which evens and smooths the fabric. Note the clips holding the selvage of the cloth (Pacific Mills, Inc.).

Adds final personality

Upon leaving the looms or machines, practically all fabrics ar in a limp, unattractive condition. Often it is difficult to believ that a flat, grayish cloth will become a beautifully colored, soft napped, highly lustrous fabric.

The fabric is first given its pretreatments to prepare it fc later processes. It may be bleached and remain white, or it ma be colored. After preparing, dyeing, or printing, the fabric ma be given many other finishes.

Specifically the term "finishing" applies to those processe through which a fabric passes after it is dyed or printed. Howeve it is customary for the trade to speak of all the operations throug which a fabric passes as "the finishing processes." These includ preparatory and coloring processes as well as the general an functional finishing processes discussed in this chapter. All fat rics—woven, knitted, or twisted—are given some finishes.

Some fabrics require many more finishes than do others. Fo example, a woolen duvetyn just off the loom is a flat, rathe loosely woven, characterless, colorless fabric. It is then dyed an napped and given other finishes so that it changes into a velvet, smooth, richly colored fabric. On the other hand, a close-textured heavy drapery damask, woven on a Jacquard loom with dyed yarns requires few fabric finishes.

Hard textured fabrics, firmly woven with tightly twisted yarns require finishing different from soft textured or napped fabric Even the same fabric may receive a soft, a firm, or a brittle finish depending upon its ultimate uses. Those woven with long, continuous filaments, as rayon, silk, or nylon, usually need to pass through fewer finishing processes than do those woven with yarns spun of short fibers.

Some operations are really preparatory processes, such as inspecting and mending and scouring or boiling the fabric. At times, the vegetable matter remaining in the wool fabric is "burned out." If a fabric has short, protruding fibers and the finished fabric is to be clear surfaced, one of the first steps is to singe off the short fibers. Early in the series of processes, many wool fabrics have the weave "set" to give the warp and filling a chance to adjust themselves. Some fabrics are softly napped, merely subduing the weave, others are deeply and heavily napped. Some have the nap laid in one direction so that it results in a lustrous, rich texture. Often such woolens pass through a process that sets the nap permanently and adds luster to the fabric. Some fabrics have controlled shrinkage. Many cottons are starched or sized; silks are weighted to replace the gum that has been removed.

The final finishing processes include straightening, pressing, and drying the fabric. Wool fabrics are usually steamed. Frequently, fabrics are sheared and brushed to remove fibers that have protruded during the finishing processes. Fabrics may be given a dull luster, a semiluster, or a high sheen. They may be smooth and shiny or have a watery or embossed surface.

Functional finishing processes. In addition to the above finishes, one or several of which are generally applied, there are others today that give added qualities to fabrics. These may or may not be applied. While some of these finishes effect only a physical change in the fabric, others impregnate the fiber and change its chemical structure.

The finishes selected for application are those necessary to give the fabric its final appearance, personality, and appeal. In the following each finishing process is described and explained briefly. The fibers to which each is applied are specified, as well as the contribution each makes to the fabric.

The processes are arranged in alphabetical order and not in sequence of application. While certain finishes normally precede others and the dry finishes are usually last, each fabric has its own definite sequence of application, often unlike that for a very similar fabric.

535

BEETLING

To give a lustrous appearance.

Applied to cottons and linens.

The fabric is wound on a wooden roll. As the roll slowly rotates in the beetling machine, large steel or wooden hammers rise and fall on the face of the goods pounding the yarns. This closes the weave and gives a soft luster and firm feel to the fabric. The cloth may be rerolled and beetled several times, thereby increasing its luster and full finish.

BRUSHING

To remove the short, loose fibers.

Applied to wool, cotton, rayon, silk, linen.

When a fabric is sheared, many of the cut off fibers remain

Brushing. Below: This machine removes all lint and loose fibers from the fabric (David Gessner Company). Left: A close up of brushing the cut-pile cords in a corduroy fabric (Bernside Mills).

on the fabric and must be removed by brushing. The machine used for this process has two large cylinders covered with strong bristles. As the fabric passes through the machine, it strikes each of these cylinders twice and the loose fibers are brushed out. At the same time, another brush cleans the lint from the back of the cloth.

The machine can be adjusted so that each cylinder gives a heavy brushing, or a light brushing, or one a light and the other a heavy brushing.

BURLING

To remove the knots in the yarns.

A contributing finish. Applied to wool.

As the yarns go through the processes of drawing, spinning, winding, warping, and weaving, irregularities are bound to occur when yarns are tied together to continue the operation or to mend breaks.

These knots or slubs are removed by hand with a small picklike instrument, called a "burling iron," and are then cut off.

"Burling and mending" is a wool term that may be all inclusive. It means pulling out and smoothing yarn knots, correcting mispicks, and threading in broken filling or warp yarns.

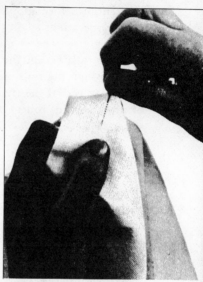

Burling and mending. Fabrics are carefully inspected, mended and corrected before they are finished (Goodall Fabrics, Inc.).

CALENDERING

To give a smooth, glazed, moiré, or embossed finish.

Applied to cotton, rayon, linen, and silk.

The purpose is to give the fabric a smooth glazed, moiré, or embossed finish. It may have a very high sheen or a softer finish, according to the method used. It is one of the last finishes given to cotton fabrics.

The fabric while damp, and often starched, is run around heavy rolls placed one above the other in a calendering machine. The rolls may be solid or hollow. One type of finish is accomplished on a swissing calender in which every other roll is covered with cotton or some other material while the others are uncovered polished metal. The fabric enters the machine from a roll, going over and under small rolls called "guide rolls" that smooth the fabric. From the guide rolls it passes around the bottom calendering roll, then around the second roll, and on through the series. Leaving the last roll, it passes over another guide roll and winds onto a shell outside the machine. Calendering gives the fabric a smooth finish, closing up the spaces left in the weave. This finish is essential before applying any of the following additional calendering operations.

A watermarked or moiré finish is made on a chasing calender, with all rolls except the first one cloth covered. After the fabric goes through the rolls once, it passes around guide rolls that carry it back to repeat the treatment several times. The heavy pressure of the clothed rollers against the fabric produces a thready, watery effect.

A highly glazed finish, as on chintz, is made on a friction calender. Only three rolls are used. The first and third are metal and the second is fabric covered. The fabric passes around the first two and out and under the third, which travels at a greater speed than the first two and thus gives a high polish or sheen to the fabric.

An embossed or crepy effect is produced on an embossing cal-

In this huge calendering machine, fabric passes over and under 43 steam heated rolls where it is pressed and dried. (The Crompton-Richmond Company).

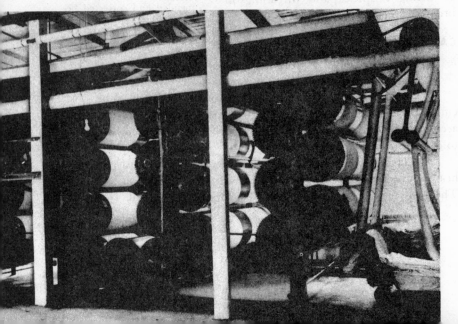

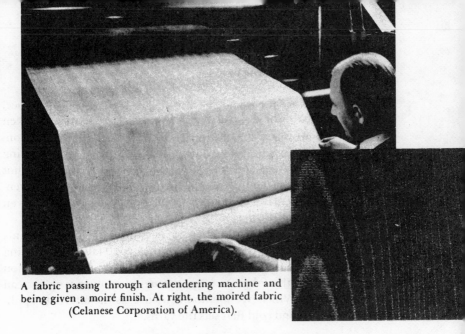

A fabric passing through a calendering machine and being given a moiré finish. At right, the moiréd fabric (Celanese Corporation of America).

ender, which has two or three rolls. If two rolls, one is covered; if three rolls, two are covered. The desired pattern is engraved on the steel roll, which is slightly heated and placed in the calendering machine with the one or two cotton rolls that have been moistened with soap and water. As the machine operates, the pattern from the engraved roll is pressed onto the fabric of the cotton covered roll, which is then dried. As the fabric is run through the machine, the ridges on the cotton covered rolls are impressed in the fabric.

If the pattern is engraved on the roll and then cut away, the result will be a lustrous pattern on a dull ground. If the ground is engraved and cut away, the result will be a dull pattern on a lustrous ground.

In the *Schreiner calender* the engraved lines are at an angle on the calender roll so they will fall parallel to the fibers in the twisted yarn, when the fabric passes through the machine.

The *Palmer unit* is a large steam heated cylinder attached to the finishing end of a tentering machine. It is used primarily for rayons but cotton and silk may be finished in the same machine. The fabric leaving the tentering machine is guided between the felt covered rollers in the Palmer unit. The heat, pressure, and smooth cloth coverings on the rollers impart to the fabric a soft and mellow finish.

CRABBING

To permanently set the weave.

Applied to wool.

Fabrics leaving the loom have a characterless appearance. Woolens usually have a reedy, loose construction, easily shaken apart. Closely woven worsteds appear to have an irregular construction. Crabbing is the operation that sets the warp and filling yarns into their permanent places in the fabrics. If this is not done, the fabric, when fulled, may shrink unevenly; or the unevenness may cause creases or weakening of the fabric when it is given further finishing treatments.

The fabric passes around a series of rollers in a machine, the first units of rollers immersing the wool fabric in boiling water and the last unit operating in cold water. For completion of the process, the fabric may go through a cooling tank instead of cold water. Whichever method is used, it is the alternate applications of heat and cold that sets the yarns.

DRY DECATING*

To set the luster.

Applied to wool, rayon, and silk.

This finish is given to wool fabrics if the luster obtained in pressing is to remain permanent. When applied to rayon and silk fabrics, it gives them a mellower, softer handle (without excess use of oil), eliminates the appearance of breaks and cracks, dulls the luster, and evens up the fabrics. All of this generally resulting in more depth and clearness of the dyed fabrics.

* For wet decating see page 561.

Crabbing. A three bowl crab machine used mostly in the production of woolens and worsteds for setting cloth in the wet state. The cloth passes through tanks of hot water and then under pressure of large squeeze rolls. The final dip is through cold water to chill the cloth before removing it from the machine (David Gessner Company).

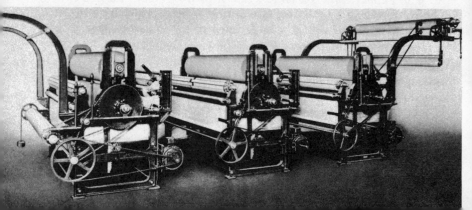

Decating. This machine is used for setting principally woolen and worsted fabrics in the dry state. It is invariably used in the production of face finish fabrics such as kersey and broadcloths. The application of heat, moisture, and pressure gives a permanent luster and reduces shrinkage
(David Gessner Company).

The fabric is wound on perforated cylinders, called "shells." Cotton fabric is sewed to both ends of the fabric to be decated. As a result, the cylinder is first covered with the cotton fabric, then with the fabric to be processed, and finally with a protective covering of cotton fabric again.

The cylinder is then sealed in a tank, the air is taken out of the tank by a vacuum pump, and steam is forced through the fabric. First the steam goes into the cylinder and through perforations through the fabric. Then the procedure is reversed, and steam is forced through the fabric from the boiler and into the perforated cylinder. All the while the cylinder is slowly rotating so that the steam permeates through the entire fabric.

Following steaming, the roll is removed to a cooling frame, and the fabric is cooled by the air and a vacuum that pumps the steam out of the cloth, after which is it unwound.

In another method, the fabric is wound onto a large, perforated drum, between layers of cotton felt. Steam is then forced through the fabric, after which it is cooled and dried by forcing air through it. The moisture and heat even and set the fabric in width and length, make permanent the luster that the wool attained from previous pressing, and soften the luster and feel of rayon and silk.

FULLING

To make compact and shrink the fabric.

Applied to wool.

Fulling, or felting, or milling as it may be called, is a finish peculiar to wools. It is the action of moisture, heat, and pressure on the wool fibers that causes the wool to shrink, the weave to close up, and the fabric to have a closer, fuller body. All worsted fabrics are fulled and many woolens are given a slight fulling to give body and softness to the fabric.

The fabric is immersed in soapy water or a weak acid solution, then pounded or pressed. The moisture makes the wool fibers more plastic and less elastic. One theory of felting is that the pressure on a bundle of fibers causes the fiber to move toward its root. As the fibers travel, they contact and intermesh with scales of other wool fibers, forming a mass of fibers in close contact. The degree of fulling determines the degree of close contact or shrinking. Fulling closes up the weave, resulting in shrinkage. Heat is used to speed up the action as it makes the wool fiber more pliable. The art or mystery of fulling is one of the aspects of woolen manufacturing that is being closely studied today. Broader scientific knowledge will mean the creation of new textures and finishes.

Two types of machines are used for fulling. One type, called

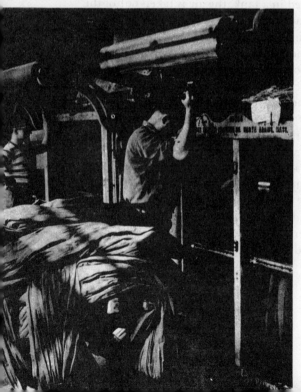

Fulling. Wool fabrics being fulled or felted in a fulling machine where they are beaten with mallets as they pass over moving rollers (American Wool Council, Inc.).

the "cylinder" or "rotary fulling mill," has two or more cylinders. The top roll is heavier than the others. Wet fabric, in a rope form, is passed into the machine and pulled round and round through the machine and under the heavy roll. The alternate compressing and relaxing of pressure causes the wool to felt and shrink. The operation is continued until the fabric has acquired the proper shrinkage and felting.

In another type, called the "hammer fulling machine," the fabric is put in the machine, which has hammers that pound the fabric tumbling about in the machine. This method is used when less fulling is desired.

GIGGING

To raise the fibers.

Applied to wool, spun rayon, spun silk, casein.

Gigging is one of the fiber raising operations, napping (see page 545) is the other. While the two operations are similar, they are used to obtain slightly different surface textures. By either method the fibers are given a slight raising that softly screens the outline of the weave, as in flannel; or more fibers are raised to completely hide the weave, as in wool broadcloth; or a sufficient number of fibers are raised to the surface to form a dense, short or deep, pile, as in fleece. These two processes also give the fabric softness and "depth." For wool, it may be undertaken before or after the fulling operation.

Gigging. Slat gigs are used for raising fibers for napped or gigged fabrics. The machines are built either as single cylinder machines or double cylinder machines. The teasles used on these machines are held stationary in slats mounted on the revolving cylinders. The cloth is run on the machine all the way from a semiwet condition to a soaking wet condition. The machine in raising the fibers attacks them gently, straightening them out and laying them parallel. All high finish face goods require gigging (David Gessner Company).

Gigging. The rolling teasle gig is used for raising the fibers on fabrics requiring a lofty pile. The teasles on this type of gig are mounted on small spindles on the large cylinder of the machine and have a rotating motion, whereas on the slat gig they are held stationary. Woolen blankets are the most typical fabrics finished on the rolling teasle gigs (David Gessner Company).

The machines for wool gigging and napping are much the same. They have wide drums or cylinders on which frames are mounted. For gigging the frames contain rows of teasels (from a thistle plant) that are like cone-shaped burrs. For napping, the frames have fine steel wires.

As the fabric passes through the machine, it comes in contact with the frames. The drums containing the frames may rotate in the same direction as the fabric but faster, and the fibers are raised and straightened. If a one-cylinder machine, both the direction of the cylinder and of the fabric are reversed as the fabric begins its second journey through the machine. A more modern machine may have several cylinders alternately rotating, the first in one direction and the next in the opposite direction, and so forth. The fabric is then turned as it goes from one cylinder to the next.

On the other hand, one cylinder may operate in the direction of the movement of the cloth and the next in the direction opposite to that of the cloth. In that case, in a machine with many

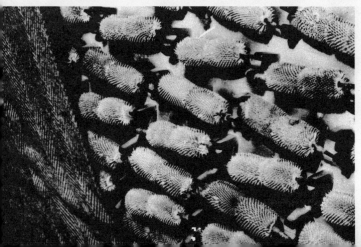

Rows of vegetable teasles mounted on gigging frames. They are used to give a light napping to such wools as this soft herringbone twill fabric (Kenwood Mills).

cylinders (each alternating in direction of rotation) the fabric is not turned. The method used is determined by the surface texture desired for the fabric. For either method, the raising of the fibers must be accomplished gradually so that the fibers are not pulled out completely, damaging the fabric.

As the fabric leaves the machine, it may be laid full width in folds, or it may be folded to half its width and laid in folds, or it may be wound on a roll.

There are two types of gigging: moist and wet. In moist gigging, the roll of fabric is immersed in a trough filled with water. From here the fabric, unwinding, goes through the machine and winds on a roll that is also immersed in water. At the same time, a soft napping may be given to the back of the fabric as the fabric contacts one cylinder or drum that has brushes in place of teasels or wires in the frames.

Wet gigging is very much like the moist gigging. The upright fibers raised by moist gigging or napping are laid down in one direction by wet gigging and a lustrous, smooth fabric, such as broadcloth, is produced.

NAPPING

To raise the fibers.

Applied to wool, cotton, spun rayon, spun silk, and casein.

Wool napping. The reasons for raising wool fibers and the description of the machine have been given under gigging (see page 543).

Napping. The single acting napper or French napper is a machine used for raising or lifting the fibers of all types of woven and knitted goods. Here there is one set of rolls, the points of which all point in one direction, that is, point the same direction as the cloth travels. The nap produced in a single acting napper has a tendency to lay in one direction (David Gessner Company).

In napping, the frames mounted on the drums or cylinders contain steel wires, which have a more powerful action than do the vegetable teasels used for gigging. Therefore, if a fabric is to be given a deep pile rather than a soft raising that subdues or conceals the weave, napping is applied instead of gigging.

There are two types of napping, moist and dry. Moist napping is applied to fabrics that have been fulled and is usually followed by wet gigging. That is, the fibers made upright by napping are laid down in one direction by wet gigging. Dry napping is usually used when an upright dense pile is required. Such fabrics are usually not fulled prior to napping.

Cotton napping. Cottons are napped for the same reasons as are wools—to give a soft filmy surface, to blend colors or screen the weave, to give a brushed and raised surface or a deep, thick pile.

The cotton napping machine is different from the wool gigging or wool napping machine. For cotton napping the machine has one large cylinder, some 4 to 5 feet in diameter, and several small cylinders or rolls, about 5 inches in diameter. The small rolls are covered with the wire cards, called "card clothing." The wires are slightly bent or set at an angle. If there are only eighteen small rolls, all of them have the wires bent in the direction the cloth is moving as it passes through the machine. If thirty-six rolls, the

The double acting napper is a machine used for giving a raised or woolly effect to all types of woven and knitted fabrics. There are two series of rolls in the napping cylinder, named pile rolls and counterpile rolls, each alternating and running counterclockwise to the cylinder and cloth. The points of the napper clothing point in opposite directions. The point of the pile rolls in the direction in which the cloth travels, and the counterpile in the opposite direction. By controlling the speeds of these two sets of rolls, different degrees of nap can be obtained. On a double acting napper the nap can be produced that does not rough or lay in either direction (David Gessner Company).

lternate eighteen have the wires bent in the opposite direction. The wires bent in the direction the fabric is traveling raise the bers so that a long nap results; those bent in the opposite direction give the short nap.

The fabric, thoroughly dry, passes around the large cylinder nd, as it travels, it comes in contact with the wires from the small ylinders, which raise the nap. Further cotton finishes may be iven napped fabrics to obtain an altogether different surface exture. The fibers may be laid in one direction and steamed to produce a smooth lustrous surface, or they may be curled to produce such fabrics as Canton flannel.

Spun rayon napping. If the spun rayon is made from rayon taple fibers that are the length of woolen or worsted fibers, 3 to 4 nches long, and the yarns are spun on woolen or worsted machines, t is napped by the wool method. If the rayon staple fibers are the ength of the cotton fibers, up to 1½ inches long, and the yarns are pun on cotton machinery, the fabric is napped on the cotton napping machine.

PERCHING

To examine and repair the fabric.
A contributing finish.
Applied to wool.

In some factories the term "perching" is used only for examining defects in the fabric, drawing a chalk mark around the imperfection, and passing it on to be mended. In other factories the inspecting and mending are done by the same person.

The fabric may be hung over a rod so that light reflects through the weave, or it may be drawn over a table with a daylight lamp reflecting on the fabric. Mending is done by highly trained workers who must not only know the technical part of weave construction but must have the skill to repair the fabric so the defect cannot be detected.

Worsted fabrics must be carefully mended as there is no subsequent finishing that will cover the mistake such as for napped fabrics. There may be a broken thread that must be woven in by hand, over and under, as are the other yarns and tied in at the beginning and end to the yarn being repaired. If a double yarn has been woven in for a short space, this must be cut and pulled out without breaking the weave.

547

PRESSING

To smooth and give luster.

Applied to wool.

Just as for pressing fabrics in the home, the three requisites are moisture, heat, and pressure. The results obtained are also the same; the fabric is ironed smooth, all wrinkles and creases are removed, and a lustrous appearance is added. Calendering, used for other fibers, is also a pressing process.

Wherever moisture and heat are used to finish fabrics, this is done for the same purpose—to make the fiber more pliable or plastic. The pressure flattens out the fibers, which permits a greater reflection of light, resulting in a higher luster.

If wool fabrics are merely pressed, the luster will soon disappear. Therefore, the fabric must be subjected to carefully regulated steam pressure. There are two methods for pressing wools. One is to lay the fabric between iron presses in long folds, with a fiber board between each fold, and apply steam and pressure to the fabric; the other is to run the fabric around cylinders, applying steam and pressure. The latter is a faster operation but does result in some stretching of the fabric.

A vertical press is used for the first method. It has two heavy metal plates, between which the fabric is placed. A sheet of fiber board is laid between each fold of the fabric, with each twelfth fiber board having metal tabs that connect with the electric current that distributes heat through the fabric.

When the required amount of fabric is placed in the press, the lower metal plate is pressed against the upper, and the fabric is thus pressed for about four hours. The fabric is then refolded

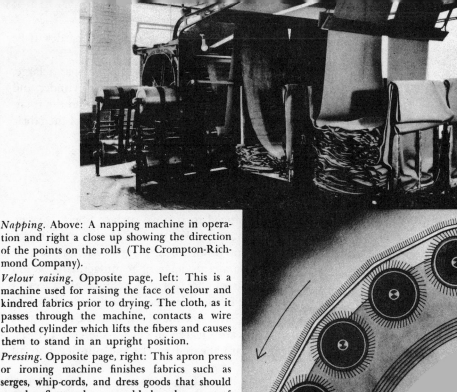

Napping. Above: A napping machine in operation and right a close up showing the direction of the points on the rolls (The Crompton-Richmond Company).

Velour raising. Opposite page, left: This is a machine used for raising the face of velour and kindred fabrics prior to drying. The cloth, as it passes through the machine, contacts a wire clothed cylinder which lifts the fibers and causes them to stand in an upright position.

Pressing. Opposite page, right: This apron press or ironing machine finishes fabrics such as serges, whip-cords, and dress goods that should not be flattened as would be the case of a regular rotary press. The cloth is carried on an endless felt apron which prevents any lengthwise stretch of the fabric (David Gessner Company).

Pressing. This hydro-pressing unit is used in the final finishing of woolens, worsteds and spun rayons. The process dampens the cloth, provides a 20-minute lag which allows the dampness to fully permeate the fabric and then presses or irons the material. The moistening before pressing gives a better press and more permanent finish than is obtained by pressing dry. The operation is continuous and is usually the final operation on fabrics requiring high finish (David Gessner Company).

so the parts that were not in contact with the fiber board will be in contact now, and the operation is repeated.

A rotary press is used for the second method. It has a large, hollow cylinder that rests in a metal bed. Both the cylinder and the bed are heated with steam. The fabric passes around the rotating cylinder. The metal bed is pressed against the face of the fabric as the cylinder carries it through the bed.

SHEARING

To cut and remove the surface fibers or fuzz.

Applied to wool, cotton, rayon, linen, and silk.

After the fabrics are singed to remove surface fibers, they go through many finishing processes that tend to eject or raise fibers, even though they may not be brushed, gigged (as is wool), or napped. To singe a fabric at this stage of manufacturing would involve further treatments of the fabric to remove all traces of singeing. Therefore, some fabrics are sheared to cut off all surface fibers and to leave a sharp, clear outline of the weave. Some fabrics with slightly raised fibers are sheared to cut off the excess fibers; other dense pile fabrics are sheared to cut the pile level so all raised fibers are the same length.

The shearing machine has a wide, spiral cylinder to which is attached the cutting blade. The fabric passes into the machine, over brushes that brush up the fibers, and then into the revolving cylinders. The fibers are propelled by the spirals to the blade, which cuts them off.

It is the adjustment of the machine that determines how much of the fabric is sheared. It must start gradually, and then become deeper as the fabric goes from one cylinder to the other.

Stripes, checks, and patterns may be obtained on pile fabrics by cutting the pile. It is the manipulation of the cutting—raising and lowering the cutting device—that cuts off the fibers according to a planned pattern.

SHRINKING

To reduce the fabric in length and width.

Applied to wool, cotton, rayon, and linen.

Up until the past few years the greatest problem concerning the upkeep of fabrics was presented by shrinkage. Cotton and linen, and particularly wool fibers, have a tendency to shrink when

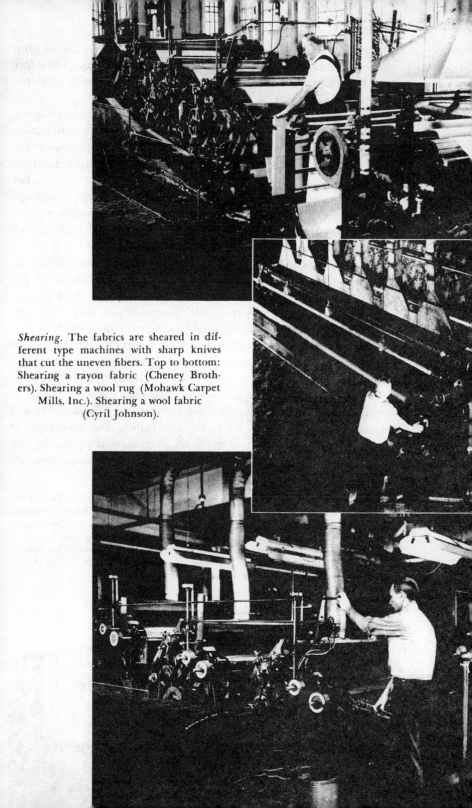

Shearing. The fabrics are sheared in different type machines with sharp knives that cut the uneven fibers. Top to bottom: Shearing a rayon fabric (Cheney Brothers). Shearing a wool rug (Mohawk Carpet Mills, Inc.). Shearing a wool fabric (Cyril Johnson).

wet. Filament rayon, silk, and nylon yarns will not shrink, but loosely woven fabrics made of any fibers may shrink. Therefore the fiber itself and the construction of the fabric both may contribute to shrinkage.

As fibers are spun into yarn, drawn out over and over again, as the warp is prepared for weaving, and as the fabric is woven, there is a constant strain and pulling. After fabrics are woven they pass through certain wet finishes, often including dyeing, which tend to bring the fibers back to normal, but only partially so. For these reasons, unless fabrics are processed to control shrinkage, many of them will shrink when subjected to laundering or wetting with water. Shrinkage may be as high as 15 to 20 per cent on such fabrics as matelassé or waffle cloth.

Shrinkage control of wool. The tendency of wool fabrics to shrink is due to the unique character of the wool fiber, which has felting qualities. Whenever heat, moisture, and pressure are applied to wool, the fibers contract, causing felting and shrinking. The wool fiber and its felting qualities are explained in Chapter 2.

Several of the wool finishes discussed in this chapter are applied by means of heat, moisture, and pressure, and in each case the wool shrinks to a certain extent. Such wool finishes as steaming, fulling, and decating are for the purpose of shrinking the wool fabric, as well as for contributing other desirable features. Up until a recent development, discussed later, for controlling the shrinkage of wool, there have been two methods in general use: one by hot steam and the other by cold water. Neither of these methods

London or cold water shrinking of woolens and worsteds. The fabrics are moistened by means of a fine spray which mechanically is made to permeate the goods, after which it passes into a sweat box which still further diffuses the moisture. The cloth continues on into a drying chamber where it is dried by cuddling on an endless apron that prevents all distortion. This process makes the cloth ready as far as shrinkage is concerned for cutting into clothes (David Gessner Company).

controls shrinkage to a point where further undesirable shrinkage will not result if the fabric is washed.

The hot steam method may be applied in three ways: open steaming and cooling; cylinder steaming; and decating. Steam shrinking is accomplished in different types of machines, according to the intended results.

For heavy overcoat fabrics and napped and pile fabrics, an open steaming and cooling machine is used and as little tension as possible is put on the fabric. For tightly and firmly woven fabrics, such as twilled suitings, and for wool mixture fabrics, a cylinder steaming machine is used. For light weight dress fabrics and for fabrics with a high sheen, such as broadcloth, the decating machine is used. The fabric is wound around a perforated cylinder. Steam is forced through the fabric from inside the cylinder, out through the perforations, and then, reversed, through the fabric into the cylinder.

The cold water method is called "London shrinking." It is used primarily for fine worsteds. A cotton or woolen blanket is dampened and laid on a platform. Folds of wool fabric and blanket are placed alternately until there is a required number of layers. A weight is placed on the pile, compressing it for several hours until the dampness from the blankets has moistened the wool fabric. The fabric is then hung in folds and dried in the air or in drying machines with hot steam pipes.

Wool shrinkage. Below: Chemically shrunk wool fabrics; comparison of the same original size treated and untreated wool fabric. The treated fabric retained practically its original size while the untreated fabric shrunk as shown. Right: Chemically shrunk wool yarns. Left to right: The original yarn, the shrinkage of the untreated yarn when washed, the approximately original size of the treated yarn when washed (Calco Chemical Division, American Cyanamid Company).

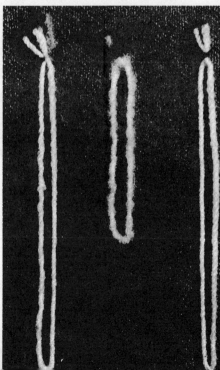

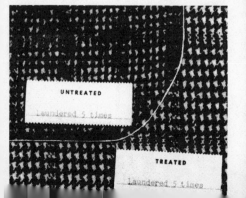

Cold-water shrinking and drying are usually followed by pressing. The fabric is folded between pressboards and a hydraulic press compresses the fabric.

A chemically controlled wool shrinkage process gives satisfactory results and in no way affects the hand or appearance of the fabric. The crease-resistance, resiliency, or warmth-giving quality is not changed. Dyes are not adversely affected. In fact, many dyestuffs have an increased fastness to laundering after this treatment. The desired finish of the fabric remains and there are no signs of frizziness. The treated fabric has a higher resistance to alkali than does untreated fabric. Testing has shown that the treated fabrics have no tendency to cause dermatitis. It may be more difficult to nap a fabric after this treatment, and there may be a fuller feel to the fabric. But a softener will assist this condition.

The fabric has less than 5 per cent length by a 2 per cent width shrinkage, and the wool shows very little tendency to felt. This means that the treatment makes possible the washing of many wool fabrics that formerly required dry cleaning. It also permits the removal of water soluble stains.

Chemically, the finish is an alkylated melamine formaldehyde.

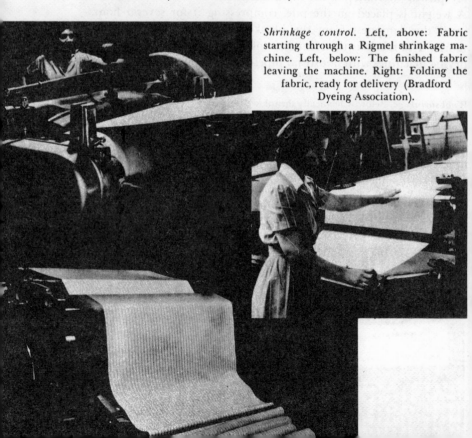

Shrinkage control. Left, above: Fabric starting through a Rigmel shrinkage machine. Left, below: The finished fabric leaving the machine. Right: Folding the fabric, ready for delivery (Bradford Dyeing Association).

t is water soluble at the time of application but condenses during heat curing to form a highly insoluble resin, part of which is within the fiber. Both woven and knitted wool fabrics of all types have been given this finish with good results.

Care is taken that the fabric is clean before treatment. Then it is run through a bath of the solution in a padder machine, squeezed through rollers, run through a tentering frame and into drying chambers, for drying and curing. The temperature for curing is usually from 280° to 300° F. and the time five to ten minutes. It is sometimes stated that temperatures as high as 290° F. will damage the wool fiber. Experiments on both wet and dry wool fabrics have shown that the wool did not deteriorate or lose its tensile strength appreciably when tested under curing conditions. Heating causes white wool to turn very slightly yellow. Loss in tensile strength seems to be caused by the resin forming on the surface during treatment, but this can be minimized with proper application of the solution.

After curing, the fabric is washed to remove the uncured resin and any remaining formaldehyde.

Shrinkage Control of Cotton, Linen, and Rayon

Fabrics made from these fibers do not have inherent felting and shrinking properties as fabrics made of wool fiber do, and yet such fabrics may shrink beyond practical usability unless given a shrinkage controlling finish. Many methods were tried, chemical and physical, and others, such as merely washing the fabric. The former were not very successful and the latter often destroyed the usual commercial surface finishes.

Some years ago technicians considered the major causes of shrinkage, such as the constant stretching of the yarns during yarn making, weaving, dyeing, and finishing under tension. Finally a method of controlling shrinkage was figured out, based on mechanical compression that restores the yarn equilibrium in a fabric, resulting in more warp and filling yarns per inch than the commercially finished fabric had before this process was applied.

As the first step in the process a sample of the fabric is measured, then washed by U. S. Textile Test Method CCC-T-191A, dried, ironed, and again measured. The percentage of shrinkage in both warp and filling is calculated and the compression shrinkage machine is adjusted so that the fabric will be shrunk in the same proportions. Water, steam, and heat are used to prepare the fabric for shrinkage.

The fabric is fed to the machine, over bars, to guide rolls that deliver it to rubber intake rolls. It is the rotation of these rolls that determines how fast the fabric is to go through the machine. From the intake rolls the fabric enters a large closed chamber, called the "sky housing," where it passes over and under idler rolls and is conditioned by water or steam. It is then sometimes passed over a heating cylinder and is finally caught by clips on a short tentering frame, where the fabric is adjusted to proper width. From the tentering frame it travels around a small feed roll to the main steam-heated cylinder. A woolen felt blanket passes around the feed roll as well as around the large cylinder. The feed roll blanket combination is sometimes called the "belt shrinker," because here is where the actual warp shrinkage takes place.

In weaving and finishing most tension is on the warp yarns, consequently, the greatest shrinkage is in the length of the cloth. The thickness of the blanket used and the size of the feed roll

The Sanforizing machine that controls shrinkage: (a) Unshrunk fabric, (b) feed-roll controls, (c) water sprays, (d) steam and water vapor applied here preparatory to shrinking, (e) cross section showing point where shrinkage is obtained, (f) width controlled here, (g) steam drum for drying, setting and shrinkage and finishing face fabric, (h) Sanforized-shrunk fabric ready for the market (Cluett, Peabody & Co., Inc.).

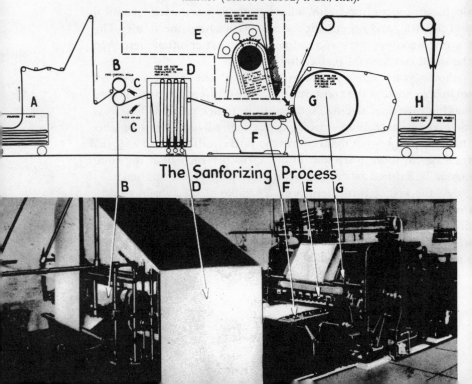

depend upon the type of fabric being shrunk and on the amount of shrinkage desired—the thicker the blanket and the smaller the feed roll the greater the shrinkage.

The fabric passes over the curve of the blanket as it travels around the feed roll, but between the blanket and cylinder as it rotates around the large cylinder. After passing over the curved surface, the blanket contacts the fabric, which is then smoothly pressed, dried, and the weave set as it passes around the steam-heated cylinder. If the other side of the fabric requires pressing and smoothing, the fabric passes around an auxiliary cylinder, with the opposite side of the fabric to the cylinder. Since shrinkage must always be the last finishing operation, the fabric is then ready for folding or rolling for market shipment.

According to technical experts, fabrics given this shrinkage process do not have more than ¾ per cent residual shrinkage in either length or width, but for practical purposes, 1 per cent is specified to the trade. There are more yarns per inch in both warp and filling than before shrinking, and this makes the fabric more closely woven and compact, which gives it greater strength and longer wearing qualities. As only water and steam are used, color and finish are in no way impaired, and the fabric is given a smooth, ironed surface texture.

Since the cost is slight, all fabrics should have controlled shrinkage today, regardless of their price. Before satisfactory shrinkage methods were developed, shrinkage of fabrics caused garments to decrease in size beyond repair, imposing great losses on consumers. Fabrics and merchandise made of fabrics that have been given controlled shrinkage treatment are so labeled. They can be laundered with no fear of further appreciable shrinkage.

A chemically controlled rayon shrinkage process has been developed. The treatment is given the fabrics in the gray goods. First, a protective colloid is padded on the goods. This prevents the caustic soda, applied later, from attacking the yarns. The fabrics are then dried and treated with a 30 per cent caustic soda solution. After this treatment the fabric is given a short dip in water and the caustic soda is neutralized in bicarbonate of soda. The goods are then boiled off and dyed and finished in the usual manner. The fabrics are slack dried after dyeing and before final tentering.

It is believed that, since the rayon yarns swell, the caustic soda changes the physical structure of the yarn. This process sets the

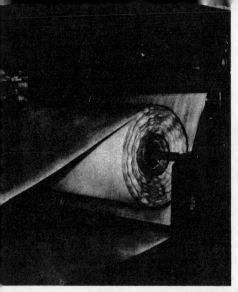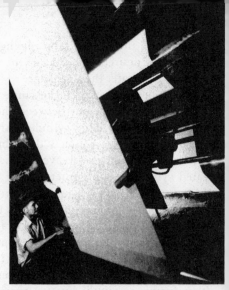

Singeing. Fabrics being singed in two types of singeing machines to remove surface fibers (Goodall Fabrics, Inc., and Pacific Mills).

yarns in the fabric and prevents further appreciable shrinkage.

It is found that fabrics, so treated, have a 15 to 20 per cent better color value and sharp, clear prints can be produced. This process has been particularly effective on cotton and spun rayon mixtures. If anything, it improves the hand of the fabric.

In addition to these specific shrinkage control finishes, there are certain special or functional finishes that also contribute some shrinkage control to the fabrics. For example, the crease-resistant finishing processes are very effective for controlling the shrinkage of such fabrics as matelassé, waffle cloth, and marquisette.

SINGEING

To remove surface fibers and lint.

Applied to wool, cotton, rayon, linen, and silk.

Fabrics that are to have a clear surface, such as clear-surface worsteds and the majority of cotton fabrics, are singed before they start through numerous other finishing processes.

The fabric passes through closed chambers in a large machine. Moving at a high rate of speed, 100 to 300 yards a minute, the fabric passes over open gas flames or over plates or rollers heated red hot. Some machines allow for singeing both sides of the fabric at the same time. This operation may be repeated two to four times. After being singed, the fabric is immediately immersed in water to avoid the danger of sparks igniting the fabric.

For cottons, this bath, or a second bath, may contain dilute sulfuric acid or an enzyme solution that will dissolve the starch used as a sizing for the yarns prior to weaving.

STARCHING

To add luster and to improve the body of the fabric.
Applied to cottons.

Starching is applied to cottons in a manufacturing plant for the same reasons as it is in the home laundry. Softeners are added to the water and the starch. These consist of various oils, chemicals, and other ingredients. If added weight is to be given the fabrics, materials like clay and chalk are included.

Starches are obtained from such sources as corn, wheat, rice, and potatoes. The starch is made much as at home. The ingredients are mixed in cold water and then the solution is heated and becomes a thick gelatinous liquid.

The starch mangle consists of a trough that holds the starch and the rolls, around which the fabric passes as the starch is being applied to it. The padder machine used for pad dyeing is frequently used for starching fabrics.

STEAMING

To shrink and to condition the fabric.
Applied to wool and spun rayon.

Regardless of when hot steam is applied to wool in the finishing processes, it has a tendency to shrink the fabric. When given a special steaming finish, it is for the purpose of shrinking the

Semi-decating or steaming. This is used for treating woven and knitted woolens, worsteds, and rayons where less severe processing is required. It is also used for refinishing London shrunk fabrics. This process reduces shrinkage and brings out the natural fiber luster. The steam is forced through the holes in the cylinder and out through the fabric which is wound around the cylinder (David Gessner Company).

fabric in both length and width. If it is applied after final decating, it is for the purpose of removing the luster from the fabric.

The steaming machine consists of a steam box with a perforated copper cover, over which is cloth. As the fabric goes over the steam box, steam is forced through the fabric.

If the fabric being steamed has been napped, the cloth may be brushed as it leaves the machine.

If shrinkage is to be more complete, the fabric on leaving the steam box may go over a heated table where the moisture is evaporated. Wool fabrics that have been given a shrinkage finish are reduced in both length and width as well as given a fuller, more compact body.

TENTERING

To straighten and to dry the fabric.

Applied to wool, cotton, rayon, linen, and silk.

During construction, dyeing, and wet finishing, the fabric has gone through numerous operations. As a result, the selvage is uneven and some yarns are irregular. The fabric now has to be dried and straightened before the dry finishing processes can be applied.

The drying and straightening is accomplished by feeding the fabric into a flat frame that straightens and carries it over steam

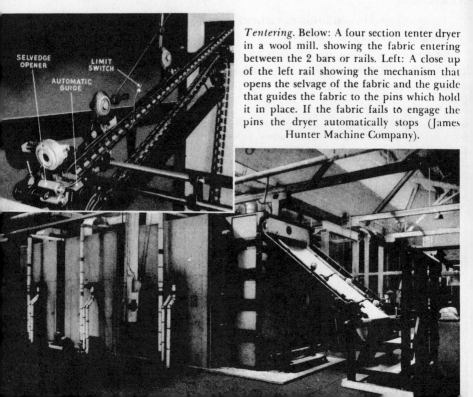

Tentering. Below: A four section tenter dryer in a wool mill, showing the fabric entering between the 2 bars or rails. Left: A close up of the left rail showing the mechanism that opens the selvage of the fabric and the guide that guides the fabric to the pins which hold it in place. If the fabric fails to engage the pins the dryer automatically stops (James Hunter Machine Company).

coils. Sometimes a room is built around the tenter frame, which may be 30 to 60 feet long. Each end of the tenter extends out of this room.

In the tenter, the fabric is caught and held by pins or clips on the tenter chains that stretch the fabric to the proper width. Some tenters have a wheel with brushes that press the edge of the fabric onto the pins; others have a small, automatic electric machine that guides the fabric. If clips are used, they automatically pick up the edge of the fabric. As the clipped or pinned fabric moves, it is systematically straightened to the required width.

As the fabric passes through the drying machine, hot air is circulated through the fabric, the air being gradually reduced in temperature from about 180° in the first chamber to room temperature as the fabric leaves the machine. The temperature is scientifically controlled to prevent damage to the wool fiber by excessive heat. The temperature in a cotton tenter oven may be 300° F. or more.

WET DECATING

To set the nap and to add luster.

Applied to wool.

This finish is applied primarily to napped wool fabrics with a high luster that are to be dyed. If the luster is to be retained through dyeing and the other finishing processes it must be made permanent.

A tentering frame with clips that hold the selvage of the fabric.

Again, heat and moisture are employed but, in place of the pressure used in other processes such as fulling and pressing, the fabric is held at a certain tension on the roll.

The wet-decating machine has a perforated cylinder mounted in a trough. The fabric is wound face down on the cylinder. Then hot water is circulated through the fabric and out through the perforated cylinder, as well as from the tank through the fabric into the cylinder. Or steam may be forced from inside the cylinder out through the perforations and through the fabric. If hot water is used, it is drawn off and cold water added to cool the fabric. In the steam process cold air is forced through the fabric.

WEIGHTING

To give body and desired appearance.

Applied to silk.

Weighting consists of adding substances to silk to replace the gum that has been boiled off prior to bleaching and dyeing. In degumming, silk loses from 20 to 27 per cent in weight, leaving the fabric soft and more or less flimsy, depending on its weave.

Excessive weighting results in seriously weakening the silk. The metals in the weighting solution cause the fabric to wear out much more quickly than they would otherwise. Sugar, salts of tin and lead, and iron have been used for centuries in weighting silk. A certain amount is necessary to produce the body and desired appearance of the fabrics. Lingerie fabrics are practically the only types that are acceptable without some weighting. With a normal amount of weighting, correctly applied, the fabric will give satisfactory wear.

It was not until the mass production of the twentieth century that silks frequently were given excess weighting. Amount and kind of weighting are difficult to detect because silk can absorb the weighting liquor up to 300 per cent or more of its original weight with no noticeable change in the fiber.

If silks are excessively weighted, the metallic salts, on exposure to air, will cause the fabric to decompose. Perspiration also causes rotting of the fibers that have excessive weighting.

A Federal Trade Commission ruling states: "Silk fabrics may be termed 'pure dye' or 'pure silk' if they do not contain more than 10 per cent of any substance other than silk with the exception of black, which may contain 15 per cent weighting."

It is not known how the term "pure dye" originated, as it has nothing to do with dyeing but refers to weighting. It may have come from "pure dyestuff," meaning that only dyes and no excess weighting have been added.

The fabric to be weighted is placed in the weighting solution bath and remains there for several hours. The fabric will absorb about 10 per cent of the weighting in each bath. If more is required, it is given a second bath, always being dried between each bath. Frequently, the liquor bath (containing the sericin) remaining after degumming, is added to the dye bath to restore a part of the weight loss. Such weighting in no way damages the fiber. The metal used and the strength of the solution is dependent on the fabric and the knowledge and integrity of the manufacturer. Care must be taken in the process so that the fabric absorbs the solution evenly throughout. After weighting, the fabric is rinsed and dried.

FUNCTIONAL FINISHING PROCESSES

It is only in the past few years that progress has been made in giving fabrics additional finishes that contribute immeasurably to their suitability for specific purposes. Perhaps one of the greatest moneysavers to consumers was the development of shrinkage control methods that prevent a fabric from shrinking appreciably after laundering, leaving the garment the correct, wearable size.

Shrinkage control has been discussed with general finishes rather than be grouped with these special finishes that give the fabrics additional functions. This is because washable fabrics demand controlled shrinkage and the time is past when such a finish should be other than a part of the normal finishes given to any washable fabric.

Other finishes have gained great importance, and their uses during the war proved so successful that their wider application to future civilian goods was assured. Light weight, durable fabrics will withstand rain and wind as will heavy and strong tent and tarpaulin fabrics. Fabrics for the uniforms of servicemen and of workers in war plants have been made fireproof. Other fabrics are made resistant to mildew, moths, and insect pests.

There are some finishes that ensure crisp fabrics remaining crisp after laundering without addition of starch; others that make fabrics resist wrinkles and creases or more readily absorb or disperse moisture. These finishes assure continued beauty as well as increased service to the fabric in use.

The development of these finishes has progressed simultaneously with the advent of new chemical compounds, the discovery of polymerization, and the knowledge of how to make synthetic fibers and plastics.

Oil cloth, waterproof raincoats, slickers, and window shades are examples of surface coatings applied to cloth. Mercerization is a finish given to cottons that creates a chemical change in the fiber, giving it added strength and luster. These and a few others constituted the entire field of functional finishes until chemists reached the point where they were ready to explore further into the realm of synthesis, and create new chemicals, and work out the methods for their transformation into yarns and finishes.

When applied. According to the type of finish it may be the first or the last finishing treatment; it may be applied at any in-between stage of finishing the fabric; or it may be before or after the fabric is dyed, or during the dyeing process.

How applied. The fabric may be immersed or dipped in the finishing liquid, the finish may be calendered on, under pressure, by running the fabric through rollers. The fabric may be immersed in the liquid in a mangle and then squeezed through rollers. The finish in a paste solution may be coated or spread on the fabric, or the finishing liquid may be sprayed on the fabric.

Results obtained. The finish may form an invisible film on the fibers or coat the yarns. Some impregnate the fiber, form a compound with the fiber, and change it chemically. Others penetrate the fiber, set the finish in the fiber, but do not change it chemically.

Some are durable and will last the useful life of the fabric, others are semidurable. Some remain after washing or dry cleaning the fabric, some can be dry cleaned but not washed, some can be washed but not dry cleaned, while others require reapplication after washing or dry cleaning.

Chemicals Used

The materials and chemicals used include a wide range. Many are highly complex and often secret formulas. Just as in the discussion of rayons and other synthetic fibers and of dyes and finishes, some knowledge of chemical terms is necessary to understand what gives fabrics these newer finishes. While it is impossible in such

564

a book as this to give a complete technical discussion of the formulas, the reader will find in Chapter 15 a brief explanation of the chemicals used in finishing.

Many of the materials used in the past are still applied today. They include asphalt and paraffin, both obtained from certain petroleums; creosote and tar, both coal-tar products; natural waxes, such as beeswax; and vegetable oils, such as linseed oil from the seed of the flax plant. There is one group that is used to coat the fabric. When applied they waterproof and mildewproof fabrics. These also may be combined with other chemical compounds to form other types of finishes.

Compounds of metallic elements form one large group. These include aluminum, chromium, copper, lead, magnesium, and tin. The elements silicon (a solid) and fluorine, hydrogen, nitrogen, and oxygen (gases) also are united with other elements to form finishing compounds. They are applied extensively to safeguard fabrics from moth damage, to make them fire- or flame-resistant, waterproof or water-repellent.

The cellulose compounds form a specialized group and include cellulose acetate, cuprammonium, and nitrocellulose, which are the same chemical compounds as used for rayons. Cellulose acetate is a cellulose ester. Cellulose ethers are better for finishes than the esters. There are four types of cellulose esters used: methyl cellulose, which is soluble in water; ethyl cellulose, which is insoluble in water; and carboxy methyl cellulose and hydroxy ethyl ether (glycol cellulose), which are soluble in caustic soda.

According to their formulas and methods of application these compounds contribute many different kinds of finishes. They may produce a fabric that is nonlinting, lustrous, crisp or soft, transparent or translucent. They may give a crisp finish to cottons, which will remain crisp after laundering without the addition of starch. The same finish at the same time may set the weave, as in crisp, specially finished marquisettes. Finishes made from these compounds may reduce slippage and give some shrinkage control to the fabric. It is this type of finish that gives permanent stiffness to the collars of men's shirts. They may make the fabric mildew-resistant and waterproof. Some impregnate the fiber while others coat the fiber.

The synthetic resins are a more recent group of finishes. Many of them consist of the same materials that make the hundreds of today's plastics. One group is similar to that from which synthetic fibers are made. These finishes have complex chemical formulas, all are compounds of some carbon derivative. They include urea-formaldehyde resins; phenol-formaldehyde resins; alkyd resins containing phthalates, which are obtained from oxidizing various benzene derivatives; acrylic acid series, containing carbon, hydrogen and oxygen; and polyvinyl or polyacryl compounds, which are polymerized ethylene derivatives.

The formaldehyde and phenol resins are initially soluble, and under the action of heat they become insoluble and will harden on the fiber. When heated, the water is eliminated and the molecules unite. The other resins listed above are always soluble and when applied in a finish they penetrate the fiber. By the application of light, heat, acids, or alkalies, the molecules in the chemical compound combine with themselves.

These resins are used to render fabrics crease- and crush-resistant, to give fullness to fabrics, and to add luster to cottons. At the same time they may contribute nonslipping qualities, greater strength, and shrinkage control. They are also used to finish water-repellent and waterproof fabrics. Urea formaldehyde is excellent for such fabrics as marquisette, matelassé, and waffle cloth.

The cation-active group is one of the latest additions to these functional finishes. Metallic ions have positive charges and are called "cations." Nonmetallic ions have negative charges and are called "anions." The chemicals used for most finishes and dyes are the negatively charged or anions that are active. In these newer finishes the positively charged or cations are active and attach themselves to the anions (see Chapter 15).

Most of the cation-active chemical compounds are composed of nitrogen, hydrogen, and oxygen. They include the nitrogen compounds, some with simple and others with very complex formulas. One group is the amines, which is ammonia (hydrogen and nitrogen) with the hydrogen replaced by a hydrocarbon radical. Another group consists of the long-chain quaternary ammonium compounds that actually combine with the fiber.

These cation-active substances are used to a great extent as water-repellent finishes. They may also render a fabric more absorbent, add a permanent softness to the fabric, and make the dyes more resistant to water, acids, and alkalies.

Waterproof finishes differ from water-repellent finishes. Waterproofing covers the fabric and closes up all the interstices or spaces between the yarns, while water-repellent finishes are applied to the individual fibers and the spaces in the fabric remain open.

Waterproof fabrics are completely rain and waterproof as the fabric is nonporous. Water will not pass through it under the normal pressure to which the fabric will be exposed. The fabric may be coated on one or both sides.

Waterproof fabrics are used extensively for such items as tents, tarpaulins, awnings, shower curtains, hospital sheetings, and rain wear. Protective work clothes may be made of fabrics that are both waterproof and chemical resistant. Since the finish renders the fabric nonporous, it does not permit air circulation or the means for the evaporation of perspiration. That is why air vents are cut in most raincoats.

Many of these waterproof finishes contribute other qualities to the fabric, either by themselves or in combination with other finishes. They may make the fabric resistant to mildew, moths, germs, perspiration, and acids. The best finishes are spot and stain, grease, oil, and dirt resistant. They will not peel, crack, dry out, harden, or stick, and are not affected by the heat of the sun. It is claimed that some give the fabric a lintless or lustrous surface, an improved hand, and a degree of shrinkage control.

Waterproof finishes coat the fiber or fabric. The materials used include rubber and synthetic elastic, linseed oil, tar, pyroxylin (cellulose nitrate solution), cellulose acetate, other cellulose esters and ethers, and synthetic resins, such as a vinyl resin. Heavy non-apparel fabrics may be waterproofed with asphalt, beeswax, paraffin, lead oleate, or tar.

Water-repellent Finishing Processes

Water-repellent fabrics are not impervious to water as the interstices or pores are not closed. The chemical compound used increases the surface tension between the water and the treated fibers, making it difficult for water to seep through the spaces.

Water-repellent fabrics, used for wearing apparel, contribute protection from rain and wind, without excessive weight. These finishes are now widely applied to children's clothes, work clothes, sportswear, curtains and draperies, and many other such fabrics.

In addition to their water-repellent quality the finishes contribute other features to the fabrics. One of the most important is that they will render the fabric resistant to nongreasy spots, stains, and perspiration, and retain no odors. If reliable finishes are properly applied, they improve the hand and feel of the fabric, often giving a soft, supple hand. They resist wilting in warm and humid weather, and prevent dirt from becoming imbedded in the fabric. Some are combined with mildew-resistant features.

The materials used include paraffin and other waxes; metallic compounds, as aluminum acetate and metallic soap; vinyl or acrylic resins; and long chain quaternary ammonium compounds. They either impregnate or coat the fiber.

One method was to treat the fabric with aluminum acetate and then in an emulsion of soap or of soap and paraffin. Another method employs aluminum acetate solution that contains the paraffin in colloidal suspension. A newer process is the quaternary ammonium compound that combines with the fiber.

Crease-resistant Finishing Processes

It was the introduction of a crease-resistant finish that opened up the field of resin finishes that impregnate the fiber. They render fabrics crease- and crush-resistant and may be applied to all types of fibers. One of their greatest contributions is making velvets, that were noted for crushing, crush-resistant.

These finishes not only make the fabric resist deep creases but speed the recovery from wrinkles, helping the fabric to retain a new appearance. They improve the body, drape, and hand of the fabric, are odor-resistant and, to a certain extent, control shrinkage. Such finishes make dyes considerably faster to washing and are used most successfully on spun rayons, linens, and cottons, especially voiles.

These finishes are of the synthetic resin group, including urea formaldehyde, phenol, and melamine resins; alkyd resins containing phthalates; and the polyvinyl compounds, all of which are formed by polymerization.

Starchlike Finishing Processes

The classification for this type of finish should be interpreted to mean that after washing, with no addition of starch, the fabric upon being ironed will return to its original crisp appearance.

These finishes are used to give a crisp hand and a transparent appearance to a fabric such as organdy. They may also give a soft hand and lustrous appearance to voile and lawn. Some types are applied to plissé crepe to give the plissé a greater degree of permanency. They are used on the linings inside collars of men's shirts to help keep them crisp without starching. They are employed on many sheer cotton and rayon fabrics used for dresses, blouses, children's clothes, and curtains.

These finishes may prevent sagging and slipping of the yarns as well as wilting in warm weather. They make the fabric lintless and the smooth surface texture resistant to snags and abrasion. They also give some shrinkage control.

The chemicals used are those of the cellulose compound group, including viscose, cuprammonium, hydroxy, ethyl cellulose, and several other types of the cellulose compound group. They coat the surface and have a high degree of durability.

Absorbent Finishing Processes

There are two of these processes that accomplish results in a different way. In one, the treated yarns soak up the moisture and break it into small molecules, which then evaporate. The other method holds the moisture by absorption, as the treated fibers take up the moisture and disperse it into the yarns.

These finishes are used for towels, underwear, corsets, sportswear, and items where absorptive qualities contribute added features. They also aid the dyeing and printing processes. They may improve the hand and feel of the fabric. The quatenary ammonium compounds are used for these finishes.

Glazed Finishing Processes

These are the finishes that give a glaze to such fabrics as chintz and fine muslin. They remain lustrous after laundering. Women's and children's dresses and aprons, curtains and draperies, and many other fabrics are now given a glazed finish. Usually more finely woven fabrics are used. This finish contributes a sleek, smooth surface that resists soil. The fabric has better draping qualities and there is some control of shrinkage.

Such resins as melamine formaldehyde, urea formaldehyde, and alkyd are used for these finishing processes.

Moth-preventive Finishing Processes

Without some treatment wool fabrics can be seriously damaged by moths, particularly so if they are stored without precautionary measures. Moths seem to thrive in a temperature of 70° and over, which is not only summer heat but the temperature of most homes in winter.

There are two types of insect pests that destroy wools. One is the black carpet beetle, the other is the moth. The moths seen flying are not those that do the damage. The moth lays her eggs in the wool and they hatch out in from six to ten days. It is the tiny larvae (worms) that before they begin to fly feed on the wool fiber, destroying the fabric.

While moth crystals and balls are still widely used as preventives, many fabrics, particularly blankets and rugs, are given special finishes that render the article resistant to these insect pests for a time at least.

The chemicals most frequently used include silico-fluoride, and cinchona alkaloids (a poisonous substance found in certain plants). Naphthalene moth balls and paradichlorbenzene (chlorine and benzene compound) are two crystals recommended for home use by the United States Department of Agriculture.

Fabrics given some of these treatments can be dry cleaned and washed several times without destroying the finish. Some are destroyed by washing but not by dry cleaning while others can be washed but not dry cleaned. Light weakens some of the finishes but has no effect on others. Labels on the merchandise should state the lasting qualities of the finish and the method of its care.

Mildew-resistant Finishing Processes

Mildew is a parasitic fungus, which, when found on fabric, is living off what is called "dead matter." It thrives in a warm, humid atmosphere.

Finishes given to prevent mildewing are odorless, with no toxic properties, and are nonirritating to the skin. They will not alter the feel or hand, the color, porosity, or tensile strength of the fabric, and may be perspiration-and odor-resistant.

Certain acids or metallic compounds may be added to starches to give a fair amount of protection. Fabrics that have been waterproofed with such materials as asphalt, creosote, other coal-tar products, and rubber, are also rendered mildew-resistant.

These finishes include metallic compounds such as chromium, copper, lead, mercury, and zinc salts, as well as certain organic compounds. Some may be applied with waterproof or fire-retardant finishing or during dyeing.

Flameproofing and Fireproofing Finishing Processes

"Flameproofing" and "fireproofing" are words used to describe two slightly different results of the processing. After flameproofing the fabric may char and glow upon being exposed to flames but will not flame or actually burn. After fireproofing, the fabric may char but will not glow after the heat is removed. In fact, most fabrics char faster with than without these finishes.

There are two types of processes—water soluble and insoluble—used for these finishes. The water-soluble process employs ammonia phosphate and ammonia sulfamate. A very simple finish that can be applied in the home is a 10 per cent solution of borax and boric acid. Being water soluble, these finishes need to be re-applied after each washing.

The insoluble compounds include chlorinated compounds such as vinyl chloride, or chorinated rubber, usually used with antimony oxide. These render the fabric weather-resistant and some of them are wash-resistant.

Most of these chemicals make the fabric stiff and harsh and add weight, features not greatly objectionable in most fabrics requiring flame or fire protection. The use of a solution like ammonium sulfamate leaves the fabric soft, but the finish can be readily washed out, although it is retained through dry cleaning.

A flame-resistant finish may or may not be toxic. It in no way impairs the quality of the fibers or fabrics; in fact, it may prolong the life of the fabric. Some processes also make the fabric resistant to odors, to rotting, and to the feeding of moths and insects. Some of the finishes cannot be detected by sight on the fabric.

Precautions necessary. Most of the above has been written from a positive point of view, because the majority of the well-known functional finishing processes applied to fabrics today are durable, reliable, and do give the described features. However, some lesser known processes may not be completely harmless, as they have not been carefully analyzed and tested, and have not withstood the supreme test of time.

THE FABRIC FINISH	WHAT IT DOES	APPLIED TO:						
Pretreatments, and General Finishes		Wool	Cotton	Linen	Silk	Spun silk	Filament rayon	Spun rayon
1. Beetling	Gives lustrous appearance		x	x				
2. Brushing	Removes short, loose fibers	x		x	x	x	x	x
3. Burling	Removes knots in yarn	x						
4. Calendering	Smooths, glazes, moirés, or embosses		x	x	x		x	x
5. Carbonizing	Burns out vegetable matter	x						
6. Crabbing	Sets the weave permanently	x						
7. Dry decating	Sets the luster permanently	x						
8. Fulling	Makes compact and shrinks	x						
9. Gigging	Raises the fibers	x						
10. Kier boiling	Prepares fabric for bleaching		x	x				
11. Napping	Raises the fiber	x	x					x
12. Perching	Examines the fabric	x	x	x	x	x	x	x
13. Pressing	Smooths and gives luster	x			x			
14. Scouring	Removes oil, sizing, and dirt	x	x	x	x	x	x	x
15. Shearing	Cuts the surface fibers	x	x					x
16. Shrinking	Adjusts width and length	x	x	x				x
17. Singeing	Removes surface fibers and lint		x	x		x		x
18. Starching	Adds luster and improves body		x	x				
19. Steaming	Shrinks and conditions	x						
20. Tentering	Straightens and dries	x	x	x			x	x
21. Wet decating	Sets the nap and adds luster	x						
22. Weighting	Replaces boiled out gum				x			
Functional Finishes								
1. Absorbent	Absorbs or disperses moisture	x	x	x	x	x	x	x
2. Crease-resistant	Resists crushing and creasing	x	x	x			x	x
3. Fire-resistant	Chars but does not burn	x	x	x	x	x	x	x
4. Germ-resistant	Destroys germs	x	x	x				x
5. Glazed	Gives a sheen		x					
6. Mildew-resistant	Resists parasitic growth		x	x				x
7. Moth-preventive	Destroys moths	x						
8. Starchless	Remains crisp without starch		x	x				
9. Waterproof	Completely protects against water	x	x	x	x		x	x
10. Water-repellent	Resists moisture, stains, and wind	x	x	x	x		x	x

Opposite page: An original hand loomed fabric combining spun silk and wool (Dorothy Liebes).

Contribute comfort and beauty

A great many fabrics, such as those for draperies, bedspreads, blankets, mattresses, rugs, and towels, are especially constructed for use in the home. Attractiveness, service, comfort, and protection are features that are considered when creating many of these specialized fabrics that add immeasurably to our American way of life. The fibers and yarns, the weaving, knitting, and twisting, and the dyes and finishes are the same as for other fabrics. In fact, some are the same fabrics as explained previously. Some have different finishes or added decorations to better equip them for home use. Others are especially designed and constructed.

BEDSPREAD FABRICS

There are three types of bedspread fabrics. One consists of fabrics decorated after they are woven. These are used almost exclusively for bedspreads. The second includes fabrics woven especially for bedspreads, their designs and colors planned before weaving, with special attention to the appearance of the spread on the bed. The third type consists of fabrics that have other uses but are also suitable for bedspreads.

Opposite page, top to bottom. Left: Thick 3-ply yarns woven to form loops that create the design; Jacquard pattern with coarse floating yarns; double woven matelassé. Right: Smooth surfaced Jacquard pattern and twill ground; Jacquard pattern, with woven-in tufts.

574

Right and wrong side of unclipped punch work combined with
clipped tufting.

Decorated Fabrics

These bedspread fabrics have single tufts, or continuous running tufting, or punch work designs. Firm, strong fabrics are used, such as 64-square sheetings (64 warp yarns and 64 filling yarns), and kraft cloth which is a plain weave fabric made of heavy yarns. Some lower priced spreads are made on sheeting of lower thread count. The fabrics may be white, natural, or colored. The decorative yarns may be white or colored.

Unclipped tufting is a continuous running stitch, which leaves uncut loops of tufting on the surface. The running stitch can be noted on the back of the fabric.

Clipped tufting is made the same as unclipped tufting, except that as the decoration is stitched on, a knife cuts the loops. The back is the same as that of unclipped tufting. Clipped tufting is often wrongly called "chenille tufting." Any fabric or decoration called "chenille" must employ chenille yarns.

Candlewick or hobnail is made the same as tufting except that there is no continuous row of stitches in the back. The tufting is spaced, and the stitches on the front are cut, so that individual tufts of yarn are spaced throughout the face of the fabric.

Unclipped punch work is accomplished differently from tufting. The low, uncut pile is in a pattern. On the back of the fabric the stitches are finer and more closely set than in tufting.

Clipped punch work is the same as unclipped punch work except that the loops are cut by hand to form a closely set, velvety pattern.

Many spreads include a combination of two or more of the

above methods of decoration and many are in contrasting colors. Variety is also given by the fineness or coarseness of the yarns, the height of loops, cut or uncut, and the thinness or thickness of the pile.

The design is first drawn on heavy duck. The sheeting is then fastened over the pattern, which is transferred to the sheeting by rubbing it with colored wax. The sheeting is then ready for applying the yarns.

Machine tufting is accomplished by a machine similar to the sewing machine. If designs or curves are being tufted, only one needle is used. If straight rows are being sewn, many needles may sew several rows at one time. There is a machine with 228 needles that may tuft as many as six spreads at one time as well as sew on interesting designs. The yarns are fed to the needles from large spools near the machine.

For punch work, the fabric, with the pattern printed, is mounted on a large round embroidery frame. Cones of yarn are hung above the frame. The operator punches in the stitches with a small, electrically driven machine that has one or two needles through which the yarn is laced.

Woven Bedspreads

These bedspreads have the pattern and color woven in. Many weaves and combinations of weaves are used. The variety is as wide as the imagination of the designer. There are plain, twill, and satin weaves, and variations of these weaves. There are the waffle, swivel, lappet, and other floating yarn weaves. There are backed fabrics and double fabrics. They may be woven on plain

Top, left to right: Candlewick or hobnail, unclipped tufting. Bottom: Both are clipped tufting; sample on right has higher and denser pile (Right and wrong sides illustrated).

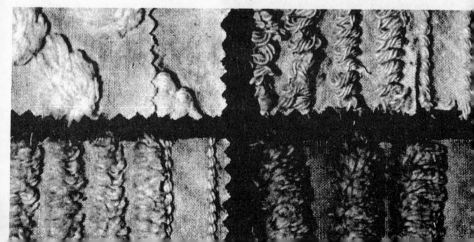

and Jacquard looms and on looms with dobby attachments. Patterns may be small or large. The bedspreads may be in one color, two colors, tones of one color, or multicolored. They are usually woven with dyed yarns.

Some are loom quilted, others are loom tufted. The loom-quilted fabrics are suggestive of intricate hand quilting. The tufts may be woven in while the fabric is being woven.

The weave contributes no more than do the yarns that are used in the fabric. They may be fine yarns woven in a satin weave; they may be coarse, tightly twisted yarns, or full, loose yarns, woven in colorful floating weaves to form the pattern.

The surface may be smooth and lustrous with rayon satin ground and Jacquard pattern. It may be loose and rough, with a homespun appearance or with a raised and quilted texture.

Fabrics Cut and Sewn

Bedspreads made of cut and sewn fabrics use a wide assortment of fabrics discussed in other chapters. They range from the sheer, crisp, and frilled, organdy bedspread for a very feminine room to the heavy velvet or corduroy tailored spread. While almost any

fabric may be adapted for a bedspread, those most frequently used include the following: the sheer, crisp dotted Swiss, marquisette, and organdy; the softer dimity and voile; and the crinkled fabrics, both plissé crepe and seersucker, which make serviceable bedspreads, easy to launder. Many daytime dress fabrics, as printed muslin, gingham, and piqué, as well as poplin with its close firm weave, are employed for bedspreads. The drapery fabrics of chintz and cretonne are often used for both drapery and bedspread.

Wool blankets of various weights and depths of nap.

578

Heavy cotton crash, homespun, sail cloth, and pile weave terry cloth are suitable for certain types of bedrooms. Corduroy, velvet, and velveteen—soft and colorful pile fabrics; plain, printed, and striped satins; and crisp, lustrous taffetas make rather elegant bedspreads.

BLANKET FABRICS

A blanket is used to give warmth. Therefore its construction and finishing should contribute warmth without excess weight. A blanket should be durable and continue to provide warmth after laundering or cleaning.

Warmth. There are no warm fabrics. They can provide warmth. They retain warmth by acting as insulators. Millions of tiny air cells or pockets between the fibers in a fabric retain body warmth and keep the cold air from penetrating. Due to the kinks and turns of a wool fiber and its resiliency, it provides more air pockets and thus is called a "warm fabric." However, tightly twisted worsted yarns can be woven into fabrics that are cool and comfortable to wear in warm weather. Napped cotton and rayon fabrics also provide air pockets which will retain warmth although to a lesser degree than wool.

The yarns. In blanket fabrics the filling yarns are napped. Therefore they should be thick, soft, and lightly twisted. For strength and added wear the filling yarns may have a stout cotton core around which is wound softly spun yarn. The warp yarns must provide strength and should be strong, tightly twisted yarns.

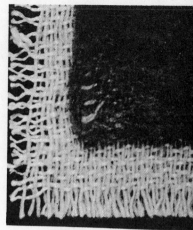

Top to bottom: A wool and cotton plain weave blanket showing the weave before napping and the fabric after napping; a rayon blanket; a cotton blanket woven in a Jacquard pattern.

579

Sheeting. Top to bottom: Muslin, carded percale, combed percale, linen.

The weave. The fabric is woven in plain, basket, twill, or herringbone twill weave, frequently with floating yarns to provide a greater napping surface. It may be woven with Jacquard patterns. Some blanket fabrics are filling backed fabrics, that is, they are woven with one warp and two fillings—a face and a back filling.

The napping. When the fabric leaves the loom it is flat. It is then napped lightly with vegetable teasels or more deeply with metal wires. Blankets are single or double napped. This does not mean that they are napped on one or both sides, as all blankets are napped on both sides. Single napping means that the nap is raised in one continuous direction. The roller in the napping machine revolves in one direction only. Double napping means that the nap is raised in two opposite directions, by certain rollers revolving in opposite direction to others. This more or less interlocks the ends of the napped fibers.

A fabric with one set of filling yarns is napped on both the face and the back of the yarn. In a fabric with two sets of filling yarns each yarn is napped on one side only.

Balance. There must be a balance between the yarns, the weave, and the napping. Since warmth is provided by the napped surface, the greater the napping, the more numerous the air cells. However, there is a saturation point in napping—the point where more napping will affect the wear of the fabric by pulling out and raising too much of the filling yarn.

A loosely woven blanket made with poor yarns can be deeply napped, hiding the weave. When washed, the pile will bunch up and even come out in spots. The weave must be strong, the warp yarns firm, and there must be just the right amount of nap for each kind of filling yarn so that it is not all napped away. Added weight means added warmth only if the fabric is well balanced in yarns, weave, and napping. The napped fibers must not be matted.

Blankets are fulled after napping. They are woven wider than the finished blanket, as fulling shrinks the fabric, giving it a closer hand and texture. They may be yarn or fabric dyed.

SHEET AND PILLOW CASE FABRICS

Sheeting is made of linen and of three types of cotton—muslin, carded percale, and combed percale. Muslin and carded percale sheets are made of carded cottons, while combed percale sheets are made of the long, combed cotton fibers.

The thread count of muslin sheets ranges from 128 to 140 threads per square inch. The government standard is 140 threads per square inch, but if carefully woven of good quality yarns, a 128 thread count will make a durable sheet at a lower price. The thread count for carded percale sheets is 180 threads and for combed percale sheets 180 or more threads per square inch.

Thread count is the total of the filling yarns plus the warp yarns in one square inch. A thread count of 144 may be 68 x 76.

While percale sheets are finer and the combed percale smoother, good quality muslin sheets may give longer wear. Most sheets are white, but there are unbleached, pastel colored, and printed fabrics. Linen makes a cool, smooth, lustrous sheet, durable and long wearing if not of too fine a quality.

COMFORTABLE FABRICS

Comfortables are used to give warmth which is created by the air pockets formed by the mass of loose fibers (usually cotton, wool, or down) that are used for the so-called filling which is quite different from the filling (crosswise yarn) of a fabric. The fabrics are used to cover and hold the filling of the comfortable in place.

The fabrics most generally selected for comfortable covers are those closely woven with a sufficiently high thread count, so that the filling remains intact, the fibers do not protrude through the fabric, and the comfortable gives durable wear.

The cotton fabrics most commonly used are heavy muslin, chintz, and sateen. These fabrics may be plain, colored, or printed. Soft cotton or rayon velveteen and rayon velvet make luxurious comfortables as do lustrous rayon satins and crisp taffetas.

Above: Herringbone twill weave ticking. Below: Twill weave ACA ticking.

MATTRESS TICKING FABRICS

Cotton fabrics for mattress and pillow tickings must be strong, closely and firmly woven of tightly twisted, durable yarns. The colors should have a good degree of fastness. They are woven in a twill weave or with a Jacquard pattern on a twill or satin ground weave and are usually made with woven-in stripes or with a floral design. The stripes may be narrow or wide or a combination of wide and narrow. They may have a Jacquard pattern or pattern and stripes may be combined.

ACA cotton ticking is the conventional type. It is made in 4-, 6-, or 8-ounce weights. This means that in a 32-inch wide material of 8-ounce weight 1 yards weighs 8 ounces. In a 32-inch wide material of 6-ounce weight 1½ yards weigh 6 ounces. In a 32-inch wide material of 4-ounce weight 8 yards weigh one pound.

Ticking fabrics may be given certain finishes that make them germ,-moisture,-and perspiration-resistant.

TOWELING FABRICS

Terry towels are made usually of cotton and have a woven-in, uncut pile. The pile may be on one side, on both sides, or in arrangements, such as all over on one side and in stripes on the

Top to bottom: Wide novelty stripe, Jacquard pattern, novelty stripe. All have satin ground weave.

opposite side. The quality of a terry weave fabric is determined by the firmness and closeness of the ground or underweave, by the quality and closeness of the pile, and by the strength of the selvage and the hems.

To examine the underweave, hold a terry towel to the light. If wide spaces of light come through the interstices or pores between the weave, the towel is probably poorly made. Only pin points of light will show through a well-made towel.

The pile may be single, double, or triple. That means, each loop may be made of one, two, or three yarns woven simultaneously. The absorbency of a towel is in the pile, which is made of the softer, more loosely twisted yarns that take up the moisture, and in the underweave, which receives and disperses the moisture.

Terry towels may be white or colored. If the latter, some are yarn dyed, and some are piece dyed. While a colored towel may be more attractive, it has less absorbency because the hollow-tubed cotton fiber, which absorbs the moisture, is impregnated or coated with dyestuff.

Most terry towels are woven with a plain weave ground. However, some few may be in a twill weave or in a twill weave effect. Terry towels are also woven on a loom with a dobby attachment or on a Jacquard loom to create geometric or more intricate patterns. Athletic, ribbed towels are woven with the pile in thickset cords or ribs, giving more energetic friction in a rubdown.

Many different weaves are used for borders. Some are heavy ribs without a pile. Some are woven with ratiné yarns forming a nubby surface texture; others, with chenille filling yarns producing a soft, velvety surface. A so-called Mitcheline border is woven with heavy yarns on a Jacquard loom. The pattern, accentuated by the heavier yarns, is raised above the ground weave.

When laundered, terry towels shrink· somewhat but become fuller and softer. If dark colored, the dye may bleed on first washing but this does not affect the color of the towel as this is the release of excess dyestuffs on the yarns.

Hand-towel fabrics are usually made of linen or cotton or a combination of the two fibers. However, rayon may be combined with either. Huck towels, woven in a huck weave, form the largest group of hand towels. The Jacquard woven or damask hand towels are frequently made of linen. Other towels range from the finer, sheer cambric guest towels to the coarse, uneven

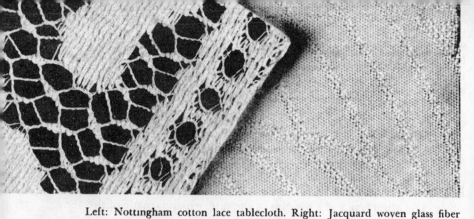

Left: Nottingham cotton lace tablecloth. Right: Jacquard woven glass fiber tablecloth.

textured, plain weave crash towels. Hand towels are usually white but may be dyed or printed.

Dish-towel fabrics are of cotton, linen, combinations of the two, or of combinations of spun rayon with cotton and linen. They are woven in the huck or plain weave. A good quality towel should be firmly but not too tightly woven. The more loosely twisted yarns are necessary for better absorption yet, if cotton, they do cause linting of the fabric.

The success of a dish towel depends on just the right twist to the yarn and the firmness of the weave, so that the moisture will be absorbed and the fabric remain as lintless as possible. Linen of course does not have the tendency to lint. Spun rayon contributes less linting than cotton as well as absorbency. It also adds brilliancy to the colors.

Glass-towel fabrics, made of all linen or linen and cotton, are recognized by their checked, striped, or bordered colored pattern. They are softer, finer, and more closely woven than most dish towels. While the firm, close weave means slower absorption, its smooth lintless surface makes it ideal for shining glassware.

Toweling fabrics are illustrated in Chapters 3 and 4.

TABLECLOTH FABRICS

Tablecloths include a wide range of fabrics. They may be made of linen, cotton, spun rayon, or glass fibers, or of filament rayon yarns. They are either woven or twisted. The weaves include for the most part the Jacquard, satin, plain, and twill, although other weaves are utilized at times.

584

Damasks are woven in the satin weave, with or without a Jacquard pattern. There are two classes, single and double. Both of the fabrics are woven single and the name applies to the type of weave. A single damask has a 5-leaf-satin weave and a double damask an 8-leaf-satin weave. The longer floats of filling yarns over 8 warp yarns produce a more lustrous fabric. According to authorities a double damask should have 180 threads per square inch and at least 50 per cent floating yarns. There may be double damasks with a lower thread count, and a lower quality double damask may not be as good as a fine single damask. However, a true double damask is superior in quality. Most damasks are white or cream colored but some are dyed, usually a pastel color.

There are numerous other types of woven tablecloths. They frequently utilize dyed yarns which may be fine or coarse. The weaves may be close or widely spaced. The fabrics have many different textures and patterns. Fabrics such as gingham, and printed piqué, twill, or crash are adapted also for tablecloths. Damasks as well as other types of tablecloths are often woven in standard sizes.

Machine made lace tablecloths are twisted on Nottingham machines. They range from a fine filet net to a coarse textured fabric with large patterns. One or more complete cloths are twisted at one time.

Left: Nottingham cotton lace tablecloth. Right: Jacquard pattern linen damask. Below: Cotton, plain and basket weave, luncheon cloth.

CURTAINS

Glass-curtain Fabrics

Glass curtains are so named because they are hung over the window glass. They are made of sheer, light weight fabrics, discussed in previous chapters, such as the crisp organdy and dotted Swiss; the softer chiffon, ninon, voile, and scrim; and the leno weave grenadine, marquisette, and madras gauze. Opaque glass curtains are made of such fabrics as pongee, casement cloth, and muslin. They may be plain colored or printed. Pongee and casement cloth are usually ecru in color. Casement cloth is finer and softer than pongee, which has more "body." Pongee is known for its elongated slubbed yarns in both warp and filling.

Lace-curtain Fabrics

Lace curtains may be sheer, closely or openly twisted, or heavier and openly made or so closely twisted that they resemble fabric. Most lace curtains are made on the Nottingham or the bobbinet machine. Nottingham curtains are recognized by their V-shaped

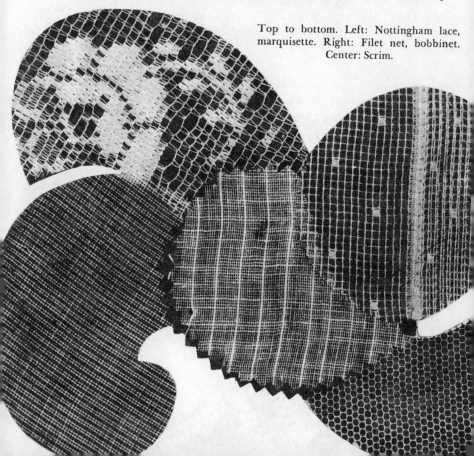

Top to bottom. Left: Nottingham lace, marquisette. Right: Filet net, bobbinet. Center: Scrim.

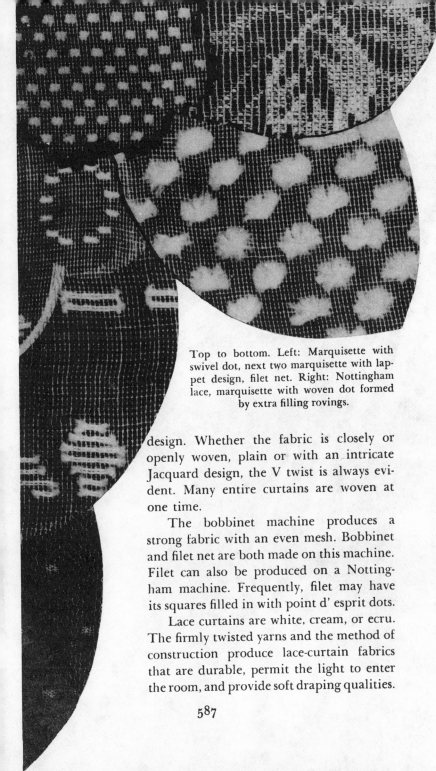

Top to bottom. Left: Marquisette with swivel dot, next two marquisette with lappet design, filet net. Right: Nottingham lace, marquisette with woven dot formed by extra filling rovings.

design. Whether the fabric is closely or openly woven, plain or with an intricate Jacquard design, the V twist is always evident. Many entire curtains are woven at one time.

The bobbinet machine produces a strong fabric with an even mesh. Bobbinet and filet net are both made on this machine. Filet can also be produced on a Nottingham machine. Frequently, filet may have its squares filled in with point d' esprit dots.

Lace curtains are white, cream, or ecru. The firmly twisted yarns and the method of construction produce lace-curtain fabrics that are durable, permit the light to enter the room, and provide soft draping qualities.

587

Above, clockwise: Print
ed muslin, Bedford cord
(c o m m o n l y called
piqué), challis.

Opposite page. Printed drapery and upholstery fabrics. Left to right. Top: Cotton and wool crash,
chintz. Middle: Sailcloth, serge, sateen, bomber
cloth. Bottom: Knitted double fabric, crash.

DRAPERY AND UPHOLSTERY FABRICS

There is one large group of gay, colorful fabrics that are typically drapery
fabrics, but they are also used for up-
holstered furniture and for slipcovers.
Chintz with its brightly printed, usually
floral, pattern is often highly glazed and
may be given a durable glazed finish
that will last the life of the fabric. Cretonne is slightly heavier
than chintz, frequently printed with a large design. A true cretonne has a printed warp and, when woven, gives a soft muted
pattern. Drapery sateen is semiheavy, woven in a lustrous sateen
weave, and often printed. Bomber cloth is a strong, durable fabric,
woven in a broken twill weave with fine warp and heavy filling
yarns. When printed, the pattern is soft in appearance.

Crash, homespun, sailcloth, and duck are all plain weave fabrics of various weights, usually printed. Monk's cloth is a basket
weave fabric, usually four-by-four basket weave. It may be white,
cream, natural, or colored, but is often made with white and tan
plied yarns, giving a mottled appearance. Jaspé is a fabric that
has a streaked color effect achieved by twisting together various
colored yarns.

589

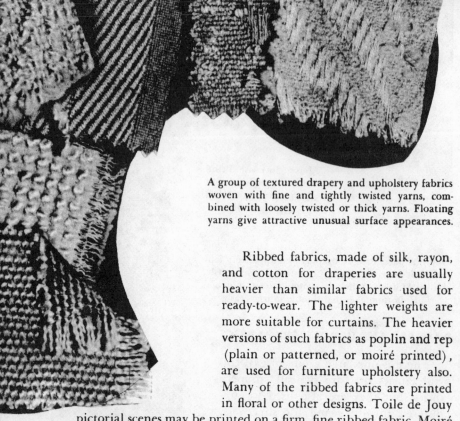

A group of textured drapery and upholstery fabrics woven with fine and tightly twisted yarns, combined with loosely twisted or thick yarns. Floating yarns give attractive unusual surface appearances.

Ribbed fabrics, made of silk, rayon, and cotton for draperies are usually heavier than similar fabrics used for ready-to-wear. The lighter weights are more suitable for curtains. The heavier versions of such fabrics as poplin and rep (plain or patterned, or moiré printed), are used for furniture upholstery also. Many of the ribbed fabrics are printed in floral or other designs. Toile de Jouy pictorial scenes may be printed on a firm, fine ribbed fabric. Moiré is a watered effect usually on a fine ribbed fabric.

The ribs of these fabrics run crosswise except rep, in which the ribs may run lengthwise or crosswise. They are constructed by warp yarns weaving over groups of filling yarns in a so-called warp rib weave even though ribs run across the fabric in the direction of a filling. Ribbed fabrics are distinguished by the fineness or coarseness and the roundness or flatness of the rib. (See definitions in Chapter 21).

Two ribbed fabrics are constructed differently from the others, piqué and Bedford cord. Yarns float on the back to form the ribs. A true piqué has a fillingwise rib, with the ribs running crosswise on the fabric. Bedford cord has a warpwise rib. Warp stuffer yarns are often introduced to raise the cord in these fabrics. Most fabrics called "piqué" are woven with lengthwise ribs.

Other dual-purpose fabrics are the sleek and smooth satins. Some types used for draperies are heavy and quite different in appearance from the satin dress fabrics. Hammered and antiqued satins may be used for upholstery.

Twilled gabardine and serge are also used for draperies. Both are heavier than similar ready-to-wear fabrics and are usually made of cotton. The wool fabrics have a left- to right-hand twill but the cotton drapery fabrics may have a right- to left-hand twill. They are often printed.

Many nubbed, slubbed, and other irregular yarns are used to make rough textured drapery fabrics. Ratiné yarns have slubs or bunches of yarns and, when woven into a ratiné fabric, produce a nubby surface texture. Bouclé yarns have tiny loops or knobs that produce a fabric with loops or knobs over the surface. Classed in this group are fabrics with nubby appearance that are woven with a very thick, loose yarn in the filling and a very fine, tightly twisted yarn in the warp. These fine, tight yarns tie in the thick yarns, producing a raised knobbed appearance.

Jacquard patterns, many most elaborate, are used for both drapery and upholstery fabrics. They are made of heavier, often coarse yarns. Damask is noted for its flat, reversible pattern. Brocade has a slightly raised surface pattern. Brocatelle has a high, rather padded patterned surface. Matelassé, a double fab-

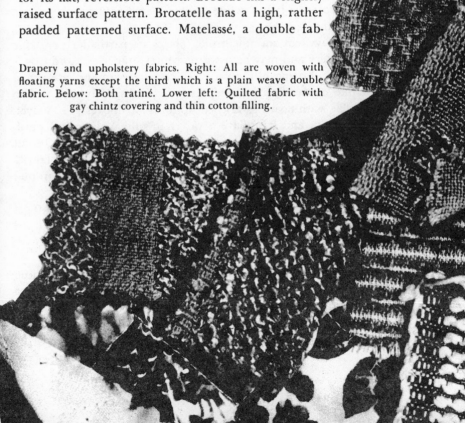

Drapery and upholstery fabrics. Right: All are woven with floating yarns except the third which is a plain weave double fabric. Below: Both ratiné. Lower left: Quilted fabric with gay chintz covering and thin cotton filling.

Above, left: Taffeta woven with dyed yarns. Right: Moiré print on a ribbed fabric.

Opposite page, top to bottom. Left: Casement cloth, printed faille, rep. Right: Faille, woven with dyed yarns, satin, poplin printed with Toile de Jouy pattern.

ric, has a quilted or puckered effect. Tapestry is multicolored with the pattern usually in pictorial design.

Quilting belongs in this fabric group. A quilted velvet makes a rich, decorative fabric, while quilted glazed chintz makes a gay, fresh looking fabric.

The heavy, cut and uncut pile fabrics are particularly suitable for upholstery and may be used for winter draperies. Corduroy is woven with floating filling yarns which, after weaving, are cut to form pile cords of various depths and widths. Velveteen is also woven flat, with floating yarns cut to form an all-over pile. Velvet, velour, plush, and frisé have a woven-in pile. Velour and plush have a cut pile, frisé usually an uncut pile, and velvet may be cut or uncut or a combination of the two. The pile of these fabrics may be wool, cotton, or rayon, but they usually have a firm ground weave of strong cotton yarns.

Other fabrics have been adapted to draperies for certain types of rooms. These include such fabrics as yarn dyed gingham, printed percale, soft, printed challis, crinkled seersucker, and such heavy twill weave, yarn dyed fabrics as denim and tickings.

In addition, there is a wide variety of novelty fabrics produced by combinations of weaves, such as a satin and twill weave or a ribbed and twill weave combination. However, a greater number

Above left: Brocatelle. Right: Brocade.

Opposite page, top to bottom. Left: Plush, imitation matelassé, tapestry. Right: Frisé, damask, cut and uncut mohair pile in Jacquard pattern.

of the fabrics that look new are made by standard weaves but with interesting yarn combinations, such as soft, loose, and thick yarns with tightly twisted yarns; lustrous filament rayon yarns with cotton yarns; or uneven yarns with smooth, sleek yarns.

WINDOW SHADE FABRICS

Window shades are made of coated fabrics. The quality of a shade cloth depends on two factors: the quality of the fabric and the quality of the finish and its method of application. A close, firmly woven cloth that does not require excess sizing to fill the interstices will give better service than a loosely woven fabric closed up and stiffened with sizing. Muslin and the finer cambric are the fabrics generally used.

The finishes, or coatings, are of different types, depending on the quality of the shade and whether it is to be made opaque or translucent. Some finishes are applied by machine while others are hand painted. A pyroxylin shade cloth has a pyroxylin coating, which is a cellulose compound, applied to an unfilled cambric fabric. This makes a fine quality window shade. There are other machine-applied finishes on both muslin and cambric.

Linseed oil, from flax seed, is used for hand-painted shades. A light coating renders the fabric translucent, a heavy coating makes it more opaque.

594

RUG AND CARPET FABRICS

Rugs and carpets are classified by their fibers and their weaves as well as by their types. Broadly, their fiber classifications are wool, rayon, cotton, linen, kraft fiber, sisal, and combinations of these fibers. Within each fiber classification rugs are made of various weaves, which are explained in the chapter on weaves as are the looms on which they are woven.

Wool rugs and carpets. Wool rugs have a wool pile surface. They are further classified as velvet (cut pile) and tapestry (uncut pile), woven on a plain loom in a velvet or woven-in pile weave; Wilton (cut pile) and Brussels (uncut pile), woven on a Jacquard loom in a Wilton or woven-in pile weave; Axminster (cut pile), woven on an Axminster loom in a woven-in pile weave; chenille (cut pile), woven on a chenille loom with chenille filling yarns; and handmade orientals.

Wilton and Brussels. Both are woven exactly the same way, the only difference being that Wilton has a cut pile and Brussels has an uncut pile surface. The ground warp is usually cotton but may include jute. There are cotton warp-stuffer yarns. The filling is heavy cotton or jute. The pile warp yarn may be woolen or worsted. The colors and patterns are woven in on a Jacquard loom. The wool pile yarns, in their various colors, are on spools in frames back of the loom. They are woven over wires, the same as for some velvet fabrics.

The rugs have a short, thick pile. Worsted yarns give a stiffer, more upright pile than do the softer woolen yarns. They also give more rows of pile per inch than do woolen yarns.

Left to right: Brussels, Wilton, chenille.

Velvet and tapestry. Both are woven exactly the same way, the only difference being that velvet has a cut pile surface and tapestry has uncut loops. These fabrics may be plain colored or with patterned effects. If a design, the warp pile wool yarns are first printed (see drum printing in Chapter 18) and then wound on large beams and used for weaving in the pile.

Velvet and tapestry rugs are made the same as any velvet fabric that has an extra warp woven over a wire to form loops. The wire may be withdrawn to form uncut loops or a knife on the end of the wire cuts the loops as it is withdrawn.

The ground warp yarns are heavy cotton and there may be cotton or jute warp-stuffer yarns. The filling yarns are heavy cotton or jute. The warp pile is wool. The pile is woven in on every other pick. On the back can be seen the warp and the filling yarns that do not carry the pile. The pile and the alternate filling yarns are on the face of the fabric. Therefore all of the wool is on the surface.

The patterned types have a short, dense pile, while the plain-colored types have a dense pile of varied height.

Axminster. These rugs and carpets are noted for their many possible colors although they may be simple in design and color. Because of the way they are woven, each tuft of pile could be a different color. Axminster is the only rug that cannot be rolled crosswise because of its stiff jute yarns.

There are two sets of cotton warp yarns, one that weaves with the filling and the other that is used for stuffer yarns. The filling is heavy jute and two yarns weave at one time. The pile yarns are wool. They weave under every other double set of jute fill-

Left to right: Axminster, velvet, tapestry.

ing yarn and are bound in by the weave. Each tuft of pile is cut.

The more tufts per inch, the better the quality of the rug, all other things being equal. They average from five to seven tufts per inch and more for better quality rugs. On the back can be plainly seen the finer, ground warp yarns and the coarse, double jute filling yarns that do not carry the pile. Between these filling yarns, slightly indented are the alternate filling yarns, covered with the underpart of the pile tuft. By holding a sample vertically, the looping of the pile around the filling can be noted.

The pile may be woolen or worsted yarn and is soft and deep. The patterns are frequently Oriental in design.

Chenille. This rug fabric is woven with chenille filling yarns (see Chapter 10), which may be woolen or worsted. A chenille blanket is first woven and then cut into strips to form the chenille yarn.

There are three sets of cotton warp yarns: the ground warp, the catcher warp that weaves in and holds the chenille yarn, and the stuffer yarns. The filling ground yarns are heavy, coarse wool or hair. On the back of the carpet can be seen the heavy, ground filling yarns, the ground warp yarn, and the catcher yarn. If the chenille yarns are ripped off the rug, a completely woven fabric remains.

The back is thick, soft, and woolly. The face has a dense, close, deep pile. When the fabric is folded, there is no line of space between the pile as in the other weaves. The pile is locked in during the weaving of the chenille blanket.

Rayon and rayon blended with wool are sometimes employed for the face pile. The rayon staple fibers are used alone or blended with wool fibers prior to spinning the yarns.

OTHER TYPES

There are many other types of carpets and rugs in addition to those discussed above. They are made of wool, cotton, linen, kraft fiber, and sisal. Some few are made of jute, grass, cocoa fiber, and palm leaves. Many are combinations of these materials such as kraft fiber combined with cotton, wool, sisal, or jute; or wool combined with linen; or sisal and jute.

They may be broadloom or from room size down to small scatter rugs, including bathroom rugs. They may be in bright or subdued colors or in pale pastels. Many are yarn dyed, others are printed. Some types are reversible and others are not, according to the way they are made.

Cotton rugs.

They may be woven, tufted, hooked, braided, knitted, or felted. Some are smooth-textured, some have a pile and others are carved. There are different ways in which the pile is attained. It may be stitched on, tufted and cut (sometimes with carved designs), knitted in, or woven in with chenille yarns.

The range of types and styles is so wide and varied that it is not possible to mention here more than the major types. Previous chapters discussed the fibers, yarns, weaves, knits, as well as the dyeing and printing processes. The explanations apply to these fabrics as well as to any other fabric.

Cotton Rugs

Cotton rugs are made from strong ply yarns or from thick rovings. The kinds of yarn and the way they are twisted contribute greatly to the final texture of the rug.

The rugs may be woven flat in a plain, twill, or Jacquard weave, or in combinations of these weaves, with thick, soft, dyed yarns.

Chenille rugs are woven with chenille filling yarns cut from a previously woven chenille blanket (see Chapter 10). The yarn may be cut off center to produce uneven pile yarns, which gives the finished rug a shaggy appearance. The rugs are woven on a plain or Jacquard loom. Usually there is a stuffer as well as a ground warp. The reversible chenille rugs are woven with two filling yarns, one that weaves through the shed to form the face pile, and the second that weaves through the shed on the next pick to form the

599

back pile. Chenille rugs may be finished to give an embossed effect.

Tufted cotton rugs are made by punching yarns into duck or canvas fabrics. Frequently, two or more fabrics are stitched together to add strength and weight. A paper on which the design has been perforated is laid over the fabric, and a colored powder or washable dye liquid is brushed over the perforations, tracing the design outline on the cloth. Then, with the use of punch work machines, the yarns are pushed through from the back, to form a pile on the face of the fabric.

Tufted rugs are also woven on Jacquard looms, which weave the ground weave and the tufts at the same time. This permits the introduction of color and design at the time of weaving.

Many textures may be achieved by tufting. Thicker or thinner yarns may be used, the pile may be short or long, the loops may be left uncut or may be cut to form a thick pile surface. The cut pile may be sheared later to form designs by variations in the height of the pile.

The raw stock or yarns are dyed previous to tufting. Frequently these rugs are incorrectly called "chenille rugs."

Knitted cotton rugs are made with dyed and heavy twisted yarns. The knitting machine is set to gauge the height of the pile which is knitted into the fabric. After knitting the rug passes through a finishing process that sets the twist and gives a firm, close body to the fabric.

Stitched-on pile is made by stitching onto a heavy fabric, such as duck, a thick fringe of ply cotton yarn. The rows of pile are about ¾ inch apart. The pile may be high or low, and the loops cut or uncut.

Braided cotton rugs are made by cutting fabrics into strips, braiding them by machine, and stitching them together. Another method is to braid heavy cotton yarns into wide and narrow strips, then stitching a wide and a narrow strip together with a crisscross stitch. With dyed yarns it is possible to achieve many color effects, such as stripes, checks, and plaids.

Rugs similar in appearance to braided rugs are produced by twisting strips of fabric and stitching them together with a crisscross stitch.

Hooked rugs are made by inserting yarn or narrow strips of fabric through a heavy jute, linen, or cotton fabric, such as canvas, burlap, or monk's cloth. The design is drawn on the back of the

Linen rugs.

fabric, and the dyed strips of cotton or wool yarn are pushed through to the face according to the pattern. They may be left as uncut loops or may be cut to form a cut pile design.

Texture surfaces are produced in a number of different ways. The rug may be given a textured effect, regardless of the fiber used or of the machine on which it is made. It may be achieved by hard twisted yarns; by combination of high and low pile; by combination of cut and uncut pile; by multitone patterns; and by hand carving. In addition, the hand-carved effect may be produced on a Wilton loom by drawing into the back of the rug certain pile yarns. The absence of these pile yarns on the surface forms the carved effect.

Linen Rugs

These rugs are made from flax grown in the United States (see the chapter on linen). The flax tow is dyed, carded, and spun into heavy yarns. The rugs are woven on plain or Jacquard looms. They have a rough texture with a typical linen luster.

Linen is also used as the backing for wool-tufted rugs.

Kraft Fiber Rugs

These fibers are made from fir or spruce wood. The logs are delivered to paper mills, where they are cut into chips, boiled

Fiber rugs.

in chemicals to remove the resins and to produce the pulp, and are then rinsed to remove the chemicals. The pulp is dyed and passed through a machine which, by rapid vibration, causes the fibers to cohere, forming long, tough sheets, known as "kraft fiber." The sheets are rolled into huge rolls and are delivered to the rug manufacturing plants. There, they are cut into strips of various width, and are tightly twisted on twisting machines into heavy yarns.

The rugs are woven in the plain, twill, and Jacquard weaves, and in variations and combinations of these weaves. They may be all kraft fiber or kraft fiber combined with cotton, sisal, or wool. The rugs are reversible. One side may be printed with a colorful stenciled design. In the trade these rugs are termed "fiber rugs."

Sisal Rugs

Rugs may be woven in the plain, twill, or Jacquard weave, or in variation or combinations of these weaves. They also are reversible, and may have one side stencil printed.

Felted Rugs

There are several types of felted rugs. For example, sisal fibers may be blended with jute fibers and felted. The rug is dyed and then printed, producing two-tone or multitone effect.

Cotton or wool fibers may be felted into rugs, then dyed or printed, or a design may be stitched on the felted fabric.

Another method of so-called felting is to apply fibers, such as jute, or jute and rayon, hair, or others to a cotton or burlap back. A bed of needles in the machine punches the fibers into the cloth until it is covered with a thick mass of fibers. The rug is then dyed and printed.

Rug Cushions. Carpet or rug cushions are made of hair, cotton, jute, rubber, or paper. New cattle hair makes the most resilient and the best cushion for most purposes. If fibers are used, the felting is usually on a fabric base. In Chapter 14 is a description of the method of felting a cattle-hair rug cushion. Some rug cushions are treated to give them added qualities, such as deodorized and made resistant to moths.

Grass Rugs

The grass used for rugs grows wild in the United States on the prairies of Minnesota and Wisconsin, and near Winnipeg in Canada. It is a tough, wirelike grass that grows about 2 feet high. After the grass dries it is cut, cured, bound into long continuous strands, and woven into rugs in a plain weave.

Grass rugs are attractive in the natural grass color or printed on one side by stenciling or by spraying. Also, different colored warp yarns may be introduced. These rugs are reversible. Most grass rugs are varnished for protection.

Sisal rugs.

SUMMARY

Decorated: Clipped and unclipped tufting. Candlewick or hobnail.
  Clipped and unclipped punchwork.
Woven.
Cut and sewn.

COMFORTABLES

BLANKETS

Wool, cotton, and rayon.

SHEETS

Linen.
Cotton: Carded percale, combed percale, and muslin.

MATTRESS TICKING

TOWELS

Terry, huck, damask, and dish.

CURTAIN, DRAPERY, AND UPHOLSTERY FABRICS

LACE CURTAINS

WINDOW SHADES

RUGS AND CARPETS

Wool: Wilton and Brussels.
       Velvet and tapestry.
       Axminster.
       Chenille.
Cotton.
Linen.
Kraft fiber.
Sisal.

RUG CUSHIONS

Opposite page: A drapery and upholstery fabric combining cotton and lustrous
rayon (Goodall Fabrics, Inc.).

FABRICS DEFINED

Each suitable for its ultimate use

Most of the fabrics, discussed in this chapter, are manufactured at this time. A few of the formerly made fabrics are included although other similar fabrics, or the same fabrics with other names, have replaced these original fabrics.

Each fabric is presented as having distinctive characteristics. The fibers, with which it is constructed, are listed. As stated in the first chapter, the assumption has been taken that no fabric is exclusively owned by any fiber. A fabric with a definite construction, finish, and similar appearance may be made with any one of a number of fibers or with a combination of two or more fibers. For this reason all fabrics should be labeled with the percentage of fiber content.

On the other hand, there are fabrics that are made of different fibers and called the same name and yet are totally dissimilar. These include such fabrics as wool broadcloth and cotton broadcloth, wool sharkskin and rayon sharkskin, and wool Cheviot and cotton Cheviot.

No distinction has been made between woolen and worsted fabrics and all have been classified as wool. Likewise, rayon is listed rather than filament rayon and spun rayon. These distinctions have been discussed in the wool and rayon chapters.

The specialty fibers are not listed as fabrics because these fibers may be used to construct various kinds of fabrics. While the synthetic fibers, other than rayon, are beginning to be made into many types of fabrics, only a few are included here because the ultimate, successful use of these fibers for certain fabrics are not fully determined at this time.

As a general rule, no merchandise terms as coatings, suitings, shirtings, diaper cloth, ski or snow cloth, towelings, or linings are included because these items determine use and may be made of various fabrics. There are a few exceptions such as balloon fabric, airplane cloth, and parachute fabric to which other names have not been given. Other exceptions include awning duck which is a particular kind of colored duck, glass toweling which is a distinctive type of fabric, mackinaw cloth which is a colorful plaid wool fabric, and ticking which is made in various weaves.

Sheeted plastics that in usage (such as for shower curtains) are taking the place of formerly woven fabrics have been omitted, except for the one fabric shown below. The advancement made by such synthetic plastics undoubtedly will show future development for an even wider use of such materials in merchandise formerly made of woven fabrics.

Some fabrics are distinguished by their yarns, many others are known by their type of construction. Some have the same construction but have been given various names because of their different finishes.

Today's fabrics are made with the fiber or fibers that are best suited to the construction, finish (including coloring), and ultimate use of the fabric. Each fabric has sufficient distinguishing characteristics to be listed as a definite type of fabric, manufactured for specific purposes. Each one should be appreciated for its contribution to our American fabrics.

Beginning on the next page, each fabric is briefly defined. They are listed alphabetically for easy reference. Most of the illustrations show a small portion of the back as well as the face of the fabric, because at times a fabric can be more readily distinguished by noting its back construction.

A plastic made from a vinyl resin and sheeted into a thin draping material. It is printed with dyestuffs that become an integral part of the plastic, making a lustrous, colorful, cleanable material.

AIRPLANE CLOTH

PLAIN WEAVE

Cotton, linen

Linen airplane cloth, used for parts of airplanes, is made from the best quality linen yarns. Cotton airplane cloth is a close textured, firmly woven fabric made of long, combed fibers that are twisted into fine, yet strong yarns. The fabric for airplanes is further treated with a coating. Untreated cotton airplane fabric may be bleached and dyed and used for apparel. The fine yarns, close, even weave, and the mercerized finish produce a durable fabric. *Uses:* The treated for airplanes, the untreated cotton fabric for men's shirtings, sportswear, boys' suits, luggage, etc.

ALBATROSS

PLAIN WEAVE

Wool, cotton

A fine, light weight fabric woven in an open plain weave. It has a pebbly surface obtained by the slight spiral twisting or creping of the yarns. It is piece dyed, often in pale shades. Cotton albatross has a slightly napped face. Albatross is sufficiently soft for infants' wear. *Uses:* Negligees, infants' wear, dresses for nuns, etc.

ARGENTINE CLOTH

See TARLETON

ART LINEN

PLAIN WEAVE

Linen

A fabric made with firm cylindrical yarns that can be easily "drawn" for needlework. It is generally white, ecru, or tan, but may be dyed in a wide range of colors. *Uses:* Table covers and center pieces, etc.

ASTRAKHAN CLOTH

PILE OR VELVETEEN WEAVE OR
KNITTED

Wool, mohair, cotton, rayon

A thick, spongy fabric with a surface of softly curled pile yarns. It is the curled yarns more than the method of construction that distin-

608

guishes astrakhan. It was first made from the strong, curly wool obtained from sheep raised in Astrakhan, Persia. Today's version imitates this fabric. The yarns are curled before weaving. Since mohair can be curled effectively, this fiber makes most attractive astrakhan. There are two methods of woven construction. One is by the velveteen weave using the curled yarns for the filling and floating them over the warp yarns. After weaving, the floating filling yarns are cut, and they curl over the surface of the fabric. They may be left uncut and, with the shrinkage of the ground yarns, the floating filling yarns are thrown up as loops. The other method is by the velvet weave. The curled yarns are used as the pile warp, which is woven over a wire. As the wire is withdrawn the loops may be left uncut or the wire may have a knife that cuts the loops, producing a curled cut pile surface. A knitted fabric constructed with previously curled yarns is also called astrakhan. If a fabric is woven, heavily napped, and then finished to give a curled effect, it is not astrakhan, but is probably chinchilla cloth. *Uses:* Coats and trimmings.

AWNING DUCK

PLAIN WEAVE

Cotton, linen

A heavy, closely woven duck, often called awning stripe. The stripe is usually woven from dyed yarns, but it may be printed or painted on one side only. Both warp and filling yarns are heavy, usually plied yarns. Heavy duck is used for better grades and drill or a lighter weight duck for less expensive fabrics. For maximum service this fabric may be given finishes that make it resistant to fire, water, and mildew. *Uses:* Awnings, beach umbrellas, hammocks, summer furniture covers, etc.

BAGHEERA

See VELVET

BALBRIGGAN

PLAIN KNIT

Cotton

A fine closely knitted fabric, usually in tubular form, made with one continuous yarn. It has a right and wrong side and much elasticity. It will "run" if a stitch is broken. *Uses:* Underwear, hose, sweaters. gloves, caps, etc.

BALLOON FABRIC
PLAIN OR BASKET WEAVE
Cotton, silk, linen

A very finely yet strongly woven fabric. The fabric is treated with a thin layer of rubber to make it waterproof. Cotton balloon cloth is made with the finest of combed yarns, usually with two or more yarns plied together. The fabric may be woven in a plain weave or in a variation of the basket weave. Cotton balloon cloth, used for wearing apparel, is often dyed. It makes a smooth, long-wearing, fine fabric. *Uses:* Treated, for balloons; untreated cotton, for shirts, dresses, etc.

BARATHEA
RIB OR TWILL WEAVE
Wool, silk, rayon, cotton

A fabric with a pebbly surface texture. Originally barathea was a registered trade name for a silk tie fabric. It is now a name applied to fabrics woven with a broken effect weave and often small, fancy, designs usually in two colors. It may be a rib or twill weave in which short broken ribs alternate to produce a coarse granulated effect. Wool barathea is made of worsted yarns and is well fulled. *Uses:* Neckties, women's dresses and suits, trimmings on men's dress suits, etc.

BATISTE
PLAIN OR JACQUARD WEAVE
Wool, cotton, rayon, silk, linen

Batiste is a sheer fabric made with any of the principle fibers spun into fine yarns. Wool batiste is a fine, smooth fabric similar to nun's veiling. Cotton batiste receives its smoothness from the long combed fibers spun into tightly twisted yarns. The fabrics are singed and the yarns or fabrics mercerized to give a lustrous appearance. Silk or rayon batistes have the smooth appearance of silk mull. Linen bastiste is a finely woven fabric, softer than most linens. A heavier batiste used for foundation garments is woven of cotton or cotton and rayon in a firm plain weave, at times with a Jacquard pattern. Batiste is bleached white or dyed a plain color, and at times embroidered. *Uses:* Cotton batiste for infants' wear, lingerie, handkerchiefs, neckwear, linings, dresses, etc. Made in other fibers, for dresses, negligees, etc. The heavier for foundation garments.

BEAVER CLOTH
See IMITATION-BEAVER CLOTH

BEDFORD CORD

BEDFORD CORD WEAVE

Wool, cotton, rayon, silk

This is the name of the weave as well as of the fabric. The pronounced lengthwise ribs are produced by interweaving the filling, in a plain or twill weave, over alternate groups of warp yarns. Stuffing warp yarn may be introduced to raise the cords. (Further explanation of the weave is on page 339.) Cotton Bedford cord is sometimes napped on the back. It is piece dyed, or if cotton, it may be printed in stripes or checks. Today the finer constructions are called piqué. *Uses:* Suits, topcoats, slacks, riding and hunting clothes, uniforms; infants' coats, etc. Heavier qualities, for draperies, upholstery, etc.

BENGALINE

RIB WEAVE

Silk, rayon, wool, cotton

A firm closely woven fabric with ribs spaced at intervals running across the fabric. The ribs are fine and slightly heavier than poplin ribs. It is woven with a warp rib weave formed by groups of filling yarns weaving as one to form the rib. There are many more warp than filling yarns. Cotton bengaline is made usually of combed yarns and mercerized. Rayon bengaline may be woven of filament or spun rayon. *Uses:* Silk, rayon, and wool, for women's dresses, suits, coats, draperies, etc. Cotton for women's dresses, men's shirts, draperies, etc.

BIRD'S-EYE

BIRD'S-EYE WEAVE

Linen, cotton, rayon

While many fabrics are called bird's-eye because they are woven to give the same effect, a true bird's-eye is woven with an appearance of a small diamond shaped design having a dot in the center. The yarns are usually loosely twisted, particularly the filling yarns. The type of loose construction, the pattern, and the kind of yarns make a soft, absorbent fabric. Bird's-eye diaper cloth is woven in the same manner, with soft, loosely twisted yarns. A worsted fabric, incorrectly called bird's-eye, is woven in a twill weave with alternate light and dark groups of warp yarn and alternate dark and light groups of filling yarns. In weaving this produces a small spotted effect. *Uses:* Diapers, reversible towels, etc.

BOBBINET

See NET

BOLIVIA

SATIN WEAVE

Wool

A soft, velvety or plushlike napped fabric. It is frequently made with the addition of some specialty fiber such as alpaca or mohair. The napped tufts are in diagonal or vertical rows rather than in an over-all pile effect. It is generally piece dyed. *Uses:* Coats, suits.

BOMBER CLOTH

BROKEN TWILL WEAVE

Cotton

A firmly woven, durable fabric with about twice as many warp as filling yarns. The warp yarns are fine and, for greater strength, weave in groups of two. The filling yarns are heavy. The fabric is woven in a four-harness, filling-faced, broken twill weave. It is usually printed. *Uses:* Draperies, etc.

BOUCLÉ

PLAIN OR TWILL WEAVE, OR KNITTED

Wool, rayon, cotton, silk, linen

Distinguished by small spaced loops on the surface of the fabric. It is a flat, irregular surfaced fabric, woven or knitted from specially twisted bouclé yarns that have small loops. An explanation of how these yarns are made is given on page 291. Bouclé has loops occurring at intervals and thus can be distinguished from astrakhan, which has a dense curled pile covering the entire surface of the fabric. Bouclé is made in textures from the light, cotton dress weights to the heavy, wool coating weights. *Uses:* Coats, suits, dresses, sportswear, etc.; draperies, etc.

BRILLIANTINE

PLAIN OR TWILL WEAVE

Silk, worsted, mohair, rayon, cotton

The most commonly known brilliantine has a cotton warp and a worsted or mohair filling. It may be given a soft calender finish or a stiff finish. Silk brilliantine is a light, loosely woven fabric made of unthrown silk in both warp and filling, and therefore the yarns slip easily. Cotton brilliantine is made of a fine warp yarn and a heavier, slightly twisted filling yarn. The fabric is given a lustrous appearance by mercerization. Glacé brilliantine has a colored cotton warp and an undyed mohair filling. *Uses:* Dresses, suits, linings, etc.

612

Wool

Cotton

BROADCLOTH

TWILL, PLAIN, OR RIB WEAVE

Wool, cotton, rayon, silk

Two entirely dissimilar fabrics, made with different fibers, weaves, and finishes: the wool and spun rayon broadcloths are woven with twill weaves; cotton, rayon, and silk broadcloths are woven with plain or rib weaves. *Wool broadcloth,* or wool mixed with spun rayon, is a fine, rather openly woven twill weave fabric, that is fulled to give it a close, uniform texture. It may be woven with woolen or worsted yarns. The fabric is napped, sheared, dampened, and the pile permanently laid down in one direction. This gives wool broadcloth a smooth, lustrous, fine, velvetlike surface texture. *Cotton or spun rayon broadcloth* has fine crosswise ribs, produced by the warp rib weave. The ribs are formed by a number of filling yarns weaving as one to form the ribs. It has the finest ribs of all the ribbed fabrics. It may be made of all carded, all combed, or combination of carded and combed yarns. The best qualities are woven with long, combed cotton woven in a high count and mercerized to produce a softly lustrous fabric. Cotton broadcloth may be bleached, dyed, or printed. *Silk broadcloth* and *filament rayon broadcloth* are woven in a plain weave, but with fine crosswise ribs obtained by heavier filling than warp yarns. Because of the ribs, the closeness of the weave, and the high twist of the silk yarns, silk broadcloth is thick and rather stiff. *Uses:* Wool and silk, for women's dresses, suits, coats. Cotton and rayon, for men's summer evening jackets, shirts, shorts, pajamas; women's dresses, hostess gowns, pajamas; children's dresses, pajamas; bedspreads, draperies, tablecloths; linings; etc.

BROCADE

JACQUARD WEAVE

Silk, rayon, cotton

A rich appearing fabric with a Jacquard woven pattern in low relief, that is, slightly raised. This distinguishes it from damask, which has a flat, reversible pattern, and from brocatelle, which has a pattern in high relief. The brocade pattern, formed by floating filling yarns, may be in one or several colors. Cotton brocade usually has the ground of cotton and the pattern of rayon or silk. Brocade was originally a heavy silk fabric, woven with a gold and silver pattern. It derived its name from the Spanish word "brocade" meaning "to figure." *Uses:* Draperies, upholstery, etc.; rayon and silk, for blouses, dresses, wraps, men's neckwear, etc.; cotton or mixtures for corsets.

BROCATELLE
JACQUARD WEAVE
Silk, rayon, linen, cotton

A tightly woven and stiff, elaborate fabric, with a Jacquard woven pattern in high relief. This distinguishes it from brocade, which has a pattern in low relief. The pattern is formed by the warp yarns. The areas not raised are backed with extra yarns. *Uses:* Upholstery, draperies, etc.

BUCKRAM
PLAIN WEAVE
Cotton, linen

The most commonly known buckram is heavily sized and stiffened. Two fabrics are glued together. One is a low-count, open weave fabric, the other is much finer. Buckram is also the name of a single, strong linen fabric, stiffened with flour paste, china clay, and glue. A stiff finished scrim is also known as buckram. *Uses:* Interlinings and stiffening in clothing. Also used in bookbinding and millinery trades, as it can be moistened and shaped.

BUNTING
PLAIN WEAVE
Cotton, wool

A loosely woven, thin fabric, usually dyed in bright shades as required for flags and banners. Wool bunting is made of tightly twisted worsted yarns and is dyed in fast colors. Cotton bunting is not as durable, and frequently the colors have a low degree of fastness. *Uses:* Flags and banners, better worsted buntings for curtains, etc.

BURLAP
PLAIN WEAVE
Cotton, jute, hemp

A very coarse, heavy fabric. It may be loosely or more closely woven, as burlap covers a wide range of weights. A burlap sack is often called a gunny sack. *Uses:* Sacks and other wrappings, backs of floor coverings, furniture covering, etc.

BYRD CLOTH

TWILL WEAVE

Cotton

A very closely and firmly woven fabric made from fine combed yarns, mercerized and dyed. Its tight weave makes it wind resistant but usually it is also given wind-resistant and water-repellent finishes. *Uses:* Aviation and ski suits, sport jackets, raincoats, etc.

CALICO

PLAIN WEAVE

Cotton

A light weight, rather coarse fabric, woven with carded yarns. Calico has a lower thread count than does percale, another printed fabric. The colors are printed on one side, either plain or patterned. The fabric is sized for crispness but washes out, requiring starching when it is laundered. So-called calico prints may be applied to muslin, cambric, chintz, and other fabrics. *Uses:* There is practically no calico today.

CAMBRIC

PLAIN WEAVE

Linen, cotton

A fine, firm, close weave fabric. A true linen cambric is very sheer, but today there are coarser fabrics that are also called cambric. Cotton cambric is a light, finely woven fabric. It may be given a soft finish with little luster and calendered on one side only. Or it may be heavily starched and calendered, obtaining a lustrous, smooth surface glaze, which is, however, removed when laundered. This latter type of cambric is frequently dyed. The soft finished fabrics are bleached white or may be printed. Cambric originated as a linen fabric in Cambrai, France, from which it received its name. *Uses:* Handkerchiefs, lingerie, aprons, shirts, dresses, table linens, etc.; glazed cambric for linings, curtains, fancy costumes, etc.

CANVAS

PLAIN WEAVE

Cotton, linen

A heavy, closely woven, firm fabric that is rather stiff. Canvas is really a heavy weight duck. There are many fabrics classified as canvas.

Perhaps the best known is the type that is made from coarse, hard twisted yarns. When woven in stripes with colored yarns or printed with stripes, it is called awning stripe canvas. A sail canvas is a very heavy fabric woven with a two-ply linen warp and a coarse cotton filling. A canvas used for needlework is made with two-, three-, or four-ply yarns, strongly woven in a more or less open plain weave, which produces a stiff, open-meshed fabric. A lighter version, woven with linen warp and cotton filling and dyed, is called a chess canvas. *Uses:* According to the weight and type, for awnings, sails, shoes, art needlework of many kinds, furniture covers, linings, etc.

CASEMENT CLOTH

PLAIN WEAVE

Cotton, rayon, silk, mohair

A light weight, rather closely woven, opaque fabric. While dyed other colors, it is usually white, cream, or tan in color. It may have leno woven designs. Better quality cotton casement cloth is mercerized and has a lustrous surface. The filling yarns are heavier than the warp yarns. *Uses:* Principally for curtains.

CASSIMERE

TWILL WEAVE

Wool

A closely woven fabric with a two-by-two twill. It is well fulled and closely sheared, giving the fabric a close, smooth and somewhat lustrous surface texture. It is frequently made of all worsted but may have a worsted warp and woolen filling. When made of all worsted it has a somewhat hard feel. It holds a crease well. *Uses:* Men's suits.

CAVALRY TWILL

TWILL WEAVE

Wool

A firmly woven fabric with a pronounced, left- to right-hand twill. It has a double twill effect. Compared to elastique, cavalry twill has heavier cords which are more widely spaced, and the twill can be noted on the back of the fabric. It is made of tightly twisted yarns and finished to give a clear, hard surface. *Uses:* Military suits and coats, civilian slacks, sportswear, etc.

CHALLIS

PLAIN OR TWILL WEAVE

Wool, rayon, cotton

An extremely soft, light weight, well-draping fabric, firmly but not too closely woven with fine yarns. Usually it is printed with a small floral design, but may be bleached or dyed a plain color. Challis is woven in a high thread count with fine twisted yarns. Originally, challis was made with silk warp and worsted filling. *Uses:* Women's dresses, blouses, negligees, pajamas, children's dresses; men's ties, shirts; comfort covers, draperies, linings, etc.

CHAMBRAY

PLAIN WEAVE

Cotton, linen

Distinguished by a white frosted appearance due to being woven with yarn dyed warp yarns and white filling yarns. Chambray is a smooth, closely woven fabric with a soft luster. It belongs to the same family as gingham and madras. Some chambrays are made with a corded stripe and some are embroidered. There is also a heavier version, made with coarser yarns into a somewhat spongier fabric. *Uses:* Finer for children's dresses, women's dresses, blouses, smocks, aprons, men's shirts, sportswear; heavier for work shirts, linings, mattress ticking, etc.

CHARMEUSE

See SATIN

CHEESECLOTH

PLAIN WEAVE

Cotton

A very loosely woven fabric, made with slightly twisted yarns, in a low thread count. It is similar to bunting but not as closely woven. The yard width is called "tobacco cloth." Cheesecloth may be unbleached, bleached, or dyed. *Uses:* Wrapping food, such as meat and cheese; curtains, fancy costumes; lower grades for cleaning cloths.

CHENILLE

ANY WEAVE

Wool, rayon, cotton

The name of the yarn or of any fabric woven with chenille yarns. These yarns, discussed on page 294, have short cut fibers or pile protruding all around. In weaving, the chenille yarn becomes the filling, forming a pile on the fabric. Chenille yarns may be constructed to weave soft, downy robes, or heavy chenille carpets and rugs. *Uses:* Rugs; women's robes, knitting yarns, trimmings, etc.

CHEVIOT

TWILL, HERRINGBONE TWILL,
OR PLAIN WEAVE

Wool, rayon, cotton

Wool or wool and spun rayon cheviot has a somewhat rough and shaggy surface texture. It is woven from fairly coarse yarns, and is similar in appearance to tweed and homespun, which, however, are woven from heavier yarns. It is this surface texture rather than the weave that distinguishes Cheviots, as they may be woven in twill, herringbone twill, or other twill weave variations. Most Cheviots are made from woolen yarns, but some better Cheviots are woven from worsted yarns and are called worsted Cheviots. It is a well fulled, close fabric, either stock or piece dyed, often in brilliant shades. *Cotton or rayon Cheviot* is an altogether different fabric made of coarse yarns in a plain weave and given a soft finish. It has yarn dyed stripes or checks. Cotton Cheviot is similar to one weight of chambray. Cheviot was originally a fabric made of the coarse wool of the sheep raised in the Cheviot hills in the North of England. *Uses:* Wool, for men's and women's coats and suits; cotton, for men's shirts.

CHIFFON

PLAIN WEAVE

Silk, rayon, wool

A light, gossamer sheer fabric. It is woven in an open weave with tightly twisted yarns. Silk chiffon is made with raw silk single yarns in both warp and filling, usually thrown forty or fifty turns but sometimes as much as seventy or eighty turns to the inch. Rayon chiffon is woven with high-twist crepe yarns. Chiffon is piece dyed or printed and may be given a soft or stiff finish. The word "chiffon" sometimes appears before the name of other types of fabrics to indicate the lightness of that fabric, as for example, chiffon velvet. *Uses:* Evening dresses, formal blouses, trimmings, scarfs, etc.

CHINCHILLA CLOTH

TWILL WEAVE, DOUBLE FABRIC, OR KNITTED

Wool, cotton

A thick, full fabric, soft and deep, with an irregular surface texture caused by curled nubs or tufts. A true chinchilla is a double fabric, with the two fabrics firmly interwoven. The face fabric is woven in a twill weave, and the back fabric may be in a plain or a twill weave, or may be an altogether different fabric such as a woven plaid or a knitted fabric. The fabric is woven with long, floating, woolen yarns and is deeply napped. The napped surface is then rubbed into small rounded and curled tufts. The knitted fabric is woven with nubby yarns and then brushed to raise the nap. The fabric may be all wool or wool and cotton. Chinchilla received its name from the resemblance of its surface texture to that of chinchilla fur. The fabric now may have textures different from the original version. *Uses:* Coats, and women's and children's hats.

CHINTZ

PLAIN WEAVE

Cotton

A printed and usually glazed fabric. The better qualities are in a firm, close weave, made with hard twisted warp yarns and coarser, slackly twisted filling yarns. Chintz is noted for bright, gay colors in flower, bird, and other designs but may be printed a plain color. Some chintz is fully glazed and others semiglazed. It cannot be called fully glazed unless it has been waxed or starched, and the ingredients of the finish pressed in with hot rollers. While this finish gives lustrous, clear colors, it will wash out when laundered. Chintz may be given a chemical glazed finish that will not wash out and will last as long as the fabric. *Uses:* Curtains, draperies, furniture covers, dressing table skirts, children's dresses, women's dresses, smocks, housecoats, sportswear, etc.

619

CORDUROY

CORDUROY WEAVE
Cotton, rayon

A fabric with raised, cut pile cords running lengthwise of the fabric. These are produced after the fabric is woven. Extra filling yarns float over a number of warp yarns. After weaving, the floating filling yarns are cut and the pile is brushed and singed to produce a clear, corded effect. The fabric is then dyed and the back is softly napped. The ribs are of different depth and width. Thick-set corduroy has from eight to eleven cords or wales per inch, fine corduroy has from sixteen to twenty-one wales per inch. *Uses:* Dresses, jackets, skirts, suits, slacks, sportswear; bedspreads, draperies, upholstery, etc.

COTTONADE

TWILL WEAVE
Cotton

A heavy, firmly woven fabric, distinguished by a speckled stripe alternating with a dark stripe. The speckled yarns are produced by twisting a colored yarn with a natural or white yarn to form a two-ply warp yarn, and weaving it with a dark filling yarn. Cottonade is similar to covert with the exception that the twill in cottonade is more prominent and the speckled effect is in stripes rather than all over the fabric as in covert. The broken stripes are produced because some warp yarns are white and colored, and others are a solid color. It may be napped on the back. *Uses:* Work clothes, linings, etc.

COUTIL

TWILL, HERRINGBONE TWILL, OR
JACQUARD WEAVE
Cotton, rayon

A firm, strong yet light weight fabric, woven with hard-twisted ply yarns. It may be woven with stripes or a small dobby or Jacquard pattern. Frequently the ground is cotton and the pattern a lustrous rayon. *Uses:* Women's foundation garments.

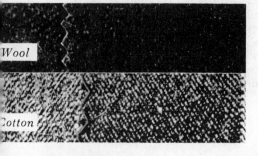

Wool

Cotton

COVERT

TWILL WEAVE
Wool, cotton, rayon

A medium to heavy weight close-ly woven fabric distinguished by

620

its mottled or flecked appearance, which is produced by dark filling yarns and white and colored (or two different colored) warp yarns. Cotton warp yarns are generally produced by a mock twist, that is, the yarns are spun from white or colored (or different colored) rovings. Wool warp yarns are produced by twisting together white and colored (or different colored) yarns. Wool covert cloth has the twill weave running from left to right, while in the cotton version the diagonal is usually from right to left. Some cotton coverts are woven in a plain weave. Generally, there are twice as many warp as filling yarns. Covert is a clear finished, hard textured fabric. *Uses:* Wool and wool versions, for men's topcoats, women's coats and suits. Cotton, for men's shirts, women's dresses, work clothes; men's and women's sportswear, uniforms; draperies, bedspreads, etc.

CRASH

PLAIN WEAVE

Cotton, linen, rayon, jute

A rather loosely woven fabric with thick regular or irregular yarns. Crash is made in various weights. It may be dyed or printed. The fabric is beetled, or the fibers otherwise flattened so that the fabric has a soft, lustrous appearance. If thick, soft yarns are used, the fabric has absorbent qualities. *Uses:* Heavier for draperies, table covers, upholstery; lighter, for sportswear, dresses, etc. One type, suitable for towels.

CREPES

PLAIN, TWILL, SATIN, JACQUARD

Silk, rayon, cotton, wool

Crepe fabrics are constructed with highly twisted yarns in either the warp or in the filling or in both. Most well-constructed crepe fabrics are woven with yarns with a right twist alternating with yarns with a left-hand twist. As the two yarns curl in opposite directions, they tend to support each other, hold the fabric in position, and prevent too great a shrinkage. Crepe fabrics range from very fine, almost smooth surface texture to a predominating, crinkled, pebbly, and mossy effect. However they can be recognized by their crinkled or crisped surface of regular or irregular appearance. Since the crepe is obtained from the twist of the yarns woven into the fabric, it is permanent. While crepes can be made in any weave, they usually are in a plain weave, which is most adaptable to highly twisted yarns. The twisting and throwing of yarns are fully discussed in Chapter 10. The following are typical crepes: Crepe de Chine, Georgette crepe, marocain crepe, and seersucker, each discussed in its alphabetical place in this chapter. Other typical crepes are:

Canton crepe. Silk or rayon. Plain weave. A soft, rather thick, slightly lustrous crepe fabric. It is similar to crepe de Chine but is heavier and rougher. It also has a more crinkled appearance due to the predominance of heavier crepe filling yarns, which give Canton crepe a slightly crosswise ribbed effect.

Chiffon crepe. Silk or rayon. Plain weave. A sheer fabric, similar to Georgette crepe but softer. The silk fabric is made with a hard twist, single, coarse warp yarn, and a regular, two- or three-ply crepe or hard twist, single, coarse filling yarn. The finishing of the cloth gives a crepe "handle."

Crepe back satin. See satin backed crepe listed under satins.

Crepe charmeuse. Silk or rayon. Satin weave. A soft, rich-draping fabric with a dull luster. It has a satin face and a crepe back. The silk fabric is made with a grenadine warp and crepe filling; or of an unthrown warp and alternate picks of crepe and loose-twist tram filling. The rayon fabric has hard twisted warp yarns and crepe filling yarns.

Crepe meteor. Silk. Twill weave. A fine, crinkled fabric, similar in appearance to crepe de Chine but more lustrous. It is made with unthrown silk warp and a three- to six-ply crepe filling as is crepe de Chine, but in a twill rather than a plain weave.

Crepon. Silk, rayon, wool. Plain weave. A heavier crepe fabric with slightly wavy, lengthwise ribs, formed by thicker, alternately twisted, crepe warp yarns. May be dyed or printed. It was originally a wool fabric.

Flat crepe. Silk or rayon. Plain weave. A smooth, soft fabric, less crinkled in texture than most crepes. The alternate right- and left-hand twisted yarns are in the filling, and are slightly lighter yarns than the warp. The yarns as well as the finish contribute to the flat appearance of the fabric.

Mourning crepe. Silk. Plain weave. It has a dull surface finish obtained by pressing with hot, engraved rollers.

Romaine crepe. Silk, rayon, wool. Plain weave. A fairly heavy crepe fabric, woven with alternate right- and left-hand twisted filling yarns.

Uses: Women's dresses, suits, evening wear, lingerie, shoes, handbags, etc.

CREPE DE CHINE

PLAIN WEAVE

Silk, rayon

A finely crinkled fabric with a smooth and lustrous surface, exceptionally soft, light weight, and well draping. The original silk fabric was woven with the warp of unthrown silk yarns and the filling of three- to six-thread crepe yarns. The crepe is obtained by alternating two filling yarns twisting in one direction with the next two yarns twisting in the opposite direction. Silk crepe de Chine is woven in the gum, and the crepy appearance is achieved during the degumming process. It may be piece dyed or printed. *Uses:* Dresses, blouses, lingerie, accessories, curtains, etc.

CRETONNE

PLAIN, TWILL, OR SATIN WEAVE

Cotton, linen, rayon

Made of well-twisted, fairly round yarns. The fabric is firmly woven and printed. It differs from chintz in that it is not glazed and the subdued or gay, bright patterns are usually, but not always, larger than those on chintz. Some cretonnes are typical examples of warp printed fabrics. These, when woven with a white or plain-color filling over the printed warp have subdued colors, and the pattern is softly outlined. Warp printed cretonnes are reversible, as the pattern shows equally on either side. For warp printing, see page 522. *Uses:* Draperies, furniture covers; lighter weights, for beach wear, etc.

CRINOLINE

PLAIN WEAVE

Cotton

A highly sized and stiff, openly woven fabric. It is the same as tobacco cloth or gauze except that these two fabrics have no sizing. Crinoline was originally made of horsehair combined with linen or cotton. The sizing on the all-cotton fabric is to give it the stiffness that was contributed formerly by the horsehair. *Uses:* Stiffenings, linings, millinery, bookbinding, basis for embroidery, etc.

DAMASK

JACQUARD WEAVE

Linen, cotton, silk, rayon, wool,
mohair

A firm, lustrous, reversible fabric with a satin ground weave and a flat Jacquard pattern. This distinguishes it from brocade, which is nonreversible and has a slightly raised pattern. On one side, the warp yarns that form the pattern float over the filling yarns. The opposite side is the reverse. It is the difference in the reflection of light on the horizontal and vertical yarns in the satin weave that gives damask its contrasting luster. If different colored yarns are used for ground and pattern, it shows the opposite coloring on the reverse side. There are three types of damask. One is made of fine yarns, bleached white or colored, such as used for tablecloths and towels. Another is made of heavier, lustrous, colored yarns and is woven into fabrics used for draperies and upholstery. A third is made of silk or rayon and may be piece dyed; it is used for evening wraps or curtains. One drapery type is made with the pattern formed with floats of ratiné, spiral, or other twisted filling yarns. One tablecloth type may be double woven, in which case the pattern is the same rather than reversed on both sides. *Uses:* Tablecloths, napkins, towels; draperies, upholstery, bedspreads; evening wraps, etc.

DENIM

TWILL WEAVE

Cotton

A stout, long wearing fabric, made with coarse, hard twisted ply yarns. It is woven with white filling yarns and colored warp yarns, usually indigo dyed blue, but it may be gray or brown. If a three-leaf warp right-hand twill, more colored warp yarns show on the face of the fabric and more white fillings are on the back. It may also be woven in a left-hand twill. In lighter weights it may have yarn dyed, woven-in stripes and checks. The compact, close weave of denim and its calendered, smooth surface texture give a durable fabric that resists water, snags, and tears. While denim is made in widths from 27 to 30 inches, it is graded and sold on the basis of the weight of the 28-inch width. A 2.20 denim designates that there are 2.2 yards of denim to the pound; a 2.45 denim designates 2.45 yards of cloth to the pound. The lower the number, the heavier the fabric. A different basis is used for the heavier weights: ounces to the yard instead of yards to the pound as employed for the lighter denims. A 9-ounce denim has 9 ounces to 1 yard. The higher the number, the heavier the fabric. *Dungaree* is the same as denim except that it is a four-leaf warp twill with colored filling yarns in place of the blue, gray, or brown warp ends of denim. *Uses:* Heavier weights, for work and military clothes, upholstery, draperies. Lighter weights, for sportswear and draperies, etc.

DIMITY

PLAIN WEAVE

Cotton, linen

A sheer, crisp, corded fabric. The yarns are fine and well twisted. Cotton dimity is made usually of combed yarns, but may be made of carded yarns. The fabric is mercerized, giving it a soft luster. The cords run lengthwise, forming stripes, or both lengthwise and crosswise, forming checks. The cords may be single or there may be double or triple groupings of cords, formed by weaving two, three, or more yarns together. Dimity may be bleached white, printed, or dyed. *Uses:* Children's dresses; women's dresses, blouses, neckwear, lingerie; boys' blouses; curtains, table runners, table scarfs, bedspreads, lampshades; art needlework foundations, etc.

DOESKIN

SATIN WEAVE

Wool, rayon, cotton

A very fine fabric. The weave is not visible on the face, which is napped and then given a rather lustrous finish under moisture. This lays down the nap and produces a very smooth, level face. The fabric takes its name from doeskin, which is soft-finished leather made from the hide of a doe. Doeskin is woven with many warp yarns set very closely together. The satin weave is warp faced. Cotton doeskin may have a twill weave. *Uses:* Men's suitings, sportswear, uniforms; women's suits, coats, sportswear, etc.

DONEGAL

PLAIN OR TWILL WEAVE

Wool, rayon, cotton

A tweed type fabric made with yarns that have colored nubs. The yarns are coarse ply yarns twisted so that thick slubs are formed (see page 289). Originally, the fabric was a homespun, woven by Irish peasants. The coarse homespun yarns were uneven, often with thick spots. Donegal is now made in imitation of these colorful hand-loomed fabrics. *Uses:* Men's sportswear; women's coats, suits, sportswear, etc.

DOTTED SWISS

PLAIN WEAVE

Cotton

A rather open weave, sheer, crisp dotted fabric. The dots may be woven by extra filling yarns with the floating ends cut between the dots, or they may be woven with swivel bobbin and shuttle, or with a lappet attachment. So-called "flock dots" may be extra fibers fastened onto the fabric; or the dots may be a paste composition printed on the fabric; or the dots may be electrocoated onto the fabric. A true dotted Swiss has the dot woven in. There may be white dots on a white background, colored dots on a white background, or colored or white dots on a colored background. *Uses:* Children's dresses and bonnets, women's dresses, aprons, blouses, neckwear; curtains, bedspreads, dressing table skirts, etc.

DRILL

TWILL WEAVE

Cotton

A heavy, firm fabric, made with a three-harness warp twill in a right to left diagonal. It is often finished in the gray but may be bleached, yarn dyed, or printed. It is sized and pressed to make a compact finished fabric. *Khaki cloth* is a drill in khaki color. *Middy twill* and *jean* are fabrics practically the same as drill. *Uses:* Middies, women's uniforms, work clothes, men's work clothes, sportswear; linings; upholstery covers, curtains, etc. It also has many industrial uses.

DUCK

PLAIN WEAVE

Cotton, linen

Closely woven, durable fabrics, made in different thread counts with ply yarns of various sizes and weights. Some are woven with single yarns combined with ply yarns. The heaviest duck is really a canvas, but most weights are lighter than canvas. It may be natural color, white, dyed, printed, or painted. The original duck was a coarse linen fabric made of two-ply warp and filling, but now duck is made almost entirely of cotton. *Uses:* Heavy, for sails, tents, awnings, upholstery, etc.; light, for sportswear, work clothes, etc. Duck also has many home and industrial uses.

626

DUNGAREE

See DENIM

DUVETYN

TWILL WEAVE
Wool, rayon, silk, cotton

A fabric with a very soft, velvetlike finish. The surface fibers are raised by emery rollers. The fabric is then sheared, singed, and brushed, to produce the rather lustrous, smooth surface texture. The back of the fabric is lightly brushed up also. Cotton duvetyn is a soft finished twill weave fabric that is napped, sheared, and finished to give a smooth, flat, downy surface with the weave concealed. The fabric was first made in Paris and the word "duvetyn" is derived from the French word "duvet," meaning "down." *Uses:* Women's coats, suits, dresses; men's sport coats.

ELASTIQUE

TWILL WEAVE
Wool

A firm, clear faced, hard textured fabric with a steep double twill. The diagonal is from left to right. The back of the fabric does not show the double twill cord. Elastique is a strong, durable fabric, made of hard twisted, worsted yarns. About the middle of the nineteenth century, elastique was the name of an overcoating fabric made from fine merino wool. *Uses:* Men's and women's suits, coats, slacks; military suits and coats.

END-AND-END CLOTH

PLAIN WEAVE
Cotton

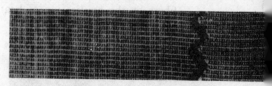

A closely woven fabric with a fine colored line or pin check, formed by alternate white and colored (or two different colors) in the warp or in both the warp and filling. For example, one warp yarn may be white, the next colored, the third white, and so on. When the white filling is woven in, there will be a fine, broken, colored line, running lengthwise of the fabric. With the exception of the colored lines an end-and-end cloth is similar to chambray. *Uses:* Men's shirts, children's wear, etc.

ÉPONGE

PLAIN WEAVE

Wool, rayon, silk, cotton

A loosely woven, spongy fabric, woven with soft, ply yarns unevenly twisted. It has an irregular surface due to the distinctive knotlike or nubby irregularities in the ratiné yarns used in the filling. See page 290 for the description of these yarns. Éponge is a French word meaning "sponge." *Uses:* Dresses, suits; draperies, etc.

FAILLE

RIB WEAVE

Silk, rayon, cotton

A soft yet firm ribbed fabric with the ribs running crosswise of the cloth. This is a warp rib weave formed by more filling yarns that weave as one. There are many more warp than filling yarns. Compared to grosgrain, faille is softer and contains larger, flatter ribs. The original silk faille was made with unthrown silk or with organzine warp and with a coarse silk or slackly twisted cotton filling. *Uses:* Women's suits, coats, dresses, blouses, robes; trimmings, curtains, draperies, etc.

FELT

FELTED

Wool, hair, cotton, rayon

A thick, compact fabric, constructed by compressing masses of loose fibers by the application of heat, moisture, and pressure. *Uses:* Hats, table covers, linings, etc.

FLANNEL

PLAIN OR TWILL WEAVE

Wool, cotton, rayon

There are many types of flannel. Wool or spun rayon flannel is a light-weight fabric with a soft napping on one side, which partially conceals the weave. It is usually in a twill weave, with the yarns more tightly twisted than for the typical cotton flannel, and it may be loosely or firmly woven. If wool, it is generally woven with woolen

628

yarns but may be made of worsted yarns. It may be yarn dyed, piece dyed, or printed. *Uses:* Women's dresses, suits, coats, skirts; girls' dresses and coats; men's and boys' suits, trousers, jackets, and shirts, etc.

Canton flannel. A twill weave cotton fabric. It may be medium or heavy and has a long, soft napping on the back of the fabric. It may be bleached, yarn or piece dyed, or printed. *Uses:* Linings, robes, sleeping garments, diapers, etc.

Outing flannel. A plain or twill weave cotton fabric, napped on both sides. It usually has stripes woven of colored yarns. However, it may be bleached, yarn or piece dyed, or printed. Domet flannels are the same but do not have the colored stripes. *Uses:* Sleeping garments, children's underwear, diapers, linings, etc.

French flannel. A very fine, twill weave fabric, softly napped on one side.

Shaker flannel. A plain weave fabric, napped on both sides. It is softer and fuller than Canton flannel.

Forestry cloth. A strong, twill weave, olive drab flannel, first used by the United States Forestry Service.

Suede flannel. A smooth fabric napped usually on both sides. After napping the fabric is sheared and the fibers pressed into the fabric giving the appearance of a close felted fabric.

FLANNELETTE

PLAIN OR TWILL WEAVE

Cotton

A soft, plain or twill weave fabric, lightly napped on one side. It may be bleached, dyed a solid color, with woven-in colored stripes, or printed. *Kimono flannel.* A plain weave, printed fabric, napped on one side, the same as flannelette. *Uses:* Sleeping garments, baby blankets, diapers, etc.

FLEECE

PLAIN OR PILE WEAVE OR KNITTED

Wool or wool and cotton

A fabric with a deep, soft, wool nap or pile. Fleece may be woven in a plain weave or knitted and then heavily napped by wire or teasel napping, which raises the surface fibers. Usually, wire napping is used as it gives a deeper nap. Or it may be woven in the pile weave, with an extra warp forming the long pile loops, which are cut to make a thick, cut pile surface. The dense pile yarns form air pockets, which ensure good insulating properties, giving warmth without excessive weight. Specialty fibers, such as cashmere, camel hair, mohair, and alpaca are often used for fleece, usually with the back or ground weave made of wool. Frequently, cotton is employed for the back of the fabric with the face pile of wool. This does not necessarily mean that the fabric is inferior. Cotton makes a strong back and at the same time allows, at a lower price, for the warmth providing wool pile. It is stock, skein, or piece dyed. *Uses:* Men's, women's, and children's coats.

FORESTRY CLOTH

See FLANNEL

FOULARD

TWILL WEAVE

Silk, rayon, cotton, wool

A light weight fabric, soft and well-draping, usually printed. It was originally a silk fabric printed with colored dots on a white ground. It now may be plain colored. *Uses:* Dresses, ties, etc.

FRIEZE

TWILL OR BROKEN TWILL WEAVE

Wool, cotton

Heavy, coarse, napped fabric. It usually has heavier warp than filling yarns and may be made to produce a mixed color effect. The nap is rough and irregular and the fabric has a more or less hard feel. Frieze is woven in a right- to left-hand twill or in a broken twill. *Uses:* Coats, sports coats, and mackinaws.

FRISÉ

PILE WEAVE

Wool, mohair, rayon, silk, cotton, linen

A heavy pile fabric with all loops uncut or some cut to form a pattern. Frisé is a durable, closely woven fabric. It may be made on a plain or a Jacquard loom. The extra warp pile yarns weave over a wire which, when withdrawn, leaves uncut loops. Or a knife on the end of the wire may cut part of the loops to produce a cut and uncut design. *Uses:* Upholstery, etc.

GABARDINE

TWILL WEAVE

Wool, rayon, cotton

A hard finished, clear surfaced, left- to right-hand twill fabric. Gabardine has a fine, steep diagonal wale, which is not evident on the back of the cloth. The steep twill is due to weaving a greater number of warp than filling yarns. Gabardine is a tightly woven, firm, and durable fabric. It is rather lustrous but may be given a dull finish. Cotton gabardine is closely woven of firmly twisted, long, combed yarns, and is mercerized to make a firm fabric with a soft luster. Gabardine may be yarn or piece dyed. *Uses:* Men's and women's suits, coats, raincoats, sportswear, uniforms, men's shirts, etc.

GAUZE

PLAIN OR HALF LENO WEAVE

Cotton, rayon, silk

A sheer, transparent fabric. Cotton gauze, woven in a plain weave, is similar to cheesecloth, but with more tightly twisted yarns woven with a higher thread count. Silk and rayon gauze may be woven in a half leno weave (sometimes called gauze weave). *Uses:* Trimmings, curtains, plain weave cotton for bandages.

GEORGETTE CREPE

PLAIN WEAVE

Silk, rayon

A sheer, loosely woven fabric. Compared to crepe de Chine it has a harder finish, less luster, and a more crepy but less crinkled surface. It is made with highly twisted yarns alternating in direction of twist in both warp and filling. Silk Georgette crepe is made with two- or three-ply yarns. They are piece dyed or printed. *Uses:* Women's dresses, blouses, negligees, lingerie; curtains, bedspreads, lampshades, etc.

GINGHAM

PLAIN WEAVE

Cotton

A light to medium weight, closely woven fabric, made with dyed yarns, in stripes, checks, plaids, and solid colors. If woven with white and colored yarns, there usually are the same number of white and colored yarns in the same sequence in the warp as in the filling. They are made of carded or long staple, combed cottons, woven with a close, fine weave. Some ginghams may have coarser yarns and less firm weaves. Ginghams are mercerized and have a soft, lustrous appearance. In addition, they are sized and calendered, which gives them a firmer body and a higher sheen. Tissue and zephyr ginghams are sheer ginghams, being woven of finer yarns in a higher thread count. *Uses:* Children's dresses and playclothes; women's dresses, robes, pajamas, sportswear; men's shirts, sportswear; curtains, draperies, spreads, dressing table skirts, laundry bags, etc.

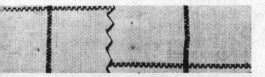

GLASS TOWELING

PLAIN WEAVE

Linen, cotton, rayon

A rather loosely woven fabric, made of highly twisted yarns. It can usually be recognized by its blue or red stripes or checks. However, other colors may be used or the towel may be white with a colored border. All linen gives the best results because of its ability to absorb moisture and its lack of fuzziness and lint. Cotton also absorbs readily and when the yarns are tightly twisted there is less linting. Some glass towels are made from linen and cotton or these fibers combined with rayon. *Uses:* Towels, kitchen curtains, table covers, etc.

GRENADINE

LENO WEAVE

Cotton, silk, rayon, wool

A loosely woven fabric with hard twisted yarns that are dyed before weaving, producing stripes or checks. Grenadine is also the name of silk yarns that have been given a certain twist (see page 287). Silk grenadine is made with a grenadine warp and given a stiff finish. Curtain grenadine often has a clipped swivel dot or dobby pattern. *Uses:* Neckwear, dresses, curtains, etc.

GROS DE LONDRES

RIB WEAVE

Silk, rayon

A closely woven, ribbed fabric with the ribs running crosswise of the fabric. Its distinguishing characteristic is its alternate heavy and fine ribs. A heavy rib may be followed by one or more fine ribs and then another heavy rib. Gros de Londres is woven in a warp rib weave, the crosswise ribs being formed by groups of filling yarns that weave as one yarn. *Uses:* Women's evening wear, dresses, blouses, curtains, etc.

632

GROSGRAIN

RIB WEAVE

Silk, rayon, cotton

A hard finished, closely woven, ribbed fabric, with the rib running crosswise. This is a warp rib weave formed by a number of filling yarns weaving as one. The ribs in grosgrain are heavier than in poplin and are more rounded than in faille. The original silk grosgrain was made with an organzine warp and a coarser filling of softly twisted tram or cotton. *Uses:* Women's coats, suits, dresses; neckties, trimmings, hatbands, ribbons, etc.

HOMESPUN

PLAIN WEAVE

Wool, rayon, cotton

Coarse fabric of the tweed type, loosely woven with irregular, lightly twisted, unevenly spun yarns, imitating the natural irregularities of homespun yarns. The open, plain weave and the thick yarns give a somewhat spongy feel to the fabric. Homespun is stock dyed and at times mixed colored yarns are introduced to give more of a hand-loomed appearance to the fabric. *Uses:* Women's coats, suits, sportswear; men's sport coats and jackets; children's coats; draperies, upholstery, etc.

HONEYCOMB (Waffle cloth)

HONEYCOMB WEAVE

Cotton, rayon, wool

This is the name of both the weave and the fabric. It is a rough textured cloth, with a raised, square, oblong or diamond-shaped pattern, produced by floating warp and filling yarns that form the ridges along the lines of the floats. *Waffle piqué* is discussed on page 645. *Uses:* Toweling, novelty dress fabrics. Heavier, colored, honeycomb fabrics are for bedspreads, etc.

HOPSACKING

BASKET WEAVE

Wool, rayon, cotton, linen

An open basket weave fabric made with coarse yarns. It is usually woven with two warps and two fillings but may be a three-by-three basket weave. Hopsacking has a hard, rather rough texture and is very durable. *Uses:* Men's and women's sportswear; draperies, etc.

HUCKABACK

HUCKABACK WEAVE
Linen, cotton

A firm, long wearing fabric, with a semirough, patterned surface. Huckaback is the name of the weave (see page 340) as well as of the fabric. While it appears to be made with a complicated weave, it is a plain ground weave with floating yarns. The warp yarns float on the surface and the filling yarns float on the back. A typical weave has two warp ends floating on the face of the fabric over five picks, and two filling yarns floating over five ends on the back. Coarser yarns produce more pronounced patterns. The floating yarns create a greater surface absorbency, making the fabric particularly adaptable to toweling. *Uses:* Towels, quilting covers, draperies, lighter for shirtings, etc.

IMITATION-BEAVER CLOTH

TWILL OR SATIN WEAVE
Wool, cotton, rayon

A heavily fulled and felted fabric made to imitate beaver fur. Fabrics made from any of the fibers are deeply napped. In the finishing processes the fabric is shrunk considerably to give it a full close body. The cotton fabric is napped on both sides. *Uses:* Coats, uniforms, sportswear, children's coats, caps, hats, linings, etc.

IMITATION-MOLESKIN CLOTH

SATIN WEAVE
Cotton

A thick, heavy, strong fabric, with a hard, smooth face and napped back that imitates the fur of the mole. It is woven with tightly twisted yarns, with about twice as many picks as ends. It may be printed to imitate yarn dyed woven fabric. The napped back gives it softness and warmth, and the close, firm weave gives it a smooth surface texture. Undyed sample shown. *Uses:* Men's winter shirts, jackets, jacket linings, work clothes, etc.

JASPÉ CLOTH

PLAIN WEAVE
Cotton, rayon

This fabric takes its name from the jaspé yarns. Different colored warp yarns are woven with single-color filling yarns, giving a series of faint or blended multicolored stripes; or light, medium, or dark shades of the same color. The yarns are hard twisted, making a firm, hard, durable fabric. *Uses:* Upholstery, draperies, covers for couches and pillows, etc.

JEAN

TWILL OR CHEVRON TWILL WEAVE
Cotton

A right- to left-hand twill fabric, similar to drill but finer. It may be bleached, dyed in solid colors or in stripes. The fabric is pressed to give it a firm clear-surfaced texture. *Middy twill* is the same as jean. *Uses:* Work clothes, sportswear, foundation garments, children's play clothes, linings, etc.

JERSEY

PLAIN KNIT
Silk, wool, rayon, cotton

A fabric knitted with one continuous, light weight yarn, usually in tubular form. It has a distinct right and wrong side: the face stitches making an up and down rib effect and those on the reverse giving a crosswise rib. It is very elastic and will "run" if a stitch is broken. *Uses:* Lingerie, dresses, sweaters, sport shirts, gloves, etc.

KERSEY

TWILL WEAVE
Wool

A heavy, highly lustered, finely napped fabric. Kersey is tightly woven, well fulled, napped, and closely sheared. The nap obscures the weave. It is similar to melton but heavier and with a more lustrous surface. Kersey was a woolen cloth made originally in Kersey in Suffolk. Its manufacture was strictly regulated by statutes as to weight, texture, etc. *Uses:* Military and civilian uniforms, overcoats, etc.

KHAKI CLOTH

See DRILL

KIMONO FLANNEL

See FLANNELETTE

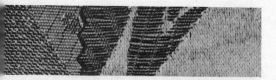

LAMÉ

**PLAIN, SATIN, OR TWILL GROUND
WEAVE WITH JACQUARD PATTERN**

Silk or rayon, with metal

A metallike or glittering fabric, with a Jacquard woven pattern in low relief. This fabric is much the same as brocade with the exception that metal threads are used for the ground, or the pattern, or as decoration for either or both. *Uses:* Evening wear, blouses, trimmings, millinery, etc.

LAWN

PLAIN WEAVE

Cotton, linen

A light weight, sheer, fine fabric, made with fine, carded or combed yarns, woven in a high count. Lawn may be given a soft or crisp finish. Usually it is a semicrisp finish, not as soft as voile and not as crisp as organdy. It may be bleached, dyed, or printed. It is sized and calendered, which results in a softly lustrous appearance. Lawn is similar to batiste, muslin, and nainsook, three fabrics that are given different finishes from lawn. Originally it was a very fine linen fabric, made in Laon, France. *Uses:* Infants' and children's wear; women's dresses, blouses, lingerie, neckwear; curtains, bedspreads, etc.

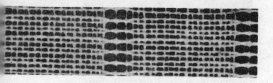

LENO

LENO WEAVE

Cotton, rayon, silk, nylon, wool

Leno is primarily the name of a weave (see page 343), but it is also often the name of a fabric. Some fabrics are woven in a leno weave, but are not called such. Exceptions are such fabrics as marquisette and grenadine. This weave is used for a wide range of fabrics, from the very sheerest to as heavy as that used to weave a wool pile rug. All of the fabric may be constructed by leno weave or only the decoration may be in this weave. The fabrics are woven with a leno device, which may be attached to a plain, dobby, or Jacquard loom.

LONGCLOTH

PLAIN WEAVE

Cotton

A soft, closely woven, bleached or printed light weight fabric. It is woven in a high count, with fine, slightly twisted yarns. Longcloth is very lightly sized and calendered and has little luster. It is most similar to nainsook, which is heavier, has more sizing, and, hence, more luster. *Uses:* Children's wear, lingerie, shirtings, etc.

MACKINAW CLOTH

TWILL WEAVE OR DOUBLE FABRIC

Wool

A thick, heavy, well-felted and napped fabric, noted for plaid or other color combinations. If a single fabric, it is similar to the heavier, usually plain colored melton. However, it may be woven double, with both sides the same, or one side in one pattern and the other plain or in another pattern. The lower priced fabrics may have a cotton warp or may be blends of wool, cotton, and rayon. *Uses:* Mackinaws and other heavy jackets, ski clothes, etc.

MADRAS

PLAIN OR JACQUARD WEAVE

Cotton

A finely woven, soft fabric, distinguished by a check, or a cord or stripe running lengthwise in the direction of the warp. A dobby or Jacquard pattern may be woven in. The fabric may be all white with white or colored stripes or checks; or may have a colored warp and white filling, with white or colored stripes or checks. A true madras is made with woven-in checks or stripes of dyed yarns. If the stripes or patterns are printed on the fabric, it is not madras. Good quality madras is made of long, combed cotton fibers and is mercerized to make it lustrous and durable. A heavier madras is used as a corset fabric. So-called curtain madras is quite a different fabric and is really a marquisette or grenadine with clipped spots. It is not madras. *Uses:* Men's shirts, pajamas, etc.

MALINE

See NET

MAROCAIN

PLAIN WEAVE

Rayon

A heavy crepe fabric with slightly wavy and rather heavy crosswise ribs, produced by thick alternately twisted crepe filling yarns. It resembles Canton crepe but is much heavier. The fabric may be white, dyed, or printed. *Uses:* Dresses, blouses, etc.

MARQUISETTE

LENO WEAVE

Cotton, rayon, nylon, silk, wool

A sheer, open weave fabric, soft or crisp. Its fineness and weight are determined by the yarns, which may be one, two, or three ply. The fabric may be bleached, dyed, or printed and may be woven with dots or with dobby or Jacquard patterns. Marquisette was originally a silk dress fabric. *Uses:* Curtains, evening wear, dresses, children's wear, etc.

MATELASSÉ

A DOUBLE-WOVEN JACQUARD WEAVE

Silk, rayon, cotton, wool

A double-woven fabric with a raised or puckered surface that has the appearance of a quilted or puckered effect. The warp yarns in both the face and back fabric are controlled by the Jacquard. The two fabrics are woven together with extra crepe yarns. Cotton matelassé has a quilted appearance while rayon and silk matelassé have puckered or blistered patterns. *Matelas* is the French word for "mattress, pad." *Uses:* Silk and rayon, for dresses, suits, evening wraps; cotton, for bedspreads, draperies, upholstery, etc.

MELTON

TWILL OR SATIN WEAVE

Wool

A thick, heavy, well-fulled or felted fabric, with a smooth surface. The fabric is napped and then very closely sheared. It is usually woven with large, soft spun, woolen yarns but may be made of finer yarns and finished with a smoother, more lustrous surface. Lower priced meltons may have a cotton warp. It is almost always in plain colors. *Uses:* Overcoats, jackets, uniforms, etc.

MESSALINE

See SATIN

MIDDY TWILL

See DRILL

MILANESE

WARP KNIT
Silk, rayon, nylon

A very sheer fabric made with many warp yarns knitted on a Milanese machine. The stitch makes a fine twill rib, running diagonally on the fabric. It has very little elasticity and is practically "run-proof." *Uses:* Lingerie.

MOIRÉ

USUALLY A RIB WEAVE
Silk, rayon, cotton

A term applied to a finish giving a watery appearance on fabrics as well as the name of the fabric. The description of the methods used, their application, and how they affect the fabric is given in Chapter 19. A moiré finish can be applied to any number of fabrics but it seems to give the best results on ribbed cloth. While most moiré finishes can be steamed or washed out, on acetate rayon moiré is permanent. *Uses:* Formal dresses and wraps, suits, negligees, pajamas, curtains, draperies, bedspreads, etc.

MOLESKIN CLOTH

See IMITATION-MOLESKIN CLOTH

MOMIE CLOTH (Mummy cloth)

CREPE WEAVE

Cotton, silk, rayon, wool, linen

A crepy, dull surfaced fabric, often called mummy cloth. Originally the fabric was made with a silk warp and wool filling and was used for mourning dress because of its lusterless surface. Today's version of momie cloth is woven on a dobby loom, with fine warp and heavy filling yarns in a crepe weave, simulating fabrics woven with heavy crepe yarns. It may be dyed or printed. A crinkled, lusterless, black fabric, with a cotton warp and wool filling is also called momie cloth as is plain woven, heavy cotton or linen fabric, used for embroidery. *Uses:* Tablecloths, towels, dresses, shirts, etc.

MONK'S CLOTH

BASKET WEAVE

Cotton

A heavy, loosely woven fabric made of coarse and rough yarns. It is natural color, with woven-in stripes or plaids, dyed solid colors, or printed, usually brown, blue, or green. Monk's cloth is a balanced basket weave. Similar fabrics are friar's cloth, usually also called monk's cloth, a four-by-four weave; and druid, an eight-by-eight weave. Monk's cloth is a durable, substantial fabric. Sometimes flax, jute, or hemp are used in the filling, but it is usually all cotton. *Uses:* Couch and furniture covers, draperies, etc.

MOQUETTE

See VELVET

MOUSSELINE DE SOIE

PLAIN WEAVE

Silk, rayon

A crisp, sheer fabric, usually dyed plain colors. It is more closely woven than chiffon and not as soft as voile. "Mousseline de soie" is French for silk muslin. *Uses:* Evening wear, trimmings, etc.

640

MULL

PLAIN WEAVE

Cotton, rayon, silk

A soft, sheer, lustrous fabric. It is similar to lawn but softer. Its finish is similar to nainsook or batiste. Originally, mull had a cotton warp and silk filling. *Uses:* Infants' and children's wear, etc.

MUMMY CLOTH

See MOMIE CLOTH

MUSLIN

PLAIN WEAVE

Cotton

A plain weave fabric ranging from light weight to the heavy qualities used for sheets, pillow cases, and for industrial purposes. Muslin is made of carded yarns in various twists in thread counts from fine to coarse. They are given from a little to heavy sizing, the latter to the poor quality fabrics. They are calendered to produce a smooth "ironed" surface texture. Muslin may be unbleached, bleached, dyed, or printed. Muslin sheeting is discussed on page 581. *Uses:* Heavier, for sheets, pillow cases, bedspreads, curtains, etc. Lighter, white colored or printed, for underwear, women's dresses, uniforms, children's wear, etc.

NAINSOOK

PLAIN WEAVE

Cotton

A light weight, soft fabric, slightly heavier than batiste. The better qualities are mercerized and have a soft luster. In addition, one side may be calendered and have a higher gloss than the other side. Nainsook is similar to batiste, dimity, cambric, longcloth, except for the finishes which differ with each fabric. It may be bleached, dyed, or printed. *Uses:* Lingerie, blouses, neckwear, infants' wear, etc.

NEEDLE POINT—WOVEN

ANY WEAVE

Wool

A fabric with a finely pebbled texture obtained by the use of nubby yarns. It is covered all over by small nubs formed by the irregularities of the yarn. The nubs may be small or more pronounced, dependent on the weight of the fabric. Woven needle-point fabric should be distinguished from needle-point lace which is made by twisting, as explained in the twisting chapter. *Uses:* Women's coats and suits.

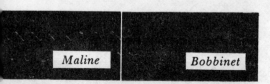

Maline   *Bobbinet*

NET

TWISTED

Cotton, linen, silk, rayon, nylon

A mesh fabric made on Levers, bobbinet, or Nottingham machines. It may be fine and sheer, or coarse and open, and have a variety of meshes, such as square or hexagonal. *Uses:* Curtains, dresses, millinery, trimmings, etc. Nets include:

Bobbinet has a fine, firm, even mesh. It is made of cotton or rayon on a bobbinet machine. It is also called *Brussels net.*

Maline has a fine, open, diamond-shaped mesh. It is made of silk or rayon on a Levers machine.

Tulle has a fine, more closely-textured, hexagonal mesh. It is made on a Levers machine of silk or rayon.

Mosquito net has a coarse, firm mesh, constructed for protection against insects. It is made of cotton on bobbinet or Levers machines.

NETTING

KNOTTED

Cotton, linen, other bast fibers

Openwork mesh fabric made by knotting with fisherman's knot. Also called *fish netting. Uses:* Fish nets, sportswear, bags, furniture coverings, etc.

NINON

PLAIN WEAVE
Silk, rayon

A very thin, smooth, rather crisp fabric, heavier and not as soft as chiffon. Originally, it was an all-silk fabric, made of low denier raw silk. *Uses:* Evening dresses, scarfs, curtains, etc.

NUN'S VEILING

PLAIN WEAVE
Wool

A very sheer, thin, and soft fabric, woven with finely twisted yarns, which give it a firm feel. It may be white or dyed plain colors. *Uses:* Women's and children's dresses, nun's veiling, etc.

ORGANDY

PLAIN WEAVE
Cotton

A crisp, thin, transparent fabric, woven with tightly twisted fine yarns. Organdy is noted for its crispness. If starched and calendered, the crispness will wash out and requires replacing after each laundering. Organdy may be given a chemical finish as a result of which the fabric will retain its crispness and transparency after laundering. Organdy may be bleached, dyed, or printed. *Organza* is the same as organdy but is made of rayon yarns that have been given an organzine twist (see page 288). *Uses:* Children's dresses, women's dresses, blouses, neckwear; curtains, bedspreads, dressing table skirts, etc.

OSNABURG

PLAIN WEAVE
Cotton

A rough, strong fabric woven in a low thread count weave, with coarse uneven yarns. It is made in light, medium, and heavy weights and may be in plain colors or printed in stripes, checks, and novelty effects. This lighter weight colored fabric may be called crash. Some Osnaburg is given a finish to produce a more lustrous and smoother surface texture. Osnaburg was originally made from coarse flax yarns. *Uses:* Lighter weights for men's and women's work clothes and sports wear; heavier, for curtains, upholstery, mattress ticking, awnings, etc. It has many industrial uses.

OTTOMAN

RIB WEAVE

Silk, rayon, cotton

A heavy, ribbed fabric, with round, prominent ribs, larger than those of most other ribbed fabrics. The silk fabric usually has a silk warp and a silk, worsted, or cotton filling. It is woven with a warp rib weave, formed by groups of filling yarns that weave as one. There are many more warp than filling yarns. The ribs may be formed at intervals to give a striped effect. *Uses:* Women's coats and suits, trimmings, etc.

OXFORD CLOTH

BASKET WEAVE (2 x 1)

Cotton

A fabric usually woven in a basket weave, with two fine warp yarns and one filling yarn equal in size to the combined two warp yarns. The warp yarns are not twisted together but are two separate yarns weaving as one. Colored yarns may be used to form stripes, and small fancy designs may be introduced or the fabric may be bleached or dyed. Oxford cloth has a variety of uses as it ranges from a lighter to a rather heavy, durable fabric. Better grades are mercerized, giving the fabric a soft luster. *Uses:* Men's shirts, pajamas, underwear; women's dresses; men's and women's sportswear, work clothes, uniforms; draperies, spreads, dressing table skirts, etc.

PANNE

See SATIN AND VELVET

PARACHUTE FABRIC

PLAIN WEAVE

Nylon, silk, rayon

A very fine, soft, lustrous, and strong fabric. While usually in a plain weave, it may be woven in a different weave. It may be white, plain color, or dyed in camouflage colors. Parachutes for humans are silk or nylon. Those for cargoes as medicine, foods, etc., are usually rayon, and the color determines the kind of cargo it will carry. *Uses:* Parachutes, lingerie, blouses, etc.

PERCALE

PLAIN WEAVE

Cotton

A firm, smooth fabric finished without a gloss. It may be white, printed in plain colors or with designs, usually small and often geometric. Percale has a higher count weave than muslin. It is very similar to cambric, being made of cylindrical yarns, but does not have the gloss typical of cambric. Percale is starched and calendered. Some are given a crepe finish and are called "crinkled sheeting." Sheeting percales are of very high thread count. Some are made of combed yarns and are called "combed percales;" others of high thread count are made of carded yarns and are called "carded percales." Percale sheetings are discussed on page 581. *Uses:* Sheets and pillow cases; crinkled, for bedspreads and curtains; printed, for children's dresses, men's shirts, boys' blouses, women's dresses, aprons, etc. Sheeting fabrics have many other household and industrial uses.

PERCALINE

PLAIN WEAVE

Cotton

A glazed or moiré finished fabric with a high sheen. It is light weight, woven finer than percale. Percaline is sized and calendered and may be bleached or piece dyed. *Uses:* Usually for linings.

PIQUÉ

BEDFORD CORD OR HONEYCOMB
   WEAVE

Cotton, rayon, silk

A group of fabrics with varied surface textures formed by raised cords or wales. Plain piqués in the United States are fine Bedford cords with lengthwise ribs. (English piqués are made with crosswise ribs.) The raised cords or wales are of varied widths and thicknesses, according to the yarns used. Piqué ranges from medium- to heavy-weight fabrics; but whatever weight, it retains its rugged appearance. It is made from combed or carded yarns and the better grades are mercerized. The combed yarns are woven in a higher thread count than are the carded yarns. They may be bleached, dyed, or printed. *Pin wale piqué* has fine raised cords. *Waffle piqué* is woven with honeycomb weave and retains the raised cord effect. There are many other novelty versions of piqué, such as embroidered piqué and "bird's-eye" piqué, not woven in the bird's-eye weave. *Uses:* Children's dresses, women's dresses, blouses, robes, neckwear, handbags; men's, women's, and children's sportswear; bedspreads, draperies, slipcovers, table runners, etc.

PLISSÉ
USUALLY PLAIN WEAVE
Cotton, rayon

It is not the weave, but plissé printing that gives this fabric its name. It has a crinkled surface in either puckered stripes or patterned effects. See page 521 for a discussion of plissé printing. The crinkle is permanent. *Uses:* Women's and children's dresses, lingerie; curtains, bedspreads, etc.

PLUSH
PILE WEAVE
Silk, rayon, mohair, cotton

A fabric with a low or high woven-in cut pile. Compared to velvet, plush has a deeper, less dense pile. Two fabrics are woven face to face, with an extra pile warp connecting them, then cut in the center to form the pile on each fabric. Plush also may be woven as one fabric, with the extra pile warp yarns woven over wire to produce the pile. As the wires are withdrawn, a knife on the end of the wire cuts the loops. *Uses:* Wraps, corset linings, upholstery, etc.

PONGEE
PLAIN WEAVE
Silk

A pale or dark ecru colored fabric woven with irregular tussah or wild silk yarns in both warp and filling. These uneven yarns give a broken crossbar effect characteristic of pongee. The silk used to weave tussah pongee is neither degummed nor thrown. The gum is usually removed after weaving. Imitation silk pongee is not made wholly of real tussah silk. Silk pongee was first woven in China on hand looms from wild silk. A rayon or cotton fabric woven with elongated slubs is not a true pongee. *Uses:* Dresses, shirts, pajamas, curtains, etc.

POPLIN
RIB WEAVE
Wool, cotton, silk, rayon

A light weight, firm, finely ribbed fabric, with the ribs running crosswise of the fabric. Next to broadcloth, poplin has the finest ribs of all ribbed fabrics. It is a warp rib weave formed by groups of filling

646

yarns that weave as one. Originally, poplin was made with a silk warp and a heavier, cotton or wool filling. Cotton poplin may be made with carded or combed yarns and is usually mercerized to give a soft lustrous appearance. Poplin may be white, yarn dyed, piece dyed, or printed. *Uses:* Coats, suits, women's dresses, men's shirts, sportswear, uniforms, etc. Cotton poplin is also used for pajamas, etc.

RADIUM
PLAIN WEAVE
Silk, rayon

A firm, closely woven fabric that is smooth and soft. Originally, it was made with a warp of unthrown (untwisted) silk, and a filling of alternate yarns of right- and left-hand twisted tram of thirty or forty twists to the inch. Usually, it is tightly woven under high tension and is finished to produce a high luster. However, some radiums have a more or less dull finish. It is woven in the gray and may be piece dyed and printed. *Uses:* Dresses, lingerie, linings, etc.

RATINÉ
PLAIN OR TWILL WEAVE OR KNITTED
Wool, rayon, cotton, silk

Ratiné is also the name of the type of yarn from which the fabric is constructed. It may be knitted or woven with ratiné yarns. When woven, usually a plain weave is used. However, it is the coarse, nubby appearance of the fabric that distinguishes it rather than its weave or knit. Ratiné yarns are discussed on page 290. It is a spongy, rather bulky fabric. The surface texture is due to the distinctive knotlike irregularities in the ratiné yarn. When woven, the ratiné yarn is used in the warp. *Uses:* Dresses, blouses, coats, suits, curtains, etc.

REP
PLAIN OR RIB WEAVE
Silk, rayon, cotton, wool

.A firmly woven fabric with prominent rounded ribs running crosswise or lengthwise of the fabric. If a warp rib weave (with ribs running crosswise of the fabric), it is formed with groups of filling yarns weaving as one to form the weave. This fabric has many more warp than filling yarns. If a filling rib weave (with ribs running lengthwise of the fabric) it is formed with groups of warp yarns weaving as one to form the rib, and there are many more filling than warp yarns. Another method of construction has both warp and filling yarns

647

arranged in a sequence of one coarse and one fine. There are usually two filling bobbins and two warp beams. Woven in a plain weave, the coarse filling passes over the coarse warp and the fine filling passes over the fine warp yarns, thus forming the lengthwise ribs. This may be reversed to form crosswise ribs. *Uses:* Men's, women's and boys' clothing; heavier, for draperies and upholstery, etc.

SAILCLOTH
PLAIN WEAVE
Cotton, linen, jute

A very heavy and strong canvas. There are many qualities and weights. Lighter weight and often printed canvas is usually called sailcloth. *Uses:* Sails, awnings, etc.

SATEEN
SATIN WEAVE
Cotton, rayon

A lustrous, smooth fabric. It may be either a filling or warp sateen. In the filling sateen a filling yarn passes under one warp yarn and then it floats over a number of warp yarns to again weave under one warp yarn, and so forth. In this weave, the sheen is crosswise of the fabric. In a warp sateen the warp passes under one filling yarn and then over a number of filling yarns, and again under one filling. A warp sateen, frequently called "satine," is usually a stronger fabric than is a filling sateen. Better-quality sateens are mercerized but may be calendered only to produce a luster. They may be bleached, dyed, or printed. *Uses:* Draperies, comfortable and mattress covers, bedspreads, upholstery; men's and women's robes, pajamas, sportswear; women's slips, linings, etc.

SATIN
SATIN WEAVE
Silk, rayon, nylon, cotton

Satin is the name of the weave (see page 337) as well as of the fabric woven in this weave. However, all satin weave fabrics are not called satin. For example, some napped wool fabrics are woven in a satin weave to permit more effective napping, such as zibeline, doeskin, and melton. If more warp yarns are on the surface, it is called a warp-faced satin, if more filling yarns, a filling-faced satin. Satin was originally an all-silk fabric, with the warp predominating on the surface. The best satin fabrics have the warp and filling yarns intersecting

648

widely but evenly spaced. There are many types of satin. The following are some of the most important silk or rayon satins.

Crepe back satin. This may also be called satin back crepe, as the fabric is reversible. One side is lustrous satin and the other side is a crinkled crepe fabric, produced by weaving alternately hard twisted filling yarns—first a right-hand twisted yarn and then a left-hand twisted yarn. This fabric has somewhat the appearance of crepe meteor, which is woven with crepe yarns in a twill weave.

Satin Canton. A soft, slightly heavier fabric, with a lustrous surface and a Canton crepe back, which has a crosswise ribbed effect due to heavier filling yarns. The crepe is produced by the filling yarns, which have alternate right- and left-hand twists. The crepe ribs on the back of the fabric give a rather pebbly appearance to the lustrous satin surface. Canton crepe is similar.

Panne satin. A satin fabric finished to be rather stiff and highly lustrous.

Ribbed satins. Some ribbed fabrics such as bengaline and faille may be woven with a satin face—the ribs giving a broken, lustrous surface to the face of the fabric. These fabrics may also be given a moiré finish.

Satin backed velvets. A lustrous satin weave may be used for velvet as well as other pile weave fabrics.

Messaline. A light weight satin fabric, woven with fine yarns. It is a five-shaft warp satin weave, rather loosely constructed.

Antique satin. A heavy, dull lustered fabric, woven with uneven yarns.

Ciré satin. A fabric given a finish that produces a very high, leather-like luster. The fabric is often dyed in metallic colors, such as gold, black, steel gray, and copper. The finish is produced by applying wax under heat and pressure. This same finish may be given to lace.

Charmeuse. A fabric with a lustrous surface and a dull back. Silk charmeuse is woven with an organzine warp and usually a spun silk filling, but the filling may be highly twisted yarns.

Uses: Extensively used for all types of women's wear, trimmings, linings; curtains, draperies, bedspreads, upholstery, etc.

SCRIM

PLAIN WEAVE

Cotton, linen

An open weave, light weight but durable fabric, woven with rather coarse yarns. It is usually white, cream, or ecru and may have woven-in white or colored stripes, checks, or designs. The better quality is woven with strong two-ply, carded yarns. It may be mercerized. *Uses:* Curtains.

SEERSUCKER

PLAIN WEAVE

Cotton, rayon

A durable fabric with permanent woven-in, crinkled stripes, running lengthwise in the fabric. The crinkle is formed by the warp yarns. There are two warp beams on the loom. Warps from one beam are woven taut and those from the other are held slack. Seersucker is woven from good quality, dyed yarns, carded or combed, in checks and stripes. It ranges from sheer to heavy weights, and may be printed in floral or geometric patterns. Plissé crepe, while similar in appearance, is quite a different fabric, produced by a printing process. Crepes woven with alternately twisted yarns, to give the creped effect, are also different, as seersucker utilizes tight and slack warp yarns rather than a high twist for creping. *Uses:* Children's play clothes, women's dresses, blouses, housecoats, lingerie; men's summer suits, pajamas; men's and women's sportswear; women's military uniforms; bedspreads, curtains, slipcovers, etc.

SERGE

Silk, wool, cotton, rayon, silk

TWILL WEAVE

A fabric with a well-pronounced diagonal rib on both face and back. The twills run from lower left to upper right on the face of the fabric and left to right on the wrong side. Cotton serge may have the twill running from right to left on the face and from left to right on the back. It is a clear, hard finished cloth. *Storm serge* is made of coarser yarns and is a wiry, rather heavy fabric. *French serge* is made of very fine, soft yarns and produces a fine twill. *Uses:* Suits, coats, dresses, skirts, trousers; draperies, upholstery, linings, etc.

SHANTUNG

PLAIN WEAVE

Silk, rayon, cotton

A fabric woven with elongated slubbed filling yarns giving an irregular surface texture. Originally it was a tussah silk fabric, woven with yarns that had irregularities, knots, and other natural imperfections. It may be dull or lustrous, soft or firm, fine or heavier. *Uses:* Dresses, suits, pajamas, blouses, sportswear; curtains, etc.

Wool

Rayon

SHARKSKIN

BASKET OR TWILL WEAVE

Rayon, wool, cotton

There are two distinctly different types of sharkskin. Both have a sleek yet delicate, pebbly surface appearance. One type is woven in a two-by-two twill. The yarns in both warp and filling are alternately white and colored. The weave ribs run from left to right and the color lines run from right to left. A second type is made of acetate rayon only. The filament yarns are tightly woven in a basket weave, giving the fabric a certain crispness. This fabric has a chalky luster peculiar to acetate rayon woven in this manner. *Uses:* Women's suits, sportswear, etc.

SUEDE FLANNEL

See FLANNEL

SURAH

TWILL OR HERRINGBONE TWILL WEAVE

Silk, rayon

A soft, light weight, firmly woven fabric made of fine soft yarns. It is frequently made in plaid design or with dyed yarns. The word "surah" is broadly used in the trade to designate any twilled silk or rayon fabric. *Uses:* Dresses, blouses, mufflers, ties, etc.

TAFFETA

PLAIN WEAVE

Silk, rayon, wool

A fabric with a slightly crossribbed effect, as the filling yarns are somewhat larger than the warp yarns. It has the same, or approximately the same, yarn count in both warp and filling, and a firm, close weave, often with yarn dyed, tightly twisted yarns. Silk taffeta usually is woven with an organzine warp and a loosely twisted tram filling. It may be bleached, woven with dyed yarns, or printed. Chiffon taffeta has good draping qualities and a soft lustrous finish, secured by passing the fabric under pressure through hot rollers. Changeable taffeta is made with the warp and filling in different colors. *Uses:* Women's dresses, evening wraps, slippers, lingerie, negligees; bedspreads, curtains, draperies, dressing-table skirts, pillows, etc.

651

TAPESTRY

JACQUARD WEAVE

Cotton, wool, silk, rayon

A fabric woven in patterns with multicolored yarns. As compared to damask and brocade, the surface texture of tapestry is much rougher. Tapestry requires two warps and all but the simplest, two or more fillings. Machine made have a somewhat smooth back but hand-made tapestries have a rough back because the bobbins are worked from the wrong side. On the loom, the back of the fabric is uppermost in the shed. The hand-weaving of tapestry has a fascinating story. The weaver follows the pattern with the painted design lying in the loom under the warp yarns. The main outlines of the pattern are traced on the warp yarns, but reference is constantly made to the illustration for the pattern and color. A distinctive tapestry pattern is pictorial and usually large. *Uses:* Draperies, upholstery, etc.

TARLATAN

PLAIN WEAVE

Cotton

A light weight, open weave, transparent fabric, woven from thin yarns and lightly stiffened. It may be highly glazed on one side and is then often called *Argentine cloth.* Tarlatan may be white or dyed plain colors. *Uses:* Fancy dress costumes, curtains, display backgrounds, lining stiffenings, etc.

TERRY CLOTH

PILE WEAVE

Cotton, linen

An uncut pile fabric. The loops may be on one or on both sides, all over the fabric in stripes, checks, or in a Jacquard pattern, and may be single, double, or triple. The loops are formed by extra, loosely twisted pile warp yarns wound on a second beam on the loom. The ground warp on one beam is held tight and weaves the underweave; the other warp is woven in loosely and forms the loops. When double or triple, two or three lose warp yarns are woven in. A complete discussion of the terry weave is on page 348. Designs may be woven in by the use of the dobby attachment or on the Jacquard machine. The fabric may be bleached, yarn dyed, piece dyed, or printed. Better quality terry cloth has a firm, close underweave and closely spaced loops. Terry cloth with its underweave construction and loops of loosely twisted pile yarns makes a most absorbent fabric. *Uses:* Towels, beach wear; draperies and bedspreads, etc.

TICKING

TWILL, HERRINGBONE TWILL, SATIN,
OR JACQUARD WEAVE

Cotton, linen

A very closely woven, stout fabric made of firm yarns. It is woven with dyed raw stock or dyed yarns. Ticking may have woven-in colored stripes, narrow to wide, or combinations of narrow and wide stripes; a Jacquard pattern (called damask ticking) or a Jacquard pattern combined with stripes. The best known ticking is white with colored stripes (see page 582). Ticking is sheared, sized, and calendered, resulting in a stiff finished fabric with a smooth, lustrous surface. It may be given special finishes to make it water-repellent, and germ-resistant. *Uses:* Mattress and pillow covers, upholstery, etc.

TOBACCO CLOTH

See CHEESECLOTH

TRICOT

WARP KNIT

Silk, rayon, cotton, wool, nylon

A fine, closely knitted fabric made with many warp yarns on a tricot machine. The fabric has a fine chain stitch, running lengthwise on one side and crosswise on the reverse side. It has little elasticity and "runs" only under strain. *Uses:* Underwear, dresses, shirts, gloves, scarfs, etc.

TRICOTINE

TWILL WEAVE

Wool, rayon

A clear finished, hard textured fabric woven with a steep, left to right, double, diagonal rib. The same fabric woven with wool is called "cavalry twill." *Uses:* Suits, slacks, uniforms, sportswear, etc.

TULLE

See NET

TWEED

PLAIN, TWILL, OR VARIATION OF
TWILL WEAVE
Wool, rayon, cotton

A wide range of rough surfaced, sturdy fabrics. Originally, tweeds included only fabrics woven with a homespun effect, usually with multicolored yarns. There have now been added fabrics that are more uniformly and closely woven, of smoother yarns. They include monotone tweeds, woven with yarns of different shades of the same color. The fabrics range from lighter weight suitings to heavy coatings. In addition to monotone and multicolor effects, they may be woven with plaids, checks, stripes or other patterned effects. Harris tweeds are made by hand in the Outer Hebrides off the coast of Scotland. The dyes in these yarns are cooked over peat and the smell of peat often remains in the fabric, particularly when it is damp. Tweed derived its name from the River Tweed in Scotland, on the banks of which the fabric was first woven. *Uses:* Suits, coats, jackets, skirts, dresses, trousers, sportswear, etc.

VELOUR

PILE SATIN, OR TWILL WEAVE
Wool, cotton

There are two types of velour. One is a heavy, pile-weave fabric. The terms "velour" and "plush" are often used interchangeably for this same fabric. The other fabric is woven in a satin or twill weave, and given a "velour finish," which is similar to the finish given to duvetyn. Since napped velour has a longer and more open nap, the weave is not concealed as in duvetyn. "Velour" is the French word for velvet. *Uses:* Pile weave, for draperies, upholstery, couch covers, etc. Twill weave, for coats, suits, etc.

VELVET

PILE WEAVE
Silk, rayon, cotton

A woven-in pile fabric, which usually has a cut pile but may have an uncut pile. The pile may be woven on a plain, twill, or satin ground by one of two methods. Two fabrics may be woven simultaneously, face to face, with an extra pile warp yarn connecting the two. As the fabric is woven, these warp yarns are cut at the center, leaving a pile on each fabric. Or a single fabric may be constructed with an extra pile warp yarn woven over a wire. The wire is withdrawn, leaving an uncut pile, or a knife on the end of the wire may cut the pile as the wire is withdrawn. Or a part of the pile may be cut, making a pattern of cut and uncut pile. Or the pile may be woven over groups of filling yarns which are later withdrawn leaving an uncut pile.

There are many different weights and types of velvet. Some are fine and almost transparent, others are medium weight, while others are heavier and stiff or very heavy. Velvet may be lustrous or dull, according to the type of pile yarns used. The following are some of the most important velvets.

Transparent velvet. A very light, soft draping, diaphanous velvet. The pile is slightly longer than in most velvets.

Chiffon velvet. A light weight, softer velvet, with a short, thick pile.

Lyons velvet. A heavier weight, crisp velvet that does not softly drape. It has a closely woven, stiff back, with an erect, short, thick pile.

Panne velvet. A rich looking velvet with the pile pressed down in one direction, by passing the fabric over rollers and subjecting it to steam and pressure. It has a satiny appearance. If the roller imprints a pattern in panne, it is called *embossed velvet.* Embossing is also used as an incidental step in the creation of a sculptured pattern. The areas of the pattern that are destined to stand highest (with the longest pile) are first laid flat by embossing; then the fabric is sheared to lower the height of the remaining pile. The embossed portion then is steamed upright so that it stands higher than the sheared portion.

Coating velvet. A heavy dress velvet, with an erect pile set so closely that when the fabric is folded no break or rib appears. It is often woven with a heavy, mercerized cotton back.

Brocaded velvet. A pattern is formed by taking out a part of the pile. It usually has a very sheer ground weave, which is often slightly stiffened. The fabric is woven as are other velvets. Then chemicals are applied in the desired pattern to the back of the fabric. This carbonizes the pile warp when heated, and it is removed from the fabric, leaving the remaining pile to form the pattern. This fabric is sometimes called *façonné velvet.*

Moquette. A velvet woven with large Jacquard patterns, used principally for upholstery.

Millinery velvet. Similar to Lyons velvet, crisp, with a short, close pile. It is made in narrow widths.

Upholstery and drapery velvets. This is the heaviest of the velvets, usually with a stiff back and very thick pile.

Bagheera velvet. A soft, fine, uncut pile velvet, with short, closely set loops. It is now seldom manufactured.

Uses: According to its type and weight, velvet has many uses. Women's dresses, suits, coats, robes, blouses, hats, handbags, shoes; children's coats, dresses, hats; men's robes; curtains, draperies, bedspreads, upholstery, etc.

VELVETEEN
VELVETEEN WEAVE
Cotton, rayon

A fabric with an all over, short, closely set pile. The fabric is woven flat with two sets of filling yarns. One weaves with the warp, to make the ground weave, and the other weaves into the warp at intervals (as in satin weaving) and then floats over a number of warp yarns. After weaving, these floating filling yarns are cut and brushed to form a pile. Velveteen pile is not entirely erect. It slopes slightly to create the sheen. Cotton velveteen is usually made of mercerized yarns to add to the luster of the fabric.

Hollow-cut velveteen can be achieved either by the arrangement of the float of the extra filling thread, or grooves can be cut in a piece of velveteen with a sufficiently deep nap. It is hard to discern the difference by eye. However, the former process leaves all erect fibers, along any one warp yarn, the same length, whereas in the latter process the groover does not always follow the direction of the warp yarn exactly, and an examination of the length of the napped fibers, along the line of a warp yarn, may show up variations.

Uses: Dresses, coats, suits, negligees; children's coats; draperies, bedspreads, upholstery, etc.

VENETIAN CLOTH
SATIN WEAVE
Wool, cotton

A smooth, strong, sleek fabric, usually made with warp-faced satin weave. It may be napped and pressed. Cotton Venetian cloth, made of mercerized yarns, usually two-ply, is similar to sateen but heavier and with a more lustrous sheen. It may be white or colored. *Uses:* Wool, for suits, and coats; cotton, for linings, dresses, draperies, cushion covers, etc.

VOILE
PLAIN WEAVE
Silk, cotton, wool, rayon

A sheer, open weave fabric, light weight, softly draping, and transparent. It is made of highly twisted yarns. Warp and filling yarns can be easily seen, as there is a tiny space between the adjacent yarns. Silk chiffon voile has grenadine yarns one way and twisted single yarns the other way. Wool voile is rather wiry, made of tightly twisted worsted yarns. Cotton voile in the best grades is made with a hard twisted, two-ply yarn. The fabric is gassed to remove surface fibers and to provide a clear surface texture. Cotton voile sometimes is woven with clipped spots, known as clipped dot voiles. Other versions are drawn work and embroidered voiles. Voile may be white, piece dyed, printed, or be woven with various stripes and checks. *Uses:* Women's lingerie, blouses, neckwear; children's clothes; curtains, bedspreads, etc.

WAFFLE CLOTH
See HONEYCOMB

WHIPCORD
TWILL WEAVE
Wool, cotton, rayon

A fabric having the heaviest rib of all the standard twilled fabrics. The twill is steep and prominent and runs from left to right, except for some cotton whipcords, which have ribs running from right to left. Many cotton fabrics have the back of the cloth slightly napped. At times, mixed colored and white warp yarns are woven with solid color filling yarns, producing a somewhat pepper and salt effect. *Uses:* Trousers, riding habits, uniforms, sportswear, etc.

ZIBELINE
SATIN WEAVE
Wool

A sleek, lustrous, velvety, soft fabric. The fabric is napped and then steamed and pressed. It is more lustrous than broadcloth. Zibeline received its name from the zibeline, a small fur bearing animal found in Siberia, a member of the sable family, with a fine black fur. *Uses:* Coats and dresses.

SUMMARY

PLAIN WEAVE
Airplane cloth
Albatross
Art linen
Awning duck
Balloon fabric
Batiste
Buckram
Bunting
Burlap
Butcher's linen
Calico
Cambric
Canvas
Casement cloth
Challis
Chambray
Cheesecloth
Chiffon
Chintz
Crash
Crepe
  Canton crepe
  Chiffon crepe
  Crepon
  Flat crepe
  Mourning crepe
  Romaine crepe
Crepe de Chine
Cretonne
Crinoline
Dimity
Donegal
Dotted Swiss
Duck
End-and-end cloth
Éponge
Flannel
  Outing flannel
  Shaker flannel
  Forestry cloth
  Flannelette
  Kimono flannel
Fleece
Gauze
Georgette crepe
Gingham
Glass toweling
Habutai
Homespun
Jaspé cloth
Lawn
Longcloth
Madras
Marocain
Mousseline de soie

Mull
Muslin
Nainsook
Ninon
Nun's veiling
Organdy
Osnaburg
Parachute fabric
Percale
Percaline
Pongee
Radium
Sailcloth
Scrim
Seersucker
Shantung
Taffeta
Tarlatan
Tweed
Voile

RIB WEAVE
Bengaline
Broadcloth
Faille
Grosgrain
Ottoman
Poplin
Rep

BASKET WEAVE
Balloon cloth
Hopsacking
Monk's cloth
Oxford cloth
Sharkskin

TWILL WEAVE
Barathea
Broadcloth
Byrd cloth
Cassimere
Cavalry twill
Cheviot
Chinchilla cloth
Cottonade
Coutil
Covert cloth
Crepe
  Crepe meteor
Denim
Donegal
Drill
Duvetyn
Elastique
Flannel
  Canton flannel

Outing flannel
French flannel
  Flannelette
Foulard
Frieze
Gabardine
Imitation-beaver
  cloth
Jean
Kersey
Mackinaw cloth
Melton
Serge
Sharkskin
Surah
Ticking
Tricotine
Tweed
Velour
Whipcord

SATIN WEAVE
Bolivia
Crepe
  Crepe charmeuse
Doeskin
Imitation-moleskin
  cloth
Melton
Sateen
Satin
Ticking
Venetian cloth
Zibeline

CREPE WEAVE
Momie cloth

**DOUBLE
WEAVE**
Chinchilla cloth
Mackinaw cloth

PILE WEAVE
Astrakhan
Fleece
Frisé
Plush
Terry cloth
Velour
Velvet

**JACQUARD
WEAVE**
Brocade
Brocatelle

Crepe
Damask
Matelassé
Tapestry
Ticking

LENO WEAVE
Grenadine
Leno
Marquisette

KNITTED
Astrakhan
Balbriggan
Chinchilla cloth
Fleece
Jersey
Milanese
Tricot

TWISTED
Net
Bobbinet
Maline
Tulle
Mosquito net

KNOTTED
Netting

FELTED
Felt

**FABRIC NAMED
FOR WEAVE**
Bedford cord
Bird's-eye
Corduroy
Honeycomb
Huckaback
Piqué
Velveteen

**FABRIC NAMED
FOR FINISH**
Moiré
Plissé

**FABRIC NAMED
FOR YARN**
Bouclé
Chenille
Lamé
Ratiné

Opposite page: Fibers, yarns, and fabrics are physically tested and dyes and finishes are chemically tested to determine acceptable standards and constantly improve fabrics (American Wool Council, Inc.).

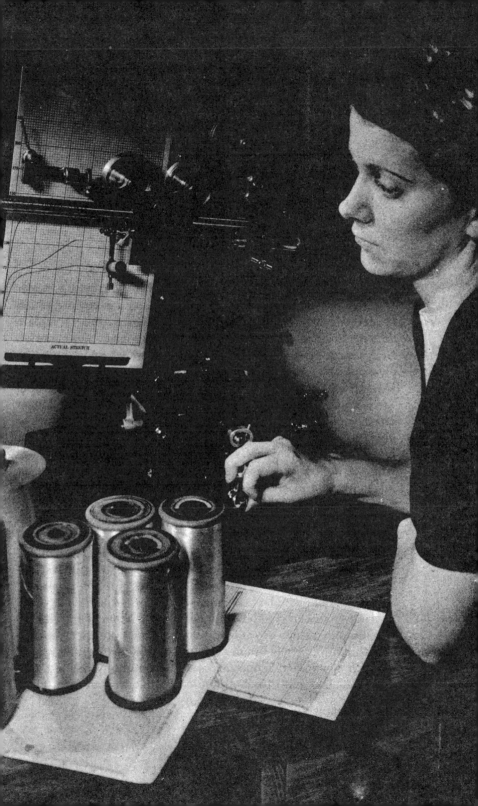

TESTING AND STANDARDS—

Constantly improve fabrics

One chapter is completely inadequate to acquaint the reader with the testing and identification of fibers and the standardization of methods and results. It would take a lengthy book to explain the tests made, the methods applied, the equipment and materials used, exactly how the tests are carried out, what results should be obtained, and what the results mean in terms of standards and qualities. Often one test is not conclusive. As an example, the burning tests, which were widely used to identify fibers, are no longer satisfactory because of the extensive blending and combining of fibers in fabrics. It may show one fiber that is present but not reveal others. Authoritative fiber content tests can only be made with the use of chemicals. Without a well-equipped laboratory it is not possible to positively identify or conclusively test a fabric for fiber content, quality, colors, or finishes.

Therefore, a book such as this can do little more than stress the need for fiber identification and for standardized quality; and discuss, in a general way, what is being accomplished in the field. National organizations, interested in tests and standards, have published excellent literature, in brief and clear form, on all acceptable tests and standards. It explains what has been done, what is being done, the methods, and the results. Many of their books and pamphlets can serve as classroom texts and reference books.

Millions of people in our country are consumers of fabrics, yet only a few thousand connected with the technical phase of fabric manufacturing and distribution are in a position to identify all fabrics and know how they will perform in use. The average consumer is not a technician, nor should he or she be expected to have the knowledge or the equipment to test fabrics.

Consumers speak of the requirements of fabrics in a nontechnical language, qualifying what they expect of each fabric they purchase. They want fabrics made of fibers and yarns that will give them the features and length of service required. They may want fabrics that are sheer, medium weight, or heavy; loosely or tightly woven; white, plain colored, striped, checked or patterned, printed or decorated; light or dark, dull or with a lustrous sheen; smooth and sleek, or coarse, or rough textured; crisp, or soft and draping; for cold, warm, or rainy weather; and giving reasonable wear under ordinary care.

The technician interprets these consumer needs into specifications and methods of constructing and finishing the fabrics. He determines the best fiber or fibers, yarn or yarns, and weave or weaves for each fabric, considering the use and the service required. He decides whether it will be white, plain color, printed, or decorated. His selection of dyestuffs is guided by the fiber or fibers used and by the ultimate fastness required. As a general example, a high degree of fastness to water or light has to be assured, or the fabric can be dyed a less fast color if it will not be greatly exposed to either. He considers the surface texture—dull or lustrous, smooth or rough—and knows which fibers, yarns, weaves, and finishes to use to obtain the desired result. He knows the fibers, yarns, and constructions that produce fabrics to be worn in heat or cold. He selects the finishes that give the fabric its final personality and that help it retain its original appearance.

All along the way he tests fibers, yarns, fabrics, dyes, and finishes. Does it have controlled shrinkage? Does the color have a high degree of fastness to light and washing and to acids and alkalies? Will the durable luster, or crispness, or crease-resistance remain through many washings? Will the sizing, when washed out, leave a weak, sleazy fabric?

These are only a few of the multitude of questions that must be answered before the fabric reaches the ultimate consumer. Each fiber, each yarn, each weave, each dyestuff, and each finish makes a certain, definite contribution to a fabric. To test and standardize one or two and not the others, means very little. A firmly and strongly woven fabric, made of low-quality fibers and yarns is no better than a poorly woven fabric constructed of the highest quality fibers and yarns. Neither will give the maximum service. A well-woven fabric, made of high-quality yarns, will not give service if the dyestuffs are not right or are improperly applied. Therefore, there must be a completeness of the right selection of fibers, yarns, constructions, dyestuffs, and finishes, considering the ultimate use of each fabric.

In the past few years, particularly since World War I, testing has advanced to the stage where it is widely practiced and becoming more and more standardized. Individual manufacturers have technical and research departments. These manufacturers have joined with others in the same industry to organize industrywide research departments. No one industry is independent but is interrelated with many allied industries. For example, the fabric manufacturer who uses cotton, wool, and rayon is dependent on the men who grow the cotton, raise the sheep, and make the rayon. He is dependent on the producers of dyes and of finishing compounds. He must rely on the designer and producer of the machines and, finally, he and the manufacturer of consumer products made of his fabrics must speak a common language.

Large chain stores, mail-order houses, store buying groups, as well as some individual stores now maintain laboratories that test merchandise and set up buying specifications and standards. The National Retail Dry Goods Association cooperates with individual and national testing and standard organizations. The National Consumer-Retailer Council was an outgrowth of the retailers' recognition of consumer participation in fabric testing and standards. Many of the women's national organizations are working with industry in an advisory capacity. Individually owned testing laboratories are functioning. As time goes on, the results of testing by manufacturers, industries, and other organizations will be reflected in better consumer goods.

To coordinate the work of all these individual manufacturers, industries, groups, and associations, national organizations have been formed. They facilitate the exchange of information, promulgate comparative tests, set up standards, and assist in the adoption of a common terminology, a definition of terms that are adopted by all allied industries and groups. Some set up definite standards that are nationally accepted; others develop tests with no effort to set standards, while still others promulgate standards and tests with the expectation that all reliable manufacturers will use them even though they are not compulsory.

Four very important national organizations are briefly discussed here, because they are valuable sources of reliable data for those who wish to study these problems further. Moreover, they are mentioned to show the extent to which standards and testing principles are now being developed. These four organizations are: United States Bureau of Standards, a bureau of the Department of Commerce; American Association of Textile Chemists and Colorists (A.A.T.C.C.); American Society for Testing Materials (A.S.T.M.); and American Standards Association (A.S.A.).

The National Bureau of Standards was established in 1901. Its function is the development, construction, custody, and maintenance of reference and working standards, and their intercomparison, improvement, and application in science, engineering, industry, and commerce. More simply stated, it serves as a clearing house for scientific and technical information. Fabrics are only a small part of its vast scope.

The bureau has two major work projects. One is scientific and technical research, the other, the development of standards. It works with the national government and with state governments and cooperates with industrial and research organizations as well

Right: Reliable manufacturers examine every yard of fabric before it leaves their plants (Celanese Corporation of America). Below: Rugs and carpets are burled (wool term) and each yard of fabric is examined and any flaws are corrected (Mohawk Carpet Mills, Inc.).

as with similar institutions abroad. Its services are available for a fee to private industry. Frequently an industry will send research associates to work in the National Bureau of Standards on projects that are of joint interest.

The bureau has made many discoveries and has made valuable contributions to methods and standards. It has no police power but offers its findings to interested industries and associations. It has a large staff of about 1,000 employees, some two-thirds of whom are technically trained. Its laboratories are well equipped for modern, scientific studies. The bureau's publications may be obtained from the Superintendent of Documents, Government Printing Office, Washington, D. C.

The American Association of Textile Chemists and Colorists is a society of textile chemists, colorists, dyers, and others. Established in 1921, it now has some 3,500 members who represent a nationwide cross section of the textile and allied industries. The organization states, "The A.A.T.C.C. does not prepare specifications to control textile materials. It supplies factual data in the form of Standard Test Methods on the basis of which such specifications can be drafted by individual manufacturers, but its policy is not to concern itself with trade specifications or informative labels, because these represent merchandising and industrial features."

Its functions cover the following: 1. To promote increased knowledge of the application of dyes and chemicals in the textile industry. 2. To conduct and sponsor research in practical problems on chemical processes and materials of importance to the textile industry. 3. To establish and maintain quality standards, whereby the textile industry is enabled to regulate itself from within, on a practical basis. 4. To provide for its members channels for the free exchange of ideas and professional experiences, for the benefit of the industry. 5. To serve as a lasting educational medium for the technical personnel of the industry, by keeping its members posted on new developments. 6. To make possible a number of publications dedicated to the above purposes.

The actual work is under the direction of the general research committee and is largely carried out by thirty-seven subcommittees formed of membership personnel, all specialists in their field of work. These subcommittees study and report on a wide range of processes and they set up quality standards. For example, they analyze fabrics and colors for resistance to water, shrinkage, dry cleaning, light, creasing, perspiration, acids and alkalies, chlorine, rubbing (crocking), insect pests, mildew, and fire. They test for fiber mixtures, permanency and effect of finishes on fabrics, and set up standard methods of chemical analysis and testing for determining the finishing materials used.

The above is only a partial list of the many finishing and dyeing processes covered by this association. Its services are "freely available to the entire industry, and all textile manufacturers and industrial organizations are invited to use them." The A.A.T.C.C. cooperates with other organizations interested in textile standards. As standards and testing methods are developed, they are adopted by cooperating manufacturers, industries, and organizations interested in the project.

The association maintains a laboratory at the Lowell Textile Institute where much of its research and development work is carried out. This organization issues a yearbook, which records the activities and accomplishments of the organization's committees. This is a useful reference book for the student as well as for the chemist and colorist, as it records in detail the latest test methods of the association and lists textile chemicals and their uses as well as a great deal of other valuable data.

Above: Examining rayon yarns under a powerful microscope to discover any flaws in the filament (Celanese Corporation of America). Below: Weighing wool to assure evenness of texture and weight (American Wool Council, Inc.).

Tiny skeins of rayon are weighed (Celanese Corporation of America).

The American Society for Testing Materials is a "national, technical society for 1. the promotion of knowledge of the materials of engineering, and 2. the standardization of specifications and the methods of testing." Established in 1898, with seventy members, it now has a membership of more than 5,300 individuals and organizations, representing all parts of the United States, Canada, South America, and many other foreign countries.

There is one committee in this society that interests itself in fabrics. It has a membership of 212, including 91 fabric producers, 51 consumers, and 70 whose interest is general. The organization believes that suitable standards can be set only by such a balance of interest as its membership gives. The membership includes leading fabric manufacturers as well as those who use fabrics to make consumer goods. Most of the large mail-order houses and chain store organizations are members, who present the store's requirements. Consumer advisory groups give the consumer's point of view. Government departments and bureaus, commercial testing organizations, certain colleges and technical schools represent different interests.

The work is done by committees and subcommittees made up of organization members. The scope is wide, and the results have been valuable to all who make use of them.

Testing methods and standards have been set up for specific

fibers; wool, cotton, rayon, glass, asbestos, linen, and other bast and leaf fibers. These include standards for the fibers, yarns, and fabrics. Further advancement has been made in defining standards that are common to all fibers. General and specific methods of testing and procedures for the use of testing machines have been worked out. A common nomenclature, definitions of words and terms, and phraseology have been carefully studied to give just the right shade of meaning, understandable to all concerned. Testing and standards for bleaching, dyeing, and finishing have been developed, and standards have been prepared for certain types of household and garment fabrics.

This is but a general summarization of the functions of the society. Each year a book is published which is a compilation of their standards. It is prepared in such a way that it is readily understood and is usable as a textbook. No standard or test is included until it has proved successful and practical in use by the fabric industry. The book gives, in detail, testing procedure for fibers, yarns, fabrics, bleaching, dyeing, and finishing. Any testing procedure must be complete in every minute detail or it is considered inadequate and is not included.

New standards and tests are being developed constantly, and all final standards are reviewed every six years. The society works closely with industry, other organizations, and institutions such as schools and colleges. It cooperates with the National Bureau of Standards, the United States Department of Agriculture, and other government departments and bureaus.

The testing methods and standards have had a marked influence on manufacturing, testing, and even the selling of fabrics. Frequently disagreements between buyer and seller are settled by the authoritative and unbiased standards that have been developed.

The demand for standard specifications and methods of testing for ultimate consumer goods and the promotion of knowledge of materials for such goods have been growing rapidly in recent years and have been reflected in the work of the organization in a number of important fields, among them fabrics. Under a recently adopted policy, this society will extend its present work in the preparation of standard tests and specifications for ultimate consumer goods, and will place in the hands of a new administrative committee the responsibility for planning and directing the development of this program.

667

An example of the wide variety of equipment required for arraying fibers according to length. The following is not an all inclusive list: (a and c) banks of combs, (b) forceps, tips padded with hard leather for transferring fibers from one set of combs to the other, (d) depressor for placing fibers in combs, (f) dissecting needle, (g) fork for scooping up fiber groups off velvet surface, (h) aluminum plate covered with velvet cloth, (i) special rule for measuring length of fiber groups, (k) smooth plate for placing fibers onto velvet surfaces, (l) wire rack for holding fiber groups wrapped in paper, (m) smooth pointed tweezers, (n) lift for raising combs in place, (o) rack for holding velvet covered plates (American Society for Testing Materials).

The American Standards Association's membership includes eighty-six national technical societies, trade associations, and government departments. Organized in 1918 it is an outgrowth of the need for standards required for World War I. To date it has built up more than 800 American standards in and out of industry.

The major function of the association is "to provide the means for arriving at national industrial standards. In doing this it coordinates and unifies the standardization activities of many groups, doing work which these groups could not do at all or could not do as effectively themselves."

The work is carried on by committees. More than 3,000 men and women, representing manufacturer, consumer, safety, governmental, and other interests are serving on these committees. They are constantly developing new standards or revising standards in use. Anyone can seek information on standards now accepted or on new standards that are needed. The hundreds of developed standards include "subjects as diverse as steel girders and the shrinkage of cotton goods."

The A.S.A. also acts as an authoritative channel for international cooperation through the newly formed United Nations Standards Coordinating Committee and through its contacts with the Latin American countries. A great degree of international standardization has been reached in many of the sciences. There is a common denominator in international weights and measures. However, much less has been accomplished for standardization in industry.

Standards for consumer goods are a part of this association's functions. These standards are for the product as a whole, using the tests and standards for individual fibers and fabrics as set up by such organizations as the A.S.T.M. and the A.A.T.C.C. For example, one project included the study of the materials and the construction used for bedding and upholstery, with a method of labeling for consumer identification. The methods for developing standards and the standards themselves are agreed upon by all groups affected. Participating as an "Advisory Committee on Ultimate Consumer Goods" are representatives of groups whose memberships include millions of consumers. The National Retail Dry Goods Association provides the retailers' point of view and five federal government bureaus actively participate.

TESTING PROCEDURES

Testing procedures used by manufacturers, industries, stores, local and national testing associations, and government bureaus have been standardized in regard to equipment used (including machines, testing samples, chemicals, and so forth), methods, and results.

By developing comparable testing methods, it is possible to obtain comparable results on fibers, yarns, and fabrics of the same general qualities. It is from these results that standards have been set.

Testing is divided into two types: physical and chemical. Physical testing is the examination and testing of fibers, yarns, and fabrics to determine their content, structure, strength, and other qualities. Chemical testing is also used for all three—fibers, yarns, and fabrics—to identify fiber content and to test the fibers, yarns, constructions, dyes, and finishes under normal use, such as their reaction to water, light, acids, and alkalies.

The purpose of the following summary is to point out the knowledge and the infinite care required for adequate testing,

Mullen tester which is used to determine the bursting strength of woven or knitted goods (Calco Chemical Division, American Cyanamid Co.).

rather than a description of the tests or the results obtained. However, it shows briefly what is being accomplished and what progress is being made to raise the standards of consumer goods.

Physical Testing

Physical testing requires the use of delicately balanced machines, with precision adjustments, and other equipment that permits the analysis of thousands of individual tiny fibers, the counting and measuring of their fineness, length, and strength.

In testing, the same as in every step along the trail of fabric manufacturing, equipment and machines carry out mechanical feats that cannot be duplicated by human handwork. The machines are results of man's creative ability, designed to perform operations essential for a closer approach to perfection.

A few of these many testing devices are illustrated here. Those who use them do not fall prey to routine but bring the devices to life by developing new methods and setting new standards.

A breaking strength and elongation machine is used to test individual strands, skeins of yarn, or fabrics. Metallic clamps hold the fabric on the machine; the skeins of yarn are on cylindrical spools, one of which can rotate. Clamps or other devices hold the individual strands being tested. To test for bursting strength, an attachment may be added to the breaking strength machine or a regular bursting tester may be used.

A thickness gauge has a weight that presses down on the fabric. A delicate dial mechanism on the machine registers the thickness.

A twist tester holds a strand between two clamps, one of which

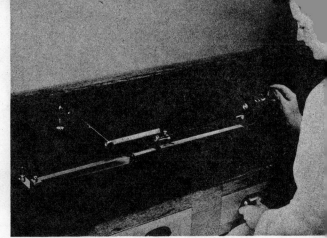

Suter twist counter which is used to determine the amount of twist in yarns (Calco Chemical Division, American Cyanamid Co.).

rotates. A counting device indicates the number of revolutions to a minimum accuracy of one turn.

The tests. If you were to visit a testing laboratory where physical tests are being conducted, you would see numerous testing machines, the microscope with its powerful lens, magnifying from 100 to 500 times; glass slides; clips for picking up the fibers; dissecting needles; devices for cutting cross sections of the fibers; fine combs for laying the fibers parallel. There may be clamps holding strands of fibers or yarns, or hundreds of tiny fibers may be lying parallel on a black velvet cloth.

The fibers are tested. Hundreds of tiny fibers from one batch may be tested. They are counted, measured, weighed, and examined. They are examined in cross section for identification, whether the batch is made up of two or more different fibers. If two or more, the approximate percentage of each is estimated. The average length is calculated and the fineness determined. They are further tested for their tensile strength and resilience.

The fineness of fibers is measured in microns, which are 1/1000 millimeter. A millimeter is 0.03937 inch. Therefore, a micron is 0.000039 inch. The gauge of the animal and vegetable fibers and the deniers of the synthetic fibers are determined in microns. The higher the number of the gauge, the finer the count; but, the lower the number of deniers the finer the count.

The yarns are tested. To completely test a yarn, the fibers of which the yarn is composed are also examined and tested. In addition, the yarn itself is tested. One of the most important factors is the twist of the yarn, whether single or made up of plied yarns

Left: Testing a carpet fabric for wear by submitting it to mechanical crushing, scuffing, and abusive friction. Right: Checking a carpet pile for height with a thickness gauge during and after wear test (Mohawk Carpet Mills, Inc.).

(more than one), and whether an S or Z twist or combinations of these twists. The number of twists per inch is also determined. The yarns are untwisted and their component parts are examined for the method of twisting, the balance or length relationship of the plied yarns, and their ability to remain twisted. These yarns range from the single, slightly twisted yarns, to the heavy, ply or core yarns. Some have nubs or slubs, or are thick and thin, or are creped. Each is tested for the qualities it is to give the finished fabric.

The number of the yarn is calculated. Each kind of fiber has an individual measuring standard (see Chapter 10). For example, the yarn number of cotton is determined by the number of 840 yard hanks it takes to make a pound. The finer the yarn, the greater the number of hanks. The breaking strength of individual strands as well as of skeins of yarn is tested to determine their lasting quality under wear and strain.

The fabrics are tested. To completely test a fabric, the yarns, and frequently the fibers also, are tested to determine the kind of yarns as well as the fiber content. In addition, the fabric itself is physically tested. The width and length are measured, and the thickness and weight of the fabric are determined. The yarn count is taken, that is, the number of warp yarns and filling yarns to an inch are counted. Some fabrics, as sheetings, require an appreciably balanced count, others, for less strenuous wear or woven differently and with yarns of different weight, may have many more yarns in the warp than in the filling or vice versa.

The breaking strength is tested in a tensile testing machine, using either a grab test sample of the fabric, a narrow strip, or a raveled strip. Fabrics are also torn warpwise and fillingwise to determine their tearing strength. Breaking and bursting tests are essential to determine the wearing qualities of a fabric and are applied particularly to those that will be subjected to strain.

Chemical Testing

Chemical testing concerns itself more especially with dyes and finishes, the processes that use chemicals for their application. It includes fastness to light, sun, water, acids, and alkalies, and other chemicals. It also tests the lasting qualities of functional finishes. It is interested in the washing, ironing, and dry cleaning of fabrics and their resistance to friction. The following is a brief resumé of the tests and standards set up for fabrics.

Fastness to light. While fabrics used for some purposes do not require a high degree of fastness to light, all those used for wearing apparel and most of those used in the home, such as curtains and draperies, do need a high resistance to fading in light.

Fabrics are light tested by two methods: exposure to the sun and exposure to artificial light. For the sun test, the fabric, under glass, is exposed to the sun in a location free from shadows, as on a roof. Exposure is made between 9 A.M. and 3 P.M. on sunny days from April 1 to October 1.

For the artificial light test the fabric is placed in a Fade-Ometer.

Taber abraser which is used to determine the resistance of fabrics to abrasion. This is often related to the wear resistance of a fabric. (Calco Chemical Division, American Cyanamid Co.).

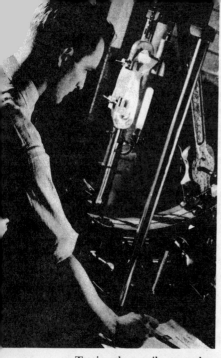

The fabric rotates about the arc of light to permit uniform exposure. Fabrics after being exposed are kept in a dark room for at least two hours and then compared with the original fabric.

Standards have been set up for the number of hours of exposure that are required for the color to fade. For example, if a fabric does not fade appreciably in five hours but does in ten hours of exposure, it is in one class. If it does not fade appreciably in 96 hours but does in 192 hours, it is in another classification. There are nine such standards, representing degrees of fastness to sun or artificial light.

Fastness to rubbing. The problem is the rubbing off or crocking of dyestuffs used to color fabrics. During wear, fabrics in garments are subjected to considerable rubbing and friction at feet, elbows, wrists, underarm, seat, neck, and so forth. It is essential that fabrics, particularly those worn next to the skin, have sufficient colorfastness to resist rubbing as well as perspiration, stain, and water.

Causes of rubbing may be poor dyes, incorrect application of the dyestuffs, or poor rinsing after dyeing. Or it may be caused by the use of wrong dyes on certain fibers.

Dyestuffs injured by rubbing the fabric may cause the color to rub off onto other fabrics and, if more severe, the fabric itself may be rubbed away. Water spots may cause the color to bleed and stain other fabrics and at the same time may produce spotty fading of the fabric, but this is not rubbing or crocking.

Testing the tensile strength of a fabric showing (below) a close up of the bursted fabric (Pacific Mills).

The fabric is tested in a crockmeter machine. The dyed fabric, dampened or dry, is placed on the machine. A bleached, soft cotton fabric is clipped over a so-called finger, which is weighted and rests upon the cloth. By turning the handle, the white cloth is rubbed on the fabric, and the amount of coloring on the white cloth is noted, as is the color of the dyed fabric.

Fastness to perspiration, acids, and alkalies. Fabrics used for wearing apparel must have a high degree of resistance to perspiration. The fabric may change color or rot if it comes in contact with perspiration, which is an acid secretion given off by the body. The acid may in time become alkaline.

The cause of fading is the action of the acid or alkali, which may change or remove the dyestuffs. Strong alkaline perspiration will ultimately rot animal fibers in fabrics.

Dyed fabrics that may be exposed to perspiration must be tested for resistance to the acid and alkalies in perspiration. Street dust may have alkali content and many washing compounds contain alkali. Also, in finishing dyed fabrics, acids may be used in carbonizing, milling, cross-dyeing, printing, and sizing. Alkalies may be used for fulling, washing and scouring, mercerizing, and sizing.

Dyed cottons are tested for fastness to acids by spotting with hydrochloric acid or acetic acid. Wool colors are usually fast to acid. For fastness to alkalies, cottons are tested by steeping for two minutes in concentrated ammonia solution, or in a sodium carbonate solution. Wool and silk are steeped for two minutes in a sodium carbonate solution or tested with calcium hydroxide paste, or exposed for twenty-four hours to concentrated ammonia. These tests are standardized as to the chemical percentage of the solution, its temperature, length of time required, and results.

Testing a fabric with a spray tester for water-repellency (The Du Pont Company).

Testing fabrics in a Fade-Ometer for resistance to fading in light and sun (American Wool Council, Inc.).

Fastness to dry cleaning. Dyed fabrics are tested for their reaction to cleaning agents. The samples, about 2 by 4 inches, are rotated in a cleaning fluid, which is usually carbon tetrachloride (or Stoddard solvent or chlorethylene) and a very small percentage of dry-cleaning soap, water, and alcohol. A second cleaning is in the solvent only. The fabric is then rinsed and dried, hand or steam pressed, or steamed, and compared with the original sample. Standards have been determined on which to judge the results.

Fastness to sea water. A color may fade when subjected to salt water. Bathing suits, beach robes, and towels that are used in sea water should be dyed with colors that are resistant to salt water. Certain dyes in wools give a better degree of fastness to salt water than do any dyes in any other fibers.

The cause of fading is the action of the sodium chloride (salt) on the chemicals in the dyestuffs. Bathing suits should always be thoroughly rinsed before drying.

If dyed cottons and linens are fast to washing, they are also fast to sea water. Therefore tests are made on silk or wool fabrics that may be exposed to sea water. Since sea water is not readily available, a water solution containing sodium chloride and anhydrous magnesium chloride is used. The dyed fabric is wrapped rather tightly with undyed fabrics and is placed in a closed vessel containing the solution. After twelve to twenty-four hours the staining on the undyed fabrics is noted.

Shrinkage of fabrics. When testing for shrinkage, a fabric is subjected to very much the same treatment as it would receive in home or laundry washing. A sample piece, at least 20 inches square, is used. The soap, the temperature of the water, and length of washing time, are all standardized. The sample is washed, dried, pressed, and then measured in length and width, and the shrinkage calculated for both length and width.

Pressing of fabrics. Other tests are made to judge the reaction of fabrics that are subjected to pressing. Standards have been set up for fabrics that change color when heat is applied. These standards are determined by the length of time that is required for the fabric to return to its normal color. Fabrics, both wet and dry, are also tested by pressing with a white pressing cloth between the iron and the fabric to find out how much of the color bleeds into the pressing cloth.

Resistance to insect pests. A great many tests and standards have been set up for determining the resistance of wool yarns and fabrics to insect pests. These pests include moths and carpet beetles. Both untreated fabrics and those given insect-resistant finishes are tested.

Larvae of the black carpet beetle or of the webbing clothes moth are used, as it is the feeding of the larvae that causes moth damage. They are placed on the fabric in a dish, box, or cage. After twenty-eight days the yarn or fabric is examined for holes present and broken and sheared fibers. The quantity of excrement deposited is noted, and the number of larvae remaining alive are counted.

If the fabric has been treated with a compound to make the fabric resistant to insect pests, the fabric is given further tests to determine the lasting quality of the compound. It is washed, dry cleaned, and pressed. It is further tested for light, abrasion, sea water, perspiration, acid, and alkali.

INDEX

The items listed in bold face type are the chapter titles and the pages included in each chapter. The items listed under chapter headings are a brief summary of subject content.

Pile attachments (on loom), 320-321
Pile looms, 320
Pillow lace, 395
Pima cotton, 71-72
Piqué fabric, 645; cotton, 91, 92-93; home fabrics, 578, 590; rayon, 199
Piqué weave, 340
pH, 431
Phenol, 444, 568; phenol-formaldehide resins, 449, 566
Photoengraving method (printing), 516
Phthalates, 443, 565, 566
Physical testing, 668-671; fabrics, 670-671; fibers, 669; yarns, 669-670
Plain circular machines, 384
Plain looms, 316-318
Plain rib, 367
Plain stitch (knitting), 365-367
Plain weave, 330-331
Plain woven fabrics, 329-334; basket weave, 332-333; plain weave, 330-331; rib weave, 333-334
Plating or plaiting, 372-373, 407
Plissé fabric, 646; cotton, 85; home fabrics, 578
Plissé printing, 521-522
Ply yarns, see twisting yarns, 289
Plush, 646; home fabrics, 592
Point d'esprit, 411
Point de Paris, 407, 412
Point de Venise, 412
Polyacryl compounds, 565
Polyamide, 226; polyamide resinoids, 450
Polymer, 226, 233, 234, 237
Polymerization, 226, 231, 234, 237, 449, 450, 514, 566, 568
Polyvinyl acetate, 450; polyvinyl chloride, 450; polyvinyl compound, 566, 568
Poplin, 646; cotton, 91; home fabrics, 578, 590; rayon, 199; wool, 44
Pongee, 646; home fabrics, 586
Positron, 428
Potassium, 440-441; potassium antimony tartrate, 441, 479; potassium bichromate, 481, 486; potassium carbonate, 441; potassium chloride, 441; potassium compounds and their uses, 440-441; potassium hydroxide, 441; potassium nitrate (salt peter), 435, 441; potassium sulfate, 441; potassium sulfide, 441; potassium tartrate, 441
Preparing cotton fibers, 80-81; blending fibers, 80; breaking up bunches, 80-81; forming picker lap, 81; opening and cleaning, 80
Pressing, 548-549; silk, 161; wool, 59
Pressure method of engraving rollers, 515
Pretreatments, chapter 16, 453-462
Bleaching all fibers, 459-461; carbonizing wool, 458-459; immunizing cotton, 456; kier boiling cotton, 454-455; mercerizing cotton, 455; scouring wool, 456-458; summary, 462
Primary colors, 488-491
Primary wall (cotton fiber), 73
Printing, chapter 18, 507-532
History, 508-511; methods, 519-521 (application, 519-520; blotch, 519-520; direct, 519-520; discharge, 520-521; resist, 521); pastes, 511-512; pigments, 512-514; roller machine, 516-519; summary, 532; types (batik, 508; block, 524-525; drum, 523-524; electrocoating, 530-531; flock, 529-530; plissé, 521-522; roller, 515-516; screen, 526-527; spray, 528; stencil, 527-528; stencil lacquer, 530; vigoreaux, 522-523; warp, 522); Western-world printing, 509-511
Printing, cotton, 101; linen, 128; rayon, 219; silk, 160
Printing methods, 519-521; direct, 519-520; discharge, 520-521; resist, 521
Printing pigments, 512, 514
Properties of substance, 423
Protein fibers, casein, 238; soybean, 242
Proton, 428
Pulled wool, 21

Punch machine, 397
Pure dye silk, 562
Purl stitch, 368
Pyroxylin, 165, 567, 594

Quaternary ammonium compounds, 566, 568, 569
Quercitron bark, 471
Quilling, 283
Quilted fabrics, 592

Rabbit, 29
Raccoon, 29
Racking, 373
Radical, 426
Radium, 647; rayon, 206
Rambouillet sheep, 9, 10, 16
Ramie, 137-139; degumming, 138-139; fiber, 138; fiber to fabric, 138-139; growing, 137-138; plant, 138; properties, 139; uses, 139; yarn, 139
Random yarn, 295
Raschel-knitting machine, 382-383
Raschel stitch, 370
Ratiné fabric, 647; cotton, 88; home fabrics, 591; rayon, 202; wool, 52
Ratiné yarn, 290
Raw silk, 155, 158-161
Raw-stock dyeing, 497-498
Rayon, chapter 7, 163-220
Cellulose, 169; change liquid to solid, 190-193; change solid to liquid, 178-185; chemicals, 188-189; cotton lint and linters, 174-177; delustering, 196-197; dyeing, 217-219; fabrics, 198-212; filament and fiber to fabric, 197; filament yarn preparation, 193-194; finishing, 213-217; history, 164-169; knitting, 197; printing, 219; spinnerets, 186-187; spun rayon, 194-196; summary, 220; twisting, 197; weaving, 197; wood pulp, 170-173
Rayon fabrics, 198-212; creped, 198; knitted, 212; lustrous, 204-205; luxurious, 206; pile surface, 210-211; ribbed, 199; sheer, 208-209; slubbed and nubbed, 202-203; twilled, 200; utility, 201
Rayon filaments, 188
Rayon processes, see filament rayon processes
Rayon shrinking, 555-557
Rayon staple fiber, history, 168-169; length of fibers, 196; manufacture, 194-196
Reducing agents, 475; reduction, 475
Rayon warp yarn preparation, 301; warp sizing or slashing, 301
Reed (on loom), 315
Reel, 296, 300
Rep, 647; cotton, 91; home fabrics, 590
Repoussé lace, 401
Reprocessed wool, 61
Réseau, 407
Resins, 449-451, 487-488
Resist printing, 521
Retarding agents, 474
Retting flax, 117-119; chemical, 119; cold water, 118; dew, 118; warm water, 117-118
Reused wool, 61
Rib circular machines, 384
Rib stitch (knitting), 367-368
Rib weave, 333-334
Ribbon lapper, cotton, 276
Richelieu, 412
Ring spinning machine, 273
Roller engraving, 515-516; pantograph, 515-516; photoengraving, 516; pressure, 515
Roller printing machine, 516, 518-519
Romaine crepe, 622; rayon, 209
Romeldale sheep, 18
Romney Marsh sheep, 17
Roughing (flax fibers), 121, 277
Roving; cotton, 82, 276-277; linen, 123, 281; spun silk, 158, 285; wool, 41, 270
Rug and carpet fabrics, 596-603
Rug and carpet looms, 324-328
Rug cushions, 419, 603

687

TECHNOLOGY AND SOCIETY

An Arno Press Collection

Ardrey, R[obert] L. **American Agricultural Implements.** In two parts. 1894

Arnold, Horace Lucien and Fay Leone Faurote. **Ford Methods and the Ford Shops.** 1915

Baron, Stanley [Wade]. **Brewed in America:** A History of Beer and Ale in the United States. 1962

Bathe, Greville and Dorothy. **Oliver Evans:** A Chronicle of Early American Engineering. 1935

Bendure, Zelma and Gladys Pfeiffer. **America's Fabrics:** Origin and History, Manufacture, Characteristics and Uses. 1946

Bichowsky, F. Russell. **Industrial Research.** 1942

Bigelow, Jacob. **The Useful Arts:** Considered in Connexion with the Applications of Science. 1840. Two volumes in one

Birkmire, William H. **Skeleton Construction in Buildings.** 1894

Boyd, T[homas] A[lvin]. **Professional Amateur:** The Biography of Charles Franklin Kettering. 1957

Bright, Arthur A[aron], Jr. **The Electric-Lamp Industry:** Technological Change and Economic Development from 1800 to 1947. 1949

Bruce, Alfred and Harold Sandbank. **The History of Prefabrication.** 1943

Carr, Charles C[arl]. **Alcoa, An American Enterprise.** 1952

Cooley, Mortimer E. **Scientific Blacksmith.** 1947

Davis, Charles Thomas. **The Manufacture of Paper.** 1886

Deane, Samuel. **The New-England Farmer,** or Georgical Dictionary. 1822

Dyer, Henry. **The Evolution of Industry.** 1895

Epstein, Ralph C. **The Automobile Industry:** Its Economic and Commercial Development. 1928

Ericsson, Henry. **Sixty Years a Builder:** The Autobiography of Henry Ericsson. 1942

Evans, Oliver. **The Young Mill-Wright and Miller's Guide.** 1850

Ewbank, Thomas. **A Descriptive and Historical Account of Hydraulic and Other Machines for Raising Water,** Ancient and Modern. 1842

Field, Henry M. **The Story of the Atlantic Telegraph.** 1893

Fleming, A. P. M. **Industrial Research in the United States of America.** 1917

Van Gelder, Arthur Pine and Hugo Schlatter. **History of the Explosives Industry in America.** 1927

Hall, Courtney Robert. **History of American Industrial Science.** 1954

Hungerford, Edward. **The Story of Public Utilities.** 1928

Hungerford, Edward. **The Story of the Baltimore and Ohio Railroad, 1827-1927.** 1928

Husband, Joseph. **The Story of the Pullman Car.** 1917

Ingels, Margaret. **Willis Haviland Carrier, Father of Air Conditioning.** 1952

Kingsbury, J[ohn] E. **The Telephone and Telephone Exchanges:** Their Invention and Development. 1915

Labatut, Jean and Wheaton J. Lane, eds. **Highways in Our National Life:** A Symposium. 1950

Lathrop, William G[ilbert]. **The Brass Industry in the United States.** 1926

Lesley, Robert W., John B. Lober and George S. Bartlett. **History of the Portland Cement Industry in the United States.** 1924

Marcosson, Isaac F. **Wherever Men Trade:** The Romance of the Cash Register. 1945

Miles, Henry A[dolphus]. **Lowell, As It Was, and As It Is.** 1845

Morison, George S. **The New Epoch:** As Developed by the Manufacture of Power. 1903

Olmsted, Denison. **Memoir of Eli Whitney, Esq.** 1846

Passer, Harold C. **The Electrical Manufacturers, 1875-1900.** 1953

Prescott, George B[artlett]. **Bell's Electric Speaking Telephone.** 1884

Prout, Henry G. **A Life of George Westinghouse.** 1921

Randall, Frank A. **History of the Development of Building Construction in Chicago.** 1949

Riley, John J. **A History of the American Soft Drink Industry:** Bottled Carbonated Beverages, 1807-1957. 1958

Salem, F[rederick] W[illiam]. **Beer, Its History and Its Economic Value as a National Beverage.** 1880

Smith, Edgar F. **Chemistry in America.** 1914

Steinman, D[avid] B[arnard]. **The Builders of the Bridge:** The Story of John Roebling and His Son. 1950

Taylor, F[rank] Sherwood. **A History of Industrial Chemistry.** 1957

Technological Trends and National Policy, Including the Social Implications of New Inventions. Report of the Subcommittee on Technology to the National Resources Committee. 1937

Thompson, John S. **History of Composing Machines.** 1904

Thompson, Robert Luther. **Wiring a Continent:** The History of the Telegraph Industry in the United States, 1832-1866. 1947

Tilley, Nannie May. **The Bright-Tobacco Industry, 1860-1929.** 1948

Tooker, Elva. **Nathan Trotter:** Philadelphia Merchant, 1787-1853. 1955

Turck, J. A. V. **Origin of Modern Calculating Machines.** 1921

Tyler, David Budlong. **Steam Conquers the Atlantic.** 1939

Wheeler, Gervase. **Homes for the People,** In Suburb and Country. 1855